THE OXFORD GUIDE
TO WRITING

THE OXFORD GUIDE TO WRITING

A Rhetoric and Handbook for College Students

THOMAS S. KANE

Editorial Consultant
Nancy Sommers

New York   Oxford
OXFORD UNIVERSITY PRESS
1983

Copyright © 1983 by Oxford University Press, Inc.

Library of Congress Cataloging in Publication Data

Kane, Thomas S.
  The Oxford guide to writing.

  Includes index.
  1. English language—Rhetoric. 2. English language—
Grammar—1950–   . I. Title.
PE1408.K2728 1983   808'.042   82–14093
ISBN 0–19–503245–4

Printing (last digit): 9 8 7 6 5 4 3 2 1

Printed in the United States of America

Acknowledgments

Many people have contributed to this book. The encouragement of my friend and colleague, Professor Leonard J. Peters, has been a necessary condition of the book's completion; his influence exists in every chapter. Professor Nancy Sommers has been enormously helpful, suggesting important revisions of the organization as well as innumerable textual improvements. Professors Miriam Baker of Dowling College, David Hamilton of the University of Iowa, Robert Lyons and Sandra Schor, both of Queens College of the City University of New York, and Joseph Trimmer of Ball State University read the manuscript, and I profited from their perceptive comments. Ms. Cheryl Kupper copyedited the manuscript with unusual thoroughness and sensitivity; her advice and emendations have greatly improved the text. Finally, I wish to thank the staff of the Oxford University Press: my editor, John W. Wright, whose patience and helpfulness exceed even the broad limits of those virtues which editors are expected to possess; Natalie Tutt, who labored over permissions; and Nancy Amy, who somehow pulled things together.

Kittery Point, Maine   T. S. K.
October 1982

To the Student

Writing, it has been said, can be learned but not taught. This paradox has truth. Writing is not a simple activity like tying a knot or riding a bicycle. Someone can show you how to make a bowline and guide your hands until you can tie it quickly and without thinking. But writing is much more complex, requiring considerable control over the most intricate achievement of the human mind—language. So vast and profound are the possibilities of language that we never master writing in any final sense; we just acquire a little more skill.

In a sense writing is not an "activity" at all, not, that is, a set of procedures you study and learn completely to control. Because it involves our very being—what we are, our beliefs and feelings and knowledge—writing must be a learning about one's self which never ends and which, as the remark has it, cannot be taught.

Yet this truth does not mean that there is no purpose or value in composition classes and instructors and books like this one. What teachers and textbooks can do is to help you learn. Not so much by supplying directions like a cookbook, as by showing you the inner workings of good prose, by explaining techniques skillful writers have developed over the centuries, and by warning of pitfalls. Such explanations and warnings are not rules. At best they are generalities about what effective prose is (and is not), but they do not define writing by settling once and for all what it must be.

Of course there are rules in a narrow sense ("conventions" might be a better word): such matters as when to use a semicolon or when to write "whom" instead of "who." These conventions are not unimportant and good writers observe them, but observing them will not in itself make you an effective writer. More important is knowing what you wish to do when you sit down to write and being able to bring to your task some of the almost infinite possibilities of diction, sentence structure, and paragraph development which our language offers. It is upon the skill with which you use language to achieve a clear purpose that readers will respond to you.

That response, however, is perhaps not the most important thing. The greatest benefit of learning to write is not pleasing your instructor or getting into graduate school or advancing your career, desirable as these goals are. It is developing as a human being. Writing forces us to explore and thereby to enlarge our capacity to think and feel and perceive. Learning to express ourselves on paper may not make us better in a moral sense (though one hopes that it does), but it will make us more complex and interesting individuals, more deeply aware of ourselves and of the world around us.

Writing, then, not only can be learned; it is worth learning.

HOW THIS TEXT IS ORGANIZED

But the learning is beset by a special difficulty. When we write we do a number of things almost simultaneously: arrange ideas, think about the reader, select words, form them into phrases and clauses and these in turn into sentences, shape the sentences into paragraphs, link the paragraphs to form larger units. There is no clear-cut logical sequence to all this. Thinking, choosing words, forming sentences all go on more or less together. Yet to study composition it is necessary to impose a "logic" of some kind. The problem is where to begin and how to proceed. Most textbooks are organized around the four elements of composition: diction (that is, words), the sentence, the paragraph, and the essay. This book will be no exception. After an introduction and a brief review of the writing process, we shall spend most of our time on the essay, the paragraph, the sentence, and diction, in that order.

These sections are the heart of the text. However, the progression from essay through paragraph and sentence to diction is not inflexible. Each part is sufficiently self-contained so that it is possible to begin, say, by studying paragraphs and then swing back to the essay or forward to the sentence. In fact any order will work, and no order is inherently superior. Each manner of proceeding has advantages and limitations. Whatever sequence of study you follow, you must, to write at all, select words, make sentences, develop paragraphs, and organize an essay.

Following the discussions of the basic elements of composition are chapters on particular kinds of writing—description, narration, argument and other forms of persuasion, the research paper, and the discussion examination answer. The book concludes with a survey of punctuation, a review of grammar, and a glossary of usage.

The first thing you should read is the Introduction. Its several chapters will acquaint with such fundamental concepts as "strategy," "style," "grammar," and "usage." But before turning to the Introduction, you might wish to leaf through the book, familiarizing yourself with its plan and, in a general way, with its contents.

Finally, a word about the correction symbols and abbreviations used throughout the book. The more important symbols are printed on the end papers, each followed by a page number referring you to the place in the text where that particular writing problem is discussed. Your instructor will use some of these symbols in marking your papers. They are an important feature of the book, enabling you to improve your writing by identifying specific mistakes and difficulties.

To the Teacher

The art of writing is endlessly fascinating, and learning and teaching that art are intellectual challenges of a high order. Yet the challenges, it seems to me, are not often met. More could profitably be said about the varieties and subtleties of good prose than is usually attempted in composition courses: how the syntax of a sentence is part of what it says, how shadings of emphasis and rhythm deepen and enrich meaning, how paragraphs are complicated by blending different means of development.

Students can respond more creatively than we may suppose to the challenge of good prose. Certainly many must begin on elementary levels, learning to write grammatically complete and uncluttered sentences or to develop simple paragraphs before moving on to the nuances of rhythm or to more intricate paragraphing. Still, composition can be taught with something of the rigor and closeness the art deserves without ignoring the widely differing needs, knowledge, and abilities of students.

In its contents this book attempts to do justice to the complexity of good prose; and, in its organization, to achieve a flexibility which allows teachers to select and arrange the material to suit the needs of their students and their own interests. For advanced and capable student-writers it explores some of the finer workings of prose. For those less advanced it stresses fundamentals in the hope that from such beginnings these students too will be stimulated to learn more about the art of writing well.

The treatments of the essay, the paragraph, the sentence, and diction (which, in that order, constitute the major parts of the text) have been divided into relatively self-contained sections, some elementary, others more advanced. Parts dealing with other aspects of composition (grammar, punctuation, usage, and such special modes of writing as persuasion, description, narration, the research paper, and the discussion answer) are similarly self-sufficient, offering focal points for the course which may be used or passed over according to the needs and interests of particular classes.

Aside from the complexity of the subject and the varied requirements and capacities of students, another problem in teaching composition is the absence of any settled ordering of the subject. Probably no two teachers approach composition in exactly the same way. One may prefer to begin by talking about the general nature of the writing process; another, to plunge right in with essays; still another, to start more slowly with paragraphs or sentences. Each method has its advantages and its limitations. There is no one place—or right place—to begin, and no one path by which to move forward.

Such diversity derives from the paradox that while writing involves several skills working together more or less simultaneously, teachers must fragment this process and present its "parts" in an order which by its very nature is arbitrary and artificial. We all know we do not write in a neat sequence of steps: selecting words; then arranging them into phrases, clauses, and sentences; and then adding these to make paragraphs. To write is to do all these things at once, and more besides. But teachers cannot present composition with that simultaneity. Willy-nilly, we most impose order, discussing one "step" at a time, falsifying in order to clarify.

Yet while some sequence or other is unavoidable, it is well for a textbook not to lock teachers into a progression congenial to its writer but not necessarily so to his or her colleagues. Accordingly, the principle of self-containment applies not only to the sections within the major parts of this book, but also to those parts themselves. While they are arranged in the order essay, paragraph, sentence, diction, that order is not essential. Teachers who prefer may begin with the paragraph or the sentence or diction. Any sequence is possible.

Contents

PART ONE   THE WRITING PROCESS

PART TWO   THE ESSAY

PART THREE   THE EXPOSITORY PARAGRAPH

Section I   Basic Structure

Section II   Methods of Development

INTRODUCTION

Truths and Misconceptions about Writing

Many people have misconceptions about writing. These misunderstandings hinder learning, making composition seem more difficult and mysterious than it is. Let's begin by explaining a few basic truths.

YOU CAN LEARN TO WRITE

The ability to write is not a rare skill which people are born with. Writing well is a craft anyone can acquire. If you wish to learn, you can compose clear English prose.

WRITING IS WORTH LEARNING

We sometimes hear that writing is no longer necessary in an age of easy vocal communication. Phones and tape recorders have taken over much of the function of letters and memos. And one can succeed in many professions without being able to write very well. But there is another side. The very advances in communications technology which support the claim that writing is no longer important have in fact increased the power of language.

If technology has reduced the need to write in some areas, it has increased the need in others. More careers in writing probably exist now than ever before—in technical reporting and journalism, for instance (even television news is written before it is spoken). And in professions where writing is not essential, the ability to express your ideas on paper is still an asset.

Beyond these practical considerations, learning to write means growing into a more complicated, more interesting person. We are human beings before we are lawyers, accountants, or teachers. Language is the essence of our humanity. Even in the modern world—especially in the modern world—writing is neither useless nor antiquated. It remains an essential skill.

GOOD PROSE IS RECOGNIZABLE

People do agree about what is well written and what is not. Of course, there are different ways of writing well, but within broad limits most of us recognize when a paragraph or essay is effective and when it isn't. The point is worth making because it is self-defeating to believe that judgments about writing are only personal opinions. Without standards of good prose, nothing could be worth learning. But standards do exist; much can be learned about how to write effectively; and it is, as we have said, worth learning.

CORRECTNESS IS NOT THE ESSENCE OF GOOD WRITING

Effectiveness and correctness are sometimes confused. The first means that a piece of writing achieves its purpose; the second, that it is written according to rules of grammar, usage, spelling, and punctuation. The confusion cuts both ways. On the one hand, some people overvalue correctness—using *ie* and *ei* properly, for instance, or distinguishing *who* and *whom* or *shall* and *will*. On the other hand, others—probably reacting against an overemphasis on grammar and mechanics—dismiss correctness as an empty "school-room" virtue.

Both extremes are foolish. To write effectively requires more—considerably more—than observing conventions of grammar and usage and punctuation. At the same time, good writers do follow these conventions—or if they do not, it is for sound reasons and then only rarely. Writing is a social activity, subject, like all such activities, to rules agreed upon by the group. In your college work, the rules are those of formal composition. They are not particularly difficult or restrictive, and most of them you probably know already. Don't overestimate their importance, but don't ignore them. Observing rules does not in itself make prose good; but carelessly disregarding them seriously reduces the effectiveness of otherwise good writing.

WRITING IS DIFFERENT FROM TALKING

Writing and talking are both forms of communication using language. Yet they differ in important aspects. The difference is easy to overlook, especially since we learn to speak long before we learn to write, and we continue all our lives to talk a great deal more than we write. Still, it is a mistake to suppose that writing is simply putting down on paper what you do when you speak. Speakers depend upon numerous

extraverbal resources of communication: tone of voice, loudness or softness, a raised eyebrow, a shrug, a deprecating or emphatic gesture.
Moreover, speakers usually share with their listeners a common ground
of interest and experience which can compensate for vagueness in language. But writers cannot depend upon so close a sharing with readers,
nor can they shout, whisper, wink, point, or grab a lapel. All of their
meaning must come through their words, and if these are vague, no
compensation is available.

Certainly writing is not essentially different from speaking. There
is even a compositional style that aims at the relaxed informal tone
we associate with conversation. And it is good when you write to imagine
that you are talking to someone, trying to explain your ideas or feelings.
In fact, however, no one *is* there. Readers must get at your meaning
solely through the words you put on paper. You must choose and arrange
those words, then, with great care.

WRITING IS MORE THAN SIMPLY
FINDING WORDS TO FIT IDEAS

Writing is not merely a process of thinking of something to say and
then selecting or looking up the words needed to say it. Writing is
more complicated. Its complexity is revealed by the ambiguity of the
word itself. On the one hand, "writing" means the total activity of
producing a composition; on the other, the physical act of forming letters. In this text we shall use "writing" in the former sense, to signify
the entire complex process, but with the understanding that the physical
act is a part of that process. An important part.

As a process, writing involves both mental and physical activity: you
think *and* you use a pencil. But these are not the clearly distinct steps
they seem. It is *not* the case that we decide what to say and then
write it down. Few of us really know, in an absolute and final sense,
what we want to say until we have said it. That is what the novelist
E. M. Forster meant when he asked, "How do I know what I think
until I see what I have to say?" We begin with a general idea of what
we would like to communicate. The struggle to translate that idea into
marks on paper pushes us in unexpected directions. Thus "writing"
in the abstract sense both feeds and is fed by "writing" in the physical
sense.

This truth ought not to discourage you. On the contrary, it should
be encouraging. Too often inexperienced writers are hamstrung because
they mistakenly suppose they have to know exactly what they want
to do before they can begin to write. Of course, the more you know,
the better. Even so, the paradox remains that you never completely
know until you use the pencil.

EVERYBODY HAS THINGS TO SAY

Anyone who isn't a carrot or a turnip has plenty to write about. You have lots to say. But you must be willing to accept a challenge. Writing throws us upon our own resources, and usually these are not quite adequate. We must stretch and strain our minds, often in discomfort and frustration. It is tempting to make the excuse, "I can't think of anything to write about." Ducking challenges, however, is self-limiting. We grow by meeting them.

EXERCISE

Do you share any of these misconceptions about writing? Can you think of others not mentioned here? Work out your feelings and ideas in a paragraph or two and be prepared to discuss them in class, raising questions or objections that the brief discussion in this chapter does not answer.

CHAPTER 2

Basic Considerations: Purpose

INTRODUCTION

In this chapter and the next two, we discuss several matters fundamental to composition: purpose, strategy, style, and the conventions (or "rules") gathered under the headings of grammar, usage, and mechanics. These things determine and limit what you do as a writer, and we shall return to them again and again in our study of the paragraph, the essay, sentence structure, and diction.

The most basic matter is purpose. The purpose of composition is to communicate. Written communication involves not only a writer but also a subject and a reader—something said and someone to whom it is said—and subject and reader are not completely independent of each other. Subjects do not exist solely in themselves. They exist in the mind and words of the writer and, potentially, in the minds of readers. Experienced writers think of a subject as something to be communicated to particular readers. More often than not, thinking about a subject in terms of the reader deepens your understanding of it. The very effort of conveying ideas to a specific kind of reader leads to new insights into the ideas themselves.

Purpose, then, involves both subject and reader.

SUBJECT

Ideally any subject you write about should interest you. It should also be within the range of your experience and skill, yet stretch that experience and skill, so that you are not merely rehashing something you already know. The subject must fit the length of your paper: subjects like the New Deal or the Vietnam War are too complicated to treat in a three- or four-page theme. On the other hand, a subject should not be trivial or simple-minded.

Don't be afraid to express your own opinions or feelings. They are an important part of the subject. No matter what the topic is, you are really writing about how *you* understand it, how *you* feel about it. Revelation of personality gives interest to writing. Readers enjoy sensing a particular mind at work, hearing a clear voice, sharing a genuine and unusual sensibility. If you have chosen a topic neither too vast nor too simple, and if genuine feeling comes through, you will be interesting. Interest lies not so much in the topic as in what you have done with the topic.

READER

Insofar as purpose means the effect you want to have upon your readers, it can take three shapes: to inform, to persuade, and to entertain.[1] These aims are not mutually exclusive. In fact it would be rare to find a piece of prose shaped purely by any one of them. But while all may enter into a specific composition, one usually predominates and determines basic strategy and style. Whichever purpose is paramount, it will succeed or fail depending upon how much thought you give to the people who are going to read your prose.

More fundamentally, you do not want to repel your readers. This does not mean that you must flatter them or be afraid to say anything they may disagree with. It does mean that you must respect your readers. Do not look down on them; if you really think yourself superior, it is wisdom to conceal the feeling. Do not take your readers' interest for granted, or suppose they possess endless patience to wade through wordy, murky prose to get at your meaning. Your job is to make the reader's task as easy as possible. Good writers take trouble to be clear so that their readers do not have to take unnecessary trouble to understand.

More specifically, you must ask yourself questions about readers: What can I expect them to know and not know? What do they believe and value? How do I want to affect them by what I say? What attitudes and claims will meet with their approval; what will offend them? What objections may they have to my ideas, and how can I anticipate and counter those objections? Are they likely to understand irony, to be put off or amused by irreverent wit?

Readers are annoyed if you assume too much about their knowledge. Tossing off unusual words or information without explanation is a put-down, a way of saying, "I know more than you." On the other hand,

---

1. There is a fourth purpose: to manipulate and bamboozle readers. Much advertising and political writing—though not all—is of this sort. The writers hope to mislead us, to hide the truth, to make wrong seem right. Such an intention is akin to persuasion, but in this book we reserve the term "persuasion" for open and legitimate attempts to change readers' ideas and beliefs. We do not discuss bamboozling.

defining the obvious also implies a low opinion of readers: don't tell them what a wheel is; they know. It is not easy to estimate your readers' level of knowledge; if you must err, it is better to say too much than too little. Nor it is easy to gauge their beliefs and values. A sense of the reader comes only gradually, with experience, and then imperfectly. Tact and respect, however, go a long way. Readers have egos too.

PURPOSE AND TYPES OF PROSE

The various effects a writer may wish to have upon his or her readers—to inform, to persuade, to entertain—result in different kinds of prose. Our primary concern is with writing that informs, which, depending upon its subject matter, we call exposition, description, or simple narration.

Exposition explains. It may explain how things work—an internal combustion engine, perhaps. It may explain ideas—a theory of economics. It may deal with facts—the frequency of divorce and the reasons people give for divorcing. Or with feelings and beliefs—what you think of college or what you plan to do in life. Your textbooks are exposition. A biography of Benjamin Franklin, an essay on ant colonies, or a piece on life at the court of Louis XIV are exposition. But whatever its subject, exposition ultimately states what a particular mind thinks, knows, or believes. Even when it deals with a factual historical subject such as the causes of the American Civil War, exposition is really a statement of what one person understands.

Because it is an expression of mind, exposition is constructed logically, in accordance with how we think and feel. It uses organizations such as cause/effect, true/false, less/more, positive/negative, general/particular, assertion/denial. Its movement is signaled by connectives like *therefore, however, and so, besides, but, not only, more importantly, in fact, for example.*

Later we shall look more closely at these categories of thought. For the moment it is enough to understand that exposition tells us what a writer thinks, feels, believes, knows; and that it takes its structure from the mind at work. Most of what you write in this course and others will be exposition. Because of its primacy, exposition receives more attention in this book than other kinds of writing.

But not all the attention. We shall also study the other two kinds of informative writing. Description deals with perceptions—most commonly visual perceptions. Its central problem is to arrange what we see in a significant pattern. The pattern, unlike the logic of exposition, is spatial, expressed in terms of above/below, before/behind, right/left, and so on. Simple narration is also a species of informative writing (though, like description, it may also serve to persuade or entertain). Its subject is a series of related events—a story—and its problem is twofold: to establish the time relationships of the events and to reveal their significance.

We shall also touch upon persuasion. Persuasion seeks to alter how readers think or believe. It may take the form of argument, which appeals to reason, offering factual evidence or logical proof. It may take the form of satire, which ridicules particular follies or evils, sometimes subtly, sometimes very broadly and coarsely. Or persuasion may be eloquent, appealing to our ideals and noblest sentiments.

Writing that is primarily entertaining receives the least attention, not because it is unimportant but because it is remote from the immediate needs of college students. Keep in mind that every kind of composition can be enjoyable. Insofar as it gives us pleasure and excites our admiration, any piece of well-written prose—expository or persuasive— entertains.

Nor are the lines between exposition and persuasion hard and fast. To persuade readers, you must often first inform them, and a skillful expository essay may well have the effect of persuading us to accept the truth of what it says. What counts, finally, is how the writer primarily wishes to affect his or her readers.

EXERCISE

1. Make a list of ten or twelve topics you might develop into short essays. Think of topics that deal not so much with things, places, or how-to-do projects as with your ideas and beliefs. The subjects should interest you and be within your experience, yet challenging. Be specific: don't write simply "College" but something like "What most impressed (or disappointed) me about college was _____."

Turn in the list to your instructor for comment. When it has been returned, improve the topics in the light of the comments and save them for future assignments, adding to the list as new ideas occur to you.

2. Selecting one of the subjects on your list, write a paragraph about the readers for whom you might develop it. Consider how you wish to affect those readers; what you are assuming about their general knowledge, values, attitudes, biases; whether they are your age or older or younger, come from a similar or different background; and how you would like them to regard you as they react to your writing.

CHAPTER 3

Basic Considerations: Strategy and Style

Purpose determines strategy and style. Strategy is how you go about achieving a purpose. It concerns the specific aspects of the topic you decide to develop and how you organize them, the words you select, the types of sentences you choose, and the way you compose paragraphs. Style is the actual language that makes the strategy work.

Think of the three concepts in terms of increasing abstractness. Style is the most immediate and obvious. It exists in the writing itself; it is the sum of all the choices you have made about words, sentences, paragraphs. Strategy is more abstract, discernible beneath the words as the immediate ends they serve. Purpose is even deeper and more abstract, lying below strategy and involving not only what you write about but how you wish to affect readers.

A brief example may help to disentangle these closely related concepts of purpose, strategy, and style. The example is a paragraph written in response to a fifteen-minute classroom exercise on one of several topics. The topics were stated broadly—"marriage," "parents," "teachers," and so on—so that each writer had to think about restricting and organizing his or her composition. This student chose "marriage":

> Why get married? Or if you are modern, why live together? Answer: Insecurity. "Man needs woman; woman needs man." However, this cliché fails to explain *need*. How do you need someone of the opposite sex? Sexually is an insufficient explanation. Other animals do not stay with a mate for more than one season; some not even that long. Companionship, although a better answer, is also an incomplete explanation. We all have several friends. Why make one friend so significant that he at least partially excludes the others? Because we want to "join our lives." But this desire for joining is far from "romantic"—it is selfish. We want someone to share our lives in order that we do not have to endure hardships alone.

The writer's purpose is not so much to inform us of what she thinks about marriage as to convince us that what she thinks is true. Her purpose, then, is persuasive, and it leads to particular strategies both of organization and of sentence style. Her organization is a refinement of a conventional question/answer strategy: the basic question ("Why get married?"); an initial, inadequate answer ("Insecurity"); a more precise question ("How do we need someone?"); a partial answer ("sex"); then a second partial answer ("companionship"); the final, most precise question ("Why make one friend so significant?"); and the final answer ("so that we do not have to endure hardships alone").

The progression of thought through the sentences of the paragraph can be diagrammed like this:

Sentences 1–2   BASIC QUESTION: Why get married?

Sentences 3–4   INITIAL ANSWER: Need

Sentence 5   THAT ANSWER INADEQUATE: Does not explain need

Sentence 6   MORE PRECISE QUESTION: How do we need?

Sentence 7   FIRST POSSIBLE ANSWER: Sex

Sentence 8   THAT ANSWER INADEQUATE: Evidence of other animals

Sentence 9   SECOND POSSIBLE ANSWER: Companionship

Sentence 10   THAT ANSWER INADEQUATE: Many friends can supply companionship.

Sentence 11   FINAL, MOST PRECISE, PHRASING OF QUESTION: Why is one friend extraordinarily important?

Sentences 12–14   FINAL ANSWER: Need for someone to help us endure hardship.

The persuasive purpose is reflected also in the writer's strategy of short emphatic sentences. These have two virtues: they carry conviction, essential in effective persuasion; and they suggest the speaking voice rather than a bookish tone, thus establishing an appropriate informal relationship with the readers.

Finally, the student's purpose determines her strategy in approaching the subject and in presenting herself. About the topic, the writer is serious without becoming pompous. As for herself, she adopts an impersonal point of view, avoiding such expressions as "I think" or "it seems to me," which, while they might suggest a pleasing modesty on other occasions, would here weaken the forcefulness of the argument.

These strategies are effectively realized in the style: in the clear rhetorical questions, each immediately followed by a straightforward answer, and in the short uncomplicated sentences, suggesting speech (there are even two sentences that are grammatically incomplete—"Answer: Insecurity" and "Because we want to 'join our lives' "). At the same time the sentences are sufficiently varied to achieve a strategy fundamental to all successful prose—to get and hold the reader's attention.

Remember several things about strategy. First, it is many-sided. Any piece of prose displays not one but numerous strategies—of organization, of sentence structure, of word choice, of point of view, of tone. In good writing these mesh and reinforce one another.

Second, no absolute one-to-one correspondence exists between strategy and purpose. A specific strategy may be adapted to various purposes. The question/answer mode of organizing, for example, is not confined to persuasion: it is often used in informative writing. Further, a particular purpose may be served by different strategies. In our example the student's solution was not the only one possible. Another writer might have organized by a "list" strategy:

> People get married for a variety of reasons. First. . . . Second. . . . Third. . . . Finally. . . .

Still another might have used a personal point of view, or taken a less serious approach, or assumed a more formal relationship with the reader.

As we examine various aspects of composition, strategy remains a constant, underlying concept. The various ways we shall study of developing paragraphs or of structuring sentences are in fact strategies. Think of composition always in this sense. Observe, as well, the strategies that direct the prose you read in this and other books. Experiment with these strategies in your own writing, adapting them to your purposes.

STYLE

Style, too, is a constant consideration in what follows. Teachers use the term in various ways. In its broadest sense "style" means the total of all the choices a writer makes in his or her words and their arrangements. In this sense style may be good or bad—good if the choices are appropriate to the writer's purpose, bad if they are not. More narrowly "style" has a positive, approving sense, as when we say that someone has "style" or praise a writer for his or her "style." In a narrow sense the word may also designate a particular way of writing, unique to a person or characteristic of a group or profession: "Hemingway's style," "an academic style."

Here we use *style* to mean something between those extremes. It will be a positive term, and while we speak of errors in style, we don't speak of "bad styles." On the other hand, we understand "style" to include many ways of writing, each appropriate for some purposes, less so for others. There is no one style, some ideal manner of writing at which all of us should aim. Style is a flexible concept, capable of almost endless variation. But one thing style is not: it is not a superficial fanciness brushed on like a coat of gilt over the basic ideas. Rather than the surface appearance of a piece of writing, style is its deep essence.

EXERCISE

Selecting one of the topics you listed at the end of Chapter 2, work up a paragraph of 150–200 words. Before you begin to write, think about possible strategies of organization and tone. By *tone* we mean (1) how you feel about your subject—angry, amused, objective, and so on; (2) how you regard your reader—in a formal or an informal relationship; and (3) how you present yourself.

When you have the paragraph in its final shape, ready to turn in, compose on a separate sheet of paper several sentences explaining what strategies you followed in organizing your paragraph and in aiming for a particular tone, and why you thought these would be appropriate. Turn in this brief discussion with your paragraph.

CHAPTER 4

Basic Considerations: Grammar, Usage, and Mechanics

Purpose, strategy, and style are determined by you. You decide what you want to do, plan how to achieve it, select the words, and compose the sentences and paragraphs. But these decisions must be made within a framework of rules, or conventions, over which you have little control. The conventions fall into three groups: grammar, usage, and mechanics.

GRAMMAR

Grammar means the rules which structure our language.[1] The sentence "She dresses beautifully" is grammatical. These variations are not:

Her dresses beautifully.
Dresses beautifully she.

The first breaks the rule that a personal pronoun is in the subjective case when it acts as the subject of a verb. The second violates the conventional order of the English sentence: subject-verb-object. (That order is not invariable and may be altered, subject to other rules, but none of these permits the pattern: "Dresses beautifully she.")

Grammatical rules are *not* the pronouncements of teachers, editors, or other authorities. They are simply the way people speak and write English, and if enough people begin to speak and write differently, the rules change accordingly.

1. "Our" includes adult, native, sane speakers. Obviously, small children, foreign-born people learning English, and people who are seriously disturbed mentally or emotionally may "use" the language oddly. But for the purpose of understanding grammar, such users are excluded from the category "speakers of English."

USAGE

Usage designates rules of a less basic and binding sort, concerning how we *should* use the language in certain situations. In college work the situation is usually formal composition. These sentences, for instance, violate formal usage:

> She dresses beautiful.
> She ain't got no dress.

Neither sentence is ungrammatical: both represent patterns used by speakers who are "adult, native, and sane." But both break rules governing how educated people write. Modern formal usage dictates that when *beautiful* functions as an adverb it takes an *-ly* ending, that *ain't* is not acceptable, and that a double negative like *ain't got no* or *haven't got no* should be avoided.

Grammar and usage are often confused. Many people would argue that the sentences above are "ungrammatical." Our distinction is more useful, however. Grammatical rules are implicit in the speech of all who use the language. Usage rules, on the other hand, stem from and change by social pressure. *Ain't,* for example, was once acceptable. The adverbial use of an adjective like *beautiful* was common in seventeenth-century prose. Both Chaucer and Shakespeare use double (even triple and quadruple) negatives for emphasis.

Sometimes usage conventions reflect social and class distinctions. Compare, for instance, *you all* and *youse.* Each attempts to make good a minor inadequacy of English: the failure to distinguish singular and plural in the second-person pronoun, a flaw which can cause confusion. *Youse* provides a remedy by analogy with other plurals, adding the *-s* which in English is the most common sign of plurality. The Southernism *you all* (which a Southerner does not use in talking to one person) makes good the deficiency in another way. But the social acceptability of the two forms differs enormously. *You all* is a perfectly allowable, if colloquial, plural. *Youse* was stigmatized and rooted out of the schools, probably because it was associated with lower-class immigrants.

The fact that usage rules are less basic than grammatical ones, however, and even that they may seem arbitrary, does not lessen their force. You would be ill-advised to write in a history paper that "Napoleon had a lousy time in Russia." So he did, but the violation of the usage rule against "lousy" in formal English would probably annoy your history professor. Keep in mind, too, that not all usage conventions are arbitrary or snobbish. Most try to maintain distinctions which contribute to clarity and economy of expression.

Moreover, usage applies to all levels of purpose and strategy, to informal, colloquial styles as well as to formal ones. For example, grammatically incomplete sentences (or fragments), frowned upon in formal usage, are occasionally permissible and even valuable in informal composition. (Witness the two fragments in the student paragraph on marriage back on page 11.) *So* is regarded in formal English as a subordinating conjunction which ought not to introduce a sentence. But in a colloquial style, it may work better than a more literary connective like *consequently* or *therefore*.

MECHANICS

In composition *mechanics* refers to the appearance of words, to how they are spelled or arranged on the paper. The fact that the first word of a paragraph is indented, for example, is a matter of mechanics. These sentences violate other rules of mechanics:

> she dresses beautifully
> She dresses beautifuly.

The conventions of writing require that a sentence begin with a capital letter and end with full-stop punctuation (period, question mark, or exclamation point). The conventions of spelling require that *beautifully* have two *l*s.

The rules gathered under the heading of mechanics attempt to make writing consistent and clear. They may seem arbitrary, but they have evolved from centuries of experience. Generally they represent, if not the only way of solving a problem, an economic and efficient way.

Along with mechanics we include punctuation, a very complicated subject and by no means purely mechanical. While some punctuation is determined by cut-and-dried rules, much of it falls into the province of usage or style. Later, when we study punctuation, we discuss the distinctions between mechanical and stylistic uses of commas, dashes, and so on.

GRAMMAR, USAGE, AND STYLE

Grammar, usage, and mechanics establish the ground rules of writing, circumscribing what you are free to do. Within them, you may select various strategies and work out those strategies in your own words, sentences, and paragraphs. The ground rules, however, are relatively inflexible. Of course you can choose to break them—just as a base-runner can cut directly from first base to third. But if he does, he is out. And if you choose to write ungrammatically or to pay no attention to usage or mechanics, readers will rule you out.

It is not always easy to draw the line between grammar and usage or between usage and style. Broadly, grammar is what you must do as a user of English; usage, what you should do as a writer of more or less formal English; and style, what you elect to do in order to work out your strategies and realize your purpose.

"Her dresses beautifully," we said, represents an error in grammar, and "She dresses beautiful," a mistake in usage. "She dresses in a beautiful manner," finally, represents a lapse in style. The sentence breaks no rule of grammar or of usage, but it is not effective (assuming that the writer wants to stress the idea of "beauty"). The structure slurs emphasis on the key term, which should close the statement. The writer knows one cannot say "She dresses beautiful," and so, afraid of adverbs, supplies a meaningless noun ("manner") as a hook upon which to hang the adjective. The noun, in turn, needs a marker ("a") and a preposition ("in") to link it to the verb. Thus three empty words clutter up the sentence and keep the stress from falling naturally on "beautifully," as it does in "She dresses beautifully."

Most of your difficulties with words and sentences probably involve errors in style. For native speakers, grammar—in our sense—is not likely to be a serious problem. Usage (which includes much of what is popularly called "grammar") and mechanics are more troublesome. But generally these require simply that you learn and observe clearly defined conventions. And having learned them, you will find that rather than being restrictive they free you to choose more effectively among the options available to you as a writer.

Style, on the other hand, is less reducible to rule, and leaves far greater room for argument. No one can prove that "She dresses in a beautiful manner" is a poorer sentence than "She dresses beautifully." (One can even imagine a context in which the longer sentence would be preferable.) Even so, it violates a principle observed by good writers: use no more empty words than you must.

You may think of that principle, if you wish, as a "rule" of style. We shall discuss and illustrate that and other stylistic "rules," but always understanding that they are generalizations about what good writers do, not laws dictating what all writers must do.

EXERCISE

These sentences contain numerous errors. Some are mistakes in grammar (as we have defined it); others in usage or mechanics. In each case try to identify both the mistake and the category into which it falls. Can you add examples of three or four other common errors?

1. The choice is between you and I.
2. This parking lot is reserved for customer's.
3. Purchasing cigarettes from this machine by minors are prohibited.
4. Its too much too expect.
5. Please deliver the package to myself.

6. The man was identified as a mister Robert Smith.
7. They gave it they're best effort.
8. He looks at life very cynical.
9. They wasn't expected until next week.
10. In his book on shakespeare prof Jones discusses *Hamlet* at great length.
11. I like Winter.
12. The women was standing all by herself.
13. Everybody shut their book and looked up.

PART ONE

THE WRITING PROCESS

INTRODUCTION

To describe writing as a process is both true and misleading. True because writing is a rational activity, a problem which the mind solves by thinking, whether the thinking be self-conscious or intuitive. Misleading to the degree that "process" suggests a neat series of well-defined steps, each of which is negotiated before the next is attempted.

The steps exist, and in the next four chapters we shall discuss them—invention, which is finding and developing topics to write about; outlining; drafting, or composing a tentative version of the whole essay; and revising. But a writer's movement up these steps is not a simple, steady climb. It is very different: a constant passage up and down and up. In a sense, one is on all the steps at once, yet at any given moment, chiefly on one.

If that sounds mysterious or even impossible, the reason is that writing, while neither mysterious nor impossible, is a very complex activity. For example: before you draft, you invent topics; you think of things to say. But as you invent and list topics, you are beginning to draft, selecting words and constructing sentences that later will appear in the essay. And when you have progressed to the point of drafting and revising, you will discover new topics: you will still be inventing.

Think, then, of writing as a process, one whose steps connect so intimately that each includes the others. No matter which phase of writing you are concentrating upon, you are involved with the process in its entirety.

CHAPTER 5

Invention: Gathering

For many of us the hardest step in writing is the first. How to get going! Some people are paralyzed at the prospect of having to fill a sheet of paper with words. Perhaps it's no wonder when we remember the misconceptions we glanced at in the first chapter, especially the false notions that the ability to write is a rare and innate talent, and that composition is hedged about with a thousand and one thickets of "correctness."

Forget about special talent; forget about correctness (for the moment); and remember that you have plenty to say.

Of course, saying it will be a struggle. Writing is hard work; even professionals complain about it. James Thurber and E. B. White, two of the finest writers of our time, once collaborated on a humorous book called *Is Sex Necessary?* In the Foreword they remarked:

> "Approach the subject in a lively spirit," we told ourselves, "and the writing will take care of itself." (It is only fair to say that the writing didn't take care of itself; the writing was a lot of work and gave us the usual pain in the neck while we were doing it.)

"The *usual* pain in the neck." Almost everybody labors to write, even those who work at the trade. So don't expect writing to be easy and then you won't get frustrated when you find it isn't. But it is not impossible; it is not beyond your ability. And remember this: writing something well repays the labor many times over.

The first step is to get started. If you have trouble, there is a remedy: to become conscious of the process called "invention" (in Latin the word means "to find, to come upon"). In composition "invention" simply means discovering and developing topics to write about. In its deepest sense invention is self-discovery. We write about ourselves, which is all any of us can do. The words we put on paper are pulled from within, as a spider draws its thread from itself.

To supply that inner substance, however, we must feed upon experience, acquiring ideas and information and extending our perceptiveness and sensibility. Moreover we must learn to think about what we acquire as raw material to be shaped into sentences and paragraphs and essays.

Invention, then, has two parts: gathering facts and ideas and perceptions, and exploring and exploiting them as material for writing. In this chapter we discuss the first of these parts.

Gathering material for composition is often a channeled procedure. An historian setting out to write a biography of Harry Truman, say, will read a specific range of books and articles and documents and interview particular people. (In Chapter 50, dealing with the research paper, we shall look at this kind of focused invention.)

In a more fundamental sense, however, the "gathering" aspect of invention concerns not so much researching a particular subject as making new ideas and feelings a part of yourself. The process goes on automatically. To breathe the atmosphere of a university is to inhale new ways of thinking and seeing and feeling. When, for example, you first study Greek tragedy and come upon the concept of *hubris*—the arrogant refusal to accept any limit to self-assertion or self-gratification—you have gained not simply a word but a sharper way of seeing and understanding. Education and invention, then, go hand in hand. If you read and listen and reflect upon what you see and hear, not only in the classroom or in conversations with your friends but in all the world around you, you will gather material for your writing.

You can do this even more effectively by keeping a commonplace book and a journal.

THE COMMONPLACE BOOK

A commonplace book is a record of the things we have read or heard and want to remember. These notes may come from books or magazines or newspapers. They may be statements we hear (or overhear) in conversation, on television, at the movies. The essential quality is that they be memorable. Usually a statement is worth remembering for a positive reason: it expresses an unusual idea or perception, or perhaps it puts a usual one extraordinarily well. Sometimes, on the other hand, a passage may be memorable in a negative sense, so silly or wrong-headed that it is worth remembering as an example of what people ought not to think.

Here is a sentence by the English novelist and essayist Virginia Woolf, which is memorable in a positive sense and which might well go into a commonplace book:

Women have served all these centuries as looking glasses possessing the
. . . power of reflecting the figure of a man at twice its natural size.

The idea is not original, but the image of the mirror is a novel and effective expression of an old truth. The sentence is worth remembering, especially by anyone interested in the changing roles of men and women characteristic of our time.

To keep a commonplace book, set aside a looseleaf notebook. When you read or hear something that strikes you, copy it down, identifying author and title. Leave space after the entry for the addition of your own thoughts later. If you accumulate a lot of entries, you may want to make an index or to group passages according to subject. (Here is the advantage of the looseleaf notebook.)

Whether or not you work this hard on a commonplace book, you will find that keeping one helps your writing in several ways. It provides a body of quotations (an occasional apt quotation adds interest to your writing). The very act of copying well-expressed sentences improves your own prose, and you will further strengthen your writing if you are stimulated to add your own thoughts to the entry. Most important, the commonplace book will be a storehouse of invention, of those elusive "things to write about."

THE JOURNAL

A journal—the word comes from French and originally meant "daily"— is a day-to-day record of what you do and see and hear and think and feel. A journal states your own experiences and thoughts rather than collecting quotations like a commonplace book. But, of course, you may combine the two. If you add your own comments to the passages you copy into a commonplace book, you are also keeping a kind of journal.

Many professional writers use journals, and the habit is a good one for anybody who wishes to learn to write, even if he or she has no professional literary ambitions. Journals store perceptions, ideas, emotions, actions—all future material for essays and stories. *The Journals* of Henry Thoreau are a famous example, as are *A Writer's Diary* by Virginia Woolf, the *Notebooks* of the French novelist Albert Camus, or "A War-time Diary" by the English writer George Orwell.

You compose journal entries for yourself. That does not mean, however, that they need not be well-written. On the contrary, you should express yourself as clearly as possible. Keeping a journal has little use as a way of learning to write unless you struggle to express your ideas and feelings precisely.

At the same time, you can forget some of the conventional requirements of formal composition. You do not have to write in grammatically complete sentences or to provide "howevers" and "therefores" and other such transitions for readers.

The following passage comes from the journal of Rockwell Stensrud, a professional writer who accompanied an old-time cattle drive staged in 1975 as part of the Bicentennial celebration:

> Very strict unspoken rules of cowboy behavior—get as drunk as you want the night before, but you'd better be able to get up the next morning at 4:30, or you're not living by the code of responsibility. Range codes more severe than high-society ideas of manners—and perhaps more necessary out here. What these cowboys respect more than anything is ability to carry one's own weight, to perform, to get the job done well—these are the traditions that make this quest of theirs possible.

This is clear and detailed. Little would be needed to turn it into conventional prose other than a few verbs and articles. Some writers use an even more stripped style in their journals. It doesn't really matter, so long as the essential thought or perception is clear.

EXERCISE

1. Make a list of four or five memorable quotations on topics that interest you. Ideally these should come from books or articles you are reading in their entirety. It is possible, however, to leaf through collections of quotations, anecdotes, aphorisms, proverbs. You do not necessarily have to agree with what you select; in fact, the value of a passage may be that it serves as a lens to focus your disagreement. In any case, look for statements that might work as jumping-off places for a composition of your own.

2. Keep a journal for several days. Describe what you see and do: books you read, movies you see, people you meet, classes, professors, friends. Though you are writing for yourself, remember that everything you put down is potentially material for a theme.

CHAPTER 6

Invention:
Exploring Ideas

The first step in invention is to find an idea, a starting place for a theme. The second is to explore the idea, looking for ways to develop it. Consider again the quotation by Virginia Woolf:

> Women have served all these centuries as looking glasses possessing the . . . power of reflecting the figure of a man at twice its natural size.

Let your mind play with this notion. How does it apply to our time? Perhaps women are no longer content to stand still and magnify the male image. If you think that is the case, you have an angle of attack. Woolf's sentence can serve as a point of departure, explaining what used to be the relationship between men and women, but is true no longer, or at least not as true.

Given this line of thinking, you have invented—that is, found—a particular way of exploring the idea. Think, for instance, about all that the image of a mirror implies. What do mirrors do? They hang upon the wall or stand in corners waiting to be used. What can happen to mirrors? They can crack, lose their reflective silvering, develop distorting irregularities. They can fall and break and bring bad luck. A human mirror may well get tired of hanging about waiting to be gazed into and refuse to shine back what its users expect to see.

What other references to mirrors can you think of? Snow White's "Mirror, mirror on the wall, Who's the fairest one of all?" Lewis Carroll's looking-glass that Alice stepped through, playing upon the old magical belief that mirrors are portals to strange worlds. The Greek myth of Narcissus, who fell in love with his own image reflected in a pool of water.

Don't stop with the mirror. What happens to the men who gaze too long into the looking glass of feminine eyes? Perhaps their vision becomes so limited that they see nothing but themselves, and themselves not truly. (Narcissus applies here.) And what a shock it must be when suddenly the dependable mirror no longer shows a man what he is accustomed to see.

What exactly was the "figure of a man" women were supposed to reflect and magnify? One image was the wise old patriarch: "Big Daddy," "Father who knows best," Daddy Warbucks in "Little Orphan Annie." Another was the macho man: strong, silent, courageous, tough and self-sufficient, sometimes with a dark sorrow in his past. The type goes back at least to Lord Byron's early nineteenth-century heroes, descending through many modern variations including the Westerner—John Wayne, Gary Cooper, and, enlisted in the war to conquer consumers, the Marlboro Man.

If you believe that many women are no longer willing to magnify this image of the dominating male, look for evidence. Is the he-man less common in movies and television shows? In serious literature an anti-hero has appeared, a type more vulnerable than the macho male, weaker, even at times cowardly and despicable. Can you find examples in popular culture? If women no longer admire the he-man, what traits do they value?

Thinking about Woolf's quotation, or any other, then, supplies you with many topics, raw material for a theme. Certainly, the ideas we have put forward, the questions we have asked, only scratch the surface. Traditionalists might take a very different approach, arguing that Woolf's image is exaggerated and misleading; or that, if it is true, it still expresses a valid ideal of the relationship between men and women. Starting from a given point, invention can proceed in very different directions, depending upon the values, experience, and imagination of the writers. The important thing is not that you explore a subject in any one way, but that you explore it in some way.

The exploring will be more productive if you use a pencil, that is, actually write down ideas. You may do this in one of two ways: by a kind of loose, free writing; or by a more systematic method of questioning, listing of topics, and outlining.

FREE WRITING

Free writing simply means getting ideas on paper as fast as you can. The trick is to let feelings and ideas pour forth. You jot down anything that occurs to you, without worrying about order or even that the ideas make much sense. Keep going; to pause is to risk getting stuck, like a car in snow. So you move the pencil ceaselessly, writing whatever pops into mind. Don't be afraid of making mistakes or of saying something foolish. You probably will. So what? You're writing for yourself, and if you won't risk saying something foolish, you're not likely to say anything wise.

A free-writing exercise on the Virginia Woolf sentence might go like this:

That's true—women have been mirrors for male vanity. And we're tired of it. Why should we be mirrors? Mirrors are passive and have no life of their own. They just hang about on a wall waiting to be used. At least a real mirror tells the truth. But when a man sees his reflection in a woman's eyes the truth is the last thing he looks for or wants. He probably wouldn't recognize it if it were there. He just wants the mirror to say, "Hi Handsome, you great big hunk." Hunk of vanity. Mirror, mirror on the wall, who's the macho-est man of all? Some women are happy being mirrors. They can't even imagine they could be anything else. But something's happened, and other women, more and more of them, aren't content to be looking glasses in which men can preen themselves like Narcissus gazing in the pool. So a lot of men no longer find John Wayne when they look into their girlfriends' eyes. They don't know what they see, and some of them are confused and some are angry and some are scared. If they could look deeply enough they might see Mickey Mouse instead of John Wayne, but I suppose if they could see that deep they wouldn't be Mickey Mouse (Mickey Mice?). Women are stepping through the mirror. But they're not stepping into a wonderland like Alice; they're stepping out of one, a wonderland made by men. And men better step out of it too because it's going to get awful lonesome in there.

Now that is far from a finished essay. Judged as a composition it's a mess, but a mess alive with possibilities. It is clay to mold.

THE ANALYTICAL APPROACH

Rather than jumping in and exploring the topic by writing anything that comes to mind, some writers prefer a more rational, analytical method of invention. What you do is to ask a series of questions. Some of these are specific to the subject. The questions about the characteristics and uses of mirrors, for instance, follow naturally from Virginia Woolf's image of the looking glass.

Other questions, more general in nature, can be put to almost any subject. Their answers suggest topics for development. The following list includes a few such questions, and you may think of others. Not all the questions will apply to every subject, but enough of them will always prove useful to generate an essay:

1. How may the subject be defined or described?
2. How may it be analyzed? What are its parts or varieties?
3. Which of these are most important for my purpose?
4. Are there examples of the subject?
5. What have other people said about it?
6. What is it similar to?
7. What is it different from?
8. What exceptions exist? What qualifications must be made?
9. What caused it?

10. What reason or purpose does it have?
11. What are its consequences and implications?
12. What is it related to?
13. What are its advantages and virtues?
14. What are its disadvantages, defects, limitations?

It is worth repeating that not all of these questions will prove useful on any given occasion. You must try them out, exploring those which open up topics and discarding those which do not.

Work with pencil and paper and follow three steps. First jot down the questions listed above (along with others that occur to you), phrasing them, of course, with reference to the particular subject and leaving a blank space below each question in which to write any answers it suggests. Put down all the questions; don't decide beforehand that this one or that cannot possibly have any bearing upon your problem. You may be right; even so, it is better to try each question before discarding it. In the blank spaces list any ideas that pop up. Work rapidly during this initial stage. Try to get through the entire list quickly, noting the first ideas or impressions the questions suggest. Don't attempt to edit your ideas at this stage; get them on paper, even if they seem off the track. If no ideas come, don't bog down trying to force them; leave that space blank and go on to the next question. During this phase of invention, try not to get deeply involved with any single idea. Try instead briefly to survey all possible avenues of development.

Here is how a few of the questions might be applied to the passage by Virginia Woolf:

How may the subject be defined or described?

Woolf talks about the domination of women by men, or, put the other way, the subservience our culture forces upon women, their restriction to roles in support of men.

What are its parts or varieties? How may the subject of male dominance–female subservience be analyzed?

This subject may be analyzed in several ways; in terms (1) of its causes and consequences; (2) of the interconnected roles of men and women; (3) of how this sexual pattern pervades our society— in the family, the economy, education, popular culture, and so on; (4) of the past and the changing present and future. These various organizations might be combined, but in a short theme it would be wise to focus only on one or two. For example, how the pattern is reinforced by popular culture—advertisements, television shows, movies.

What caused the dominance of men?

Probably the ultimate cause was simply the greater size and strength of the male animal together with the maternal role of the female. These conditions created a patriarchal family structure, which civilization elaborated and rationalized.

As a second step in the analytical method of invention, go over your listings carefully, selecting topics which seem promising and discarding those which do not. Now you want to give further thought to any spaces you left blank and to abandon them if no ideas come.

Finally, sort out the topics you think important and write down each at the top of a separate piece of paper. List all the ideas you can think of to support it—definitions, examples, causes, purposes, consequences, implications, comparisons, contrasts, exceptions, qualifications.

Suppose, for instance, that you decide to develop an essay based on Virginia Woolf's remark and discussing how the traditional male-female roles are reflected in popular culture. Ask yourself what are the categories of popular culture and which ones you should focus on. Write these down and, under each list, the supporting examples. Be alert also to exceptions; admitting these and trying to neutralize them will strengthen your argument:

1. Advertising
   A. The Marlboro Man. The western hero peddling cigarettes. No women; all he needs is a horse and a cigarette.
   B. Beer ads. Some focus exclusively on macho men—athletes, tough guys. Women are conspicuous by their absence. In other ads women do appear, but in subservient roles. They drink beer to show it is not an exclusively masculine habit. But their principal function is to look pretty and to gaze admiringly at their ruggedly handsome men.
   C. Exceptions. There is advertising which exploits the liberated woman (or Madison Avenue's conception of her). Virginia Slims, for example. But even these give to their slender heroines a hard, masculine self-sufficiency, a kind of transsexual version of the macho ideal. Nor do they promote a mature ideal of the relationship between men and women. If they dignify women at all, they do it at the cost of reducing men to tyrannical boobs, which is not the truth either.
2. Television shows
   A. Action shows. In most of these, women play only supporting roles. The protagonists are men. They may be given a little more sensitivity than the tough cops of the past, but they are still variations on that type—"The Streets of San Francisco" or "Maverick."

      Even in a program like "Charlie's Angels," where women play the leads, there is a shadowy male master, and the women act out masculine roles.
   B. Exceptions. In some situation comedies women are allowed to be women—"One Day at a Time," for instance, or "The Jeffersons." These are a sign of changing attitudes, but they are not the dominant note of television.

3. Movies

Here there are more signs of change—*An Unmarried Woman, The Good-bye Girl,* foreign films like *City of Women* or *Wife-Mistress.* However, it is questionable whether such films reflect popular attitudes or those of a small sophisticated audience.

Drawing up lists of topics in this manner combines some of the advantages of free writing and of analysis. In effect you are jotting down ideas freely as they occur to you, but within the framework of categories that will structure your essay.

CONCLUSION

The two approaches to invention do not exclude one another. In fact, they can prove complementary. If you find that thinking analytically about a subject does not produce topics, try free writing as a first step and simply let ideas flow without trying to arrange them. Once you have filled a sheet or two, analyze what you have written. Irrelevant topics can be discarded. Relevant topics can be grouped to bring similarities together, and the groups arranged to form a rudimentary organization.

EXERCISE

1. Select one of the memorable passages you collected in the exercise at the end of the last chapter (or, if you prefer, one of those listed below). Fill one or two pages with free writing. Put down *everything* it suggests to you. You do not have to agree with the idea; in fact, you may find it stupid. The vital goal is to get your own thoughts and feelings on paper without worrying about organization or precision. You need these qualities for a good theme, but you can work at them later.

A. Beware of all enterprises that require new clothes.

Thoreau

B. Know thyself.

Greek maxim

C. "Know thyself?" If I knew myself I'd run away.

Goethe

D. In love always one person gives and the other takes.

French proverb

E. Sex is something I really don't understand too hot. You never know *where* the hell you are. I keep making up these sex rules for myself, and then I break them right away.

J. D. Salinger

F. One is not born a woman, one becomes one.

Simone de Beauvoir

G. In high school and college my sister Mary was very popular with the boys, but I had braces on my teeth and got high marks.

Betty MacDonald

H. So this gentleman said a girl with brains ought to do something else with them besides think.

Anita Loos

I. Woman's virtue is man's greatest invention.

Cornelia Otis Skinner

J. When women kiss it always reminds me of prizefighters shaking hands.

H. L. Mencken

K. No man but a blockhead ever writes, except for money.

Samuel Johnson

L. My language is a universal whore whom I have to make into a virgin.

Karl Kraus

M. Children begin by loving their parents. After a time they judge them. Rarely, if ever, do they forgive them.

Oscar Wilde

N. If the rich could hire people to die for them, the poor could make a nice living.

Yiddish proverb

O. He's really awfully fond of colored people. Well, he says himself, he wouldn't have white servants.

Dorothy Parker

P. God gives himself to men as powerful or as perfect—it is for them to choose.

Simone Weil

Q. If God lived on earth people would throw rocks at his house.

Yiddish proverb

R. The Churches grow old but do not grow up.

Doris Langley Moore

S. Whoever takes up the sword shall perish by the sword. And whoever does not take up the sword (or lets it go) shall perish on the cross.

Simone Weil

T. If we wanted to be happy it would be easy; but we want to be happier than other people, which is almost always difficult, since we think them happier than they are.

Montesquieu

U. [College is] four years under the ethercone breathe deep gently now that's the way to be a good boy one two three four five six get A's in some courses but don't be a grind. . . .

John Dos Passos

V. If a thing is worth doing, it is worth doing badly.

G. K. Chesterton

2. Selecting a second memorable sentence, explore it for topics in a more systematic, analytical way. First decide how to analyze the subject. Then, under each, list possible topics—examples, comparisons and contrasts, reasons, qualifications, and so on.

3. With your instructor's approval, try one or both of the above exercises in cooperation with several classmates. Often a small group of people working together on a problem of invention can trade ideas and stimulate each other.

7

Outlining

INTRODUCTION

An *outline* is a plan view of an essay. By means of numbered and lettered headings, it lays out the main parts and the subdivisions of each.

Not all writers like outlines. Some prefer to let organization evolve from the process of writing, as jazz musicians like to play without written arrangements. There is something to be said for this. Writing is not strictly a linear process with a series of clear-cut steps: inventing topics, arranging these into an outline, drafting an essay from the outline, then revising the draft into the finished form.

The steps do exist, but not in watertight compartments. They interconnect and reciprocate. Invention, for instance, does not end with outlining or drafting. As you draft you continue to invent; new ideas occur and you work them in. And even when you first explore a subject for ideas you are drafting. Whether you invent by free writing or by the more analytical method, you will use words, phrases, sentences that will later appear in the essay.

Depending upon your habits of mind, you may prefer to develop a composition by applying the steps sequentially, or to work more loosely—writing, shaping, rewriting. Neither method is absolutely better, and each demands time and thought. They are simply two approaches to solving the same problem—creating a clear, well-organized piece of writing. Whether we start from one side or the other, we wind up, if we succeed, doing more or less the same thing: we must think about what we want to write, are writing, or have written; we must be alert to seize upon fresh insights as these occur and to get rid of what is redundant or irrelevant; we must remember the reader; and we must work constantly to organize what we say into a coherent whole.

Outlining is one useful tool in building that coherence.

An outline takes one of two forms, depending upon its purpose: the scratch outline and the formal outline. The first is solely for the writer's use. It corresponds to the quick sketch a carpenter might make on the back of an envelope to indicate the dimensions and rough shape of the finished work. A scratch outline may be sloppily done, with frequent scratchings-out and additions (though it will be easier to use if it is neat). It needs to be clear only to the person who made it; it may not be clear to anyone else, but then a scratch outline is not intended for anyone else.

The formal outline is for others. It corresponds to the blueprint an architect draws for a contractor. As a form of communication, a formal outline should be neat, legible, and composed according to rules agreed upon by writer and reader. While the writer uses a scratch outline and then throws it away, he or she usually turns in a formal outline with the essay. It is like the table of contents of a book; indeed, tables of contents are a kind of formal outline.

THE FORMAL OUTLINE

The rules of outlining vary somewhat from one teacher to another. In general, however, they are concerned either with the format, or appearance, of the outline, or with its logic.

RULES OF FORMAT

An outline is a plan, and like all plans it should be visually clear and accurate. A formal outline ought to be arranged on the page so that anyone familiar with the conventions of outlining can tell at a glance what the major divisions of the subject are, how many subdivisions exist in each, and so on.

1. *Adopt a conventional system of numbering and lettering and maintain it consistently.*

The most common system is this:

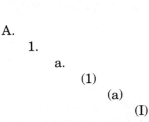

Beginning with capital roman numerals for the major headings, this method preserves a strict alternation between numerals and letters.

2. *Indent subheadings under their primary headings to indicate that they are parts of those topics.*

Except for the initial Roman numerals, the terms "subheading" and "primary heading" are relative. Any heading A is a subhead relative to I or II, but a primary head to the 1 or 2 included under it. Similarly, 1, a subhead to A, is a primary heading for a and b.

Indention has a logical significance, indicating the relationship of part to whole. As a practical matter, indent enough to make that relationship clear, but not so much that by the time you get to step (1) you have only three-quarters of an inch into which to squeeze the topic. Keep the indentation to the minimum necessary for clarity, and indent the same distance in all cases.

3. *Align equal headings vertically.*

All Roman numerals should be under one another in a straight line parallel to the edge of the paper. All the As and Bs fall in a second vertical line a few spaces to the right, and so on.

Ignoring the two rules of indentation and vertical alignment leads to an outline like this:

```
I.
A.
1.
        a.
          (1)
    (2)
              b.
    2.
  B.
```

No reader can grasp this organization at a glance. See how clear the plan becomes with the headings properly indented and aligned:

```
I.
    A.
        1.
            a.
                (1)
                (2)
            b.
        2.
    B.
```

4. *Begin the first word of each heading with a capital.*

It is not necessary, however, to capitalize any other words in a heading (unless they would normally begin with capitals).

5. *Cast all headings consistently as grammatically complete sentences or as fragments.*

Your instructor may disagree here. Some teachers do not care if you mix fragments and complete sentences in an outline; some prefer all fragments; others all sentences. Fragments have the advantage of reducing the number of words in the outline. Here, however, let your instructor be your guide.

RULES OF LOGICAL ANALYSIS

The second set of rules governing the formal outline is logical. Outlining is essentially a technique of analysis, of breaking down a series of topics, step by step, into their immediate constituents. An outline works from the general to the particular, the degree of generality or particularity being signaled by the number or letter assigned to the heading and by its indentation.

6. *Never use only one subdivision under any heading.*

Put another way, this rule states that every I demands at least a II, every A at least a B. The rule holds, of course, not merely for Roman numerals and capital letters but for all degrees of "one-ness" and "A-ness."

The rule of at least two subdivisions is logically necessary. An a indented under a 1 means that the smaller topic is part of the greater. If a is a part, then 1 must have other parts, and these have to be designated by other letters.

The rule does *not* mean that you can have only two subdivisions. The other parts may be two or three things, so that you may have c and d as well as a and b. Nor does the rule mean that you must have two or more subdivisions under every heading; this would make an outline endless. The rule means simply what it says: if you write a 1 you must also write *at least* 2; if you write a you must write *at least* b.

Perhaps you cannot think of a second subdivision after you have made a first. If so, don't make up a meaningless topic just to have something to label 2 or b. Instead, combine the topic you listed as 1 or a with the heading above it, placing a colon or dash between them. You may even wish to go further and eliminate the original main topic altogether, replacing it with what had been the solitary subheading. In either case you will wind up with *no* subdivision and so will have preserved the rule of no 1 without a 2, no a without a b.

7. *Place topics of equal weight in equal headings, topics of unequal value in appropriately unequal headings.*

The form of an outline implies that headings of the same degree (A, B, and C, for instance, or 1, 2, and 3) are of roughly equal weight (or importance, or logical value). This brief outline of ships in the U.S. Navy during World War II violates that implicit logic:

    I. Ships of the line
        A. Battleships
        B. Aircraft carriers

    C. Cruisers
    D. *U.S.S. Morgan*

Heading D violates the rule of logical equality. A, B, and C all designate categories of vessels; D names a specific ship. The proper heading here is "Destroyers." It would be equally illogical to include "Destroyers" as heading 1 under "C. Cruisers."

    8. *Make headings meaningful.*

This rule stretches the meaning of "logical" a little, but it is important. A heading should convey something specific to the reader. Meaningfulness is determined not simply by the heading itself, but by the heading plus everything subsumed under it.

Consider this very short outline:

Books
  I. Books I like
 II. Books I don't like

The headings tells us nothing except that the writer will talk first about books he likes and second about books he doesn't like. But what kind of books these are, what reasons the writer has for preferring this variety to that—of all this we learn nothing.

To say that the essay itself will inform us of these matters says nothing. For one thing, part of the function of a formal outline is to inform. Obviously an outline cannot and should not say everything the essay says. Still it ought to carry more meaning than this. For another thing, so brief and superficial an outline warns of a similarly thin and insubstantial theme. To have meaning, these headings must have further analysis:

Books
I. Books I like
    A. Travel books
        1. Especially about remote and unusual places
           a. *Hashish,* by Henry de Monfreid
           b. *Travels in Arabia Deserta,* by Charles M. Doughty
        2. Also travel books about Europe
           a. *The Path to Rome,* by Hilaire Belloc
           b. *Travels with a Donkey,* by Robert Louis Stevenson
    B. Biographies
        1. Prefer those which are scholarly and historical
           a. *Hitler,* by Alan Bullock
           b. *Thomas Jefferson,* by Duman Malone
        2. And so on. . . .

This is enough to show how the heading "Books I like" can become meaningful. Now the outline tells the reader something and suggests that the paper itself will present some satisfying detail.

9. *Keep a reasonable proportion in the parts of an outline.*

The main headings of an outline should be analyzed to roughly the same degree. There is something wrong, for instance, in an outline that carries the major headings I and II into the fourth subdivision but merely lists heading III—presumably of comparable importance—with no analysis at all.

A rule of thumb will keep you from such gross errors of proportion: carry the analysis of any heading to within one step of the analysis of other headings of the same degree. For example, if headings I and II go to (1) and (2), topic III ought to be analyzed at least to a and b. Or III could be carried one step below, to (a) and (b).

Obviously equality of analysis cannot be applied rigidly. In analyzing any subject you will find some parts more important than others. Usually, however, the rule of one-step equality is enough to allow for such differences. Any time you find that you have grossly violated this rule consider whether you have neglected something important or whether the topic you have slighted is less important than you first thought and might better be placed in a more subordinate position or perhaps dropped altogether.

10. *Be sure the outline is accurate.*

A formal outline is generally turned in with the essay. It should promise nothing the essay does not deliver. Nor should it fail to indicate any major topic developed in the essay.

Since most people find that as they write they move away from a working outline in one spot or another, check your formal outline against your essay and amend it where necessary to make it an accurate plan of what you have actually written.

THE SCRATCH OUTLINE

The rules do not apply in any legal sense to the scratch outline. This is a working outline, for your eyes alone. Even so, it simplifies matters to follow the formal rules.

Once you have invented topics (see the previous chapter), it is only a step to a working outline. As you fit the topics into outline form, proceed with the analysis one division at a time. Decide first upon the primary sections of the essay; next analyze these into their principal subdivisions; then carry each subdivision one stage further; and so on.

Working this way has several advantages. You are less likely to lose sight of your whole plan and become embroiled in details at too early a stage. You will find it easier to maintain a rough equality of analysis, so that the outline does not become disproportionate. And you will be able to cut off the analysis at any stage and still have a complete outline suitable for a theme.

For brief compositions an outline should go through at least the second subdivision:

I.
    A.
        1.
        2.
        3.
and so on.

Whether you need to subdivide further into a, b, (1), (2) depends upon subject and occasion. Longer papers require these more detailed stages; they are probably unnecessary for the typical freshman theme. It is, however, less a fault in outlining to do too much than too little.

A SAMPLE OUTLINE

An outline drawn up for an essay on Virginia Woolf's remark might look like this:

  I. Beginning
    A. Virginia Woolf's remark
    B. People claim the world is changing, which may be true
    C. But change is only beginning
        1. Our culture still expects women to act subservient roles
        2. Evidence can be seen in popular culture, especially in television, both advertising and programs
 II. Television ads
    A. Those depicting women in traditional housewife role
    B. Ads aimed at men
        1. The Marlboro Man
        2. Miller Lite Beer
III. Television shows
    A. Police action programs
        1. "Streets of San Francisco," "Kojak"
        2. Variations of the tough cop
    B. Jiggle television
        1. Women as sex symbols
        2. Heroines who never think
IV. Things are better at the movies
    A. Decline of the John Wayne hero
    B. Films that take women seriously
        1. American movies—*An Unmarried Woman, Private Benjamin*
        2. Foreign films—*Wife-Mistress, City of Women*

V. Closing
   A. Films a sign of change
   B. Still a long way to go
   C. But women will get there

FROM OUTLINE TO ESSAY

A good outline is more than just an essay plan. It contains a good deal of the substance. Headings are the germ of topic sentences and of much of the supporting detail. Moreover, the outline suggests, for short papers at least, a tentative paragraph structure. To find this, estimate how many paragraphs your paper is likely to consist of. No absolute rule exists for such an estimate because expository paragraphs vary considerably in length. A fair average is 120 words. That is very approximate, however, and you ought really to base any estimate upon the average length of the paragraphs you write (the average may change as you acquire more experience in developing them).

Once you have decided roughly how many paragraphs your paper will have, study the outline to find the most likely places for paragraph breaks. Count all the headings of each order—all the roman numerals, capital letters, arabic numerals, and so on—and tentatively locate new paragraphs at the level where the count most closely corresponds to the estimated number of paragraphs. In a short paper of 400 words, for example, developed from an outline having three major headings, the paragraphs would probably begin at I, II, III. In an 800-word version of the same subject, at the As, Bs, Cs; and in a still longer treatment, at the 1s, 2s, 3s.

Of course, any calculation of paragraph structure should be flexible. It would be foolish to bind paragraphing rigidly to an outline. A section you thought would make a single paragraph may require two; another may turn out not to warrant a paragraph to itself and to be better combined with another topic. An outline is a guide, but a guide you should leave behind as the actual writing opens up unforeseen possibilities.

EXERCISE

1. Choosing one of the topics on a particular subject which you invented in the exercise following the previous chapter, work it into a formal outline.

2. Trade outlines with a classmate and check his or her work against the rules of outlining. Look especially for vague, meaningless headings and for As without Bs and 1s without 2s.

3. Outline a short essay you have read or a section or chapter from one of your textbooks. Did the exercise force you to read more carefully? Can you see any places where the organization might have been improved?

8

Drafting and Revising

DRAFTING

A *draft* is an early version of a piece of writing. Most of us cannot compose a paragraph or an essay well on the first attempt. We must write, rewrite it, do it again. These initial efforts are called drafts, in distinction from the final version. As a rule the more you draft, the better the finished essay.

About drafting, the best general advice is the same as for free writing: keep going and don't worry about small mistakes. A draft is not the final product; it is by nature rough and imperfect. It is impossible to compose an essay one perfect sentence at a time. You get lost in the quest for perfection, becoming so involved with detail that you lose sight of the whole composition.

A portrait painter does not start near the top of the blank canvas, paint the subject's hair in complete and finished detail, drop to the forehead, the eyes, and so on. Neither should writers. You must rough out the full composition, then gradually develop and refine, keeping the total effect always in mind.

In a draft, then, accept imperfections. Don't linger over small problems. If you can't remember a spelling, spell the word the best you can with a question mark after it and check it later. If you can't think of an exact term or phrase, put down the closest words you can think of (again with a query). Your primary goal is to develop ideas, to work out a structure. Don't lose sight of these major goals by pursuing minor ones—proper spelling, conventional punctuation, the exact word. These can come later.

There is a limit, however, to the similarity between drafting and the kind of invention we discussed as "free writing." Free writing is exploration and discovery; your pencil should move wherever your mind takes it. A draft is more purposeful. Presumably you know, more or less, what you want to do, and the draft is an early version of an organized composition. Therefore you are not as free as in exploratory writing. In the latter you should never feel the need to go back and start over because you can go anywhere. But in drafting you will occasionally enter blind alleys; then you must stop, back out, and set off in a new direction. The mistake will not be completely unproductive if it tells you where you don't want to go.

Most people find it easier to write drafts with a pen or pencil, though some can work successfully on a typewriter. If you draft in longhand, skip every other line and leave adequate margins: you will need the space for revision. If you type, double space. Use only one side of the paper, reserving the other for major revisions or additions. It is a good idea, when you number the pages of your draft, to put a brief identifying title on each sheet: "First draft, p. 1," "Second draft, p. 3."

In a composition of any length, consider stopping every so often at a convenient point. Read over what you've written, making whatever corrections or improvements you can; then type what you've done. Seeing your ideas in print will usually be reassuring. If you don't have a typewriter, copy the section neatly in longhand; the effect will be much the same. Turn back to the draft; work out the next section; stop again and type. The alternation between drafting and typing will relieve the strain of constant writing and give you a chance to pause and contemplate what you have accomplished and what you ought to do next.

In all this advice, however, we would not be dogmatic. People vary enormously in their writing habits; what works for one fails for another. The best rule is to find a time and a place for writing that enable you to work productively and to follow a procedure you find congenial. You may like to draft in green or purple ink, to listen to music as you write, to compose the entire draft of a ten-page essay and then retype the whole thing instead of doing it section by section. Do what works for you.

A SAMPLE DRAFT

Virginia Woolf remarked that "women have served all these centuries as looking glasses possessing the . . . power of reflecting the figure of a man at twice its size." There can't be much argument that this has been the case, at least in Western societies. People claim that the world is changing and that women are being liberated from their traditional subservient role, that the mirror has cracked. Well, maybe it has, but if there is any real change, it is no more than a beginning. There is plenty of evidence that our culture still generally expects women to serve men. The evidence can be seen in advertising and television shows, which reflect and reinforce the real values of our world.

Television commercials for soaps and foods and household goods depict women in traditional roles—cleaning the bathroom, washing clothes and dishes, cooking. Now and then a woman appears in the male role of engineer or business executive, but not very often.

Even worse are the ads aimed at men. The macho male who plugs Marlboro cigarettes represents a kind of male ideal—the man on horseback, dominant, untrammeled by females, king of all he surveys. Of course, he's usually alone. The ads for Miller Lite Beer focus on athletes, active or retired, tough guys seen in exclusively male settings. The message is that "men drink beer," but impressionable adolescents will get another message: "This is how men act." If women appear in beer ads at all, it is only in subservient roles. They drink beer to show that it is not an exclusively masculine habit, even right out of the bottle or can. But their most important function is to gaze admiringly at their ruggedly handsome men.

Not only the ads on television but many of the shows reinforce women's traditional role of supporting men. Consider the popularity of police-action shows—"The Streets of San Francisco," for instance, or "Kojak," or even "The Rockford Files." In recent years the heroes may have been allowed a little more sensitivity than in the past, and they may pay lip service to liberal values, but they are still variations of the tough cop. They are still kings of a man's world, which women enter only as true-blue sweethearts or wives or, occasionally, in the role of female criminal.

On so-called "jiggle" television, women, who are invariably bosomy and curvy, are exploited as sex symbols. Their lives center on sex. They spend their time either scheming to trap the man they want or fending off men they don't want. They remind me of a character in an Anita Loos story: "So this gentleman said a girl with brains ought to do something else with them besides think." Even if they have brains these girls don't think and they don't learn. They just look sexy, pant after men, and make wisecracks.

Things are a bit better in movies. The John Wayne type is no longer so dominant, though Clint Eastwood and Burt Reynolds still play the macho game. A new male type has emerged, not so big or physically dominating, less tough, more vulnerable. More importantly, some movies do take women and their rebellion against male oppression seriously—*An Unmarried Woman, The Good-bye Girl, Private Benjamin,* for instance. Some foreign films especially are sensitive to the changing attitudes of women. In Fellini's *City of Women,* for example, women tell it like it is.

Perhaps such films are a hopeful sign, signaling a change in popular attitudes. Still the change is hardly begun. Films, especially foreign imports, are not popular in the same sense that beer ads and television sitcoms are. Fellini is still ahead of middle-class attitudes and prejudices. We still have some distance to go before women can stand up on CBS and ABC and NBC and tell it like it is.

But I think the looking glass is cracking, and women, like Alice, will eventually struggle out of a man-made wonderland. Men had better come out too because it will be awfully lonesome in there.

REVISING

Revising is not simply correcting and polishing your draft. Unless you have done an exceptional job on your first effort, revising involves more fundamental improvements. Nor is it a completely separate activity which follows drafting in a neat "step-two" sequence. The two relate more intimately. Most of us are constantly making small revisions even as we draft, either in our minds or by literally scratching out one word and writing another. The earlier suggestion, for instance, about typing a section of your draft and then going back to drafting the next part involves revision undertaken in the midst of drafting.

Yet despite this close back-and-forth relationship, drafting and revising differ in emphasis. Drafting is active creating; revision, a more thoughtful and critical form of creativity. As the writer of a draft you must keep going and not worry about small mistakes or imperfections. As a reviser you become a reader, a demanding reader who expects perfection. When you write you see your words from inside; you know what you want to say and easily overlook lapses of clarity obvious to readers. When you revise you try to put yourself in the reader's place. Of course you cannot get completely outside your own mind, but you can think about what readers know and do not know, what they believe and consider important. You can ask yourself if what is clear to you will be equally clear to them.

To revise effectively, force yourself to read slowly. Some people do this by holding a straight-edge so that they can read only one line at a time, one word at a time if possible. Others read their work aloud. This is more effective, though obviously you cannot do it on all occasions. But when you can, reading aloud not only slows you down, it distances you from the words, putting you nearer to that objectivity which successful revision requires. Moreover, it brings another sense to bear. You hear your prose as well as see it, and ears are often more trustworthy than eyes. They detect an awkwardness in sentence structure or a jarring repetition of sound that the eyes pass over. Even if you're not exactly sure what's wrong, you can hear that something is, and you can tinker with the sentences or the words until they sound better. Usually they will be better.

It helps to get someone else to listen to or to read your work and respond. Such help, however, must be used honestly. The other reader should serve only as a critic, pointing out what seems awkward or confusing. It is up to you to solve the problem.

Keep a pencil in hand as you revise (some like a different color). Mark your paper freely. Strike out imprecise words, inserting more exact terms above them (here is the advantage of skipping lines). If you think of another idea or of a way of expanding a point already present, write a marginal note to yourself, phrasing it precisely enough so that when you come back to it in an hour or a day it will make sense. If a passage isn't clear, write "clarity?" in the margin. If you sense a gap between paragraphs or between sentences within a paragraph, draw an arrow from one to the other with a question mark. Above all, be ruthless in striking out what is not necessary. A large part of revision is chipping away excess verbiage.

You will become more aware of what to look for when you revise, but we shall mention a few basics here. Most fundamental is clarity. If you suspect a sentence may puzzle a reader, figure out why and revise it. Almost as important is emphasis. Strengthen important points by expressing them in short or unusual sentences. Learn to position modifiers so that they interrupt a sentence and throw greater weight on important ideas. Look for unsupported generalizations. Even when it is clear, a generalization gains value from illustrative detail.

Sharpen your diction. Revise awkward repetitions of the same word. Replace vague abstract terms with precise ones having richer, more provocative connotations. Watch for failures of tone: don't offend readers and don't strike poses.

Be alert for errors in grammar and usage and in spelling and typography. Make sure your punctuation is adequate and conventional but no more frequent than clarity or emphasis requires. Guard against mannerisms of style. All of us have them: beginning too many sentences with "and" or "but"; interrupting the subject and verb; writing long, complicated sentences. None of these is wrong, but any word or sentence pattern becomes a mannerism when it is overworked. One "however" in a paragraph may work well; two attract a reader's notice; three will make him or her squirm.

A SAMPLE REVISION

Virginia Woolf once remarked that

"women have served all these centuries as

looking glasses possessing the . . . power

of reflecting the figure of a man at twice

its size." There can't be much argument

that this has been the case, at least in

Western societies. People claim that the

world is changing and that women are

being liberated from their traditional

subservient role, that the mirror

                                                               But

These changes improve has cracked. Well, maybe. ~~it has, but~~ if
emphasis.

there is any real change, it is no more than

a beginning. There is plenty of evidence

that our culture still generally expects

Clarifies that primarily television advertising is meant.

women to serve men. The evidence
is especially strong in television, both
~~can be seen in advertising and television~~
advertising and shows,
~~shows,~~ which reflect and

reinforce the real values of our world.

Deadwood. Context makes clear that what follows are examples.

T
~~A good example from~~ television

~~advertising are the~~ commercials for soaps

and foods and household items~~, which~~

depict women in traditional roles—cleaning

the bathroom, washing clothes and dishes,

The paragraph is underdeveloped. The addition of illustrative details reinforces the point.

cooking. It is pathetic to see how happy the

housewives are in these jobs, how proud and

pleased that their floors shine, that their

dishes reflect the envious faces of their

friends, that Tom's, Dick's, or Harry's shirts

Wordy. And it is unwise even to seem to imply that an engineer or business executive is a "male role."
Begin new sentence for emphasis.

are finally rid of "ring around the collar."

Now and then a woman appears
     as an
~~in the male role of~~ engineer or business
     . But
executive~~, but~~ not very often.

The rephrasing is more economic and more precise.

Even worse are the ads aimed at men.

A more exact word. Also the similarity of sound points up the contrast between dominance and domesticity.

cowboy
The ~~macho male~~ who plugs Marlboro
the macho
cigarettes represents ~~a kind of male~~

ideal—the man on horseback, dominant,
domesticity
untrammeled by ~~females~~, king of all he

Delete this sentence. The idea is not really connected with the topic of the paragraph.

surveys. ~~Of course, he's usually alone.~~ The

ads for Miller Lite Beer focus on athletes,

active or retired, tough guys seen in

exclusively male settings. The message

Use capitals for emphasis.

MEN
is that "~~men~~ drink beer,"

More accurate.

but impressionable adolescents will
an additional
get ~~another~~ message: "This is how
MEN
~~men~~ act." If women appear in beer ads at

all, it is only in subservient roles. They

drink beer to show that it is not an

exclusively masculine habit, even

"However" avoids the mannerism of too many initial "buts"; putting the word in an interrupting position creates a stronger sentence and one of varied movement.

T
right out of the bottle. ~~But~~ their most
, however ,
important function is to gaze admiringly

at their ruggedly handsome men.

Not only the ads on television but many

of the shows reinforce women's traditional

role of supporting men. Consider the

popularity of police-action shows—"The

Streets of San Francisco," for instance, or

"Kojak" or even "The Rockford Files." In

recent years the heroes may have been

allowed a little more sensitivity than in

the past, and they may pay lip service to
                                       and feminist
liberal‸values, but they are still

Add for precision.

variations of the tough cop. ~~They are~~
                               S
~~still~~ still kings of a man's world, which women

Dropping the subject and
verb turns this sentence
into a fragment and
makes it stronger.

enter only as true-blue sweethearts or
                                       as
wives or, occasionally, ~~in the role of~~
                        bitch
~~female~~ criminals.

Wordy. The revision also
avoids the word "role,"
which should not be
overused.

A stronger word. A bit
shocking, but striking for
that reason.

    On so-called "jiggle television," women,

who are invariably bosomy and curvy, are

exploited as sex symbols. Their lives center
                                , even at work,
on sex. They spend their time‸either

Add for emphasis.

scheming to trap the man they want or

fending off men they don't. ~~want.~~
                   are like
"Want" can be supplied
from the context.

They ~~remind me of~~ a character in an Anita

Loos story: "So this gentleman said a girl

with brains ought to do something else with

This late introduction of
the first-person pronoun
is awkward in an essay
which has so far been
impersonal. A first-
person point of view
should be established in
the beginning.

them besides think." Even if they have

The separate, short sentence is more emphatic.

                 T

brains these girls don't think. ~~and t~~hey

don't learn. They just look sexy, pant after

men, and make wisecracks.

Placing "in movies" first more clearly signals the change of topic.

   In movies t

Things are a bit better ~~in movies~~.

"Prevalent" is more precise; also don't overuse "dominant."

The John Wayne type is no longer so
prevalent
~~dominant~~, though Clint Eastwood and Burt

Reynolds still play the macho game. A new

male type is emerging, not so big or

physically dominating, less tough, more

vulnerable. More importantly, some movies

do take women and their rebellion against

male oppression seriously—*A Woman*

Under the Influence, An Unmarried

The dash makes clear that these are examples.

Woman, Private Benjamin, ~~for instance~~.

Some foreign films especially are sensitive

to the changing attitudes of women. In

Fellini's *City of Women,* for example,
speak from their minds and hearts.

Avoid cant phrases such as "tell it like it is."

women ~~tell it like it is~~.

   Perhaps such films are a hopeful sign,

signaling a change in popular attitudes.

Still the change is hardly begun. Films,

"Foreign imports" is redundant.

                  ones
especially foreign ~~imports~~, are not popular

There are too many "stills" in this paragraph. This and the following one can be dropped with no loss of meaning.

in the same sense that beer ads and television sitcoms are. Fellini is ~~still~~ ahead of middle-class attitudes and prejudices. We

"Considerable" is more exact. Add "real women" for emphasis.

considerable
~~still~~ have ~~some~~ distance to go before
, real women,
women can stand up on CBS and ABC and NBC ~~and tell it like it is~~.

Again an awkward shift to the first person.

                                                                    T
But sooner or later they will. ~~I think~~ ~~t~~he looking glass is cracking, and women,

Add for clarity. For Alice, entering Wonderland was a positive act.

in reverse
like Alice, will eventually struggle out of a man-made wonderland. Men had better

A new sentence is not only stronger here, it also makes a better closing.

                                                  It
come too. ~~because it~~ will be awfully lonely in there.

Probably you would not write such extensive marginal notes to yourself, but those in the example suggest how you should be thinking. The revisions, observe, are toward precision, emphasis, and economy.

How many drafts and revisions you go through depends upon your energy, ambition, and time. Most people who publish feel they stopped one draft too soon. Provided you do it neatly, you may revise even your final copy, changing a word or a sentence before turning in an assignment. Check with your instructor, however, before you correct submitted work. Composition teachers are usually willing to accept revisions, so long as they are not so numerous or messy that they interfere with reading. Some do want clean copy—that is, pages with no corrections, additions, or deletions.

FINAL COPY

Whether or not you are allowed to revise it, your final copy should always be neat and legible. Keep margins of an inch or more. If you type, use standard typing paper and type on only one side. Double space and correct typos by erasure or tape, not by overstriking. Keep the keys clean and invest now and then in a new ribbon. If you write in longhand, use conventional, lined composition paper. Unless the instructor says otherwise, skip every other line and write only on one side. Leave adequate margins for corrections and comments. Take time to write legibly. No one expects a beautiful copperplate hand, but it is fair to ask for readability.

CONCLUSION

Drafting and revising are closely related: the systole and diastole of composition. The draft surges from thought into words; the revision draws back into thought. Without a draft you have nothing to revise, but without revision your draft remains shapeless and incoherent. Both are necessary for clear, effective writing.

When you draft, write more than you pause: produce words. When you revise, pause more than you write: think, be critical, read as a stranger to your own ideas.

EXERCISE

1. The following paragraph reworks a description of a shipwreck by a famous American author. The rewriting has deliberately cluttered up the original with unnecessary words (about half again as many). The organization of the paragraph has not been changed. But in some places the wordier version divides into several sentences ideas that were stated as one, or joins in a single sentence ideas that were separately stated.

Revise the paragraph, removing what is repetitious or clearly implied. Try to get out 50 or 60 words. See also if you can improve the places where sentences were awkwardly broken or joined.

When you have done, your instructor can read or show you the original paragraph.

> The ship was a brig. (A brig is a sailing vessel that has two masts, both of which are square-rigged.) It was named the *St. John,* and it came from Galway, Ireland, and it was carrying a load of Irish emigrants to the United States. The ship was wrecked on Sunday morning. Now it was Tuesday morning, two days later. The sea was still breaking violently on the rocks that lined the shore. There were eighteen or twenty of the same large boxes I have mentioned in my paragraph above, that were lying on a green hillside, which was a few rods from the water. The boxes were coffins. They were surrounded by a crowd of people. The dead bodies which had been recovered from the sea were twenty-seven or eight in all. They had been collected there on the hillside near the coffins. Some of the crowd of people were nailing down the lids of the coffins. Others were carting the boxes away, and others were lifting the lids of the coffins which were yet loose in order to peep under the cloths. For each body, with such clothing as it still had, was covered loosely with a white sheet. I witnessed no signs of grief. But there was a sober despatch of business which was affecting, and one man was seeking to identify a particular body. An undertaker or carpenter was calling to another man in order to know in what box a certain child had been put by the people who recovered her body.

2. Draft a short essay of 500–600 words based on the outline you drew up at the end of the preceding chapter. Type or use an ordinary lead pencil or blue or black ink. Skip every other line and leave adequate margins.

Next, using a red pencil or pen, revise your draft. Cross out what is unnecessary or unclear. Where it is required, add words in the spaces between lines. In the margins explain briefly why you made the changes.

PART TWO

THE ESSAY

INTRODUCTION

Essay is a difficult word to define. As a noun it means basically an "attempt" or "effort" to do something. In this meaning *essay* was applied in the sixteenth and seventeenth centuries to compositions that discussed a general question from a limited and often personal point of view. Such writings were "essays" ("tries" we might say today) at treating their subjects and they made no claim to thoroughness or authority.

Something of that original sense of a tentative effort remains in the modern use of *essay*: a relatively short, self-contained piece of prose dealing with matters of fact, belief, or opinion. It is difficult to pin down the term beyond this loose sense, and probably not very profitable to try. The very looseness of the label may be a virtue, allowing us to include a great variety of things.

Most college writing—freshman themes, critical studies in literature classes, term papers for history or sociology—is conventionally said to be in essay form. Like the paragraph and the sentence, the essay is a unit of composition. Although more difficult to describe and define, it is an entity with a beginning, a middle, and an end. The paragraphs that compose these parts must tie together so that the essay is an integral whole and not a random collection of paragraphs.

Our discussion of the essay falls into three parts. First we look at its basic sections: beginning, middle, and closing. Next we examine three aspects of the essay which pervade its entire structure: point of view, persona, and tone. Finally we analyze the differences between tight and loose organization.

CHAPTER 9

Structure of the Essay: Beginning

INTRODUCTION

Readers approach an essay with a set of questions: What is this about? What does the writer plan to do (and not do)? Will it interest me? What kind of person is the author?

To begin effectively you must answer these questions, one way or another. From the writer's point of view, beginning means announcing and limiting the subject, indicating the plan of the paper, engaging the reader's attention, and establishing an appropriate tone and point of view.

Not all of these matters have equal importance. Announcing and limiting the subject are essential. Laying out the plan of the paper and angling for the reader's interest, on the other hand, depend upon your purpose and audience. Tone and point of view are inevitable: whenever you write, you imply them. In the beginning, then, you must establish a tone and point of view conducive to your purpose.

In discussing these aspects of beginning, we frequently speak of the "beginning paragraph." In fact, the length of a beginning depends upon the length and complexity of what it introduces. In a book the opening might take an entire chapter with dozens of paragraphs. In a short theme it might take only a sentence, not even a separate paragraph. But whatever their length, all effective openings fulfill the same functions.

Begin.
announce.

ANNOUNCING THE SUBJECT

In announcing a subject you have two choices: (1) whether to be explicit or implicit, and (2) whether to be immediate or to delay.

EXPLICIT AND IMPLICIT ANNOUNCEMENT

In explicit announcement you literally state in some fashion or other, "This is my subject." The philosopher Alfred North Whitehead begins *Religion in the Making* like this:

It is my purpose to consider the type of justification which is available for belief in the doctrines of religion.

The words "It is my purpose" make this an explicit statement of purpose. The opening would have been implicit had Whitehead written:

Belief in the doctrines of religion may be justified in various ways.

This beginning does not literally tell readers what the subject is, but it plainly implies it.

Because of its clarity, scholars and scientists writing for their colleagues often use explicit announcement. In less formal situations explicit announcement, when used at all, should be kept simple, avoiding words like *purpose, intention, this theme, paper, essay*. For instance, the following plain, informal statement of subject (a domestic murder in California) works very well:

This is a story about love and death in the golden land, and begins with the country.

Joan Didion

On the whole, however, it is best to avoid explicit announcement in the work you do in composition class. "The purpose of this paper is to contrast high school and college" makes a heavy-handed theme opening. It is better to establish a subject by implication: "High school and college differ in several ways." Readers will understand you; don't hit them over the head.

Implicit announcements can appear as rhetorical questions. The English essayist Hilaire Belloc begins an essay on "The Historian" like this:

What is the historian?
The historian is he who tells a true story in writing.
Consider the members of that definition. . . .

Thus Whitehead might have begun:

How may belief in the doctrines of religion be justified?

Or the theme on high school and college might have opened:

In what ways do high school and college differ?

In the last case, however, the question sounds mechanical. That is a disadvantage of opening with a rhetorical question. While better than no announcement at all, better even than the clumsiness of "The purpose of this theme is," it is not very original. Phrase your announcement as a rhetorical question only when you can do it with originality or when all other alternatives are less attractive.

The same advice holds for beginning with a dictionary definition, another way of announcing a subject implicitly. Nothing is inherently wrong in starting off with a quote from a reputable dictonary, but it has been done so often that it is trite. Of course a clever or unusual definition makes a good opening. John Dos Passos's description of college as "four years under the ethercone" is certainly more novel and provocative than the dictionary definition and might make a fine opening.

When the purpose of an essay is to define a word or idea, it is legitimate to start from the dictionary. But these qualifications admitted, the dictionary quotation, like the rhetorical question, has been overworked as a way of implicitly announcing the subject.

IMMEDIATE AND DELAYED ANNOUNCEMENT

Your second choice involves whether to announce your subject immediately or to delay. This opening line of an essay called "Selected Snobberies" by the English novelist Aldous Huxley falls into the first category:

All men are snobs about something.

Letting readers know your subject at once is a no-nonsense, business-like procedure. But an immediate announcement may not hold much allure. If the subject is of high interest, or if the statement is startling or provocative (like Huxley's), it will catch a reader's eye. Generally, however, immediate announcement is longer on clarity than interest.

So you may want to delay identifying the subject. Delay is usually achieved by beginning broadly and narrowing until you get down to the subject. The critic Susan Sontag, for instance, uses this beginning for an essay defining "Camp" (a deliberately pretentious style in popular art and entertainment):

Many things in the world have not been named; and many things, even if they have been named, have never been described. One of these is the sensibility—unmistakably modern, a variant of sophistication but hardly identical with it—that goes by the name of "Camp."

Less commonly the subject may be delayed by focusing out, that is, opening with a specific detail or illustration and broadening to arrive at the subject. Aldous Huxley opens an essay on "Tragedy and the Whole Truth" in this manner:

There were six of them, the best and bravest of the hero's companions. Turning back from his post in the bows, Odysseus was in time to see them lifted, struggling, into the air, to hear their screams, the desperate repetition of his own name. The survivors could only look on, helplessly, while Scylla "at the mouth of her cave devoured them, still screaming, still stretching out their hands to me in the frightful struggle." And Odysseus adds that it was the most dreadful and lamentable sight he ever saw in all his "explorings of the passes of the sea." We can believe it; Homer's brief description (the too-poetical simile is a later interpolation) convinces us.

Later, the danger passed, Odysseus and his men went ashore for the night, and, on the Sicilian beach, prepared their supper—prepared it, says Homer, "expertly." The Twelfth Book of the *Odyssey* concludes with these words: "When they had satisfied their thirst and hunger, they thought of their dear companions and wept, and in the midst of their tears sleep came gently upon them."

The truth, the whole truth and nothing but the truth—how rarely the older literatures ever told it! Bits of the truth, yes; every good book gives us bits of the truth, would not be a good book if it did not. But the whole truth, no. Of the great writers of the past incredibly few have given us that. Homer—the Homer of the *Odyssey*—is one of those few.

It is not until the third paragraph that Huxley closes in on his subject, of which the episode from the Odyssey is an example.

Delayed announcement has several advantages. It piques readers' curiosity. They know from the title that the opening sentences do not say exactly what the essay will be about, and they are drawn on to see what it is about and how the opening connects with it. Curiosity has a limit, however, and you cannot tease readers for long.

Then, too, the opening sentences can be used to clarify the subject, whether by applying some of the background necessary for understanding or by offering a particular example. And finally, delayed announcement can be interesting in its own right. There is a pleasure like that of watching a high-wire performer in observing an accomplished writer close in on his or her subject.

More immediate announcement, on the other hand, is called for in situations where getting to the point is more important than angling for readers or entertaining them. How you announce your subject, then, as with so much in writing, depends upon purpose, that is, upon your reason for addressing your readers.

LIMITING THE SUBJECT

*Begin.
limit*

In most cases a limiting sentence or clause must follow the announcement of the subject. Few essays (or books, for that matter) discuss *all* there is to say; they treat some aspects of a subject but not others. As with announcement, limitation may be explicit or implicit. The first—in which the writer says, in effect, "I shall say such and so"—is more common in formal, scholarly writing. The grammarian Karl W. Dykema begins an article entitled "Where Our Grammar Came From":

The title of this paper is too brief to be quite accurate. Perhaps with the following subtitle it does not promise too much: A partial account of the origin and development of the attitudes which commonly pass for grammatical in Western culture and particularly in English-speaking societies.

In less formal situations one should limit the subject less literally, implying the boundaries of the paper rather than literally stating them:

Publishers, I am told, are worried about their business, and I, as a writer, am therefore worried too. But I am not sure that the actual state of their affairs disturbs me quite so much as some of the analyses of it and some of the proposals for remedying what is admittedly an unsatisfactory situation.

Joseph Wood Krutch

Without literally saying so, Krutch makes it clear that he will confine his interest in the problems publishers face to criticizing some of the attempts that have been made to explain and solve those problems.

Besides being explicit or implicit, limitation may also be positive or negative (or both). The paragraphs by Dykema and Krutch tell us what the writers *will* do; they limit the subject in a positive sense. In the following case the English writer and stateman John Buchan tells what he will *not* do (the paragraph opens the chapter "My America" of his book *Pilgrim's Way*):

The title of this chapter exactly defines its contents. It presents the American scene as it appears to one observer—a point of view which does not claim to be that mysterious thing, objective truth. There will be no attempt to portray the "typical" American, for I have never known one. I have met a multitude of individuals, but I should not dare to take any one of them as representing his country—as being that other mysterious thing, the average man. You can point to certain qualities which are more widely distributed in America than elsewhere, but you will scarcely find human beings who possess all these qualities. One good American will have most of them; another, equally good and not less representative, may have few or none. So I shall eschew generalities. If you cannot indict a nation, no more can you label it like a museum piece.

Some limitation—explicit or implicit, positive or negative—is necessary at the beginning of most essays. Term papers, long formal essays whose purpose is to inform, technical and scholarly articles, all may have to engage in extensive boundary fixing to avoid misleading or disappointing the reader. The usual freshman theme, however, does not require very much limitation. Readers learn all they really need to know by an opening sentence like this:

College is different from high school in several ways—especially in teaching, homework, and tests.

The final phrase conveys the limitations, following the announcement in the first clause of the sentence. The subject is a contrast between college and high school, the focus is on college, and the contents are limited to three specific points of difference. That is limitation enough for a brief, informal essay, and the writer can get on with the discussion without a heavy statement like this:

I shall limit the contrast to teaching methods, homework, and tests.

No one can give you a rule by which to test whether you have limited a subject sufficiently. All you can do is to put yourself in the reader's place and ask if it is clear (whether by direct statement or by implication) what the essay will do and what it will not do.

INDICATING THE PLAN OF THE ESSAY

Another function of the beginning, though not an invariable one, is to clarify how the essay will be organized. The writer has the plan in mind when he or she composes the beginning paragraph (or at any rate revises it). The question is: Should the plan be revealed to the reader?

Writers often do consider it necessary. Harold Mattingly begins his book *Roman Imperial Civilization* with this paragraph:

The object of this first chapter is to give a sketch of the Empire which may supply a background to all that follows: to explain what the position of Emperor from time to time was, how it was defined in law, how it was interpreted by the subjects; then, around the Emperor, to show the different parts of the State in relation to one another and to him. Later chapters will develop particular themes. We shall have to consider at the close how far the constitution of the Empire was satisfactory for its main purposes, how much truth there is in the contention that imperfections in the constitution were a main cause of Decline and Fall.

The paragraph indicates not only the plan of the first chapter and that of the whole book, but also how the opening chapter fits into the larger organization.

Even with subjects less complex and grand than the Roman Empire, writers may wish to tell us how they intend to develop their essays:

I want to tell you about a woodsman, what he was like, what his work was, and what it meant. His name was Alfred D. Teare and he came originally from Nova Scotia, but all the time I knew him his home was in Berlin, New Hampshire. Probably the best surveyor of old lines in New England, he was—in his way—a genius.

Kenneth Andler

This straightforward paragraph not only announces and limits the subject but also reveals something about the organization of the essay.

Readers are prepared for a three-part structure: Teare as a person, the nature of his work, and the significance of that work. They assume that the writer knows his craft—as in this case he does—and that the order in which he mentions these aspects of his subject reflects the order in which he treats them. The plan has been clarified implicitly but nonetheless effectively.

Establishing your plan in the beginning has several virtues. It eases the reader's task. Knowing where they are headed, readers can follow the flow of ideas. An initial indication of the organization also simplifies some of the writer's problems. When a writer can assume that readers understand the general scheme of the essay, he or she finds it easier to move from point to point. Well-prepared readers are less likely to need elaborate transitions.

As with limiting the subject, one cannot set down clear-cut rules about when to reveal the plan. Generally it is wise to indicate something about the organization of compositions which are relatively long and which fall into several well-defined parts—term papers, for example. The shorter, simpler English essay less often requires that its plan be established in the opening paragraph.

When you must indicate your plan, do so as subtly as you can. The imaginary theme about high school and college that begins:

> College is different from high school in several ways—especially in teaching methods, homework, and tests. . . .

clearly implies the three parts of the essay and their order. In longer, more scholarly work you may occasionally feel it desirable to explain your organization explicitly. Be sure that your subject is substantial enough and your purpose serious enough to support such a beginning.

Begin.
interest

INTERESTING THE READER

Sometimes you can take reader interest for granted. Scholars and scientists writing for learned journals, for instance, do not have to make much overt effort to catch their readers' eyes. More commonly a writer's audience includes at least some readers whose interest must be deliberately sought. Several strategies for doing this are available.

STRESSING THE IMPORTANCE OF THE SUBJECT

Treat the reader as a reasonable, intelligent person with a desire to be well informed and say, in effect: "Here is something you should know or think about." The American poet and critic John Peale Bishop begins an essay on Picasso with this sentence:

> There is no painter who has so spontaneously and so profoundly reflected his age as Pablo Picasso.

And the historian J. E. Neale introduces a discussion of the politics of Elizabethan England by relating them to our own time:

> We are living in an age of ideological conflict and are troubled by the strains it imposes on society. The totalitarian state spreads alarm; we fear doctrinaires with their subversive organizations; we suspect fellow-travellers; we endure the Cold War; we think of quislings and the fifth column as instruments of foreign conquest. The fanatic's way of life we know to be dynamic; and though we say "It shall not happen here," we are not inclined, after our experience of the last twenty years, to boast that it cannot happen here. We are at odds about the policy we should pursue. Passion breeds passion; and unless we feel deeply about our own ideals, inevitably we are at a disadvantage against the enemies of society. Moderation, which is a liberal virtue, takes on a watery appearance. It seems uninspired and inglorious, prone to defeat.
>
> In such a dilemma it may be useful to turn to history, which is the treasury of recorded experience. History never repeats itself, but it offers analogies. Just as the historian, consciously or unconsciously, uses the present to understand the past, so there is a reverse process. It is the most weighty of the justifications for the writing and the study of history; and a nation which is historically minded is more likely to be fortunate in its affairs than one which is not.
>
> For an analogy with our own times we cannot do better than turn to the Elizabethan period in English history. . . .

AROUSING CURIOSITY

This is usually a more effective strategy than stressing the importance of the subject. You may play upon curiosity by opening with a short factual statement which raises more questions than it answers. W. Somerset Maugham begins a biographical sketch:

> I knew he was drunk.

"Who," we ask, "is 'he'? How did Maugham know he was drunk?"

The astronomer Sir Arthur Eddington begins a chapter in his book *The Philosophy of Physical Science*:

> I believe there are 15,747,724,136,275,002,577,605,653,691, 181,555,468,044,717,914,527,116,709,366,231,425,076,185,631,031, 296 protons in the universe and the same number of electrons.

It would be a curiously incurious reader who would not boggle at this and read on to learn how the writer arrived at so precise a figure.

A short step from such interest-arousing factual openings is the cryptic beginning, that is, a mysterious or not quite clear statement. Charles Lamb opens an essay:

> I have no ear.

We soon learn that he means "no ear for music," but for a moment we are startled.

Here is another example, the beginning of an essay on the abdication of the British king, Edward VIII, in 1936:

> This is the second attempt.
> I had heard him over the air. I had heard an English king explain to his people his renunciation of the British throne.
>
> Arthur H. Little

We must read on to discover that the writer means his "second attempt" to convey how deeply he was affected by the abdication speech.

To be effective a cryptic opening must not simply be murky. It must combine clarity of statement with mystery of intent. We know *what* it says, but we are puzzled about *why*. The mystery has to be cleared up rather quickly if the reader's interest is to be retained. For most of us curiosity does not linger; without satisfaction it goes elsewhere.

Carrying mystification a little further, you may open with a rhetorical paradox—a statement which appears to contradict reality as we know it. Hilaire Belloc begins his essay "The Barbarians":

> It is a pity true history is not taught in the schools.

Readers who suppose true history is taught may be annoyed, but they are likely to go on.

Sometimes mystification takes the form of a *non sequitur*, that is, an apparently nonlogical sequence of ideas. An enterprising student began a theme:

> I hate botany, which is why I went to New York.

The essay revealed a legitimate connection, but the seeming illogicality fulfilled its purpose of drawing in the reader.

AMUSING THE READER

Aside from arousing their curiosity, you may attract readers by amusing them. One strategy is to open with a witty remark, often involving an allusion to an historical or literary figure. Professor Colin Cherry begins the book *On Human Communication*:

> Leibnitz, it has sometimes been said, was the last man to know everything.

Francis Bacon opens his essay "On Truth" with this famous sentence:

> *"What is truth?"* said jesting Pilate and would not stay for an answer.

A contemporary writer alludes both to Pontius Pilate and to Bacon by adapting that beginning for her essay "What, Then, Is Culture?":

"What is truth?" said jesting Pilate, and would not stay for an answer.
"What is culture?" said an enlightened man to me not long since, and
though he stayed for an answer, he did not get one.

Katherine Fullerton Gerould

Another variety of the entertaining opening is the anecdote. Anecdotes have a double value, attracting us once by their intrinsic wittiness and then by the skill with which writers apply them to the subject. In the following opening Nancy Mitford describes the history of the French *salon,* a social gathering of well-known people who discuss politics, art, and so on:

"What became of that man I used to see sitting at the end of your table?" someone asked the famous eighteenth-century Paris hostess, Mme. Geoffrin.

"He was my husband. He is dead." It is the epitaph of all such husbands. The hostess of a salon (the useful word salonnière, unfortunately, is an Anglo-Saxon invention) must not be encumbered by family life, and her husband, if he exists, must know his place.

The salon was invented by the Marquise de Rambouillet at the beginning of the seventeenth century. . . .

Mitford's story is amusing, in a cynical fashion. More important, it leads naturally into her subject. *Naturally*—that is important, for an opening anecdote fails if forced upon the subject from the outside.

Still another entertaining opening strategy is the clever and apt comparison. It may be an analogy, as in the following passage by Virginia Woolf, the first part of the opening paragraph of her essay "Reviewing":

In London there are certain shop windows that always attract a crowd. The attraction is not in the finished article but in the worn-out garments that are having patches inserted in them. The crowd is watching the women at work. There they sit in the shop window putting invisible stitches into moth-eaten trousers. And this familiar sight may serve as an illustration to the following paper. So our poets, playwrights, and novelists sit in the shop window, doing their work under the eyes of reviewers.

Notice, incidentally, the skill with which Woolf focuses down upon her subject.

A comparison calculated to arouse interest may be a simile or metaphor. G. K. Chesterton wittily begins an essay "On Monsters" with this metaphorical comparison:

I saw in an illustrated paper—which sparkles with scientific news—that a green-blooded fish had been found in the sea; indeed a creature that was completely green, down to this uncanny ichor in its veins, and very big and venomous at that. Somehow I could not get it out of my head, because the caption suggested a perfect refrain for a Ballade: A green-blooded fish has been found in the sea. It has so wide a critical and philosophical application. I have known so many green-blooded fish on the land, walking about the streets and sitting in the clubs, and especially the committees. So many green-blooded fish have written books and criticism of books, have taught in academies of learning and founded schools of philosophy that they have almost made themselves the typical biological product of the present age of evolution.

Chesterton uses "green-blooded fish" as a metaphor for all self-centered, dehumanized people, and the metaphor attracts us by its originality.

Title

A WORD ABOUT TITLES

The title of an essay precedes the beginning and should clarify the subject and arouse interest. The title, however, does not take the place of a beginning paragraph. In fact it is good practice to make an essay self-sufficient so that subject, purpose, plan (if needed) are all perfectly clear without reference to a title.

As to titles themselves, they should ideally be both informative and eye-catching. It is difficult in practice to balance these qualities, and most titles come down on one side or the other: they are informative but not eye-catching, or unusual and attractive but not especially informative. In most college work it is more important that a title clearly indicate your intentions than that it hook your reader. In either case a title ought to be concise.

If you start your essay with a title in mind, be sure it fits the theme as it actually evolves. In the process of composition essays have a way of taking unexpected twists and turns. For this reason some teachers suggest that you do not decide upon a final title until you have done the paper and can see what you have actually written.

CONCLUSION

When composing beginnings, inexperienced writers are likely to err at either of two extremes: doing too little or doing too much. In doing too little they slight the opening, jumping too suddenly into the subject and piling ideas and information in front of the reader before he or she has time to settle back and see what all this is about.

In doing too much they make the beginning a précis of the essay and anticipate everything they plan to do. The function of an opening is to introduce an essay, not to be a miniature version of it. To make it so is to act as inappropriately as the master of ceremonies at a banquet who introduces the main speaker by anticipating everything he or she is going to say.

The effective beginning stays between those extremes. It lets readers know what to expect, but it leaves them something to expect.

EXERCISE

1. In about 100 words, compose a beginning paragraph either for the theme you outlined at the close of the preceding chapter or for another on another topic of interest. Make sure that readers understand your general subject, the limitations of your treatment, and your organization. But convey this information implicitly: do *not* write, "The subject will be . . ."; "This paper will be limited to . . ."; "The plan to be followed is . . ." Try to interest your readers and to establish a point of view and a tone appropriate to your purpose.

2. In conjunction with the exercise above, answer these questions, devoting several sentences or a brief paragraph to each:

   A. What strategy did you use to interest your readers?
   B. What tone were you seeking to establish—specifically, how did you feel about the subject, how did you wish readers to view you, and what kind of relationship did you hope to establish with them? Explain also how these aspects of tone led you to choose certain words in your beginning paragraph.

3. Study the beginning of an expository essay you like. First copy the beginning; then analyze the writer's techniques, using examples from his or her writing to support the analysis.

CHAPTER 10

Structure of the Essay: Closing

Clos. Like the opening of an essay, the closing should be proportional to the length and complexity of the paper. In a long essay the closing may be several paragraphs; in a short one, a sentence or two, perhaps not even spaced separately but simply ending the final paragraph of the discussion. For most freshman themes the closing ought to be a new paragraph but a brief one. Whatever their length, closings have well-defined functions.

Clos. sum. ## SUMMARY AND CONCLUSION

These functions include, though not invariably, summarizing and drawing a conclusion in the sense of a final judgment or inference. To do this, one usually uses an explicit phrase such as *in summary, to sum up, summing up, in short, in fine, to recapitulate.* The label may be more subtle: "We have seen, then, that. . . ." In writing that aims at relaxed informality, the label should be no more obtrusive than clarity requires. Generally, however, clarity does require some signal of summation.

Whether or not a summary is needed at all depends upon the length of the paper and the complexity of the subject. In a term paper, especially if it is argumentative, a summary is advisable. In most short, uncomplicated freshmen themes it probably is not.

Clos. conc. A *conclusive closing,* on the other hand, may be necessary, even in a short essay. Certain subjects make such a judgment obligatory. Mark Twain ends his essay "Saint Joan of Arc" with this considered opinion:

> Taking into account, as I have suggested before, all the circumstances— her origin, youth, sex, illiteracy, early environment, and the obstructing conditions under which she exploited her high gifts and made her conquests in the field and before the courts that tried her for her life—she is easily

and by far the most extraordinary person the human race has ever produced.

Any essay on Saint Joan must arrive at some sort of conclusion: Saint or fool? Traitor or patriot? Dupe or heroine? Divinely inspired or victim of hallucinations? The questions inhere in the subject.

Similarly the journalist Samuel Hopkins Adams concludes an account of Warren Gamaliel Harding (the twenty-ninth President, from 1921 to 1923) by drawing a conclusion:

> The anomaly of Warren Gamaliel Harding's career is that without wanting, knowing, or trying to do anything at all unusual, he became the figurehead for the most flagrantly corrupt regime in our history. It was less his fault than that of the country at large. Maneuvered by the politicians, the American people selected to represent them one whom they considered an average man. But the job they assigned him is not an average job. When he proved incapable of meeting its requirements, they blamed him and not themselves.
>
> That is the tragedy of Harding.

Sometimes judgments are expressed as rhetorical questions. Joan Didion ends an essay about Newport, Rhode Island, a summer resort for the very wealthy:

> Who could fail to read the sermon in the stones of Newport? Who could think that the building of a railroad could guarantee salvation, when there on the lawns of the men who built the railroad nothing is left but the shadows of migrainous women, and the pony carts waiting for the long-dead children?

Or a final conclusion may be supported by a quotation, as in this closing passage of an essay about the American Civil War:

> It was the bad luck of the South that the only great poet who commemorated the "strange sad war" was Walt Whitman of Manhattan. But Timrod's lines on the Confederate dead in Magnolia Cemetery at Charleston may serve as a final text:
>> In seeds of laurel in the earth
>> The blossom of your fame is blown,
>> And somewhere, waiting for its birth,
>> The shaft is in the stone.
>
> The laurels have grown; the shafts are all around us. They cannot be too numerous or too high.
>
> D. W. Brogan

On occasion it may not be wise, or even possible, to round off an essay with a neat final judgment, even when the subject seems to call for one. The novelist Joseph Conrad once remarked that the business of the story teller is to ask questions, not to answer them, and the point applies sometimes to the essayist, who may wish to suggest a judgment rather than to formulate it. The strategy is called an *implica-*

tive closing. The writer stops short, allowing the reader to infer the conclusion. In effect the final paragraph then opens a door rather than closes one. Here, for instance, is how a student ended a theme about a teen-age hangout:

> The old lady who lives across the street from the place says that the most striking thing is the momentary silence which, now and then, breaks up the loud, loud laughter.

On certain occasions a judgment, whether stated or implied, is not enough. When an adequate response to the subject demands action, the closing involves telling readers what to do. In the following paragraph the lexicographer Eric Partridge not only expresses an opinion about clichés, but also issues advice, making what is called a *directive closing*:

> Perhaps we might summarize the pros and cons for and against clichés in some such way as this. In the interplay of conversation, a cliché is often redeemed by a *moue* or a shrug or an accomplice-smile: "There! I've used a cliché. Very careless and humdrum of me, I suppose. But at least you know what I mean." Intonation, pauses, emphasis, these and other means can invest a cliché-ridden sentence, or set of sentences, with humor and wit, and with realism and trenchancy. In writing, we lack these dramatic, these theatrical, these extraneous aids: we *stand or fall alone*. In writing, the battered simile and the forgotten metaphor may well be ludicrous or inept or repellent; the hackneyed phrase so commonplace that it offends, the idiom so weak that it enfeebles the argument or dulls the description or obscures the statement; the foreign phrase either so inadequate or so out of place that it sets up a misgiving, a doubt, a dissent; the quotation so mauled by the maudlin, so coy in the mouths of the prim, so bombastic in the speech of the pompous, as to be risible, so very common as to lose all distinction, so inept as to fail.
> "If in doubt, don't!"

TERMINATION

Summaries and conclusions are not always essential functions of a closing. Some essays require neither. One function, however, is essential: to terminate, to say, in effect, "The End." When they are present, summaries and conclusions are in themselves signals of termination because they come normally at the close of an essay. But because they are not always present, you must know how to use other strategies which signify that you have no more to say.

Clos. t. w.

TERMINAL WORDS

The simplest strategy is to employ a word or phrase that literally says "ending." Ernest Hemingway begins the last two paragraphs of

A Moveable Feast (reminiscences of life in Paris in the 1920s) like this:

That was the end of Paris.

Equally obvious—though more mechanical—are the signposts *in conclusion, concluding, finally, lastly, in the last analysis, to close, in closing,* and so on. Adverbs that show a loose consequential relationship can also signal closing: *then, and so, thus.* These expressions are like flags on the eighteenth green. How obvious they need to be depends, as with signposts in general, upon how well the reader can see. Do, however, keep terminal words as unobtrusive as possible. In writing the best technique hides itself.

CIRCULAR CLOSING

This strategy works on the analogy of a circle, which ends where it began. The final paragraph repeats an important word or phrase prominent in the beginning. The expression may be a metaphor or simile. Or it may be a striking, unusual word. In any case it must be important, something the reader will remember. You cannot hang a sign on it— "Remember this for I shall return to it in my closing." But obviously if the strategy is to work the reader has to recognize the key term. In an essay of more than several pages, you may want to repeat the expression unobtrusively once or twice as you go along to keep it before the reader's eyes. And occasionally you might wish to emphasize the fact of completion by saying something like, "We return, then, to. . . ."

In a sketch of a famous aristocrat, Lady Hester Stanhope, the biographer Lytton Strachey opens with this paragraph:

> The Pitt nose has a curious history. One can watch its transmigrations through three lives. The tremendous hook of old Lord Chatham, under whose curves Empires came to birth, was succeeded by the bleak upward-pointing nose of William Pitt the younger—the rigid symbol of an indomitable *hauteur*. With Lady Hester Stanhope came the final stage. The nose, still with an upward tilt in it, had lost its masculinity; the hard bones of the uncle and grandfather had disappeared. Lady Hester's was a nose of wild ambitions, of pride grown fantastical, a nose that scorned the earth, shooting off, one fancies, towards some eternally eccentric heaven. It was a nose, in fact, altogether in the air.

And here are the final three sentences of Strachey's sketch:

> The end came in June, 1839. Her servants immediately possessed themselves of every moveable object in the house. But Lady Hester cared no longer: she was lying back in her bed—inexplicable, grand, preposterous, with her nose in the air.

The return to the key phrase "nose in the air" is not the only reason this is an effective closing, but it is an important one. Notice that Strachey's phrase not only latches the end of his essay to its beginning, but is also an image essential to his attitude toward Lady Hester Stanhope. The expression that completes the circle necessarily looms large in the reader's mind, and it must be genuinely important or the emphasis it receives from repetition will prove misleading.

Clos.
rhythm

RHYTHMIC VARIATION

Prose rhythm is a complex matter. For the moment it is enough to understand that, however it works, rhythm is an inevitable and important aspect of prose. Because of this, you can use it to signal closing by varying the movement of the final sentence or sentences.

Rhythmic variation often takes the form of a slowing down and regularizing. The closing sentence is likely to be long and broken into clearly marked parts; its stressed syllables, especially at the very end, are more regularly spaced or clustered. A famous example is the last paragraph of Lewis Carroll's *Alice in Wonderland*:

> Lastly, she pictured to herself how this same little sister of hers would, in the aftertime, be herself a grown woman; and how she would keep, through all her riper years, the simple and loving heart of her childhood; and how she would gather about her other little children, and make *their* eyes bright and eager with many a strange tale, perhaps even with her dream of Wonderland of long ago; and how she would feel with all their simple sorrows, and find a pleasure in all their simple joys, remembering her own child life, and the happy summer days.

The rhythm becomes as regular as that of poetry in the final phrase:

```
x    x    ,  x    , x      ,
and the happy summer days.
```

Another example is the conclusion of an essay on Jonathan Swift (who died insane) by the novelist William Thackeray:

> He was always alone—alone and gnashing in the darkness, except when Stella's sweet smile came and shone upon him. When that went, silence and utter night closed over him. An immense genius; an awful downfall and ruin. So great a man he seems to me, that thinking of him is like thinking of an empire falling. We have other great names to mention—none, I think, however, so great and so gloomy.

The pause in the middle slows Thackeray's final sentence, and the interrupters "I think" and "however" break its second half into units. The very last phrase is an example of a rhythmic pattern called a "cursus," used in Latin prose as a conventional signal of closing and adapted by many writers of English. More specifically the phrase belongs to the variety *cursus planus,* in which the stress falls upon the second syllable from the end and upon the fifth:

```
x   ,    x  x   ,  x
```
so great and so gloomy.

Instead of a regular meter or a cursus, the rhythmic variation that signals closing may appear as a cluster of stressed syllables. The novelist Truman Capote ends his report of a famous mass murder (*In Cold Blood*) by describing a visit to the graves of the victims:

Then, starting home, he walked toward the trees, and

```
                                           x    , x
```
under them, leaving behind the big sky, the whisper

```
x   ,   , x  x  x    ,   ,      ,
```
of wind voices in the wind-bent wheat.

Aside from the final three successive stresses, Capote's passage is memorable for its sound: the alliteration of *w,* and the repetitions of *n, d, t,* and the short *i.*

Occasionally writers take the opposite tack and close with a short quick sentence rather than a long, slow, regular one. Such a closing is most effective played off against a longer statement, as in this passage, which ends Joan Didion's essay "On Morality":

Because when we start deceiving ourselves into thinking not that we want something or need something, not that it is a pragmatic necessity for us to have it, but that it is a *moral imperative* that we have it, then is when we join the fashionable madmen, and then is when the thin whine of hysteria is heard in the land, and then is when we are in bad trouble. And I suspect we are already there.

Failing to use a brief sentence as a way of ending sometimes wastes a potentially good closing:

At last the hard-working housewife is ready to watch her favorite television program, but before fifteen minutes are up she is sound asleep in her chair and before she realizes it the 6:30 A.M. alarm is going off and it is time to start another day.

As an ending this would be considerably improved if the final clause ("it is time to start another day") were treated as a separate sentence set off against the longer statement.

Clos. nat.   **NATURAL POINT OF CLOSING**

A final way of signaling the end is to close the essay at a natural point, one built into the subject. For example, in a biographical sketch of someone who is dead, the obvious place to end is with the death scene, as in the passage quoted earlier by Lytton Strachey about Lady Hester Stanhope. Another instance is this paragraph, the end of Llewelyn Powys's essay "Michel de Montaigne":

> On 13 September, 1592, Michel de Montaigne, having distributed certain legacies to his servants, summoned his parish priest to his bedside, and there in his curious room with the swallows already gathering on the leaden gutters outside, he heard Mass said for the last time in the company of certain of his neighbors. With due solemnity the blessed sacrament was elevated, and at the very moment that this good heretical Catholic and Catholic heretic (unmindful for once of his nine learned virgins) was raising his arms in seemly devotion toward the sacred morsel which in its essence—*que sçais-je*—might, or might not, contain a subtle and crafty secret, he fell back dead.

Here the effectiveness of closing with the death scene is reinforced by the careful construction of the last sentence, which does not complete its main thought until the very final word. "Dead" falls into place like the last piece of a puzzle.

Natural closings are not restricted to death-bed descriptions. In an essay about your daily routine, for instance, you will end with some variation of the phrase that diarist Samuel Pepys made famous: "And so to bed." Even when a subject does not have a built-in closing, a comparison or figure of speech can provide one. The American historian W. H. Prescott concludes *The Conquest of Peru*:

> With the benevolent mission of Gasca, then, the historian of the Conquest may be permitted to terminate his labors—with feelings not unlike those of the traveller, who, having long journeyed among the dreary forests and dangerous defiles of the mountains, at length emerges on some pleasant landscape smiling in tranquility and peace.

CONCLUSION

The various strategies of closing do not exclude each other. More often than not, good writers combine several. Nor are they the only ways of ending an essay. Inventive writers tailor their closings to to subject and purpose. The poet Dylan Thomas, for instance, wittily concludes his essay "How To Begin a Story" by doing what inexperienced writers should *not* do—simply stopping in midsentence:

> I see that there is little, or no, time to continue my instructional essay on "How To Begin a Story." "How To End a Story" is, of course, a different matter. . . . *One* way of ending a story is. . . .

And Virginia Woolf closes an essay called "Reading" with this sentence:

Some offering we must make; some act we must dedicate, if only to move across the room and turn the rose in the jar, which, by the way, has dropped its petals.

It is difficult to say why this works. The rhythm is important. But so is the image. The flower that has dropped its petals is a kind of metaphor of closing. Perhaps the seeming irrelevancy of the final clause is also a signal, a gracious closing of a conversation by lightly changing the subject. In any case, the passage succeeds, ending the essay neatly and unmistakably. That is the important thing.

EXERCISE

In your next composition pay particular attention to your closing paragraph. Try to make it about 75 or 100 words. Whether you include a summary and a conclusive judgment must depend upon the subject. In any case, begin with a phrase of termination, and experiment with varying the speed and rhythm of the final sentence. If your opening paragraph permits, attempt a circular closing.

CHAPTER 11

Structure of the Essay: Organizing the Middle

Just as an essay must have an effective beginning and closing, so its middle, the body of the composition, must have a clear organization, evolved from the outline. How much help readers need to understand that organization varies with their experience. Always, however, it is an important part of the writer's task to assist them in following the flow of thought.

This can be done in two ways. One is by *signposts*—words, phrases, sentences (occasionally even a short paragraph) which tell readers what you have done, are doing, will do next, will do later, or even will not do at all. The second is by *interparagraph transitions,* that is, words and phrases that tie the beginning of a new paragraph to the one that precedes it.

Sign.

SIGNPOSTS

The most common signpost consists of a broad topic statement, including words establishing the general plan, which sets up a section of an essay (or, in a short composition, the entire discussion). For instance, the scientist J.B.S. Haldane organizes a five-paragraph section of a long essay like this:

> Science impinges upon ethics in at least five different ways. In the first place . . .
> Secondly . . .
> Thirdly . . .
> Fourthly . . .
> Fifthly . . .

Sequence may be signaled by actual numbers or letters—usually enclosed in parentheses—rather than by words like *first, second, in the first place, in the second place.* Thus the poet W. B. Yeats explains why he believes in magic:

> I believe in the practice and philosophy of what we have agreed to call magic, in what I must call the evocation of spirits, though I do not know what they are, in the power of creating magical illusions, in the visions of truth in the depths of the mind when the eyes are closed; and I believe in three doctrines, which have, as I think, been handed down from early times, and been the foundation of nearly all magical practices. These doctrines are—
>
> 1. That the borders of our mind are ever shifting, and that many minds can flow into one another, as it were, and create or reveal a single mind, a single energy.
>
> 2. That the borders of our memories are as shifting, and that our memories are a part of one great memory, the memory of Nature herself.
>
> 3. That this great mind and great memory can be evoked by symbols.

Numbers, however (and number words like *first, second, third*), lose their value if overused. Instead of clarifying the organization they are likely to confuse it, leading the reader into a labyrinth of (1)s and (2)s and (a)s and (b)s.

It is better to set up a three- or four-part analysis, employing key terms which designate the major ideas and repeating these words as subsequent paragraphs begin. For example, the television critic Edith Efron, discussing soap operas, writes:

> Almost all dramatic tension and moral conflict emerge from three basic sources: mating, marriage and babies.

She begins the next paragraph by picking up the key word "mating":

> The mating process is the cornerstone of the tri-value system.

And the following paragraph by using the loose synonym "domesticity" to link to "marriage and babies":

> If domesticity is a marital "good," aversion to it is a serious evil.

Whether you use numbers or key terms to label the parts of your discussion, once you begin this strategy you must follow through. Writers sometimes make the mistake of starting off with something like this:

> There were three reasons why the Treaty was not satisfactory. First. . . .

But then they fail to introduce the next two reasons with the obligatory *second* and *third* (or variants such as *secondly* and *finally*). The absence of signals bewilders the reader, who may not recognize when the writer passes from reason one to reason two.

Aside from setting up a group of paragraphs, signposts may also anticipate future sections of an essay or make clear that something will not be treated. In composition it is generally difficult to analyze any subject into watertight compartments. As you develop one point, you often touch upon another that you do not plan to discuss fully until later or perhaps not to discuss at all. When this happens, you may want to give a warning.

Signposts may also point backward, reminding readers of something treated earlier that bears upon the current topic. Thus a writer may say: "(see page 8)"; or integrate his reminder more neatly into the text: "As we saw in Chapter 7. . . ."

The signposts we have looked at are *intrinsic,* that is, they are actually a part of the writer's text. There are also *extrinsic* signposts, ones which stand outside the actual discussion yet clue the reader to its organization. An outline or table of contents is such an extrinsic signal. So too are chapter titles, subtitles of sections, and running heads like that on the top of this page. Sometimes a writer even begins each chapter with a brief outline or a synopsis of the material covered.

Typography and design convey other extrinsic indications of organization: the indentation of quotations, extra spacing between lines to signal a new major section, and occasionally Roman numerals centered above the division of an essay.

Philosophical and scientific writing often uses a more elaborate numbering system. A two-part number is assigned to each paragraph, the first digit designating the chapter, the second, the paragraph. The numerals 3.7 set before a paragraph signify that it is the seventh paragraph of the third chapter. Such a number helps readers refer to earlier material, though it does not really help them follow the organization of what they are currently reading.

Several variations of this numbering technique are more helpful. In a triple-numbering systems, the number in the second position designates the section or paragraph group within the chapter: 3.3.7, for example, means chapter 3, section 3, paragraph 7. Every time the second number changes, the reader knows a new section has begun. Another variation keeps the two-part number but uses the second one to designate a paragraph group rather than a single paragraph: 3.2, for instance, might include four or five paragraphs.

Trans.

INTER-PARAGRAPH TRANSITIONS

Transitions link a paragraph to what has immediately preceded it. They occur at or near the beginning of a paragraph because the new paragraph represents a turn of thought, a minor new beginning which has to be linked to what has gone before. When readers have to change tracks, the transition acts like a switch, smoothing and easing the turn. Without a transition readers may be able to make the switch, but they have to bump along for themselves.

REPETITION

The simplest transition uses the repeated word. At the beginning of the new paragraph, you pick up a key term or phrase occurring at the end of the preceding one. Writing about the Louisiana politician Huey Long, Hodding Carter ends one paragraph and begins the next with the following link (the italics are added in this and all following examples, unless noted otherwise):

> Behind Huey were the people, and the people wanted these things.
> And with *the people behind him, Huey* expanded ominously.

The historian D. W. Brogan, writing about the American Civil War, uses the same technique, except that he combines the transition and the new topic in one fragmented rhetorical question:

> It was touch and go, but Lincoln, who supported the long ordeal of Grant's bloody failures, saved the Union as the neurotic Davis lost the Confederacy its last and only chance.
> *Grant's failures?* I know how much I am running against the tide of current historical opinion when I use such terms. Yet what other name can we give to the campaign of 1864?

The repeated word makes a strong and simple link. It works well when the key term leading into the new paragraph occurs naturally at the end of the preceding one. It is awkward and artificial, however, when the term is forced into the final sentence merely to set up the transition.

THE QUESTION-AND-ANSWER TRANSITION

A second method of linking paragraphs is to ask and answer a rhetorical question. Usually the question occurs at the end of the preceding paragraph and the answer at the beginning of the following one. Nancy Mitford, commenting upon the apparently compulsive need of tourists to travel, concludes one paragraph and begins the next like this:

> *Why do they do it?*
> *The answer is* that the modern dwelling is comfortable, convenient, and clean, but it is not a home.

Less often the question appears at the opening of the new paragraph, as in this discussion of the ultimate defeat of the Crusades:

> With want of enthusiasm, want of new recruits, want, indeed, of stout purpose, the remaining Christian principalities gradually crumbled. Anti-

och fell in 1268, the Hospitaler fortress of Krak des Chevaliers in 1271. In 1291, with the capture of the last great stronghold, Acre, the Moslems had regained all their possessions, and the great crusades ended, in failure. *Why? What went wrong?* There was a failure of morale clearly. . . .

Morris Bishop

The question-answer transition makes a very strong tie, but, as with the rhetorical question generally, it is too obvious a strategy to be called upon very often. In a short paper, one use is enough.

Trans.
Sum.

THE SUMMARIZING TRANSITION

In this type of transition you begin with a phrase or clause that sums up the point of the preceding paragraph and then move to the main clause, which introduces the main topic. *If-* and *while*-clauses frequently carry such transitions:

> If I went through anguish in botany and economics—for different reasons—gymnasium was even worse.

James Thurber

> But while Bernard Shaw pleasantly surprised innumerable cranks and revolutionists by finding quite rational arguments for them, he surprised them unpleasantly also by discovering something else.

G. K. Chesterton

In each case the opening clause refers back to the discussion in the paragraph before, while the main clause points forward to the new topic. Unless idiom prohibits it, the clauses should be in that order: summary, new topic.

Long summarizing transitions like these tend to be formal in tone. On informal occasions it may be better to avoid a full *if-* or *while*-clause and effect the summary more briefly. Innumerable variations are possible. Here, for example, a student moves from the topic of college teaching methods to that of personal responsibility:

> Because of *these differences in teaching methods,* college throws more responsibility upon the student.

And in the following case Fritz J. Raddatz, writing a biography of Karl Marx, shifts from the theme of Marx's desire for personal privacy to that of his isolation from society:

> The counterpart to *Marx's isolation of himself from other men* was his isolation in society.

The summarizing transition may take even briefer form, using pronouns like *this, that, these, those,* or *such* to sum up the preceding topic. The historian J. Fred Rippy uses this device to move from the severe geographical conditions of South America to a discussion of its resources:

These are grave handicaps. But Latin America has many resources in compensation.

The journalist Hodding Carter swings us from a discussion of the political traditions of Louisiana to the birth and early years of Huey Long:

Such was the political Louisiana in which Huey Pierce Long was born in 1893, in impoverished Winn Parish in north Louisiana, a breeding ground of economic and political dissenters.

Although the pronouns are perfectly clear in these examples, they can be ambiguous when used as the subjects of sentences, especially when they refer to the whole of a preceding idea. If you do employ them in this way, be sure that a reader can understand what they refer to. Should there be a doubt, make them adjectivals modifying a word or phrase that fairly sums up the point from which you are moving: for example, "These handicaps are . . ."; or "Such political traditions were. . . ."

Trans.
log. conn.

LOGICAL CONNECTIVES

Finally, you may link paragraphs by words showing logical relationship: *therefore, however, but, consequently, thus, and so, even so, on the other hand, for instance,* and many, many more. In the following passage the historian and political scientist Richard Hofstadter is contrasting "intelligence" and "intellect." In the first paragraph he defines "intelligence." By placing the transitional phrase "on the other hand" near the beginning of the second paragraph, he signals the other half of the contrast:

. . . intelligence is an excellence of mind that is employed within a fairly narrow, immediate, and predictable range. . . . Intelligence works within the framework of limited but clearly stated goals. . . .

Intellect, *on the other hand,* is the critical, creative, and contemplative side of mind.

Here is another example—a discussion of *Hamlet*—in which *moreover* indicates that the new paragraph will develop an extension of the preceding idea:

If I may quote again from Mr. Tillyard, the play's very lack of a rigorous type of causal logic seems to be a part of its point.

Moreover, the matter goes deeper than this. Hamlet's world is preëminently in the interrogative mode. . . .

Maynard Mack

Logical connectives seldom, if ever, provide the only link between paragraphs. Rather, they work in conjunction with word repetitions, summaries, pronouns. In fact, all the various transitional strategies we have looked at commonly occur in some combination. But whatever its form, an inter-paragraph transition should be clear and unobtrusive, shifting readers easily from one topic to the next without jolting them.

EXERCISE

1. In your next theme underline the transitional words linking your paragraphs. Be prepared to explain and, if necessary, to defend your transitions.

2. With your instructor's permission exchange themes with a classmate. Study how he or she has linked paragraphs. If you find places where the links seem inadequate, suggest how they could be strengthened.

12

Tight and Loose Organization

INTRODUCTION

While all essays must have a beginning, a middle, and a closing, their organization can vary from tight to loose. In the first case the writer follows a carefully conceived plan which he or she has probably set down in outline (though some people can do this sort of organizing in their heads). In loose structure the essay unfolds in a more leisurely and associational manner: the writer may develop a point, leave it for a moment, come back to it again, and so on.

Tightness and looseness are not absolute categories into which every actual essay fits. Rather they are extremes on a scale. Most essays fall somewhere between, organized more or less tightly, more or less loosely. Nor should you think of tight and loose structure as value terms. Neither is necessarily superior to the other; each may be done well or done badly; each works for some purposes but not for others.

The advantage of tight organization (assuming it is well carried out) is clarity. Everything is neatly analyzed—a place for each idea, each idea in that place. For readers whose primary concern is to gather information or to achieve understanding as quickly and easily as possible, such clear arrangement is a blessing.

Loose organization seems more personal. It creates the illusion of a mind thinking and feeling rather than of a mind that has already arrived at its conclusions. Readers feel more involved with the *process* of the writer's consciousness, not simply its product. But while loose structure allows for a more intimate and immediate relationship between writer and reader, it sacrifices surface clarity. Ideas are not as neatly categorized, and a loose essay is more difficult to reduce to outline. One cannot skim it as quickly for ideas and information.

TIGHT ORGANIZATION

As a brief example of tight organization, study this Prologue to the first volume of his autobiography by the philosopher Bertrand Russell:

> Three passions, simple but overwhelmingly strong, have governed my life: the longing for love, the search for knowledge, and unbearable pity for the suffering of mankind. These passions, like great winds, have blown me hither and thither, in a wayward course, over a deep ocean of anguish, reaching to the very verge of despair.
>
> I have sought love, first, because it brings ecstasy—ecstasy so great that I would often have sacrificed all the rest of life for a few hours of this joy. I have sought it, next, because it relieves loneliness—that terrible loneliness in which one shivering consciousness looks over the rim of the world into the cold unfathomable lifeless abyss. I have sought it, finally, because in the union of love I have seen, in a mystic miniature, the prefiguring vision of the heaven that saints and poets have imagined. This is what I sought, and though it might seem too good for human life, this is what— at last—I have found.
>
> With equal passion I have sought knowledge. I have wished to understand the hearts of men. I have wished to know why the stars shine. And I have tried to apprehend the Pythagorean power by which number holds sway above the flux. A little of this, but not much, I have achieved.
>
> Love and knowledge, so far as they were possible, led upward toward the heavens. But always pity brought me back to earth. Echoes of cries of pain reverberate in my heart. Children in famine, victims tortured by oppressors, helpless old people a hated burden to their sons, and the whole world of loneliness, poverty, and pain make a mockery of what human life should be. I long to alleviate the evil, but I cannot, and I too suffer.
>
> This has been my life. I have found it worth living, and would gladly live it again if the chance were offered me.

In his very first sentence Russell lays out his plan—the three-fold structure of what will follow. Each part is developed in a separate paragraph. Each paragraph is explicitly tied to what precedes it. Thus the key term "love" links the second paragraph to the first. "With equal passion" provides the transition from paragraph two to three. At the beginning of the fourth paragraph the linkage is longer: the opening sentence, referring back to the two topics Russell has just discussed, leads into the following statement, introducing the new topic. Finally, the brief summarizing fifth paragraph begins with the link word "this," which points both backward to the preceding topic and forward to the concluding statement.

The tight organization also shows in Russell's individual paragraphs. The second one is a good example. It supports its topic by advancing three reasons. Each is prefaced by an organizational term blazing the trail of the writer's thought: "first," "next," "finally."

Such prose is well-crafted. The writer has taken the time and trouble both to think through what he wants to say and to provide readers with the means of following him. Anyone who pays reasonable attention to Bertrand Russell's passage knows exactly what he is going to do, what he is saying, what he has said, and finds it relatively easy to outline the writer's ideas and to answer questions about them.

LOOSE ORGANIZATION

This brief essay by Anatole Broyard provides an example of loose organization:

FIREPLACE

Our first conversational impulses exhausted, our minds and bodies lulled by food and drink, we gaze, my guests and I, into the fire.

Each of us sees different things. I watch my friends' faces grow thoughtful as the fire absorbs their vacant eyes. Some of them look melancholy, as if they are reflecting on the fact that all life is only a burning up of energy, that, in a sense, we all go down in flames. For others, the fire is like a Rorschach test: they see in it whatever is uppermost in their minds.

After dinner, there is a spiritual belching. We leave the séance of the table with its associations of family and festivities and retire back into ourselves to wait for our second wind. Temporarily dispersed, we need the fire for a focus. It may be the only thing we have in common at this moment.

Thousands of years ago fire helped to civilize men, brought them together around a common center, softened them with its influence. It is not so very different now. Two hundred years ago this same fireplace cooked the meals of the people who lived here. While I use this for a living room, a place to cook up conviviality, it was a kitchen to them, the guts of their way of life.

For us now, fire is not a necessity, but a ritual, one that has lost much of its grandeur and terror. We no longer sacrifice virgins to fire to propitiate cruel gods, nor do we burn witches. Yet there is something numinous about fire which hypnotizes us. If we have lost the habit of worship, perhaps we still hover on the edge.

In my own mood of post-prandial melancholy, I ask myself whether even this mild obeisance to the fire may not be obsolete, for few of us here have kept that old feeling for hearth and home. We are more like what Spengler called "intellectual nomads." Of the dozen guests in the room, I count six who have already abandoned one conjugal hearth for another.

Yet not one of them would buy a house without at least one fireplace. What is the country without a fireplace? they ask, as if we were Neolithic men and women. It is more important even than a swimming pool. Things have come to such a pass that we are always either heating up or cooling down our errant spirits.

When I was a bachelor, my Greenwich Village apartment boasted a brick fireplace and I used to buy bundles of logs from an ancient Italian and carry them up four flights of stairs. Even then, a modest evening's fire cost $20, or four $5 bundles. This was not counting kindling, which sometimes came in the surrealistic shape of women's high heels. Rejects from shoe factories, these hardwood heels burned very well, though their symbolism left something to be desired. A fire in New York City was almost as dear as love itself. One night I expended $60 worth of wood in honor of a young woman who was neither a virgin nor a witch.

A fire burns well only on the ashes of many former fires, and here is a profound metaphor for those prepared to pursue it. If I recall correctly, Freud saw flames as phallic, but I think that the feminine principle has at least an equal claim. A man building a fire is almost as pleased with himself as when he is kindling a woman, and a woman being warmed hisses and crackles like a fire. Good fire-builders are looked upon as macho, and those whose fires die out must suffer the opposite inference.

I find firelight more flattering than candles. Only Camilles or the dead look dramatic by candlelight, while in the glow of my fire, everyone is flushed with promise, flickering with change, with a shadow-play of mixed feelings. There is a heady hint of cannibalism in the room, as if we might like to consume one another. Why else is that woman warming her haunches on the hearth?

Conversation sputters and bursts into flame. We have thawed our cool thoughts—so much more heat-resistant than our bodies—to room temperature. I feel a blaze of affection for my friends and for my home. The fire has dried out the dampness in our souls, the winter in our hearts.

Broyard's essay gathers itself around the central theme of the importance of fire in human society. The looseness is reflected in the occasional discontinuities between paragraphs, paragraphs 7 and 8, for instance. The idea developed in the eighth paragraph extends the topic of the preceding one (an example of how far urban dwellers will go to maintain a fireplace). Yet the connection, unlike those in Russell's passage, is left implicit: no "for example" or "for instance" offers readers a handhold. Even looser are the connections between paragraphs 8 and 9 and 9 and 10. Here the writer appears simply to be musing randomly upon fire. Still, a pattern exists, a movement from the specific and personal to the general and philosophical.

It would be a mistake, then, to take this looseness for formlessness. The essay is organized. It moves in a kind of circle—from the outer world of the social group gathered after dinner around a fireplace to the inner world of the writer's sensibility, finally returning to its beginning, ending with the group circled around the fire. In a subtle way the pattern suggests the dual capacity of fire when captured and contained within the hearth: the paradox that it both unifies the group and turns each person inward upon himself or herself.

Broyard's piece is an example of what we call the familiar essay. In familiar essays the topic is less an idea or event or problem than it is the writer's individual perception of, response to, feelings about an idea or event. The structural principle is not a logical or temporal sequence existing in the topic itself; it is rather the evolving consciousness of the writer. The value of the familiar essay lies less in information or ideas (though it contains these) than in the unfolding of a sensibility, the revelation of a unique and unusual mind actively engaged upon a common human experience—sitting before a fireplace, sailing a boat, sleeping late on Sunday morning.

The difference between tight and loose organization comes down to this: in the former the writer subordinates himself to an objective principle of organization contained within the subject, the logic of cause and effect, say, an historical progression of events, or the manner in which a complex entity divides into its parts. The writer's purpose is first to understand this objective structure and second to make it clear to readers. In loose organization the consciousness and sensibility of the writer dominates the apparent topic and imposes itself upon that topic. The organization is thus a more immediate reflection of a particular mind. Ultimately, of course, it is this mind which is the real subject.

EXERCISE

Compose two short essays on similar but not identical subjects. If you wish, go back to the topics suggested on pages 32–34 following the chapter on invention. Attempt a tight organization in one essay; strive for a looser structure in the other.

CHAPTER 13

Point of View,
Persona, and Tone
in the Essay

P. V.

POINT OF VIEW

Point of view relates to how you present a subject. There are two possibilities: you may reveal your role as writer by using "I," "me," "my"; or you may avoid all explicit reference to yourself. The first is called a *personal* point of view, the second *impersonal*. The difference is not that in a personal point of view the subject is the writer, while in the second it is something else. Every subject involves the writer. The difference is a question of strategy.

However, you do not have equal freedom on every occasion to choose either point of view. Subject and purpose should determine the choice. Some topics so intimately involve the personality of the writer that they require a first-person presentation. It would sound silly to describe your summer vacation impersonally. Do not be afraid, then, to use "I" if it fits your subject and purpose, if it sounds pompous and artificial to hide behind such false beards as "this observer," "your reporter," "the writer," and so on.

On the other hand, at times a personal point of view is simply inappropriate. A scientist, writing professionally, usually tries to keep his or her personality in the background, and properly so: scientific subjects are best treated objectively.

And of course many topics can be presented from either point of view, though the two approaches will result in different essays. In such cases you must consider occasion and reader, especially the degree of formality you want. An impersonal point of view is more formal, a personal one less so.

Whichever point of view you select, establish it in the opening paragraph. It is not necessary to say, "My point of view will be personal (or impersonal)." Simply use "I" if you intend to write personally, or avoid the first-person pronoun if you do not. And maintain your point of view consistently. Don't jump back and forth between personal and impersonal presentation. At the same time, you can make small adjustments. You may expand "I" to "we" when you wish to imply "I the writer and you the reader." Whether writing personally or impersonally, you may address readers as individuals by employing "you," or shift to "one," "anyone," "people," and so on when you refer to no one specifically.

But such shifts in point of view should be compatible with the emphasis you desire, and they should be slight. Radical changes in point of view, nine times in ten, are awkward. It is good practice, then, (1) to select a point of view appropriate to the subject; (2) to establish that point of view, implicitly but clearly, in the opening paragraph; and (3) to keep to it consistently.

PERSONA

Persona derives from the Latin word for an actor's mask (in the Greek and Roman theaters actors wore cork masks carved to symbolize the character they were portraying). As a term in composition, *persona* means the writer's presence in the writing.

This derivation from "mask" may be a little misleading. Persona should not be taken to imply a false face, a disguise, behind which the real individual hides. In one sense a writer's persona is always "real." It exists, there in the prose, inseparable from the language. The words you choose, the sentence patterns into which you arrange them, even the kinds of paragraphs you compose and the way in which you organize your essay—all contribute to a personality which is, for that particular piece of writing, you.

Of course, you may say, a persona is not really the person who writes. (*Person,* interestingly enough, comes from the same Latin source as *persona.*) And of course that is true, as it is true that the same person may assume, on different occasions, quite different personas. Still, the only contact readers generally have with a writer is through his or her words. For readers the persona implicit in those words is the real, existent fact about the writer.

The question to ask about any persona is not is this really the writer? The questions are, is it really how the writer wants to appear? and is it how he or she should appear? To put the matter another way: Is the persona authentic and appropriate?

Authenticity concerns whether the personality your words convey corresponds to the personality you want to present in a particular piece of writing. To say that a persona is authentic does not necessarily mean that it is truly you. We are all many different people: we show one face to friends, another to teachers, still another to parents. Authenticity here means simply that how you appear in what you write is how you wish to appear.

But authenticity is not enough. A persona must also be appropriate. The kind of person you represent yourself to be should help you achieve the end you are aiming at, or at the very least it ought not to get in the way.

Persona is most immediately and directly revealed when a writer discusses himself or herself. For example, a clear personality emerges in the following passage from Benjamin Franklin's *Autobiography.* Franklin is explaining that when he was educating himself as a youth he learned to drop his habit of "abrupt contradiction, and positive argumentation" and to become more diffident in putting forward his opinions. (He is, of course, talking about the same thing we are—persona.)

> [I retained] the habit of expressing myself in terms of modest diffidence, never using when I advance any thing that may possibly be disputed, the words, *certainly, undoubtedly,* or any others that give the air of positiveness to an opinion; but rather I say, *I conceive,* or *I apprehend* a thing to be so or so, *It appears to me,* or *I should think it so for such & such reasons,* or *I imagine it to be so,* or *it is so if I am not mistaken.* This habit I believe has been of great advantage to me, when I have had occasion to inculcate my opinions & persuade men into measures that I have been from time to time engag'd in promoting. And as the chief ends of conversation are to *inform,* or to be *informed,* to *please* or to *persuade,* I wish wellmeaning sensible men would not lessen their power of doing good by a positive assuming manner that seldom fails to disgust, tends to create opposition, and to defeat every one of those purposes for which speech was given us, to wit, giving or receiving information, or pleasure: for if you would *inform,* a positive dogmatical manner in advancing your sentiments, may provoke contradiction & prevent a candid attention. If you wish information & improvement from the knowledge of others and yet at the same time express your self as firmly fix'd in your present opinions, modest sensible men, who do not love disputation, will probably leave you undisturb'd in the possession of your error; and by such a manner you can seldom hope to recommend your self in *pleasing* your hearers, or to persuade those whose concurrence you desire.

Franklin strikes us as a discerning and candid man, sensitive to how he affects people, but sensitive in an unabashedly egocentric way. His advice about not coming on too strong—still worth heeding—is based not so much upon concern for the feelings of others as upon a clear-eyed assessment that modesty is the way to get on in the world. Yet the very openness and ease with which Franklin urges that advice washes away its taint of self-serving manipulation.

We sense a very different personality in the passage by Bertrand Russell from his *Autobiography* (see page 85). Russell is more emotional than Franklin. His attitude toward knowledge and toward other people is less worldly and more passionate. Russell is driven to knowledge not because it serves his ambition but because of a compulsive desire to know (in fairness, it should be noted that Franklin too had much of this disinterested drive for knowledge). He sees other people not as helps or hindrances in his career, but as fellow humans, for whose sufferings he can feel compassion and sorrow.

But there is more to Russell's persona than the obvious emotion conveyed by his words. His emotion is constrained within a rational framework. As we saw, the organization of his paragraphs is tightly analytical. Here is someone who not only feels intensely but whose intellect can impose order upon his feelings and give them an even sharper focus. We sense an unusually powerful and complex mind, in which emotion and reason are not at war but are reinforcing allies. Russell's passionate response to life gains intensity because it is shaped and heightened by a reasoned, analytical structure.

Persona, as you can see, is a function of the *total* composition. It emerges not only from the meanings of the words but also from the more abstract, less obviously expressive patterns of sentences and paragraphs and from overall organization.

While it is most obvious in autobiographies, persona is not confined to such writing. It exists in all composition, even when the writer uses an impersonal point of view (that is, avoids the pronouns "I," "me," "my").

In compositions about subjects other than oneself, persona is most strikingly revealed by commentary. We catch our clearest glimpses of a personality when a writer reacts to his or her subject. For instance, in the following passages an historian discusses questions of dress and personal cleanliness in the Middle Ages:

> Hemp was much used as a substitute for flax in making linen; the thought of hemp shorts curdles the blood.

> In the thirteenth-century French romance *L'Escoufle* Sir Giles, beside the fire, removes all his clothes except his drawers to scratch himself. (Fleas, no doubt.)

> Morris Bishop

Such comments reveal writers as personalities with their own ways of looking at the world—in Bishop's case with a pleasantly cynical humor.

Often commentary is expressed subtly, blending with the discussion so smoothly that it may pass unnoticed by inattentive readers. Barbara Tuchman, describing the building where the British House of Lords sits in session, writes:

> Between the windows, statues of the barons of Magna Carta, inadvertent founders of the parliamentary system, looked down a little grimly on what they had wrought.

The phrase "a little grimly," while it describes the expressions on the statues, is more significant in conveying a complex response by the writer, an ironic sense of the difference between political myth and political reality. In myth the English barons who in 1215 forced King John to sign the Great Charter recognizing their rights were fighting for the sacred cause of freedom; to them the English people owe the beginning of their great tradition of individual rights protected by law from the encroachments of the powerful. That is the myth. The facts are probably a bit different. The barons cared little about freedom in the abstract or protection under law; they fought for no one's freedom but their own. They wanted more power for themselves and were not at all scrupulous about how they used it to exploit those beneath them. And very likely—as Tuchman implies—they would not approve of the democratic freedoms of modern England.

This complicated set of ideas and evaluations is all carried by the words "a little grimly." What is more to our point, the phrase reveals a mind alert to the ironies of history, tough enough to accept those ironies and even to find a kind of pleasure in contemplating them, and witty enough to compress that sense of irony into a trenchant phrase and project it subtly into the scene.

A persona stands behind even relatively faceless writing. Here is Charles Darwin describing the mouth of a duck:

> The beak of the shoveller-duck (Spatula clypeata) is a more beautiful and complex structure than the mouth of a whale. The upper mandible is furnished on each side (in the specimen examined by me) with a row or comb formed of 188 thin, elastic lamellae, obliquely bevelled so as to be pointed, and placed transversely to the longer axis of the mouth.

Darwin's is an observant, precise mind. He carefully refrains from saying anything more than his facts allow: notice the qualification "(in the specimen examined by me)." It would not be accurate to say that Darwin never allows emotion to infuse his words: he speaks of the duck's mouth as a "beautiful" structure, for example. Still, his feelings—unlike those of Bertrand Russell in the passage from his *Autobiography*—are very much in the background. The persona is that of a sober, objective, painstaking observer, which, for Darwin's purpose, is exactly what it should be.

TONE

Tone is a web of feeling stretched throughout an essay. It has three main strands: the writer's attitudes toward subject, reader, and self. Persona is the complex personality we sense behind the writing; tone is the more specific set of feelings from which that particular personality emerges.

Each of the three determinates of tone is important, and each has many variations. Writers may regard themselves very seriously, or with ironic or amused detachment (to suggest only three of numerous possibilities). They may treat readers as intellectual inferiors and lecture them (usually a poor tactic), or as friends with whom they are talking. They may be angry about their subject, or laugh at it, or discuss it objectively and unemotionally. Given the three variables of subject, reader, and self, together with many ways of regarding each, the possibilities of tone are almost endless.

Like persona, tone is unavoidable. You imply it in the words you select and in how you arrange them. It behooves you, then, to express an appropriate tone. You must not appear pompous or offend readers by being flippant about a subject that demands seriousness. Always remember that tone of one kind or another you are bound to have. Make it a tone that helps rather than hinders.

Let us look at a few examples of skillful writers using tone to create a set of attitudes.

ATTITUDE TOWARD SUBJECT

Many attitudes toward one's subject are possible. Often tone is simple objectivity. Here, for instance, are two paragraphs from a book about science:

> Physical science is that department of knowledge which relates to the order of nature, or, in other words, to the regular succession of events.
> The name of physical science, however, is often applied in a more or less restricted manner to those branches of science in which the phenomena considered are of the simplest and most abstract kind, excluding the consideration of the more complex phenomena, such as those observed in living beings.
>
> James Clerk Maxwell

Maxwell's purpose is to frame a definition of physical science, not to express his feelings about it. His language, accordingly, is denotative and his tone objective and unemotional.

In contrast, the writer of the following paragraph is angry:

> *The Exorcist* is a menace, the most shocking major movie I have ever seen. Never before have I witnessed such a flagrant combination of perverse sex, brutal violence, and abused religion. In addition, the film degrades the medical profession and psychiatry. At the showing I went to, the unruly audience giggled, talked, and yelled throughout. As well they might. Although the picture is not X-rated, it is so pornographic that it makes *Last Tango in Paris* seem like a Strauss waltz.
>
> Ralph R. Greenson, M.D.

In the next example an angry tone is expressed more subtly, behind a mask of irony. The writer is describing the efforts of nineteenth-century laborers to improve their working conditions:

> . . . as early as June 8, 1847 the Chartists had pushed through a factory law restricting working time for women and juveniles to eleven hours, and from May 1, 1848 to ten hours. This was not at all to the liking of the manufacturers, who were worried about their young people's morals and exposure to vice; instead of being immured for a whole twelve hours in the cozy, clean, moral atmosphere of the factories, they were now to be loosed an hour earlier into the hard, cold, frivolous outer world.
>
> Fritz J. Raddatz

Irony may be employed more humorously to create a very different tone, as in this pleasantly satiric comment on universities:

> In the colleges of Canada and the United States the lectures are supposed to be a really necessary and useful part of the student's training. Again and again I have heard the graduates of my own college assert that they had got as much, or nearly as much, out of the lectures at college as out of the athletics or the Greek letter society or the Banjo and Mandolin Club.
>
> Stephen Leacock

TONE TOWARD READER

Writers may also view their readers in widely different ways. Some tend to be assertive and dogmatic, regarding readers as a passive audience to be instructed. The playwright and social critic George Bernard Shaw attacks the evils of the capitalism in such a manner:

> Just as Parliament and the Courts are captured by the rich, so is the Church. The average parson does not teach honesty and equality in the village school: he teaches deference to the merely rich, and calls that loyalty and religion.

Shaw's tone is that of a lecturer, though certainly a witty and entertaining lecturer. He expects his readers to sit meek and silent and to learn what things are really like.

At the opposite extreme a writer may try to establish a more intimate, face-to-face tone, as though he or she were talking to a friend. In the following case the writer discusses the problems of being the "other woman" in a married man's life, of having to share him with his wife:

> One or the other of you is going to spend the night with him, the weekend with him, Christmas with him. (I've tried all three of us spending it together. Doesn't work.) One or the other of you is going to go on trips with him.
>
> Ingrid Bengis

Bengis' informal conversational tone depends upon several things. For one, she addresses her readers directly ("you"), acknowledging their presence and bringing them and herself into a more intimate, and seemingly more equal, relationship. For another she cultivates a colloquial style, one suggesting a friendly speaking voice: the contractions ("I've," "Doesn't") and the fragment ("Doesn't work").

Here is another example of a colloquial, friendly tone:

> Adventure can mean, as well, grabbing a bucket and going out some lovely summer afternoon to pick berries.
>
> Doesn't sound awfully adventurous? Well, if berrying is something you haven't done since you were a kid, or haven't ever done, and if doing it breaks your routine and gets you outside into the natural world, which is a place you don't go at all, at least not without two hundred dollars' worth of equipment on your back, and if, in spite of all this, you just up and do it, I'd say that was adventurous.
>
> Anyway, that's what this book is about—the adventure of grabbing a bucket and going. Any bucket. Any going.
>
> Ruth Rudner

Here again we find "you," contractions, fragments ("Any bucket. Any going"). Such a colloquial style, however, is *not* the result of directly copying speech. People don't really talk this way; our speech is not nearly so well-ordered and effective. Rather than imitating actual talk, the writer wishing to achieve a colloquial tone must create the *illusion* of speech, a very different thing. The long statement that makes up the bulk of Rudner's second paragraph, for instance, is a complicated periodic sentence, using parallelism and interrupted movement to delay its main point to a climactic final position. But the illusion of someone talking is successfully maintained by the diction—by the absence of formal, literary words and by the presence of idiomatic expressions associated with speech ("grabbing," "awfully," "well," "kid," "you," "haven't," "don't," "just up and do it," "anyway"). Notice how the tone is subtly altered if we substitute for the colloquial "anyway" a more bookish connective like "in any event."

A friendly, informal attitude toward your reader is obviously appropriate when dealing with a topic like Rudner's—getting personally involved with the natural world. But it need not be restricted to such subjects. Indeed, in much contemporary exposition, even of a scholarly sort, writers often relax the older convention of maintaining a formal distance between themselves and their audience and unbend to address readers directly. For example, in this sentence a well-known Shakespearean scholar is writing about *Hamlet*:

> Great plays, as we know, do present us with something that can be called a world, a microcosm—a world like our own in being made of people, actions, situations, thoughts, feelings, and much more, but unlike our own in being perfectly, or almost perfectly, significant and coherent.
>
> Maynard Mack

"As we know" not only acknowledges the existence of the readers but draws them into a closer union with the writer and flatters their intelligence and sophistication.

An even subtler instance of addressing the reader is seen in the following passage, in which an engineer defends his profession against the charge that it has been heedless of damaging the environment:

> I do not want to overstate the case. The articles, editorials, newsletters, conferences, and the like, hardly constituted a crusade on the part of the engineering profession to save the environment. More could have been done. All right, more *should* have been done. But at least there is evidence that genuine efforts were made by the profession.
>
> Samuel C. Florman

Florman's next-to-last sentence tacitly says to readers: "I hear your objection that a few articles and editorials about protecting the environment were not very much." Thus in a subtle manner the writer encourages the readers' participation in the communication, giving them a kind of input and responding to it.

Florman's sentence also illustrates another way in which writers create the illusion of a talking voice. The use of italics suggests the emphasis and pitch by which we draw attention to important words. Barbara Tuchman does this very effectively in the following passage (she is arguing that concern for freedom of speech does not require that we accept any and all pornography):

> The cause of pornography is *not* the same as the cause of free speech. There *is* a difference. Ralph Ginsburg is *not* Theodore Dreiser and this is *not* the 1920s.

Used sparingly, in that way, italics are a useful device of emphasis and a way of creating a voice with which readers can connect. But note the caution: used *sparingly*. Italicization for emphasis easily becomes a mannerism and then it loses its effectiveness and turns into an annoyance.

TONE TOWARD SELF

Toward himself or herself a writer can adopt an equally great variety of tones. Objective, impersonal exposition involves a negative presentation of the writer, so to speak. By avoiding personal references or idiosyncratic comments, he or she becomes a lucid transparency through which we observe a set of facts or ideas. For example, a British writer, discussing the Battle of Anzio in Italy during World War II, begins like this:

> The full story of Anzio, which was originally conceived as a minor landing behind enemy lines but evolved through many ups and downs into a sepa-

rate Italian front of major importance, needs a history to itself. Within the scope of the present work it is possible only to summarize the main events and their significance in so far as they affected the main front at Cassino.

Fred Majdalany

On the other hand a writer may be more self-conscious and deliberately play a role. In exposition, for instance, it is often a good tactic to present yourself a bit deferentially, as Benjamin Franklin suggests in the passage quoted earlier (page 91). An occasional "it seems to me" or "I think" or "I believe" goes a long way toward avoiding a tone of cocksureness and restoring at least a semblance of equality to the unavoidably one-way communicative street that runs between writer and reader. Thus a scholar writing about Chaucer's love poetry avoids dogmatism by a qualifying phrase:

His early love complaints are less conventional than most and have the unmistakable ring, or so it seems to me, of serious attempts at persuasion.

John Gardner

A writer's exploitation of a self-image may go considerably beyond an occasional "I think." Humorous writers, for example, often present themselves as slightly ridiculous figures. The Irish writer Sean O'Faolain begins an essay like this:

When I was a boy in school, reading my Latin texts with one finger on the word and one finger in the Notes, I did not get much fun out of it. Anyway they made us read the wrong sort of authors, respectable authors like Cicero and Livy and the dull parts of Virgil.

The attitude toward self—or rather, boyhood self—defines O'Faolain's tone: an amusing image of a bored and struggling schoolboy.

S. J. Perelman, a magnificently funny writer, often casts himself in a comic part:

Every so often, when business slackens up in the bowling alley and the other pin boys are hunched over their game of bezique, I like to exchange my sweatshirt for a crisp white surgical tunic, polish up my optical mirror, and examine the corset advertisements in the New York *Herald Tribune* rotogravure section and the various women's magazines. It must be made clear at the outset that my motives are the purest and my curiosity that of the scientific research worker rather than the sex maniac.

Such role-playing is not quite the same thing as a persona. A writer's persona is reflected in all aspects of a composition, not simply in a self-caricature designed to amuse us or the guise of a deferential friend hoping to charm us. Beyond any momentary character the writer may be playing is the shrewd and witty creator of that image. It is that creator, that intelligence and sensibility, as we apprehend it in the style, which constitutes the persona.

EXERCISE

Study one of your recent themes—perhaps one you wrote in the preceding exercise—for point of view, persona, and tone. Compose a paragraph describing the tone and persona you sense in your writing. Be specific, anchoring the description in particular words and phrases.

If you prefer, you might do this exercise by exchanging themes with a classmate and commenting upon his or her point of view, persona, and tone.

PART THREE

THE EXPOSITORY PARAGRAPH

INTRODUCTION

The expository paragraph deals with facts, ideas, beliefs. It explains, analyzes, defines, compares, illustrates. It answers questions like What? Why? How? What follows? Like what? Unlike what? It is the kind of paragraph you are most often called upon to write in reports, term papers, and discussion tests.

Our survey of the expository paragraph falls into three sections. First we discuss its basic structure; second, the variety of ways in which paragraphs may be developed; and finally, on a slightly advanced level, a few of the more complex techniques of unifying and developing expository paragraphs.

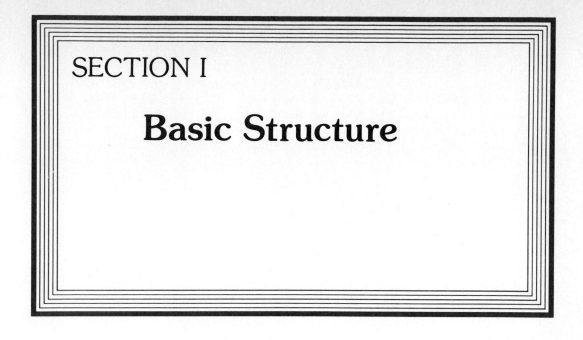

SECTION I

Basic Structure

14
The Sentence in the Paragraph

INTRODUCTION

The term *paragraph* has no simple definition. On rare occasions, when strong emphasis is wanted, a single sentence or word may serve as a paragraph. But generally in composition a paragraph is a group of sentences all relating to and developing a common idea, called the topic.

An expository paragraph is essentially an enlargement of a simple subject/predicate pattern like "Dogs bark." The subject, or topic, is simply more complicated and needs to be expressed in a sentence. This is called the *topic sentence* or *statement,* and it usually appears at or near the beginning of the paragraph. The predicate—that is, what is asserted about the topic—requires several sentences. These compose the bulk of the paragraph, developing or supporting the topic in various ways. We shall study some of those ways in subsequent chapters.

The subject/predicate pattern can vary. A paragraph may have two topics, sometimes of equal importance, sometimes of a major/minor nature. Paragraphs also differ in length and complexity, depending upon such factors as fashion, occasion, and individual preference. Today the fashion is for shorter paragraphs in expository writing than one would have used fifty or a hundred years ago. Occasion—why, where, and for whom you write—is an even more significant cause of variation. Journalists, for instance, prefer shorter paragraphs than do writers of scholarly essays. Narrative and descriptive paragraphs differ from expository.[1] In a brief composition paragraphs are likely to run shorter and simpler than in a longer paper. Finally, personal style plays a part: some writers like fewer but longer paragraphs, others numerous short ones.

No one, then, can say how long a paragraph should be. As a rough

1. We consider descriptive and narrative paragraphs in Chapters 46 and 47.

rule of thumb—very rough—think of paragraphs for your English compositions and other course papers in terms of 120 to 150 words. If most of your paragraphs fall below 100 words—50 or 60, say—they probably need more development. If your paragraphs run well over 150 words—200 or 300—consider dividing them. Numerous brief paragraphs are liable to be disjointed and underdeveloped. Too many long ones fatigue readers. But remember—we are talking about *average paragraph length*. An occasional short paragraph of 10 or 15 words may work very well, and so may one of 300.

THE TOPIC SENTENCE

A good topic sentence is concise and emphatic. It is no longer than the idea requires, and it stresses the important word or phrase. Here, for instance, is the topic statement which begins a paragraph about the collapse of the stock market in 1929:

> The Big Bull Market was dead.
>
> Frederick Lewis Allen

Notice several things. Allen's sentence is brief. Not all topics can be explained in six words, but whether they take six or sixty, they should be phrased in no more words than are absolutely necessary. The sentence is clear and strong: you know exactly what Allen means. It places the key word at the end: "dead." In the final position, a word or phrase gets heavy stress and leads naturally into what follows. Of course, if a topic sentence ends on the key term, it must do so naturally, without violating any rules of word order. Finally, the sentence opens the paragraph. The topic sentence generally belongs here—at or near the beginning.

In order to attract attention, topic sentences sometimes appear in special forms, as rhetorical questions or fragments. A rhetorical question differs from a literal question in that it does not request information. Either it asserts something emphatically ("Have we not had enough of such men?"="We certainly have had enough of such men!"), or it sets up an idea for development (that is, the writer asks the question in order to answer it). In either case a rhetorical question is a real question and must be closed with a question mark. Here are examples of rhetorical questions used as topic statements; each also opens its paragraph:

> What then is the modern view of Joan's voices and messages from God?
> George Bernard Shaw (about Joan of Arc)
> What did Lincoln's Emancipation Proclamation accomplish?
> J. G. Randall

The rhetorical question makes it easy to generate a paragraph. In some ways, too easy, tempting you to employ it too often. In a short theme one rhetorical question as a topic statement will work well; two will seem a mannerism.

Unlike the rhetorical question, the fragment—a grammatically incomplete sentence—does not automatically generate a paragraph. Its virtue is that it is unusual and hence emphatic and eye-catching, as in the second paragraph of this passage:

> Approaching the lake from the south, spread out, high up in a great V, was a flock of Canada geese. They did not land but continued on their way, trailed by the brass notes of their honking.
> *Spring.* How perfect its fanfare. No trumpets or drums could ever have so triumphantly announced the presence of royalty. I stood marveling in their wake until, cold, I returned to the firs to see what else I could summon up.
>
> Ruth Rudner (italics added)

In the following paragraph about the Great Depression of the 1930s, the topic statement closely combines a fragment and a rhetorical question:

> *And the workingman?* Already there are six million unemployed. Factory employment is down 20 per cent and payrolls are off 29 per cent. The man who still has a job can read the future in his shrinking pay envelope and dwindling bank savings. Even if he doesn't want to look, he can see the breadlines forming, and he is not unmindful that winter is ahead. He is feeling the pinch hard already; he will feel it more.
>
> Wallace Stegner (italics added)

Like rhetorical questions, fragments should be called upon only rarely as topic statements. Overused, they become common and lose their one virtue. Moreover, because fragments are easily abused, many instructors are reluctant to countenance them at all. Before you employ a fragment, for a topic sentence or any other purpose, ask your teacher's advice.

SENTENCES AS ANALYTICAL ELEMENTS

The sentences of a good expository paragraph reflect a clear, rational analysis of the topic. Here is a brief example, a paragraph by the British philosopher Bertrand Russell (the sentences have been numbered for convenience):

> (1) The intellectual life of the nineteenth century was more complex than that of any previous age. (2) This was due to several causes. (3) First: the area concerned was larger than ever before; America and Russia made important contributions, and Europe became more aware than formerly of Indian philosophies, both ancient and modern. (4) Second: science, which had been a chief source of novelty since the seventeenth century, made new conquests, especially in geology, biology, and organic chemistry. (5)

Third: machine production profoundly altered the social structure, and gave men a new conception of their powers in relation to the physical environment. (6) Fourth: a profound revolt, both philosophical and political, against traditional systems in thought, in politics, and in economics, gave rise to attacks upon many beliefs and institutions that had hitherto been regarded as unassailable. (7) This revolt had two very different forms, one romantic, the other rationalistic. (8) (I am using these words in a liberal sense.) (9) The romantic revolt passes from Byron, Schopenhauer, and Nietzsche to Mussolini and Hitler; the rationalistic revolt begins with the French philosophers of the Revolution, passes on, somewhat softened, to the philosophical radicals in England, then acquires a deeper form in Marx and issues in Soviet Russia.

Russell's nine sentences correspond to his steps in analyzing his topic. The first states the topic: the increasing complexity of nineteenth-century intellectual life. The second indicates how the paragraph will develop: it will list several causes of the complexity. Sentence 3 relates the first of these causes; number 4 explains the second; and 5 the third.

Thus far a one-to-one correspondence exists between sentence and cause. When he comes to the fourth reason, however, Russell uses not one but four sentences. Even so, the thought determines the division. The first of these four sentences (number 6 in the text) expresses the point in general terms; the next analyzes it into two aspects; the parenthetical statement qualifies a key term; and the last develops more specifically the two aspects set up earlier.

In Russell's paragraph, then, sentences correlate with thought something like this:

Sentence	Idea
1	Topic (increasing intellectual complexity)
2	Plan (list several causes)
3	First cause (larger area)
4	Second cause (science)
5	Third cause (machine production)
6	Fourth cause (stated generally)
7	(analyzed into two forms)
8	(qualification)
9	(specification of the two forms)

Showing how the sentence structure of a paragraph correlates with its ideas is called *conceptual analysis,* and you can use it to test your own paragraphs. The correspondence need not be as exact as in Russell's case, but if you cannot make a simple outline establishing reasonably clear relationships, your paragraph is probably confused and confusing.

The fact that an outline of a paragraph like Russell's reveals a coherent logical structure does not necessarily mean that the writer actually worked from such an outline. One can proceed this way, but in an essay of any length it is tedious and time-consuming to plot out every paragraph. Experienced writers adjust sentences to thought intuitively, without constantly thinking about when to begin a new sentence. Those with less experience must remain more conscious of the problem, and working up individual paragraphs from outlines provides good practice. But whether the result of careful outlining or of intuition, a well-constructed expository paragraph follows this principle: analysis of the topic determines the number of sentences.

In Russell's paragraph the topic is supported by a list of causes explaining an effect. One can support topics in many other ways: by listing effects derived from a cause, by offering examples, by specifying details, by restating, defining, qualifying, comparing, contrasting. In following chapters we shall study these means of developing paragraphs. First, however, we must look at something vital to the effectiveness of all paragraphs.

This is paragraph unity, and it occupies us in the next chapter.

EXERCISE

1. Indicate the topic sentence of each of the following paragraphs. Judging it by the criteria of clarity, concision, and focus, do you think each topic statement is effective? Does each paragraph stay within the limits established by the topic sentence?

A. *The Exorcist* is a menace, the most shocking major movie I have ever seen. Never before have I witnessed such a flagrant combination of perverse sex, brutal violence, and abused religion. In addition, the film degrades the medical profession and psychiatry. At the showing I went to, the unruly audience giggled, talked, and yelled throughout. As well they might: although the picture is not X-rated, it is so pornographic that it makes *Last Tango in Paris* seem like a Strauss waltz.

Ralph R. Greenson, M.D.

B. Consumerism as applied to women is blatantly sexist. The pervasive image of the empty-headed female consumer constantly trying her husband's patience with her extravagant purchases contributes to the myth of male superiority: we are incapable of spending money rationally; all we need to make us happy is a new hat now and then. (There is an analogous racial stereotype—the black with his Cadillac and his magenta shirts.) Furthermore, the consumerism line allows Movement men to avoid recognizing that they exploit women by attributing women's oppression solely to capitalism. It fits neatly into already existing radical theory and concerns, saving the Movement the trouble of tackling the real problems of women's liberation. And it retards the struggle against male supremacy by dividing women. Just as in the male movement, the belief in consumerism encourages radical women to patronize and put down other women for trying to survive as best they can, and maintains individualist illusions.

Ellen Willis

2. Selecting one of the general subjects below, write ten topic sentences, each on a different aspect of the subject. Work for clarity, emphasis, concision. Experiment with

placing key terms at the end of the sentence, with the rhetorical question, and—given your instructor's approval—with the fragment. Select topics that you could develop into expository paragraphs of 150 to 220 words; you might be asked to do so.

School (homework, teachers, classmates, for example)
The economic future as you see it
Teen-age culture
Popular entertainment
Sports
Hobbies

3. Make a conceptual analysis of each of the following paragraphs. Use an outline form like that on page 108, listing the sentences by number on the left, and on the right their logical relationships to what has gone before. After each statement of relationship, sum up what the sentence says. The analysis of the first paragraph might begin like this:

Sentence	Idea
1	Topic (a paradox with regard to grammar)
2	Specification: first part of the paradox (people regard grammar as dull)

A. A curious paradox exists in regard to grammar. (2) On the one hand it is felt to be the dullest and driest of academic subjects, fit only for those in whose veins the red blood of life has long since turned to ink. (3) On the other, it is a subject upon which people who would scorn to be professional grammarians hold very dogmatic opinions, which they will defend with considerable emotion. (4) Much of this prejudice stems from the usual sources of prejudice—ignorance and confusion. (5) Even highly educated people seldom have a clear idea of what grammarians do, and there is an unfortunate confusion about the meaning of the term "grammar" itself.

W. Nelson Francis

B. (1) Marriage is an immemorial institution which, in some form, exists everywhere. (2) Its main purpose always was to unite and to continue the families of bride and groom and to further their economic and social position. (3) The families, therefore, were the main interested parties. (4) Often marriages were arranged (and sometimes they took place) before the future husbands or wives were old enough to talk. (5) Even when they were grown up, they felt, as did their parents, that the major purpose of marriage was to continue the family, to produce children. (6) Certainly women hoped for kind and vigorous providers and men for faithful mothers and good housekeepers; both undoubtedly hoped for affection, too; but love did not strike either of them as indispensable and certainly not as sufficient for marriage.

Ernest van den Haag

4. Compose a single paragraph from each of the following outlines. Use the topic suggested, but otherwise express your own ideas:

A. Sentence	Idea
1	Topic ("At first college is a strange place.")
2	Restatement of topic
3	First example
4	Second example

B. Sentence   Idea

Sentence	Idea
1	Topic ("Many young people feel differently about marriage than their parents do.")
2 } 3 }	{ Comparison: parents' view { young people's
4	First reason for difference
5	Second reason
6	Third reason

15

Paragraph Unity

INTRODUCTION

Paragraph unity involves two related but distinct aspects: coherence and flow. *Coherence* means that the ideas fit together. *Flow* means that the sentences link together so that readers are not conscious of gaps.

While they are closely connected, coherence and flow are not the same. Flow is a matter of style and appears in the surface of prose, in specific words and grammatical patterns which tie one sentence to another. Coherence belongs to the substructure of the paragraph, to the relationships of thought, feeling, and perception expressed by the words.

A paragraph can have one quality without the other. There may be a genuine coherence of idea, for example, yet a lack of flow so that the sentences never come close enough together to form a smooth bridge carrying the reader from idea to idea. Instead they resemble a series of awkwardly placed stepping stones across a small stream: the reader can get from one to another but only by considerable effort and always in danger of losing his or her footing.

COHERENCE

A coh.

To be coherent a paragraph must satisfy several criteria: *relevance*—every idea must relate to the topic; *effective order*—ideas must be arranged in a way that clarifies their logic or importance; and a negative requirement, *inclusiveness*—nothing vital must be omitted.

RELEVANCE

A coh. rel.

When you write a topic sentence, you make a promise, and the paragraph that follows should accord with that promise. Do not wander from the topic. No matter how attractive an idea may be, let it go if you cannot fit it into the topic you have staked out or cannot reasonably revise the topic to include it. Nothing mars coherence more easily than irrelevance. Here is an example:

(1) College is very different from high school. (2) The professors talk a great deal more and give longer homework assignments. (3) This interferes with your social life. (4) It may even cost you a girlfriend. (5) Girls don't like to be told that you have to stay home to study when they want to go to a show or dancing. (6) So they find some other boy who doesn't have to study all the time. (7) Another way in which college is different is the examinations. . . .

The paragraph begins well. The first sentence establishes the topic and the second supports it. Then, however, the writer begins to slide away from his subject. Sentences 3 and 4 might be allowed if they were subordinated. But 5 and 6 lose contact. Some people indeed do not like to take second place to books, but that is not pertinent to a paragraph contrasting college and high school. In sentence 7 the writer tacitly acknowledges that he has wandered, throwing out a long transitional lifeline to haul us back to the subject. Rid of irrelevance, the paragraph might begin:

College is very different from high school. The professors talk a great deal more and give longer homework assignments, which interfere with your social life. College examinations, too, are different. . . .

A Coh.
order

ORDER OF THOUGHT

Relevance alone is not enough to establish coherence. All the ideas in a paragraph can relate to the topic and yet be badly arranged.

Arrangement often inheres in the subject itself. A paragraph about baking a cake or preparing to water ski is committed to following the steps of the process it describes. Likewise a logical structure in the subject determines the order of its paragraph, as in this example about the value of opposition in politics:

The opposition is indispensable. A good statesman, like any other sensible human being, always learns more from his opponents than from his fervent supporters. For his supporters will push him to disaster unless his opponents show him where the dangers are. So if he is wise he will often pray to be delivered from his friends, because they will ruin him. But, though it hurts, he ought also to pray never to be left without opponents; for they keep him on the path of reason and good sense.

Walter Lippman

There is a necessary order of thought here: first the assertion, next the reason supporting it, and then a return to the original claim, now in the form of a conclusion introduced by "So."

Inherent order may also be chronological, following the sequence of events, as in this paragraph from an account of the execution of Mary, Queen of Scots:

> Her last night was a busy one. As she said herself there was much to be done and the time was short. A few lines to the King of France were dated two hours after midnight. They were to insist for the last time that she was innocent of the conspiracy, that she was dying for religion, and for having asserted her right to the crown; and to beg that out of the sum which he owed her, her servants' wages might be paid, and masses provided for her soul. After this she slept for three or four hours, and then rose and with the most elaborate care prepared to encounter the end.
>
> James Anthony Froude

In each of these cases the topic determines the arrangement. The writers' problem is simply to understand and follow the structure inherent in their subjects. Sometimes, however, order of thought is less a function of the subject itself than of the writer's perception or analysis of it. For instance, in composing a paragraph about the three things which most surprised you in college, you might find no necessary logical or temporal relationships among the things themselves. In terms of logic or chronology, you could use any one of the possible arrangements. Which arrangement to choose depends upon you.

One solution is to place ideas in order of relative importance. The most significant may come first and the least significant last. But a climactic progression from the least to the most important is usually better. If you cannot discern any differences in importance, consider which order best connects with what has gone before or best prepares for what is to come. Should you see no basis at all for arranging ideas within a paragraph, then of course any order is legitimate. But this is not likely to happen very often. Most of the time a proper or at least a most effective way of placing ideas does exist.

Paragraph coherence, finally, has to do with what we may loosely call the logic of the paragraph. Ask about each idea: (1) Is it relevant to the topic? and (2) Is it in the best possible place?

PARAGRAPH FLOW

¶ flow

Coherence belongs to the substructure of a paragraph, to the relationship of its ideas. Flow appears in the surface, visible in the explicit words and in the similarities of grammatical pattern that link sentence to sentence. There are two basic ways of maintaining flow, though these are not incompatible and a paragraph may employ both. The first is to establish a master plan at the beginning of the paragraph and to introduce each subsequent idea by a word or phrase which fits it into the plan. The second concentrates upon linking sentences successively as the paragraph develops, making sure that each statement connects with the one that precedes it. Try to visualize the difference like this (the rectangles represent sentences; their size, number, and position are merely schematic, implying nothing about actual sentences):

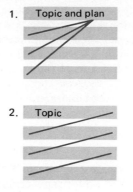

SETTING UP A MASTER PLAN

In this method of establishing flow, the opening of the paragraph makes clear not only the topic but also how it will be developed. The paragraph by Bertrand Russell, quoted on pages 107–8, is an example; it begins:

> The intellectual life of the nineteenth century was more complex than that of any previous age. This was due to several causes.

Russell's first sentence establishes the topic, his second the plan. More often these will be combined in a single sentence. Once the plan has been announced you need only fit each succeeding sentence or sentence group into its place with an appropriate word or phrase, as Russell does with "First. . . . Second. . . . Third. . . . Fourth. . . ."

Here is another example:

> There are three kinds of book owners. The first has all the standard sets and best-sellers—unread, untouched. (This deluded individual owns woodpulp and ink, not books.) The second has a great many books—a few of them read through, most of them dipped into, but all of them as clean and shiny as the day they were bought. (This person would probably like to make books his own, but is restrained by a false respect for their physical appearance.) The third has a few books or many—every one of them dog-eared and dilapidated, shaken and loosened by continual use, marked and scribbled in from front to back. (This man owns books.)
>
> Mortimer J. Adler

Both Russell and Adler work climactically, ending with the aspect of the topic they consider the most important or most interesting.

Sometimes, instead of words, numerals or letters introduce the parts of a paragraph:

For the majority of situations in which a dictionary is consulted for meaning, words may be roughly divided into three groups: (1) Hard words which circumstances make immediately important: "The doctor prescribed synthesized *cortisone.*" "*Recidivism* is a serious criminal problem in some urban communities." "*Existentialism* is a subjective philosophy." (2) Words frequently seen, usually understood loosely, but suddenly and recurrently unstable (for the individual): *synthesize, urban* and *subjective* in the preceding sentences. (3) Common familiar words which unexpectedly need to be differentiated (*break* vs. *tear, shrub* vs. *bush*) or specifically clarified, such as *fable, adventure, shake, door, remainder, evil.* Most people get by without having to clarify these common words in the third group until they become an issue. Without an issue definitions of these common words are frequently jumped on because the word looks easy to the uninitiated, although in practice they are usually more difficult than hard words to define.

Philip B. Gove

Numbering the parts of a paragraph—whether with words or figures—has the advantage of simplicity and clarity. It suits only topics which can easily be analyzed, however. It has the further disadvantage of mechanicalness. In a short essay one paragraph using this technique is enough.

The habit of relying upon numerical words or figures has a worse danger. Overindulged, it becomes self-defeating. When a subject is divided into "first" and "second" and each of these in turn analyzed into "1s" and "2s" or "a's" and "b's," readers may well think themselves in a maze. What was intended to clarify serves only to confuse.

You can avoid some of the obviousness of "first," "second," "third" by introducing various aspects of the subject into the topic sentence and then fulfilling the plan by repeating each key term as that aspect is developed:

We are controlled here by our confusion, far more than we know, and the American dream has therefore become something much more closely resembling a nightmare, on the *private, domestic,* and *international* levels. *Privately,* we cannot stand our lives and dare not examine them; *domestically,* we take no responsibility for (and no pride in) what goes on in our country; and, *internationally,* for many millions of people, we are an unmitigated disaster.

James Baldwin (italics added)

One way of creating flow, then, is to announce your plan and explicitly fit each unit into that plan. It is not a technique confined to paragraphs. You may use it to organize a portion of a long paragraph (which is what Baldwin does), or you may expand it to organize a short theme, in which case the units would be individual paragraphs rather than sentences. But it is, as we said, a mechanical mode of organization to be employed with restraint.

LINKING SUCCESSIVE SENTENCES

A flow link

The second way of maintaining flow is not to announce any master plan along with the topic, but simply to connect sentences as you go. Less obvious than "first," "second," "third," the method makes the paragraph evolve more naturally. Furthermore the technique can accommodate more complex relationships among ideas; it is not confined only to topics capable of being broken into numbered series.

Sentences can hook up in several ways.

1. *Repeating key words*

The most obvious hook-up is by verbal repetition, as in this short paragraph about Saint Patrick:

> We know that among the marks of holiness is the working of miracles. Ireland is the greatest miracle any saint ever worked. It is a miracle and a nexus of miracles. Among other miracles it is a nation raised from the dead.
>
> Hilaire Belloc

The repetition need not be so literal. Variant forms of a word may be used, or synonyms, as in this paragraph from *Democracy in America* by Alexis de Tocqueville.

> Those who cultivate the sciences among a democratic people are always afraid of losing their way in visionary speculation. They mistrust systems; they adhere closely to facts and study facts with their own senses. As they do not easily defer to the mere name of any fellow man, they are never inclined to rest upon any man's authority; but, on the contrary, they are unremitting in *their* efforts to find out the weaker points of their neighbor's doctrine. Scientific precedents have little weight with them; they are never long detained by the subtlety of the schools nor ready to accept big words for sterling coin; they penetrate, as far as they can, into the principal parts of the subject that occupies them, and they like to expound them in the popular language. Scientific pursuits then follow a freer and safer course, but a less lofty one.

Some of Tocqueville's repetitions are identical ("they . . . they"); some use variations of the same word ("sciences . . . scientific"); and some state the same idea differently ("afraid of losing their way in visionary speculation . . . mistrust systems").

Repeated words may occur in a variety of positions in the sentence.[1] The most useful patterns for establishing flow, however, are to begin successive sentences with the same words; to end them that way; or, strongest of all, to open a new sentence with the term or phrase that closed the preceding one:

---

1. For a more complete discussion of the varieties of repetition, see pages 305–7.

No man of note was ever further separated from life and fact than Lindbergh. No man could be more reluctant to admit it.

John Lardner

Charles R. Forbes went to jail. Albert B. Fall went to jail. Alien Property
Custodian Thomas W. Miller went to jail.

Samuel Hopkins Adams

Such plants to operate successfully had to run at capacity. To run at capacity they needed outlets for their whole output.

Thurman Arnold

A special case of repetition involves the use of pronouns and demonstratives. The personal pronouns and such words as *one, another, some,
the former, the latter, the first, the second, the third,* and so on link
sentences by substituting for an earlier word or phrase. Especially useful
in this way, *this* and *that* (along with their plurals, *these* and *those*)
may be employed either as true pronouns or as demonstrative adjectives.
This passage illustrates both functions:

> The blind in particular seem to become indifferent to climatic extremes;
> and there must be in everyone's cognizance two or three immovable sight-
> less mendicants defying rain and chill. . . .
> *This* insensitiveness to January blasts and February drenchings may
> be one of the compensations that the blind enjoy. Whatever else happens
> to them they never, perhaps, catch cold. And *that* is more than something.
>
> E. V. Lucas (italics added)

A vague this

There is a danger, however, in using *this* and *that* as subjects. A
connection clear to the writer does not always jump at the reader.
The risk increases when the antecedent of *this* or *that* is not a single
word, or even a phrase, but the total idea expressed by an entire sentence
or paragraph. For this reason it is sometimes better to use these words
not as nouns but as adjectives modifying a more precise subject that
fairly sums up the preceding point: "this fact," "this danger," "that
truth." As demonstratives the words still hook the sentence to what
went before, but now without risk of confusion.

2. *Connective words*

You can also link sentences by connective words and phrases. These
conjunctive, or transitional, adverbs establish a relationship between
the ideas expressed in successive sentences. The relationship may be
one of time (*presently, meanwhile, at the same time*); of space (*above,
below, in front*); or of logic (*therefore, however, as a result*).

In the following example the English critic F. L. Lucas connects his
sentences by transitional words (here italicized) in a passage answering
the claim that as a poetic device metaphor has no place in prose:

> The truth seems that metaphor too is older than any literature—an
> immemorial human impulse perhaps as much utilitarian as literary. *For*
> there appears little ground for assigning poetic motives to the first man

who called the hole in a needle its "eye," or the projections on a saw its "teeth." *In fine,* metaphor is an inveterate human tendency, as ancient perhaps as the days of the mammoth, yet vigorous still in the days of the helicopter. Why *then* should it be banned from prose?

"For . . . In fine . . . then" establish the logical framework of Lucas's argument:

Assertion . . . . Sentence 1
Reason . . . . "For," sentence 2
Assertion restated . . . . "In fine," sentence 3
Conclusion . . . . "Then," sentence 4

Transitional adverbs are best placed at or near the beginning of the sentence, where they guide the reader most effectively. Readers do not know where you are going. They are groping down a dark passage, and an important part of your job as writer is to show them the way. Connective words tell readers what to expect. *However* flashes "contradiction ahead"; *in fact* signals "here comes a strong restatement of what has just been said"; *therefore* warns "a conclusion or consequence approaching."

One can overuse such words, insulting readers by laboring the obvious. Good writers vary considerably in their use of terms like *nevertheless, even so, furthermore.* Some like to make the flow of their thought as explicit as possible and employ conjunctive adverbs frequently. Others use these words rarely, preferring to leave the logical structure of their sentences and paragraphs implicit.

There is room for both styles. Much depends upon the degree of help your readers require. But if you must err in using words such as *however* and *therefore,* too many are better than too few. Annoying readers by over-explicitness is a lesser fault than confusing them.

Acquiring a working set of these words and phrases is not difficult.[2] English is rich in them. Just to show some sort of opposition or contradiction, for example, we have *but, however, still, yet, nonetheless, nevertheless, though, instead, on the other hand, on the contrary, notwithstanding, even so*; and this list is not complete. These words are not simply equivalents, they convey subtle differences of meaning and tone. *Nevertheless,* for instance, is a more formal word than *though.* Because of slight but important distinctions in meaning and tone, good writers have ready to hand a number of transitional adverbs. A writer who can call only upon *but* or *however* cannot communicate the nuance expressed by *yet* or *still* or *though.*

And and *but* present a special case. Most often they act as coordinating conjunctions, joining words, phrases, or clauses within a sentence;

2. For a list of the more common ones, see Appendix 1.

but they can also function adverbially. Some people object to them as conjunctive adverbs, and you may have been warned, "Never begin a sentence with *and* or *but.*" The fact is that good writers frequently do start off with these words:

> I come finally to the chief defiler of undergraduate writing. And I regret to say that we professors are certainly the culprits. And what we are doing we do in all innocence and with the most laudable of motives.
>
> Willard Thorp

> Is not indeed every man a student, and do not all things exist for the student's behoof? And, finally, is not the true scholar the only true master?
>
> Ralph Waldo Emerson

> Natural philosophy had in the Middle Ages become a closed chapter of human endeavour. . . .
>
> But although the days of Greek science had ended, its results had not been lost.
>
> Kurt Mendelssohn

As openers the two conjunctions *and* and *but* are useful. *But* is less formal than *however,* while *and* is less formal and ponderous than *furthermore* or *moreover* or *additionally.* Don't be afraid of initial *and*s and *but*s; use them moderately, however, and in appropriate contexts.

3. *Syntactic patterning*

A third way of linking sentences within the paragraph is by syntactic patterning. This means repeating the same basic structure in successive or near-successive sentences. In a simple form writers often use this technique to hold together the parts of a comparison or contrast:

> In bankless Iowa City eggs sell for ten cents a dozen. In Chicago the bread-lines stretch endlessly along the dirty brick walls in windy streets.
>
> Wallace Stegner

> That New York was much more dry [non-alcoholic] on Sunday during that summer is true. That it was as dry as [Theodore] Roosevelt believed it—"I have, for once, absolutely enforced the law in New York"—is improbable. That it was dry enough to irritate the citizenry to new heights of indignation is clear.
>
> Henry F. Pringle

Syntactic patterning may be more extensive, working throughout most of a paragraph:

> It is common knowledge that millions of underprivileged families want adequate food and housing. What is less commonly remarked is that after they have adequate food and housing they will want to be served at a fine restaurant and to have a weekend cottage by the sea. People want tickets to the Philharmonic and vacation trips abroad. They want fine china and silver dinner sets and handsome clothes. The illiterate want to learn how to read. Then they want education, and then more education, and then they want their sons and daughters to become doctors and lawyers.

It is frightening to see so many millions of people wanting so much. It is almost like being present at the Oklahoma land rush, except that millions are involved instead of hundreds, and instead of land, the prize is everything that life has to offer.

Samuel C. Florman

While reusing the same sentence pattern often involves the iteration of specific words, it is the repeated syntax which most strongly ties things together. However, one cannot impose such repetition upon just any group of sentences. It works only when the thought itself is repetitive, as in Florman's paragraph where the sentences list a series of humanity's rising expectations. In such cases the similarity of pattern does what really all sentence structure should do: the form reinforces the sense.

EXERCISE

1. List the transitional devices that link the sentences in the following paragraph: repetition of key words or ideas, conjunctive adverbs, and sentences of similar construction:

Above the beginner's level, the important fact is that writing cannot be taught exclusively in a course called English Composition. Writing can only be taught by the united efforts of the entire teaching staff. This holds good of any school, college, or university. Joint effort is needed, not merely to "enforce the rules"; it is needed to insure accuracy in every subject. How can an answer in physics or a translation from the French or an historical statement be called correct if the phrasing is loose or the key word wrong? Students argue that the reader of the paper knows perfectly well what is meant. Probably so, but a written exercise is designed to be read; it is not supposed to be a challenge to clairvoyance. My Italian-born tailor periodically sends me a postcard which runs: "Your clothes is ready and should come down for a fitting." I understand him, but the art I honor him for is cutting cloth, not precision of utterance. Now a student in college must be inspired to achieve in all subjects the utmost accuracy of perception combined with the utmost artistry of expression. The two merge and develop the sense of good workmanship, or preference for quality and truth, which is the chief mark of the genuinely educated man.

Jacques Barzun

2. Each of the paragraphs below lacks unity. The problem may be merely inadequate links between sentences, or it may go deeper, involving incoherence of thought. Rewrite each paragraph, staying as close as possible to the original wording, but making the changes required to unify the sentences:

A. There are several kinds of test. Quizzes deal with only a small amount of material, usually that covered in the preceding week or two. Pop quizzes are given without any announcement. Students often miss them and have to arrange make-ups. Examinations are longer and cover more ground. The mid-term comes about the sixth or seventh week and in some courses is the only grade the teacher has for the mid-semester mark. It is important. The final comes at the end of the course and is a large part of one's grade. Students work hard preparing for finals.

B. Hats have served a number of different functions. One is comfort. People living in the hot areas of the Southwest like the ten-gallon hat because its wide brim shelters them from the sun and the large air space insulates the head and keeps it cool. Ten-gallon hats come in a smaller size (called the five-gallon); both kinds are made of felt, and both have spread to other sections of the country because of the popularity of the cowboy in American culture. (Americans are cowboy-conscious; they see cowboys on television and in the movies from childhood on.) Hats also offer protection. Football players and racing-car drivers wear helmets made of a hard plastic that completely encloses the head, which is cradled inside in a kind of webbing (which reduces impact). Construction workers wear helmets of steel or plastic in case tools or bolts are dropped on them. Hats are also decorative. Many people wear hats simply to increase their attractiveness (they hope), especially women. Frequently women's hats are silly, and sometimes they are annoying, especially if you get stuck behind one in a movie.

3. Here are ten words or phrases used to indicate logical relationships of thought. In your own words explain what each signals to a reader and whether each is relatively formal or informal in tone:

so	furthermore	anyhow
yet	besides	on the other hand
though	at all events	on the contrary
		in sum

SECTION II

Methods of Development

16

Paragraph Development: Illustration

INTRODUCTION

In this and the following seven chapters we shall study how to develop topics in an expository paragraph, focusing on one technique at a time, beginning with illustration. Such an approach oversimplifies matters because writers often combine techniques in composing paragraphs: not illustration *or* comparison *or* reasons, but illustration *and* comparison *and* reasons. But walking comes before running, and here we concentrate upon relatively simple paragraphs, saving for Chapter 25 those paragraphs that develop in more complicated patterns.

Methods of paragraph support may be sorted into three loose groups. The first includes those which stay strictly within the topic: they offer examples of it or simply say it over again in various ways, but they do not move from the topic to other related ideas. In the second group are techniques which involve a second subject, introduced for comparison or contrast. This second subject may be subordinate to the main topic, serving to clarify or emphasize some aspect of it, or it may have equal importance. The third class of techniques are more logical. These explore the meaning or ramifications of the topic—defining, qualifying, explaining causes and effects.

We begin with the simplest method of development—illustrating the topic.

ILLUSTRATION

A dev. illus.

Giving a specific example is one of the easiest ways of supporting a generalization. Good writers do this frequently, following an abstract idea with one or more illustrations:

Some of those writers who most admired technology—Whitman, Henry Adams, and H. G. Wells, for example—also feared it greatly.

Samuel C. Florman

But an effect can become a cause, reinforcing the original cause and producing the same effect in an intensified form, and so on indefinitely. A man may take to drink because he feels himself a failure, and then fail all the more completely because he drinks.

George Orwell

Illustrations show that you are not talking through your hat. Thus Florman gives us names, grounding his assertion in facts and enabling us to check that assertion against our own knowledge. Illustrations have a second virtue: they anchor an abstraction in particulars, translating an idea difficult to grasp into everyday terms. This is what Orwell does.

Brief examples like those from Florman and Orwell do not make paragraphs, of course. But examples can be extended to provide the substance of an entire paragraph. Sometimes the paragraph consists of a single instance worked out in considerable detail:

Some of the most abstract terms in the language are really faded metaphors. On examination it turns out that an earlier meaning, now forgotten, is often lively in the extreme. Hence an obvious means of invigorating our jejune vocabulary is to fall back on those lively older meanings. True enough, the average speaker does not know that they ever existed. He is not *reminded* that "express" once meant, literally and physically, "to press out." But he can learn it instantaneously from a context. It may be that only the archaic literal sense is intended, or it may be that both the physical and the metaphorical are to be grasped simultaneously. In any event, the impact of the divergent use on an attentive reader forces him to a new experience of the word, without sacrificing comprehension. An example of the use of "express" in this revivified fashion will be found in Emily Dickinson:

Essential Oils—are wrung—
The Attar from the Rose
Be not expressed by Suns—alone—
It is the gift of Screws—

Margaret Schlauch

On the other hand an illustration paragraph may consist of a number of brief examples, as in this passage about the change in modern modes of eating and drinking:

As far as the home is concerned, the biggest change in what P. G. Wodehouse called "browsing and sluicing" is probably not the decline in huge, formal meals, or shorter menus, but the odd form our food is in nowadays when we buy it. Coffee comes as a powder. Fish arrives as a frozen rectangular block. Soup, stiff with preservatives, comes in a tin or as a powder. Potatoes no longer wear their jackets but arrive pale and naked in an impenetrable plastic bag. Embryonic mashed potato comes in little dry lumps, like cattle-feed pellets. Bread, untouched by human baker, arrives wrapped and sliced in a soft lump, the "crust" seemingly sprayed on. Beer, urged upward by gas, emerges from a steel dustbin.

Frank Muir

Make sure the reader understands that examples are examples. The simplest solution is explicitly to introduce an illustration with some such expression as *for example, for instance, an example* (or *illustration*) *is, as a case in point*. The words *say, thus, consider, take, suppose* may also serve as introductions. Vary your use of these expressions; do not introduce every illustration with *for example*. With multiple examples it is not necessary to label every subsequent instance after the first has been introduced. Do not always place the phrase in the opening position. A *for instance* is equally effective between the subject and verb, where it is still near the beginning but seems less mechanical.

When the illustrative function of a detail is obvious, you can safely dispense with an introduction. Orwell does not write, "For example, a man may take to drink . . .", nor does Muir literally label his examples of the "odd forms" of modern food. He depends upon the reader's common sense. No infallible rule tells you when *for example* is superfluous and when its absence will confuse a reader. In this, as in so much of writing, you must try to imagine yourself in the reader's position. If an illustration seems even a bit confusing without an introduction, put one in.

Whether introduced or not, examples are most effective when they are specific. In Muir's paragraph the abstract expression "the odd form our food is in" is given shape and heft by "frozen rectangular block," "pale and naked in an impenetrable plastic bag," "little dry lumps like cattle-feed pellets." One can, of course, get carried away and pursue an illustration too far, so that the tail wags the dog and what was intended to support a generalization ends up supplanting it. But this is a much less common mistake than not using illustrations often enough and not developing those which are used with sufficient detail.

EXERCISE

1. Make a conceptual analysis (as explained on page 108) of each of the following paragraphs. In which paragraph(s) is the example explicitly introduced? Do you think the illustrations adequately support the topics? Why or why not?

A. Primitive peoples often build much of their religious and cultural behavior on this belief in the natural relationship of word and thing. For example, they believe that to know the name of an object, person, or deity is to gain a certain control over it: in "Ali Baba and the Forty Thieves," the words "Open Sesame!" cause the stone doors of the cave to move aside. Conversely, certain powers in the universe are thought to dislike the use of their names by mortals. Words are therefore tabooed, or euphemisms and descriptive phrases are invented such as *the little people* instead of *fairies*. The Greeks came to call those vengeful mythological creatures whose "real name" was *Erinyes* (or Furies) the *Eumenides* (or "good-tempered ones").

W. Nelson Francis

B. In the typical American family, a girl is trained from babyhood to be what the culture defines as feminine. Everyone encourages her to act cute and charming and flirt with her father, her uncles, and little boys. When she announces that she wants to be a fireman, her mother laughs: girls can't be firemen; you'll be a mother, like me. Or a nurse. Or a teacher. When she roughhouses, parents brag, "She's as tough as a boy." Yet at the same time they warn, "Someday you'll have to stop acting like a boy and be a lady." Most likely her brothers are free to play while she helps with the dishes, and her parents are more tolerant of their noise, dirt, and disobedience—after all, boys will be boys.

Ellen Willis

2. Compose a paragraph of about 150 words developing a single illustration in detail to support one of these topics:

A. A college catalogue is not easy to understand.
B. Most situation comedies on TV are ridiculous, far removed from the lives most of us live.
C. All parents have hang-ups.
D. It is surprising how many adults continue to play with toys.

3. Beginning with one of the topic sentences listed below, develop a paragraph using numerous brief illustrations. Make them clear enough so that you can dispense with an introduction:

A. One function of teen-age slang is to serve as a language which adults cannot understand.
B. Every national group has its favorite foods.
C. Courtship practices vary widely from culture to culture.
D. People make hobbies out of collecting almost anything.

CHAPTER 17

Paragraph Development: Restatement

SIMPLE RESTATEMENT

A dev. restate.

In its simplest form restatement involves nothing more than repeating the topic idea. As a way of emphasizing an important point within a paragraph, restatement is common:

> 1964 threatens to be the most explosive year America has witnessed. *The most explosive year.*
>
> Malcolm X (italics added)

Sufficiently extended, restatement may provide the substance of an entire paragraph, as in this passage about why American men are unlikely to cry (the paragraph expresses attitudes of our culture, not the writer's own beliefs):

> American men don't cry, because it is considered unmasculine to do so. Only sissies cry. Crying is a "weakness" characteristic of the female, and no American male wants to be identified with anything in the least weak or feminine. Crying, in our culture, is identified with childishness, with weakness and dependence. No one likes a crybaby, and we disapprove of crying even in children, discouraging it in them as early as possible. In a land so devoted to the pursuit of happiness as ours, crying really is rather un-American. Adults must learn not to cry in situations in which it is permissible for a child to cry. Women being the "weaker" and "dependent" sex, it is only natural that they should cry in certain emotional situations. In women, crying is excusable. But in men, crying is a mark of weakness. So goes the American credo with regard to crying.
>
> Ashley Montagu

Repeating what you have just said is both an easy and a difficult way of developing a paragraph. Easy because you do not have to search for examples or comparisons, or think about reasons and consequences.

Difficult because you must repeat yourself without being monotonous. Because of this difficulty, restatement passages are generally brief.

The risk of monotony is increased since restatement demands a closer degree of similarity in sentence structure than usual. Sentences that say the same thing are often cast in the same mold. A good example of repeated syntax in restatement appears in this passage by the poet John Masefield about the prevalence of piracy in the early part of the seventeenth century:

> It is difficult for one accustomed to the law and order of the present day to understand the dangers which threatened the Jacobean traveller. The seas swarmed with pirates; so that few merchantmen dared put to sea without arms; while very few came home without some tale of an encounter. There were pirates in the Atlantic, to intercept the ships coming home from the Newfoundland fisheries. There were pirates in the West Indies, roving for Spanish treasure-ships. There were pirates in the Orkneys, preying upon the Iceland trades. There were pirates near Ireland, especially in the south and west, ranging over the Channel, and round these coasts. But there were, perhaps, more pirates in the Mediterranean than in all the other waters put together. In the Mediterranean they had the most part of the trade of Europe for their quarry; while the coasts of Africa, and the islands of the Archipelago, provided obscure harbours (with compliant Governors) for the recruiting of the companies after a cruise.
>
> John Masefield

Aside from knowing when to stop, success in handling such repetition depends on sufficiently varying the form or content of the sentences. Masefield, for example, keeps to the same pattern for four successive sentences: "There were pirates in" + a verbal phrase. But each sentence differs in content.[1] At the same time the similarity of the sentences reinforces the point that much the same condition existed everywhere. Masefield also uses the repeated sentence structure to build toward his main topic: piracy in the Mediterranean. In the last of the five "there were" sentences he signals the climactic value of this idea by varying the pattern: opening with "But," placing "perhaps" in an interrupting position, and changing completely the pattern of the second half of the sentence.

NEGATIVE-POSITIVE RESTATEMENT

A dev.
neg. pos.

Restatement may take two special forms. One is negative-positve reassertion. You begin by saying what is *not* the case, then assert what is (the order may be reversed):

> I am not thinking of philosophy as courses in philosophy or even as a subject exclusive of other subjects. I am thinking of it in its old Greek sense, the sense in which Socrates thought of it, as the love and search

1. The last of the four has an interrupting phrase between the main clause and the verbal construction.

for wisdom, the habit of pursuing an argument where it leads, the delight in understanding for its own sake, the passionate pursuit of dispassionate reasonableness, the will to see things steadily and to see them whole.

Brand Blanshard

SPECIFICATION

A dev.
spec.

The second special type of restatement moves from general to particular, repeating an abstract idea not literally but rather by enumerating all that it entails. Brief specifications are often found within individual sentences as a way of giving substance to an abstraction:

Bound to the production of staples—*tobacco, cotton, rice, sugar*—the soil suffered from erosion and neglect.

Oscar Handlin (italics added)

The great reform measures—*the reorganization of parliament, the revision of the penal code and the poor laws, the restrictions placed on child labor, and other industrial reforms*—were important factors in establishing English society on a more democratic basis.

Albert C. Baugh (italics added)

A more extended instance of specification occurs in this paragraph about politics in Louisiana. It develops by particularizing all that is included in the phrase "the same political pattern":

Throughout the years the same political pattern prevailed. The city dominated the state: New Orleans, the nation's mecca of the fleshpots, smiling in not altogether Latin indifference at its moral deformities, and, like a cankered prostitute, covering those deformities with paint and lace and capitalizing upon them with a lewd beckoning to the stranger. Beyond New Orleans, in the south, French Louisiana, devoutly Catholic, easy-going, following complacently its backward-glancing patriarchs, suspicious of the Protestants to the north. And in central and northern Louisiana, the small farmers, principally Anglo-Saxon; bitter, fundamentalist Protestants, hating the city and all its evil works, leaderless in their disquiet and only vaguely aware that much of what they lacked was in some way coupled with the like-as-like office seekers whom they alternately voted into and out of public life.

Hodding Carter

While specification resembles illustration, it differs in an important way. An illustration is one of numerous possible cases; specification covers in detail the entire ground implied in the topic idea. In the sentence by Professor Handlin "tobacco, cotton, rice, sugar" are not simply examples of the staple crops of southern agriculture: they *are* the staple crops. Similarly Hodding Carter, beginning with the abstract phrase "political patterns," specifies that pattern in its entirety, rather than illustrating it by mentioning only a single part. Since it moves from general to particular, specification is less likely to seem monotonous, and writers use it more frequently than simple restatement.

EXERCISE

1. In each of the following paragraphs which words in the opening sentence establish the topic? Which expressions elsewhere in the paragraph restate the topic?

A. There is another paradox in man's relationship with other creatures: namely, that those very qualities he calls animalian—"brutal," "bestial," "inhuman"—are peculiarly his own. No other animal is so deliberately cruel as man. No other creature intentionally imprisons its own kind, or invents special instruments of torture such as racks and thumbscrews for the sole purpose of punishment. No other animal keeps its own brethren in slavery; so far as we know, the lower animals do not commit anything like the acts of pure sadism that figure rather largely in our newspapers. There is no torment, spite, or cruelty for its own sake among beasts, as there is among men. A cat plays with its prey, but does not conquer and torture smaller cats. But man, who knows good and evil, is cruel for cruelty's sake; he who has a moral law is more brutal than the brutes, who have none; he alone inflicts suffering on his fellows with malice aforethought.

Susanne K. Langer

B. An act of the self, that's what one must make. An act of the self, from me to you. From center to center. We must mean what we say, from our innermost heart to the outermost galaxy. Otherwise we are lost and dizzy in a maze of reflections. We carry light within us. There is no need merely to reflect. Others carry light within them. These lights must wake to each other. My face is real. Yours is. Let us find our way to our initiative.

Mary Caroline Richards

C. Lord Chesterfield's gentleman was a man of the world; but it was, after all, a very hard and empty world. It was a world that had no eternal laws, only changing fashions. It had no broken hearts, only broken vows. It was a world covered with glittering ice, and the gentleman was one who had learned to skim over its dangerous places, not caring for those who followed him.

Samuel McChord Crothers

2. Beginning with one of the following topic sentences, compose a brief restatement paragraph. Construct your sentences to resemble one another, though with enough variety to avoid being monotonous:

A. There was no doubt about it—Harry was ugly.
B. It was the dullest party I have ever been to.
C. She was very attractive.
D. I studied very hard for the exam.

3. Specification, as in the paragraph by Hodding Carter (page 131), begins with a broad statement of the topic and then repeats it in detail. Compose such a paragraph on one of these subjects:

large cities
television
professors
students

CHAPTER 18

Paragraph Development: Comparison and Contrast

A dev. comp. Here we turn to the three types of paragraphs which bring together two or more subjects: comparison, contrast, and analogy. *Comparison* deals with the similarities between several subjects, and *contrast* with the differences. *Analogy* is a special type of comparison in which a subject of secondary importance and often of a very different nature is introduced to clarify or justify some aspect of the main topic. While comparison concentrates upon similarities and contrast upon differences, they work in much the same way, and we consider them together. Analogy receives separate treatment in the next chapter.

FOCUSING A COMPARISON OR CONTRAST

Because they involve at least two subjects and offer the possibility of doing several things, comparison and contrast pose problems of focus, organization, and development. First you face the question of whether to deal only with points of likeness or of difference or to treat both. You must be clear about the focus in your own mind and make it equally clear to the reader. These topic sentences show such clarity:

> It is a temptation to make a comparison between the nineteen 'twenties and the nineteen 'sixties, but the similarities are fewer than the differences.
>
> Russell Lynes

> The difference between a sign and a symbol is, in brief. . . .
>
> Susanne K. Langer

> Shakespeare's England and the Athens of Pericles were alike in several ways.
>
> Student

A second problem of focus concerns the subjects. Will you concentrate upon one subject or treat both as equally important? If, for instance,

133

you are writing about high school and college, you have three possibilities of focus: on high school, on college, or on both. The focus must be made clear to the reader, but without the obviousness of a sentence like the following:

× In this paragraph I shall be chiefly concerned with high school.[1]

Work with a lighter hand, constructing the topic sentence so that the key term or terms function as its subject and thus naturally indicate your focus. For example, if you wish to concentrate upon high school:

In many ways high school is like college.

If upon college:

In many ways college is like high school.

And if upon both:

College and high school are alike in many ways.

In the following paragraph notice how the historian J. G. Randall keeps his focus constantly before us. (He is comparing the failure of Reconstruction after the Civil War and the refusal of the U.S. Senate to accept President Wilson's League of Nations policy after World War I.)

In the case of *both Lincoln and Wilson* the soldiers did their part and so did the Executive, but *in each case* partisanship and narrow-mindedness wrecked the program. *Under Lincoln and Johnson, as under Wilson,* there was failure of high-minded unity behind the plan of peace that bore promise of success. *In each case,* instead of needful co-operation, there was stupid deadlock between President and Congress. There was *in each case* a fateful congressional election whose effect was felt far down in later years: 1918 may be matched against the "critical year" 1866. *In each case* the President's plan failed in the sense that it failed to be adopted; the opposite plan *in each case* failed miserably by being adopted.   (italics added)

ORGANIZING A COMPARISON OR CONTRAST

A second problem posed by comparison and contrast involves organization. When you compare any two subjects—call them A and B—you must do so with regard to specific points—1, 2, 3, and so on. You may organize your material in two ways—either:

A.
    1.
    2.
    3.

1. Here and throughout the book the symbol × before a word or sentence means that it is stylistically awkward. The symbol differs from an initial asterisk, which means that the word or expression is wrong, violating a rule of grammar, usage, or mechanics.

B.
  1.
  2.
  3.

or:

1.
  A.
  B.
2.
  A.
  B.
3.
  A.
  B.

In a paragraph about high school and college you could devote the first half to high school and discuss such specifics as teachers, lessons, and homework, and spend the second half on college, treating the same points, preferably in the same order. Or you could organize around the particular similarities or differences. In the first part of the paragraph you would discuss teachers, in high school and then in college; in the second part, lessons; and in the third, homework.

Neither method of organizing is inherently better. Proceeding by A and B stresses each subject in its totality. Organizing by 1, 2, 3 emphasizes specific likenesses or dissimilarities. But while neither method is absolutely superior, one will probably serve your purpose better on any specific occasion. In the following passage, for example, the writer chose to organize around the broad subjects of Western civilization and Eastern:

> Americans and Western Europeans, in their sensitivity to lingering problems around them, tend to make science and progress their scapegoats. There is a belief that progress has precipitated widespread unhappiness, anxieties and other social and emotional problems. Science is viewed as a cold mechanical discipline having nothing to do with human warmth and the human spirit.
>
> But to many of us from the nonscientific East, science does not have such repugnant associations. We are not afraid of it, nor are we disappointed by it. We know all too painfully that our social and emotional problems festered long before the age of technology. To us, science is warm and reassuring. It promises hope. It is helping us at long last gain some control over our persecutory environments, alleviating age-old problems—not only physical but also, and especially, problems of the spirit.
>
> F. M. Esfandiary

Here, on the other hand, an historian contrasting Catholics and Protestants in the sixteenth century organizes not around the broad categories of Roman and Reformer, but rather around the specific differences that separated them:

> The Catholic believed in the authority of the Church; the Reformers, in the authority of reason. Where the Church had spoken, the Catholic obeyed. His duty was to accept without question the laws which councils had decreed, which popes and bishops administered, and so far as in him lay to enforce in others the same submission to an outward rule which he regarded as divine. All shades of Protestants on the other hand agreed that authority might err; that Christ had left no visible representative, whom individually they were bound to obey; that religion was the operation of the Spirit on the mind and conscience; that the Bible was God's word, which each Christian was to read, and which with God's help and his natural intelligence he could not fail to understand. The Catholic left his Bible to the learned. The Protestant translated the Bible, and brought it to the door of every Christian family. The Catholic prayed in Latin, and whether he understood his words or repeated them as a form the effect was the same; for it was magical. The Protestant prayed with his mind as an act of faith in a language intelligible to him, or he could not pray at all. The Catholic bowed in awe before his wonder-working image, adored his relics, and gave his life into the guidance of his spiritual director. The Protestant tore open the machinery of the miracles, flung the bones and ragged garments into the fire, and treated priests as men like himself. The Catholic was intolerant upon principle; persecution was the corollary of his creed. The intolerance of the Protestant was in spite of his creed. In denying the right of the Church to define his own belief, he had forfeited the privilege of punishing the errors of those who chose to differ from him.
>
> James Anthony Froude

DEVELOPING THE COMPARISON OR CONTRAST

Closely related to the question of organization is a final problem: the units in which to develop the comparison or contrast. The simplest plan is to spend a paragraph or a portion of a paragraph on one subject and a roughly equal passage on the other. That is what Esfandiary does in discussing the differences between Western and Eastern attitudes toward science and progress (see page 135). The eighteenth-century English essayist Joseph Addison uses a similar method (though in a single paragraph) contrasting "true" and "false" happiness, spending the first half of his paragraph on the former, the second half on the latter:

> True happiness is of a retired nature, and an enemy to pomp and noise; it arises, in the first place, from the enjoyment of one's self; and, in the next, from the friendship and conversation of a few select companions. It loves shade and solitude, and naturally haunts groves and fountains, fields and meadows: in short, it feels everything it wants within itself, and receives no addition from multitudes of witnesses and spectators. On the contrary,

false happiness loves to be in a crowd, and to draw the eyes of the world upon her. She does not receive any satisfaction from the applause which she gives herself, but from the admiration which she raises in others. She flourishes in courts and palaces, theaters and assemblies, and has no existence but when she is looked upon.

On the other hand, you may build up a comparison or contrast in a series of shorter units. These may be pairs of sentences, each expressing half the comparison:

The original Protestants had brought new passion into the ideal of the state as a religious society and they had set about to discipline this society more strictly than ever upon the pattern of the Bible. The later Protestants reversed a fundamental purpose and became the allies of individualism and the secular state.

Herbert Butterfield

Or the contrast may be contained within a single sentence:

At first glance the traditions of journalism and scholarship seem completely unlike: journalism so bustling, feverish, and content with daily oblivion; the academic world so sheltered, deliberate, and hopeful of enduring products. It is true that both are concerned with the ascertainment and diffusion of truth. In journalism, however, the emphasis falls on a rapid diffusion of fact and idea; in academic work it falls on a prolonged, laborious ascertainment.

Allen Nevins

In such cases the total effect comes from a stringing together of sharply focused specific contrasts or comparisons. In the following instance the writer is pointing out the differences between himself and his white friends and their black schoolmates in Brooklyn:

There were plenty of bad boys among the whites—this was, after all, a neighborhood with a long tradition of crime as a career open to aspiring talents—but the Negroes were *really* bad, bad in a way that beckoned to one, and made one feel inadequate. *We* all went home every day for a lunch of spinach-and-potatoes; *they* roamed around during lunch hour, munching on candy bars. In winter *we* had to wear itchy woolen hats and mittens and cumbersome galoshes; *they* were bare-headed and loose as they pleased. *We* rarely played hookey, or got into serious trouble in school, for all our street-corner bravado; *they* were defiant, forever staying out (to do what delicious things?), forever making disturbances in class and in the halls, forever being sent to the principal and returning uncowed. But most important of all, they were *tough*; beautifully, enviably tough, not giving a damn for anyone or anything. To hell with the teacher, the truant officer, the cop; to hell with the whole of the adult world that held *us* in its grip and that we never had the courage to rebel against except sporadically and in petty ways.

Norman Podhoretz

Notice that when a comparison or contrast develops sentence by sentence or clause by clause, it is kept sharp by constructing the units according to the same pattern, placing key ideas in the same positions and giving them the same grammatical functions.

The development of a comparison is related to how it is organized. The paragraph to paragraph (or half-paragraph to half-paragraph) method deals better with each subject in its entirety. Proceeding by balanced sentences or clauses is better if you wish to focus upon particular points of similarity or difference.

CONCLUSION

Writing a comparison or contrast requires that you think carefully about what you want to do and about how you can best focus, organize, and develop the material to achieve your purpose. The problem is further complicated by the fact that none of the choices we have discussed is absolute. A paragraph does not have to compare *or* contrast: it can do both. It does not have to maintain only one focus: a skillful writer can effectively shift among several. And extended comparisons can and do alter their mode of development.

An example of such complications is the following paragraph by the philosopher Bertrand Russell. He is writing about William Gladstone (Liberal Prime Minister of Great Britain four times during the latter part of the nineteenth century) and Nikolai Lenin (leader of the Bolshevik Revolution of 1917 and founder of modern Soviet Russia):

> Lenin, with whom I had a long conversation in Moscow in 1920, was, superficially, very unlike Gladstone, and yet, allowing for the difference of time and place and creed, the two men had much in common. To begin with the differences: Lenin was cruel, which Gladstone was not; Lenin had no respect for tradition, whereas Gladstone had a great deal; Lenin considered all means legitimate for securing the victory of his party, whereas for Gladstone politics was a game with certain rules that must be observed. All these differences, to my mind, are to the advantage of Gladstone, and accordingly Gladstone on the whole had beneficent effects, while Lenin's effects were disastrous. In spite of all these dissimilarities, however, the points of resemblance were quite as profound. Lenin supposed himself to be an atheist, but in this he was mistaken. He thought that the world was governed by the dialectic, whose instrument he was; just as much as Gladstone, he conceived of himself as the human agent of a superhuman Power. His ruthlessness and unscrupulousness were only as to means, not as to ends; he would not have been willing to purchase personal power at the expense of apostasy. Both men derived their personal force from this unshakable conviction of their own rectitude. Both men, in support of their respective faiths, ventured into realms in which, from ignorance, they could only cover themselves with ridicule—Gladstone in Biblical criticism, Lenin in philosophy.

Russell's paragraph covers both differences and similarities. In the opening portion he focuses equally upon both statesmen; in discussing atheism he concentrates upon Lenin; in the final portion of the paragraph he widens his focus again to include both. Yet these shifts and alterations are carefully controlled and signaled so that the paragraph never slides into confusion.

EXERCISE

1. Study the following paragraphs of comparison or contrast. Answer these questions about each:
   A. Is the writer comparing, contrasting, or combining the two?
   B. Is the focus primarily upon one topic or upon both?
   C. How is the comparison (contrast) organized and developed?

I. Let's compare the U.S. to India, for example. We have 203 million people, whereas she has 540 million on much less land. But look at the impact of people on the land.

The average Indian eats his daily few cups of rice (or perhaps wheat, whose production on American farms contributed to our one percent per year drain in quality of our active farmland), draws his bucket of water from the communal well and sleeps in a mud hut. In his daily rounds to gather cow dung to burn to cook his rice and warm his feet, his footsteps, along with those of millions of his countrymen, help bring about a slow deterioration of the ability of the land to support people. His contribution to the destruction of the land is minimal.

An American, on the other hand, can be expected to destroy a piece of land on which he builds a home, garage and driveway. He will contribute his share to the 142 million tons of smoke and fumes, seven million junked cars, 20 million tons of paper, 48 billion cans, and 26 billion bottles the overburdened environment must absorb each year. To run his air conditioner we will strip-mine a Kentucky hillside, push the dirt and slate down into the stream, and burn coal in a power generator, whose smokestack contributes to a plume of smoke massive enough to cause cloud seeding and premature precipitation from Gulf winds which should be irrigating the wheat farms of Minnesota.

Wayne H. Davis

II. Winter and summer, then, were two hostile lives, and bred two separate natures. Winter was always the effort to live; summer was tropical license. Whether the children rolled in the grass, or waded in the brook, or swam in the salt ocean, or sailed in the bay, or fished for smelts in the creeks, or netted minnows in the salt-marshes, or took to the pine-woods and the granite quarries, or chased muskrats and hunted snapping-turtles in the swamps, or mushrooms or nuts on the autumn hills, summer and country were always sensual living, while winter was always compulsory learning. Summer was the multiplicity of nature; winter was school.

Henry Adams

III. In anthropological studies of different cultures the distinction between those which rely heavily on shame and those that rely heavily on guilt is an important one. A society that inculcates absolute standards of morality and relies on men's developing a conscience is a guilt culture by definition, but a man in such a society may, as in the United States, suffer in addition from shame when he accuses himself of gaucheries which are in no way sins. He may be excessively chagrined about not dressing appropriately for the occasion or about a slip of the tongue. In a culture where shame is a major sanction, people are chagrined about acts which we expect people to feel guilty about. This chagrin can be very intense and it cannot be relieved, as guilt can be, by confession and atonement. A man who has sinned can get relief by unburdening himself. This device of confession is used in our secular therapy and by many religious groups which have otherwise little in common. We know it brings relief. Where shame is the major sanction, a man does not experience relief when he makes his fault public even to a confessor. So long as his bad behavior does not "get out into the world" he need not be troubled and confession appears to him merely a way of courting trouble. Shame cultures therefore do not provide for confessions, even to the gods. They have ceremonies for good luck rather than for expiation.

   True shame cultures rely on external sanctions for good behavior, not, as true guilt cultures do, on an internalized conviction of sin. Shame is a reaction to other people's criticism. A man is shamed either by being openly ridiculed and rejected or by fantasying to himself that he has been made ridiculous. In either case it is a potent sanction. But it requires an audience or at least a man's fantasy of an audience. Guilt does not. In a nation where honor means living up to one's own picture of oneself, a man may suffer from guilt though no man knows of his misdeed and a man's feeling of guilt may actually be relieved by confessing his sin.

                                                                              Ruth Benedict

   2. Write a paragraph of contrast on one of the subjects listed below. Confine yourself to three or four points of difference and organize around the two topics—that is, discuss all the points with regard to A before going on to treat the same points with regard to B:

   A. Psychology and sociology
   B. Football and baseball
   C. A jazz musician and Lawrence Welk
   D. New York and Chicago (or any two cities you know well)
   E. A sports car and the family sedan
   F. Studying alone and studying with one or two friends
   G. Life at home and life in a dormitory

   3. Now compose another paragraph on the same subject you selected in exercise 2, but this time organize not around A and B but around the three or four specific points of difference.

   4. Still working with the same topic, write a third paragraph beginning like this:

   Yet despite these differences __A__ and __B__ are alike in several ways.

Paragraph Development: Analogy

INTRODUCTION

A dev.
ang.

Analogy is a special type of comparison in which a second subject is introduced to show a similarity which explains or justifies something about the main topic. For example, the American writer Flannery O'Connor, addressing a class in creative writing, remarks:

> I understand that this is a course called "How the Writer Writes," and that each week you are exposed to a different writer who holds forth on the subject. The only parallel I can think of to this is having the zoo come to you, one animal at a time; and I suspect that what you hear one week from the giraffe is contradicted next week by the baboon.

O'Connor's main subject is writing, and her analogy is visiting the zoo, or rather having the zoo visit you. The analogy enables the writer to present herself with comic self-deprecation and, more seriously, to suggest something about the limitations of courses in creative writing.

Analogies differ from usual comparison in several ways. First, they focus always upon one topic, the analogy being secondary, serving to clarify or emphasize. Second, the analogical subject may have—and usually does have—a very different nature. Comparison typically involves things of like sort—a Ford and a Chevrolet, high school and college, New Orleans and San Francisco. The value of an analogy, on the other hand, is increased by finding similarities in unlike things, such as a course in writing and visiting a zoo.

ANALOGY AS CLARIFICATION

In exposition the chief function of analogy is to translate an abstract or difficult idea into more concrete or familiar terms. That is the purpose of O'Connor's analogy, as it is of the one in the following passage by the English novelist and essayist Virginia Woolf. Woolf argues that, as art, movies suffer from having developed, in a technical sense, too rapidly. Her point is abstract and so she puts it into an image more easily grasped:

> For a strange thing has happened—while all the other arts were born naked, this, the youngest, has been born fully-clothed. It can say everything before it has anything to say. It is as if [a] savage tribe . . . had found scattering the seashore fiddles, flutes, saxophones, trumpets, grand pianos by Erard and Bechstein, and had begun with incredible energy, but without knowing a note of music, to hammer and thump upon them all at the same time.

Such translations of unfamiliar ideas or experiences into commonplace terms make analogies an important mode of communication. To take one more example, Sir Arthur Eddington, an astronomer and philosopher of science, is explaining the methods of science:

> Let us suppose that an ichthyologist is exploring the life of the ocean. He casts a net into the water and brings up a fishy assortment. Surveying his catch, he proceeds in the usual manner of a scientist to systematize what it reveals. He arrives at two generalizations:
>
> 1. No sea-creature is less than two inches long.
> 2. All sea-creatures have gills.
>
> These are both true of his catch, and he assumes tentatively that they will remain true however often he repeats it.
>
> In applying this analogy, the catch stands for the body of knowledge which constitutes physical science, and the net for the sensory and intellectual equipment which we use in obtaining it. The casting of the net corresponds to observation; for knowledge which has not been or could not be obtained by observation is not admitted into physical science.
>
> An onlooker may object that the first generalization is wrong. "There are plenty of sea-creatures under two inches long, only your net is not adapted to catch them." The ichthyologist dismisses this objection contemptuously. "Anything uncatchable by my net is *ipso facto* outside the scope of ichthyological knowledge, and is not part of the kingdom of fishes which has been defined as the theme of ichthyological knowledge. In short, what my net can't catch isn't fish." Or—to translate the analogy—"If you are not simply guessing, you are claiming a knowledge of the physical universe discovered in some other way than by the methods of physical science, and admittedly unverifiable by such methods. You are a metaphysician. Bah!"
>
> Sir Arthur Eddington

ANALOGY AS PERSUASION

Aside from clarifying the unfamiliar, analogies are often used for their persuasive force. In this function do not confuse the "rhetorical" analogies we are discussing here with "logical" analogies. In logic analogy is a special and rigorous type of proof. A rhetorical analogy, on the other hand, is *never a form of proof.* By itself it cannot demonstrate the truth of any conclusion. At best a "weak form of reasoning," it suggests that because A resembles B in some respects, it resembles B in other ways. But because the resemblance between A and B is never total, everything that is true of one cannot necessarily be applied to the other.

For instance, some political thinkers have used the "similarity" of a state to a ship to justify an authoritarian society. They argue that just as a ship can survive storms only when complete power lies in the hands of a master seaman who demands unhesitating obedience from the officers and crew, so a state can survive only if its citizens submit unquestioningly to an absolute ruler. But, of course, states and ships are *not* identical, and therefore what may be true of the latter cannot be assumed to apply to the former.

Analogies like that of ship and state which draw forced or partial parallels to "prove" unwarranted conclusions are called "false" or "unfair." The English poet and essayist W. H. Auden points to another example of a misleading political analogy:

> A favorite analogy for the state among idealist political thinkers is with the human body. This analogy is false. The constitution of the cells in the body is determined and fixed; nerve cells can only give rise to more nerve cells, muscle cells to muscle cells, etc. But, in the transition from parent to child, the whole pack of inherited genetic characters is shuffled. The king's son may be a moron, the coal heaver's a mathematical genius. The entire pattern of talents and abilities is altered at every generation.
>
> Another false analogy is with the animal kingdom. Observed from the outside (how it appears to them no one knows), the individual animal seems to be sacrificed to the continuance of the species. This observation is used to deny the individual any rights against the state. But there is a fundamental difference between man and all other animals in that an animal which has reached maturity does not continue to evolve, but a man does. As far as we can judge, the only standard in the animal world is physical fitness, but in man a great many other factors are involved. What has survival value can never be determined; man has survived as a species through the efforts of individuals who at the time must often have seemed to possess very little biological survival value.

To the degree that any rhetorical analogy is seriously put forward as a form of proof, it is of course "false." Generally, however, we reserve the term for analogies that are glaringly wrong (or, a cynic might say, for those you disagree with). But even a "fair" rhetorical analogy, remember, does not prove anything in a rigorous logical sense.

Even so, good analogies can be effective devices of persuasion, like this one by Abraham Lincoln in a speech opposing the spread of slavery into territories outside the South:

If I saw a venomous snake crawling in the road, any man would say I might seize the nearest stick and kill it; but if I found that snake in bed with my children, that would be another question. I might hurt the children more than the snake, and it might bite them. Much more, if I found it in bed with my neighbor's children, and I had bound myself by a solemn compact not to meddle with his children under any circumstances, it would become me to let that particular mode of getting rid of the gentleman alone. But if there was a bed newly made up, to which the children were to be taken, and it was proposed to take a batch of young snakes and put them there with them, I take it no man would say there was any question how I ought to decide. That is just the case. The new territories are the newly made bed to which our children are to go, and it lies with the nation to say whether they shall have snakes mixed up with them or not. It does not seem as if there could be much hesitation what our policy should be.

Lincoln's argument simply assumes that slavery is wrong and does not prove it. But few in his audience would have needed proof, and the analogy makes a forceful, convincing statement of why the spread of slavery should be opposed even though it is difficult or dangerous to do anything about slavery where it already exists. (Lincoln took a more moderate position on abolition at this time—1860—than he did a few years later.)

Analogies can be especially persuasive when they are used for ridicule. In the following example a mocking comparison makes us laugh at prudes who would censor Greek and Latin poetry:

. . . a man, who, exposed to all the influences of such a state of society as that in which we live, is yet afraid of exposing himself to the influence of a few Greek or Latin verses, acts, we think, much like the felon who begged the sheriff to let him have an umbrella held over his head from the door of Newgate [prison] to the gallows, because it was a drizzling morning and he was apt to take cold.

Thomas Babington Macaulay

CONCLUSION

Whether intended to explain or to persuade, then, the analogy is always introduced in support of the main topic, and it should be kept in check, never allowed to develop on its own past the point where it has served its purpose. Usually an analogy appears after the main subject has been identified, as it does in the Macaulay example just quoted or in the paragraphs by Flannery O'Connor (p. 141) and Virginia Woolf (p. 142). It may or may not be specifically labeled an analogy, but when it follows the main topic, its supportive function is clear.

Now and again an analogy may more effectively precede the principal subject. Lincoln begins by talking about snakes before he applies the analogy to slavery (above), for example. This progression creates suspense, drawing in the reader or listener, who wonders what snakes have to do with the matter. In working out an analogy, think a bit about the strategy of placing the analogy before or after you introduce the main topic.

EXERCISE

1. Identify the analogies in the following passages. What purpose does each serve? Do you think they are effective?

> A. I am an explorer, then, and I am also a stalker, or the instrument of the hunt itself. Certain Indians used to carve long grooves along the wooden shafts of their arrows. They called the grooves "lightning marks," because they resembled the curved fissures lightning slices down the trunks of trees. The function of lightning marks is this: if the arrow fails to kill the game, blood from a deep wound will channel along the lightning mark, streak down the arrow shaft, and spatter to the ground, laying a trail dripped on broadleaves, on stones, that the barefoot and trembling archer can follow into whatever deep or rare wilderness it leads. I am the arrow shaft, carved along my length by unexpected lights and gashes from the very sky, and this book is the straying trail of blood.
>
> Annie Dillard

> B. The work of the statistician has been likened to that of the map maker who presents a traveler with a sketch of the important highways, the locations of towns, and the major geographical features. The interesting details and beautiful scenery are deliberately omitted. Towns are dots and rivers are lines, and all features of human interest which constitute the traveler's real goals are missing. But just as the map is an aid to reaching these goals, so the statistical facts of a science provide an orientation for the workers in the science. The statistician gives averages, trends, variabilities, correlations, and the particular details are lost in the abstract summary. When we turn to statistical methods, therefore, we do so to gain a general orientation, and not to explain a particular event.
>
> George A. Miller

2. It is difficult to suggest topics for analogies. They are personal things that must grow out of one's own experience and values. However, here are a few possibilities; if none appeals to you, devise your own. In any case, compose a single paragraph. You may begin either with the main topic (which comes first in the pairs listed below) or with the analogy, as it suits your purpose:

> A. Reading a difficult book and climbing a mountain
> B. A library and a cemetery
> C. College and a cafeteria
> D. A person's or a nation's conception of reality and the wearing of glasses

CHAPTER 20

Paragraph Development: Cause and Effect

The methods of paragraph development we study in this and the following three chapters are logical in a loose sense. They deal with the causes or consequences of the topic; clarifying what it means (definition); dividing it into its parts or placing it in its appropriate category (analysis and classification); or pointing out the limits of its truth or applicability (qualification). These methods of development are the heart of exposition, that is, writing that explains.

CAUSE

A dev.
Cause

We often support a topic by discussing its cause(s) or reason(s).[1] Exposition frequently must answer the question "Why?" What caused the Great Depression or World War II? What reasons lie behind your personal decision to study medicine or law or business or physical education? Why does a potential new product have a market? What reasons can be advanced to defend a man accused of a crime, or to prove his innocence?

One cannot write for very long without having to explain why something is true or not true, why an event happened or did not happen. Many strategies for developing reasons are available. The simplest is just to ask the question and then supply the answer. One scholar does this in discussing why American English is different in many ways from the English spoken in the British Isles:

> If, then, the language of the original colonists was merely the English of England, why does ours differ somewhat from theirs today? Three reasons can be offered.
> First, the people of Great Britain in the seventeenth century spoke differ-

1. *Cause* and *reason* are not strict synonyms: the former is the more general and includes the latter. In talking about composition, however, we use the terms interchangeably.

ent local dialects. What we now consider to be standard English for England developed from the language of London and the near-by counties. But the settlers of America came not only from that region but also from many others. New England was settled largely from the eastern counties. Pennsylvania received a heavy immigration from the north of Ireland. English as it came to be spoken in New England and much of Pennsylvania thus naturally was not the same English that developed as the standard in England. For instance, in what we consider typical British English of today, the final *r* has been lost. It is, however, partially preserved in General American, possibly because the Scotch-Irish of the eighteenth century preserved that sound, as they still do in Ireland.

A second cause for the difference between the two countries lies in mere isolation. Language is always changing. When two groups of people speaking the same language are separated and remain in comparative isolation, change continues in the language of both groups, but naturally it does not continue in the same direction and at the same rate with both of them. The languages thus tend to become different.

Third, the language in the United States has been subjected to various influences that have not affected the language in Great Britain—the environment, the languages of other early colonists and of the later immigrants.

George R. Stewart

A paragraph devoted to reasons may develop more subtly. Instead of using a question-answer strategy and explicitly announcing reasons, a writer may leave the causal relationship tacit. The connection exists in the substructure of the ideas, but the diction does not label it heavily. In the following paragraph, for instance, only the "for" in the opening sentence indicates the reasons pattern:

The cult of beauty in women, which we smile at as though it were one of the culture's harmless follies, is, in fact, an insanity, for it is posited on a false view of reality. Women are not more beautiful than men. The obligation to be beautiful is an artificial burden, imposed by men on women, that keeps both sexes clinging to childhood, the woman forced to remain a charming, dependent child, the man driven by his unconscious desire to be—like an infant—loved and taken care of simply for his beautiful self. Woman's mask of beauty is the face of the child, a revelation of the tragic sexual immaturity of both sexes in our culture.

Una Stannard

ONE REASON OR SEVERAL

In supporting a topic by reasons you may work with only one, treating it in detail; or you may develop two, three, or more. In this paragraph Thoreau uses only a single reason to explain why he built his solitary cabin at Walden Pond, but he restates it in various ways:

I went to the woods because I wished to live deliberately, to front only the essential facts of life, and see if I could not learn what it had to teach, and not, when I came to die, discover that I had not lived. I did not wish to live what was not life, living is so dear; nor did I wish to practice resignation, unless it was quite necessary. I wanted to live deep and suck out all the marrow of life, to live so sturdily and Spartan-like as to put to rout all that was not life, to cut a broad swath and shave close, to drive life into a corner, and reduce it to its lowest terms, and, if it proved to be mean, why then to get the whole and genuine meanness of it, and publish its meanness to the world; or if it were sublime, to know it by experience, and be able to give a true account of it in my next excursion. For most men, it appears to me, are in a strange uncertainty about it, whether it is of the devil or of God, and have *somewhat hastily* concluded that it is the chief end of man here to "glorify God and enjoy him forever."

Henry David Thoreau

On the other hand a writer may choose to discuss several reasons as Professor Stewart does in the passage quoted on pages 146–47. Here is another example, a paragraph about the advantages of marriage in the nineteenth century:

A hundred years ago there was every reason to marry young—though middle-class people seldom did. The unmarried state had heavy disadvantages for both sexes. Custom did not permit girls to be educated, to work, or to have social, let alone sexual, freedom. Men were free but since women were not, they had only prostitutes for partners. (When enforced, the double standard is certainly self-defeating.) And, though less restricted than girls shackled to their families, single men often led a grim and uncomfortable life. A wife was nearly indispensable, if only to darn socks, sew, cook, clean, take care of her man. Altogether, both sexes needed marriage far more than now—no TV, cars, dates, drip-dry shirts, cleaners, canned foods—and not much hospital care, insurance, or social security. The family was all-important.

Ernest van den Haag

The choice between treating one reason or several is not often a free option. Usually the subject determines it, or the amount of time or space you have, or what you are required to do. Asked in a history exam to discuss the three principal causes of the Depression, you cannot elect to develop only one.

ORDER OF REASONS WITHIN THE PARAGRAPH

When you must work with several reasons, you face the problem of arranging them in a significant order. If the reasons are logically serial—that is, if the topic idea is caused by A, and A in turn by B, and B by C—the organization is predetermined. To handle such a chain of cause and effect in any but the natural sequence of Topic: A-B-C would be awkward and confusing.

But the various reasons may be parallel—that is, they have no causal connection and relate only in contributing to the same result. Then you have more latitude. If the causes have an order in time, you should follow that. If they do not have a chronological pattern, you will probably have to rank them according to relative importance. Generally a climactic strategy is best: begin with the least important and conclude with the most important:

> I doubt if the English temperament is wholly favourable to the development of the essayist. In the first place, an Anglo-Saxon likes doing things better than thinking about them; and in his memories, he is apt to recall how a thing was done rather than why it was done. In the next place, we are naturally rather prudent and secretive; we say that a man must not wear his heart upon his sleeve, and that is just what the essayist must do. We have a horror of giving ourselves away, and we like to keep ourselves to ourselves. "The Englishman's home is his castle," says another proverb. But the essayist must not have a castle, or if he does, both the grounds and the living-rooms must be open to the inspection of the public.
>
> A. C. Benson

Reversing the order of Benson's two reasons would not impair the logic of his paragraph. However, it would disarrange the climactic structure. While Benson nowhere says that he considers the second reason more important, he gives it more than twice the space and repeats it three times.

EFFECTS

A dev. effect

Effects or consequences are handled much the same as reasons.[2] But now the topic idea is regarded as causing the consequences discussed in the remainder of the paragraph. The paragraph may develop only a single effect, as in this passage about how the moon affects tides:

> If the moon were suddenly struck out of existence, we should be immediately appraised of the fact by a wail from every seaport in the kingdom. From London and from Liverpool we should hear the same story—the rise and fall of the tide had almost ceased. The ships in dock could not get out; the ships outside could not get in; and the maritime commerce of the world would be thrown into dire confusion.
>
> Robert Ball

A more intricate example of the development of a single effect is this paragraph concerning the increasing use of large sums of money ("capital") in the late Middle Ages and early Renaissance:

> The discovery of the advantages of the possession and use of capital caused important changes. Of these the most important was a gradual redistribution of power. A new class emerged, imbued with a new spirit and without the traditions and prejudices of the old feudal nobility. Those who owned money found themselves with a new prestige and a new power

2. The terms *effect* and *consequence* we shall also use interchangeably.

which they used to remake the political organ so as to facilitate their transactions. In this way, the feudal organization of Europe, which was incompatible with the new economy, was destroyed and replaced by a new political structure favorable to trade and commerce. A society in which small local lords were little better than bandits and every tiny autonomous domain could exact its tax on passing goods was replaced by one in which the king's judges enforced order and the king's highways made the flow of trade easy and rapid. The old medieval organization of economy on a municipal and local-unit basis gave way to a large-scale, unified economic structure, and trade, having come to be regarded by the Crown as one of its principal sources of strength, was everywhere favored and encouraged by monarchs. Thus, through an alliance of the new money power with the old royal power, was born the national territorial state, the instrument of capitalistic advance.

W. T. Jones

While Jones's paragraph is longer and more complicated, it too treats only one effect. It analyzes that effect closely, however, to show the conditions involved. The logical structure can be diagrammed like this:

CAUSE: Growing power of capital

EFFECT: Strengthened the position of the rich.

EFFECT: Their greater leverage enabled them to increase their advantage still more by removing those aspects of the medieval economy which blocked the use of capital. Thus capitalists gradually created a society more advantageous to themselves.

SECONDARY CAUSE: As kings realized the possibility of using commerce to widen their own power base, they allied with capitalists to destroy all vestiges of medieval society in which power existed independent of the monarchy.

EFFECT: The political alliance between king and capitalist, each seeking to enhance his own power, overthrew the feudal world and ultimately produced . . .

FINAL EFFECT: the modern state.

MULTIPLE EFFECTS

A paragraph may consist of several effects rather than of only one. In the following case the writer is concerned with what the automobile has done to our society:

Thirdly, I worry about the private automobile. It is a dirty, noisy, wasteful, and lonely means of travel. It pollutes the air, ruins the safety and sociability of the street, and exercises upon the individual a discipline which takes away far more freedom than it gives him. It causes an enormous amount of land to be unnecessarily abstracted from nature and from plant life and to become devoid of any natural function. It explodes cities, grievously impairs the whole institution of neighborliness, fragmentizes and destroys communities. It has already spelled the end of our cities as real cultural and social communities, and has made impossible the construction of any others in their place. Together with the airplane, it has crowded out other, more civilized and more convenient means of transport, leaving older people, infirm people, poor people and children in a worse situation than they were a hundred years ago. It continues to lend a terrible element of fragility to our civilization, placing us in a situation where our life would break down completely if anything ever interfered with the oil supply.

George F. Kennan

Kennan does not label the logical structure of his paragraph. The pattern topic →consequences is implicit, not even indicated by a brief connective like *therefore* or *and so*. But the absence of such words is not a fault. The sentence structure keeps the logic clear. Sentence after sentence begins with the subject-verb nexus, which is, of course, essentially a cause-effect relationship: "It [the 'private automobile'] is . . . It pollutes . . . It causes . . . It explodes . . . It has . . ." The repetition of this pattern supports and clarifies the logic—a fine example of how sentence structure may be used to create paragraph unity.

CAUSE AND EFFECT

Thus far we have seen paragraphs which develop reasons in support of a topic and paragraphs which develop effects. Often, however, causes and consequences are more intimately related. Sometimes a thing may be simultaneously both cause and effect, as when the result you expect an action to have is the reason you perform it. In Kennan's paragraph the effects of the automobile are the reasons he worries about it. The journalist Pete Hamill expresses much the same point in the following paragraph, explaining that what the car has done to our society makes it "one of our jailers":

In fact, the automobile, which was hailed as a liberator of human beings early in this century, has become one of our jailers. The city air, harbor-cool and fresh at dawn, is a sewer by 10. The 40-hour week, for which so many good union people died, is now a joke; on an average day, a large number of people now spend three to four hours simply traveling to those eight-hour-a-day jobs, stalled on roads, idling at bridges or in tunnels. Parking fees are $5 to $10 a day. The ruined city streets cost hundreds more for gashed tires, missing hubcaps and rattled engines.

Frequently cause and effect form a chain: A gives rise to B, B to C, and so on. In such a linkage, B is both a consequence of A and a cause of C. This paragraph by the journalist A. J. Liebling about the effect of television upon boxing (which he calls "the Sweet Science") develops a series of causes and effects:

> The immediate crisis [of boxing] in the United States, forestalling the one high living standards might bring on, has been caused by the popularization of a ridiculous gadget called television. This is utilized in the sale of beer and razor blades. The clients of the television companies, by putting on a free boxing show almost every night of the week, have knocked out of business the hundreds of small-city and neighborhood boxing clubs where youngsters had a chance to learn their trade and journeymen to mature their skill. Consequently the number of good new prospects diminishes with every year, and the peddlers' public is already being asked to believe that a boy with perhaps ten or fifteen fights behind him is a topnotch performer. Neither advertising agencies nor brewers, and least of all the networks, give a hoot if they push the Sweet Science back into a period of genre painting. When it is in coma they will find some other way to peddle their peanuts.

Liebling treats both reasons and consequences. The initial cause is the use of television to sell products, the ultimate effect is the deterioration of prizefighting. But linking these are several specific conditions, each the effect of a preceding cause and the cause of a subsequent effect:

INITIAL CAUSE: The hucksterism of television

EFFECT: Too many free prizefights

EFFECT: Disappearance of the small fight club

EFFECT: Inadequate training of young boxers

FINAL EFFECT: Deterioration of professional boxing

All this is clearly conveyed with only a single transitional adverb ("consequently"), used to signal the chief result.

EXERCISE

1. Analyze the cause-effect (or effect-cause) pattern in each of the following paragraphs by drawing a rough diagram like the one following the paragraph by A. J. Liebling:

A. It has been a cruel decade for the magazine business. Rising production costs, postal increases and soaring paper prices have made it much more difficult to turn a profit. Television has proved a tough competitor for advertising and audience, and many of the mass circulation giants, among them *Life, Look* and *The Saturday Evening Post,* have floundered or failed in the contest.

Nancy Henry

B. Of all the miracles of television none is more remarkable than its power to give to so many hours of our experience a new vagueness. Americans have become increasingly accustomed to see something-or-other, happening somewhere-or-other, at sometime-or-other. The common-sense hallmarks of authentic first-hand experience (the ordinary facts which a jury expects a witness to supply to prove he actually experienced what he says) now begin to be absent, or to be only ambiguously present, in our television-experience. For our TV-experience we don't need to go out to see anything in particular. We just turn the knob. Then we wonder while we watch. Is this program "live" or is it "taped"? Is it merely an animation or a "simulation"? Is this a rerun? Where does it originate? When (if ever) did it really occur? Is this happening to actors or to real people? Is this a commercial? A spoof of a commercial? A documentary? Or pure fiction?

Daniel J. Boorstin

C. American women are not the only people in the world who manage to lose track of themselves, but we do seem to mislay the past in a singularly absentminded fashion. A century ago, observers as different as Mark Twain and Henry James noticed that our identities appeared to fit loosely and be readily subject to change. In part, this is because we are Americans, geographically and socially mobile, fellow countrymen of the Henry Ford who said, "History is bunk." In part, it is because we are women, people whose lives are recurrently jolted away from continuity. Look at our personal histories, and you will find old scars of passage from one state of being to another: tomboy to junior miss; drum majorette to sociology student; art historian to computer programmer; candidate for an M.A. to harried wife-and-mother; priestess of the Feminine Mystique to divorcée; housewife to financial analyst; and older woman who has suffered the feared and fracturing shift from busy middle life to lonely, widowed age.

Elizabeth Janeway

D. So the hurricane passed,—tearing off the heads of the prodigious waves, to hurl them a hundred feet in the air,—heaping up the ocean against the land,—upturning the woods. Bays and passes were swollen to abysses; rivers regorged; the sea-marshes were changed to raging wastes of water. Before New Orleans the flood of the mile broad Mississippi rose six feet above the highest watermark. One hundred and ten miles away Donaldsonville trembled at the towering tide of the Lafourche. Lakes strove to burst their boundaries. Far-off river steamers tugged wildly at their cables,—shivering like tethered creatures that hear by night the approaching howl of destroyers. Smoke-stacks were hurled overboard, pilot-houses torn away, cabins blown to fragments.

Lafcadio Hearn

2. Compose a single paragraph developing three or four reasons to support one of the following topics. Consider carefully the order of the reasons and provide them with adequate links. Feel free to use an illustration, a restatement, comparison, and so on, but give the bulk of the paragraph to the reasons:

A. The advantages of living at home while going to college (or going to college away from home)
B. The expansion of professional sports in the last twenty years
C. The contemporary mania for exercise
D. The popularity of folk music

    E. Racial (or sexual, religious, social) bias
    F. The advantages of taking a hike alone (or of jogging, bicycling, sailing alone)

3. Work out a single reason to support one of these subjects:

    A. Arguments between parents and children (or husbands and wives)
    B. The prevalence of the macho image among men
    C. Why many women (or blacks, Hispano-Americans) feel that society discriminates
       against them
    D. Poor driving
    E. Failing out of college
    F. The change in American sexual mores in the past generation

4. Selecting one of the topics below, write a single paragraph treating several effects. Arrange the effects in the best sequence and tie them together to make a unified paragraph, not a laundry list:

    A. Final examinations
    B. Television and the family
    C. Owning a car
    D. Drug addiction
    E. Taking a course in sociology (or psychology, economics, philosophy, literature,
       anthropology)
    F. Living by yourself

5. Compose a brief essay treating both cause(s) and effect(s) about any of the topics in exercises 2, 3, or 4 that you have not already used for a paragraph. The essay should be about 600 to 700 words long (excluding any words in an outline if your teacher asks for one) in five or six paragraphs.

CHAPTER 21

Paragraph Development: Definition

The essayist G. K. Chesterton said that "the only way to say anything definite is to define it." Definition is at the heart of exposition. You cannot write very long without having to define something. Perhaps it is only a single word which readers cannot be expected to know:

> Huge "pungs" (ox- or horse-drawn sledges), the connecting links between ocean commerce and New England farms, are drawn up in Dock Square three deep. . . .
>
> Samuel Eliot Morison

Perhaps it is a complex, many-sided idea. Sometimes, in fact, the whole point of a composition is to develop a definition. What do you, as an individual, or we, as a society, understand "democracy" to mean? One could write a long book defining that.

In its most basic sense *to define* means "to set limits or boundaries." That seems simple enough. But to draw the lines for so vast an abstraction as "democracy" is far more difficult than surveying a building lot. The process of definition is beset with difficulties, as any textbook of logic makes clear. These philosophical problems need not concern us, however. Our more superficial interest lies in the practical compositional problems of defining.

KINDS OF DEFINITION

Before we turn to those problems we should say a word about some of the basic kinds of definitions and about some of the purposes definitions serve.

NOMINAL AND REAL DEFINITIONS

An elementary distinction exists between the definition of a word and that of the thing, whether material object, concept, emotion, or whatever, the word signifies. Definitions of words are called *nominal* (for example, the dictionary definition you look up). Those of things are called *real* definitions (this does *not* imply that nominal definitions are somehow false). In practice the distinction between nominal and real definitions often does not matter very much. But sometimes it does. You should always be clear in your own mind whether you are primarily concerned with the word or the thing, and make that concern clear to the reader. If you want a reader to understand that you are defining, say, the term *democracy,* write "the term *democracy,*" placing the word in quotes or underlining it (the typewriter equivalent of italic type). In the following paragraph, for example, the writer wishes to make clear how the word *history* is commonly used:

> By its most common definition, the word *history* now means "the past of mankind." Compare the German word for *history—Geschichte,* which is derived from *geschehen,* meaning to happen. *Geschichte* is *that which has happened.* This meaning of the word *history* is often encountered in such overworked phrases as "all history teaches" or "the lessons of history."
>
> Louis Gottschalk

Professor Gottschalk uses a nominal definition. The essayist G. K. Chesterton, on the other hand, defines the institution (the thing) rather than the word in this passage about marriage:

> Marriage is not a mere chain upon love as the anarchists say; nor is it a mere crown upon love as the sentimentalists say. Marriage is a fact, an actual human relation like that of motherhood, which has certain human habits and loyalties, except in a few monstrous cases where it is turned to a torture by special insanity and sin. A marriage is neither an ecstacy nor a slavery; it is a commonwealth; it is a separate working and fighting thing like a nation.

CONSENSUAL, STIPULATIVE, AND LEGISLATIVE DEFINITIONS

A *consensual* definition, the most common, clarifies what people commonly use a word to mean, or understand a thing to be. A *stipulative* definition states a special meaning given to a word or concept by a writer for a particular purpose. It differs from the usual (consensual) definition, but the difference is legitimate so long as the writer explains it and so long as he or she uses the word in its special sense consistently. A *legislative* definition also differs from the conventional sense of a word or idea; it is put forward as what the word *ought* to mean or the thing *ought* to be. It differs from the stipulative definition in that the writer does not simply say, "For convenience I shall use X to mean such and so," but rather asserts "I shall use X to mean such and so; this is its proper meaning, and you should use it this way too."

INCIDENTAL AND PRIMARY DEFINITIONS

An *incidental* definition relates in a secondary way to your main point. The definition is not your chief concern but simply something you must do in order to get on with your topic. A *primary* definition, on the other hand, *is* the main point. You use it when the purpose of the paragraph or theme is to establish what a word or idea means. Incidental definitions tend to be brief; primary ones, longer and more thorough.

MODES OF DEFINING

Definitions can be developed in a variety of ways—modes of defining we shall call them. Though for convenience we consider them one at a time here, they do not exclude one another and in practice are often combined.

OSTENSIVE DEFINITION

The quickest way to define a common object is to point to it, as one might define *apple* by pointing out or picking up a piece of that fruit. This is *ostensive* definition, and it is how children acquire their first words. The writer, however, cannot point. He or she can use a picture or diagram (notice how many pictures appear in your dictionary), but then he or she ceases to be a writer. In any case, drawing pictures is seldom convenient and often impossible; many things cannot be depicted or, for that matter, pointed to. Aside from an occasional diagram or schematic drawing, then, a writer has to depend upon methods of defining which employ words.

GENUS-SPECIES DEFINITION

The most common mode of defining is that of *genus-species*. The thing, word, or concept being defined (called the *definiendum*) is placed into its genus, or class, and then distinguished from other members of that class. Thus we might define football by first classifying it as a team sport and then showing how it differs from other members of the genus "team sport," baseball, basketball, hockey, and so on. In the following brief definition of history Voltaire begins by setting it into the category "recital of facts"; then he differentiates it from the other member of that class, which he calls "fable":

> History is the recital of facts given as true, in contradistinction to the fable, which is the recital of facts given as false.

The bulk of a genus-species definition usually goes to differentiation, either implicitly or explicitly. In the first case you do not actually contrast the definiendum (football, for example) with the other things in its category (baseball, basketball, and so on), but you describe it so completely that in effect you distinguish it from them. In the case of explicit differentiation you do literally contrast the definiendum with the other members of its class, as Voltaire contrasts history with fable. Obviously a class of any size at all makes complete explicit differentiation impractical. It would require many pages, for instance, literally to distinguish football from all other team sports. Practically, explicit differentiation is generally selective rather than total.

However you differentiate the thing you define, you must distinguish between attributes which are unique to it and attributes which are not. For example, football is played in stadiums (commonly outdoors) before large crowds but so are baseball and soccer. That is not a unique feature. On the other hand, the dimensions and markings of a football field are unique. The unique qualities are what distinguish a definiendum. This does not mean, however, that you can ignore nonunique attributes. If you were explaining football to an English friend, you ought to say something about where and when it is played.

The following explanation of a map illustrates these various aspects of defining by classification and differentiation:

> A map is a conventional picture of an area of land, sea, or sky. Perhaps the maps most widely used are the road maps given away by the oil companies. They show the cultural features such as states, towns, parks, and roads, especially paved roads. They show also natural features, such as rivers and lakes, and sometimes mountains. As simple maps, most automobile drivers have on various occasions used sketches drawn by service station men, or by friends, to show the best automobile route from one town to another.
>
> The distinction usually made between "maps" and "charts" is that a chart is a representation of an area consisting chiefly of water; a map represents an area that is predominantly land. It is easy to see how this distinction arose in the days when there was no navigation over land, but a truer distinction is that charts are specially designed for use in navigation, whether at sea or in the air.
>
> Maps have been used since the earliest civilizations, and explorers find that they are used in rather simple civilizations at the present time by people who are accustomed to traveling. For example, Arctic explorers have obtained considerable help from maps of the coast lines showing settlements, drawn by Eskimo people. Occasionally maps show not only the roads, but pictures of other features. One of the earliest such maps dates from about 1400 B.C. It shows not only roads, but also lakes with fish, and a canal with crocodiles and a bridge over the canal. This is somewhat similar to the modern maps of a state which show for each large town some feature of interest or the chief products of that town.
>
> C. C. Wylie

First the writer places "map" in its genus ("a conventional picture of an area of land, sea, or sky") and illustrates it ("road maps"). Then he distinguishes "map" from the other member of its class ("chart"). Finally, in the third paragraph, he gives us information about maps which, although not essential to the definition, is nevertheless interesting and enlightening.

In working out a genus-species definition, then, the essential questions to ask yourself are these:

1. To what class does it belong?
2. What unique qualities distinguish it from other members of that class?
3. What other qualities—even though not unique—are important if readers are to understand my meaning?

SYNONYMOUS DEFINITION

A briefer and simpler method of defining involves the use of synonyms. Synonymous definitions generally appear within a sentence and explain a key term. They are necessary when you must use a word you cannot reasonably expect readers to know, or even an ordinary word to which you are giving a special sense:

> Societies, knowing that young people can change rapidly even in their most intense devotions, are apt to give them a moratorium, a span of time after they have ceased being children, but before their deeds and works count toward a future identity.
>
> Erik H. Erikson

> The questions Mr. Murrow brought up will rise to plague us again because the answers given are not, as lawyers say, "responsive"—they are not the permanent right answers, although they will do for the day.
>
> Gilbert Seldes

> Love takes off the masks that we fear we cannot live without and know we cannot live within. I use the word "love" here not merely in the personal sense but as a state of being, or a state of grace—not in the infantile American sense of being made happy but in the tough and universal sense of quest and daring and growth.
>
> James Baldwin

There is no sure guide to when you need to define a word. Certainly you should give a definition when you use a technical term in a passage aimed at nontechnical readers: lawyers do not have to be told what *responsive* means in a legal sense, but the rest of us do. And a definition is necessary when you employ an ordinary word in a special or personal sense, as Baldwin does with *love*. On the other hand, you insult readers by defining commonplace words used conventionally. Between those extremes you must settle the question of when to define for yourself based on what you can reasonably expect your readers to know or to look up for themselves.

USING ILLUSTRATIONS IN DEFINITION

Examples are useful in defining, especially when you deal with abstractions. For instance, "heroism" can perhaps be defined easiest by citing how a hero acts (or does not act). In the following paragraph an anthropologist is explaining to American readers what "self-respect" means to the Japanese. She contrasts the Japanese conception of the quality with the American. The heart of her definition, however, lies in the examples of how the Japanese behave to maintain self-respect:

> In any language the contexts in which people speak of losing or gaining self-respect throw a flood of light on their view of life. In Japan "respecting yourself" is always to show yourself the careful player. It does not mean, as it does in English usage, consciously conforming to a worthy standard of conduct—not truckling to another, not lying, not giving false testimony. In Japan self-respect (*jicho*) is literally "a self that is weighty," and its opposite is "a self that is light and floating." When a man says "You must respect yourself," it means, "You must be shrewd in estimating all the factors involved in the situation and do nothing that will arouse criticism or lessen your chances of success." "Respecting yourself" often implies exactly the opposite behavior from that which it means in the United States. An employee says, "I must respect myself (jicho)," and it means, not that he must stand on his rights, but that he must say nothing to his employers that will get him into trouble. "You must respect yourself" had this same meaning, too, in political usage. It meant that a "person of weight" could not respect himself if he indulged in anything so rash as "dangerous thoughts." It had no implication, as it would in the United States, that even if thoughts are dangerous a man's self-respect requires that he think according to his own lights and his own conscience.
>
> Ruth Benedict

METAPHOR AND SIMILE IN DEFINING

Metaphors and similes, which express a kind of comparison,[1] sometimes help to convey the meaning of a word or concept. In a famous passage, a seventeenth-century Anglican clergyman named Jeremy Taylor defined prayer by a series of metaphors, culminating in the image of a lark:

> Prayer is the peace of our spirit, the stillness of our thoughts, the evenness of recollection, the seat of meditation, the rest of our cares, and the calm of our tempest; prayer is the issue of a quiet mind, of untroubled thoughts, it is the daughter of charity and the sister of meekness; and he that prays to God with an angry, that is, with a troubled and discomposed spirit, is like him that retires into a battle to meditate, and sets up his closet in the outquarters of an army, and chooses a frontier-garrison to be wise in. Anger is a perfect alienation of the mind from prayer, and therefore is contrary to that attention which presents our prayers in a right line

1. For a more complete discussion, see pages 393–99.

to God. For so have I seen a lark rising from his bed of grass, and soaring upwards, singing as he rises, and hopes to get to heaven, and climb above the clouds; but the poor bird was beaten back with the loud sighings of an eastern wind, and his motion made irregular and unconstant, descending more at every breath of the tempest than it could recover by the libration and frequent weighing of his wings: till the little creature was forced to sit down and pant, and stay till the storm was over; and then it made a prosperous flight, and did rise and sing, as if it had learned music and motion from an angel as he passed sometimes through the air about his ministries here below.

NEGATIVE DEFINITION

Negative definition explains the meaning of something in terms of what it is not. In the following paragraph miserliness is defined by reference to the opposite concept of thrift:

> Thrift by derivation means thriving; and the miser is the man who does not thrive. The whole meaning of thrift is making the most of everything; and the miser does not make anything of anything. He is the man in whom the process, from the seed to the crop, stops at the intermediate mechanical stage of the money. He does not grow things to feed men; not even to feed one man; not even to feed himself. The miser is the man who starves himself, and everybody else, in order to worship wealth in its dead form, as distinct from its living form.
>
> G. K. Chesterton

PAIRED AND FIELD DEFINITIONS

In composition, words can be defined only by other words. Sometimes, however, the meaning of a word or idea will be intimately tied to that of another, or even to the meanings of three, four, or more terms which comprise a field of meaning. In such cases the definition of one word invariably involves that of the other or others. For example, the terms designating commissioned rank in the United States Army form a field of meaning: *captain* cannot be defined without reference to *first lieutenant* and *major*—the ranks on either side—and these in turn imply *second lieutenant* and *lieutenant colonel* and so on through the entire series of grades.

Paired and series definitions use contrast, but they double or multiply definitions—that is, the contrast works in both directions so that each term or concept is understood with reference to the other:

> Sentiment (so far as literature is concerned) may be defined, I suppose, as the just verbal expression of genuine feeling; it becomes sentimentalism when the feeling is not genuine, or when the expression strikes the reader as laid on with too thick a pen. . . .
>
> John Galsworthy

A longer, more complicated example occurs in this paragraph defining the two kinds of source material with which historians must deal:

> Written and oral sources are divided into two kinds: primary and secondary. A *primary source* is the testimony of an eyewitness, or of a witness by any other of the senses, or of a mechanical device like the dictaphone—that is, of one who or that which was present at the events of which he or it tells (hereafter called simply *eyewitness*). A *secondary source* is the testimony of anyone who is not an eyewitness—that is, of one who was not present at the events of which he tells. A primary source must thus have been produced by a contemporary of the events it narrates. It does not, however, need to be original in the legal sense of the word original— that is, the very document (usually the first written draft) whose contents are the subject of discussion—for quite often a later copy or a printed edition will do just as well; and in the case of the Greek and Roman classics seldom are any but later copies available.
>
> Louis Gottschalk

ETYMOLOGY AND SEMANTIC HISTORY

Another way of getting at the meaning of a word is through its etymology or semantic history, that is, by discussing the original meaning and the changes in significance the word has undergone. Cardinal Newman, for instance, defines the idea of a university by going back to an older name for the institution and discussing the earliest sense and implications of that term:

> If I were asked to describe as briefly and popularly as I could, what a University was, I should draw my answer from its ancient designation of a *Studium Generale* or "School of Universal Learning." This description implies the assemblage of strangers from all parts in one spot;—*from all parts*; else, how will you find professors and students for every department of knowledge? and *in one spot*; else, how can there be any school at all? Accordingly, in its simple and rudimental form, it is a school of knowledge of every kind, consisting of teachers and learners from every quarter. Many things are requisite to complete and satisfy the idea embodied in this description; but such as this a university seems to be in its essence, a place for the communication and circulation of thought, by means of personal intercourse, through a wide extent of country.
>
> John Henry Newman

While proven and easy, the strategy of discussing the etymology of a word and quoting its dictionary definition has limitations. For one thing, the approach is overused. Readers are likely to react negatively to essays on definition that begin "The dictionary defines X to mean . . ."; they have seen it too often.

For another, you must use the information you garner from a dictionary cautiously.[2] The etymology of a word is not necessarily its "true"

---

2. Dictionaries and their use are discussed in Chapter 45, pages 429–37.

or "proper" definition. Word meanings change: *aggravate,* for example, comes from Latin roots meaning "to make heavy," which helps to explain its modern sense of "to annoy, to exasperate," but one cannot argue that the contemporary meaning is wrong because it has strayed from the original. Nor do dictionary definitions tell the whole story. No matter how thorough and sensitive, they must necessarily exclude subtleties and overtones of meaning.

Certainly consulting a suitable dictionary is a required step in writing a definition paragraph or essay. But it ought not to be a substitute for thinking about the problem on your own. Nor should you quote extensively from dictionaries (unless, of course, an assignment specifically requires it). Avoid, in particular, hackneyed formulas like "Webster's says. . . ."

EXERCISE

1. Study the following examples of definition. Which are "nominal," which "real"? What methods of defining does each use? Do you think the definitions are clear? If not, why not?

A. The Seraglio is variously referred to as the Sarail, Le Grand Serai, the Seray or the Harem, though this latter is incorrect, to describe the whole palace, since it applies only to the women's quarters, the core, within the whole. The word "Harem" derives from the Arabic *haram,* forbidden, unlawful. A certain area of land centered round the Holy Cities of Mecca and Medina was considered as set apart, inviolate, and so described as *haram.* The word came to be applied, in its secular sense, to the women's quarters of a Moslem household—it was their *haram,* or sanctuary, territory apart, inviolate to all but the master of the household. The Selamlik, or men's quarters, derived from the word *selam,* a greeting, the Selamlik being the one part of the house where it is permitted to receive visitors.

Lesley Blanch

B. If we discard the authority of rules and of "reputable" writers, to what can we turn for a definition of "correct" English? At the outset it must be acknowledged that there can be no absolute, positive definition. "Correct English" is an approximate term used to describe a series of evaluations of usage dependent upon appropriateness, locality, social level, purpose, and other variables. It is a relative term, in decided contrast with the positive nature of (1) *reputability,* the determination of good usage by reference to standard authors; (2) *preservation,* the obligation to defend and maintain language uses because they are traditional, or are felt to be more elegant; (3) *literary,* the identification of good usage with formal literary usage. By discarding these traditional conceptions, and turning to the language itself as its own standard of good usage, we may find the following definition adequate for our present needs. Good English is that form of speech which is appropriate to the purpose of the speaker, true to the language as it is, and comfortable to speaker and listener. It is the product of custom, neither cramped by rule nor freed from all restraint; it is never fixed, but changes with the organic life of the language.

Robert C. Pooley

2. Using as many of the techniques of definition as the problem requires, compose a one-paragraph definition of one of the following. Your task is to make clear the nature of the *thing* or *idea* designated by the word:

scholarship   honor
courage   patriotism
love   evil

3. Define one of the following words in a paragraph or two. You may consult a dictionary, but ultimately the work should be your own and not a rehash of dictionary definitions; it should be a small essay which makes clear what the word means to contemporary users:

"philosophy"   "Christian"
"science"   "poetry"
"teen-ager"   "rock and roll"

CHAPTER 22

Paragraph Development: Analysis or Classification

In a broad sense all expository paragraphs are analytical. To write about any subject you must analyze it into particulars (whether reasons or comparisons, illustrations or consequences) and then organize these into a coherent whole. More narrowly, however, "analysis" refers to the specific technique of developing a topic by distinguishing its components and discussing each in turn.

The basic strategic questions of analysis are: "What parts does my subject have? What kinds of things does it include?" The questions are slightly different. The first applies to tightly structured subjects. Then analysis is really a kind of dissection, cutting a complex whole into components that do not exist separately, as an arm does not "exist" apart from the rest of the body. The second question applies more readily when the subject is a loosely unified collection of separate elements classed together simply for sharing some common quality. G. K. Chesterton, for example, analyzes the category "people" like this:

> Roughly speaking, there are three kinds of people in this world. The first kind of people are People; they are the largest and probably the most valuable class. We owe to this class the chairs we sit down on, the clothes we wear, the houses we live in; and, indeed (when we come to think of it), we probably belong to this class ourselves. The second class may be called for convenience the Poets; they are often a nuisance to their families, but, generally speaking, a blessing to mankind. The third class is that of the Professors or Intellectuals, sometimes described as the thoughtful people; and these are a blight and a desolation both to their families and also to mankind. Of course, the classification sometimes overlaps, like all classification. Some good people are almost poets and some bad poets are almost professors. But the division follows lines of real psychological cleavage. I do. not offer it lightly. It has been the fruit of more than eighteen minutes of earnest reflection and research.

Chesterton develops his paragraph by asking, in effect, "What kinds of people are there?" Notice that he uses the terms *class* and *classification* several times. This strategy of paragraph development is sometimes called classification. In strict terms analysis and classification are not the same. The first begins with the general and works into particulars; the second starts with the particulars and sorts them into the categories or classes to which they belong. But practically speaking, the difference is not very significant. Both are concerned with a class and a number of particulars, and the problem is to make clear that the former encompasses the latter. Thus in Chesterton's humorous analysis the broad category "people" is composed of the particular groups "People," "Poets," and "Professors."

We see a more arithmetic kind of analysis in the following passage. The writer had spent several weeks of 1934 (during the Depression) in a transient workers' camp in California and had observed and thought about the unemployed and unmarried men the camp had been set up to accommodate:

I began evaluating my fellow tramps as human material, and for the first time in my life I became face-conscious. There were some good faces, particularly among the young. Several of the middle-aged and the old looked healthy and well preserved. But the damaged and decayed faces were in the majority. I saw faces that were wrinkled, or bloated, or raw as the surface of a peeled plum. Some of the noses were purple and swollen, some broken, some pitted with enlarged pores. There were many toothless mouths (I counted seventy-eight). I noticed eyes that were blurred, faded, opaque, or bloodshot. I was struck by the fact that the old men, even the very old, showed their age mainly in the face. Their bodies were still slender and erect. One little man over sixty years of age looked a mere boy when seen from behind. The shriveled face joined to a boyish body made a startling sight.

My diffidence had now vanished. I was getting to know everybody in the camp. They were a friendly and talkative lot. Before many weeks I knew some essential fact about practically everyone.

And I was continually counting. Of the two hundred men in the camp there were approximately as follows:

Cripples	30
Confirmed drunkards	60
Old men (55 and over)	50
Youths under twenty	10
Men with chronic diseases, heart, asthma, TB	12
Mildly insane	4
Constitutionally lazy	6
Fugitives from justice	4
Apparently normal	70

(The numbers do not tally up to two hundred since some of the men were counted twice or even thrice—as cripples and old, or as old and confirmed drunks, etc.)

Eric Hoffer

ANALYSIS OF ABSTRACTIONS

Sorting out concrete topics, whether people or varieties of apples, is the easiest kind of analysis. The technique can also be used to explain abstractions, the organization of a club, a working group, a military unit, and so on. The basic questions shifts a little, more "How is it structured?" than "What are its parts?"

The analytical method remains the same, however. Study this passage explaining how the watches were arranged aboard a nineteenth-century sailing ship. (The term *watch* applies both to the two divisions of the crew who alternated in working the vessel and to the periods of the twenty-four-hour day when each group was on duty.)

> The crew are divided into two divisions, as equally as may be, called the watches. Of these, the chief mate commands the larboard, and the second mate the starboard. They divide the time between them, being on and off duty, or, as it is called, on deck and below, every other four hours. The three night watches are called the first, the middle, and the morning watch. If, for instance, the chief mate with the larboard watch have the first night watch from eight to twelve, at that hour the starboard watch and the second mate take the deck, while the larboard watch and the first mate go below until four in the morning, when they come on deck again and remain until eight. As the larboard watch will have been on deck eight hours out of the twelve, while the starboard watch will have been up only four hours, the former have what is called a "forenoon watch below," that is, from eight A.M. till twelve A.M. In a man-of-war, and in some merchantmen, this alternation of watches is kept up throughout the twenty-four hours, which is called having "watch and watch;" but our ship, like most merchantmen, had "all hands" from twelve o'clock till dark, except in very bad weather, when we were allowed "watch and watch."

Richard Henry Dana

ANALYSIS OF A PROCESS

A process is a complex set of operations occurring one after another and all directed toward achieving a specific end. The steps involved in knitting a sweater, for example, are a process, from buying the pattern and wool to the final blocking and shaping. So is the election of a political candidate, or registering for college.

In most cases the steps are relatively well defined. The writer's job is first to understand the process, separating its stages in his or her own mind, and second to explain those stages clearly enough to enable readers to perceive them. The process itself may involve making something like a sweater or a table, or it may be more abstract, as it is in the following paragraph, a brief explanation of the steps in the methodology of science:

> . . . there is a fairly clear pattern of the operation of the scientific method. First, regularities are recognized as such and recorded. Then, a formulation is sought which, preferably in the simplest and most general way, contains these regularities. This then has the status of a law of nature. The newly formulated law may, and usually will, predict further regularities which were previously unknown. Finally, the objective is a combination of two or more of these laws into a still more general formulation. For instance, the great significance of Einstein's theory of special relativity is due to the fact that it provides a combination of the electromagnetic laws with those of mechanics.
>
> Kurt Mendelssohn

One needs to pay attention to the relationship between compositional units (sentences and paragraphs) and ideas in all writing. This is especially important in analyzing a process. Thus Mendelssohn begins a new sentence for each of his steps. (In the case of steps two and three he uses several sentences for a more detailed analysis.) Moreover he carefully identifies each step with the words "First . . . then . . . finally. . . ."

The next example concerns a less abstract process: directions for making bread, specifically, how to knead the dough. The writer carries the visual separation of his compositional units even further to help readers follow the process step-by-step:

KNEADING THE DOUGH

The kneading surface, board or table should be at a height on which your hands rest comfortably when you are standing straight (mid-thigh). Keep the surface floured sufficiently to prevent the dough from sticking during kneading. The purpose of kneading is to get the dough well-mixed, of a smooth, even texture, and to further develop the elasticity of the dough.

Beginning with a lump of dough not entirely of a piece, somewhat ragged and limply-lying, commence kneading.

Flour your hands.

Picking up far edge of dough, FOLD dough IN HALF toward you, far side over near side (Figure 14),[3] so that the two edges are approximately lined up evenly.

Place your hands on NEAR SIDE of dough so that the top of your palms (just below fingers) are at the top front of the dough (Figure 15).

PUSH DOWN AND FORWARD, centering the pushing through the heels of the hands more and more as the push continues (Figure 16). Relax your fingers at the end of the push. Rock forward with your whole body

---

3. In the original, simple line drawings illustrate the text; these are not reproduced here.

rather than simply pushing with your arms. Apply steady, even pressure, allowing the dough to give way at its own pace. The dough will roll forward with the seam on top, and your hands will end up about ⅔ of the way toward the far side of the dough. Removing your hands, see that the top fold has been joined to the bottom fold where the heels of the hands were pressing (Figure 17).

TURN the dough ¼ turn (Figures 16 and 17) (clockwise is usually easier for right-handed persons). Fold in half towards you as before (Figure 17) and rock forward, pushing as before (Figures 18 and 19).

TURN, FOLD, PUSH. Rock forward. Twist and fold as you rock back. Rock forward. Little by little you will develop some rhythm. Push firmly, yet gently, so you stretch but do not tear the dough.

Add FLOUR to board or sprinkle on top of dough as necessary to keep dough from sticking to board or hands. As you knead, the dough will begin stiffening up, holding its shape rather than sagging; it will become more and more elastic, so that it will tend to stretch rather than to tear. It will stick to hands and board less and less until no flour is necessary to prevent sticking. The surface will be smooth and somewhat shiny.

Before you finish kneading, SCRAPE THE BREAD BOARD (Figure 20) and rub dough off hands and incorporate these scraps into the dough.

Place the dough (Figure 21) in the OILED BREAD BOWL smooth side down, and then turn it over so the creases are on the bottom (Figure 22). Oiled surface will keep a crust from forming on the dough.

COVER the dough with a DAMP TOWEL and set it in a warm place.

Edward Espe Brown

Here the typography greatly clarifies the analysis: separate paragraphs for each step; first lines of each paragraph moved to the left; key terms set in capitals. While expository essays do not often use such typographical aids, they help in directions—whether for cooking, putting together a bicycle, or any other mechanical process. Keep in mind, however, that these devices do not substitute for clear and proper analysis of the process. In fact, typography works effectively only after an analysis has been made.

EXERCISE

1. Choosing any two of the following topics, compose two separate analysis paragraphs. Begin with a topic sentence like Chesterton's (see page 165) and unify the paragraph with appropriate connecting words.

    A. Types of students (or professors)
    B. Kinds of dancing
    C. The basic varieties of detective fiction
    D. Small sailboat rigs
    E. Used-car salesmen
    F. The ''lines'' employed by romantically inclined young men (or women)
    G. The basic types of hammer (or of any other tool)

2. Write a set of directions for some simple activity (hitting a golf ball, repotting a plant, any such process). The problem—and it is not easy—is to analyze the process into its steps and to lay these out so clearly that a novice can perform the action. Use brief paragraphs if you wish, and any other techniques of layout that will help the reader, but treat these as aids to analysis, not as substitutes for it.

CHAPTER 23

Paragraph Development: Qualification

*A dev.
qual.*

In developing ideas one often finds it necessary to admit some limitation to their truth or applicability. The subjects with which composition deals are seldom absolutely true or absolutely false. The process we use to deal with this is called *qualification*.

Qualification does pose a problem: the danger of blurring the focus of the paragraph. Suppose, for example, that someone writes a criticism of college football. He or she begins with the flat statement "College football is a semiprofessional sport." The point is clear and emphatic, but it is not true: the issue is just not that simple. Now suppose that, recognizing the complexity of the topic, the writer adds a second sentence:

> College football is a semiprofessional sport. Some universities do play a purely amateur game.

No longer so vulnerable to the charge of oversimplifying, the writer has gained protection at the cost of confusing readers, who do not know what to expect. Will the paragraph be about universities which subsidize football or about those which do not? Has the writer presented a contradiction or two true assertions, each limited in some way?

As this suggests, qualification involves at least the appearance of contradiction. The trick is to qualify without confusing readers as to the main point. Actually it is not difficult to do, once you understand a few basic principles:

1. *Whenever possible, subordinate the qualification.*

> College football is a semiprofessional sport, although some universities do play a purely amateur game.

This makes better sense. By expressing the qualification in the adverbial *although*-clause, the writer reduces its importance. The thought, however, still progresses awkwardly. Placing the qualification last leaves it uppermost in the reader's mind, which, considering the writer's intention, is not desirable. This brings us to a second principle.

2. *When you can, place the qualification first and wind up on the main point.*

> Although some universities do play a purely amateur game, college football is a semiprofessional sport.

3. *Use qualifying words and phrases.*

> Although a few universities play a purely amateur game, big-time college football is, in general, a semiprofessional sport.

The addition of such expressions as "a few," "big-time," and "in general" further establishes the limits of the writer's assertion.[1] So phrased, the sentence has sufficient qualifications to forestall easy challenges from those who disagree with it. Yet it remains clearly focused.

4. *When a qualification must be expressed in a seperate sentence, begin it with a word stressing its obviousness and follow it by repeating the major idea.*

One cannot always work a qualification into the same sentence as the main topic. When a qualification has to be stated separately, introducing it with an admission of its truth tends to disarm it. "Of course" (or "certainly," "admittedly," "true"), you write, "such and such is the case." This tells the reader that you are well aware of the fact and implies that it doesn't matter very much. Then, with the qualification completed, you return to your main point with a strong signal of contradiction ("but," "however," "still," "yet," "even so," "nevertheless") and repeat that point, whether in a new sentence or in a clause coordinated with the qualification:

> Big-time college football is essentially a semiprofessional sport. Of course, a few universities play a purely amateur game. But these are only a few; on the whole, the game is subsidized.

Study the more extensive example in this paragraph:

> Male assumptions about women are clouded by stereotypes of thought pertaining to "women as a group." Men see other men as individuals, each with distinguishable character traits. They see women as a set of people who share certain character traits which can be summed up under the label "feminine"; and before they can think about a woman as an individual, they have to fight their way through the stereotypes. Feminine human beings are expected to be passive, emotional, indecisive, devious and intuitive rather than rational. Some good traits are allowed to them, of course: warmth, sensitivity and concern for others. But these are felt to be good

---

1. English has many qualifiers: *some, many, most, usually, commonly, occasionally, mostly,* and so on.

only in special situations—at home, within the family milieu of intimacy, and not in the business world.

Elizabeth Janeway

Janeway's next-to-last sentence concedes that men occasionally admit positive qualities into their negative image of women. The qualification, however, is neutralized by "of course" and by the immediately following assertion introduced by "but."

At times a qualification requires more extended treatment which may take several sentences or even an entire paragraph. For example, George R. Stewart, arguing that the American colonists constituted an essentially homogenous culture, writes:

> With few exceptions the colonists of European stock were of northwestern European origins, and there can have been, racially, only negligible differences among them. Even in their cultural backgrounds they differed little. They were heirs of the European Middle Ages, of the Renaissance, and of the Reformation. They were Christians by tradition, and nearly all were Protestants.
>
> Naturally the groups differed somewhat, one from another, and displayed some clannishness. They were conscious of their differences, often more conscious of differences than of resemblances. Thus a Pennsylvania governor of 1718 was already voicing the cry that the American conservative has echoed ever since. "We are being overwhelmed by the immigrants!" he said in effect. "Will our country not become German instead of English?"
>
> Nevertheless, from the perspective of two centuries and from the point of view of the modern world with its critical problems of nationality and race, the differences existing among the various colonial groups fade into insignificance. We sense, comparatively speaking, a unified population. In the political realm, indeed, there were divergences that might lead even to tarrings and featherings, but racially and socially and religiously the superficial differences were much less important than the basic unity.

Professor Stewart's second paragraph elaborately qualifies the point he makes in the first, admitting that despite their sameness the colonists did differ. Its greater length does not alter the technique of qualification, however. The paragraph begins with "naturally," stressing the truth of the concession and thereby removing its sting. In the third paragraph the writer swings back to the main point, signaling the return with the emphatic "nevertheless." And throughout he uses such qualifiers as "few," "only negligible," "nearly all," "somewhat," and "comparatively speaking."

CONCLUSION

Although learning how to qualify without blurring your topic is necessary in writing good exposition, do not overqualify. Some people cannot bring themselves to state as a fact what plainly is a fact. They feel compelled to hedge all bets and write, "It would seem as if it were in the vicinity of twelve o'clock," instead of "It's noon."

But such excesses do not mean that you should avoid qualification. They mean rather that you must know when you should not qualify as well as when you should. Reality being what it is, you will find that qualification is called for not infrequently.

EXERCISE

1. The final sentence of the passage by George R. Stewart (page 173) contains a minor qualification. Identify it and show how it illustrates the techniques we have been discussing.

2. Explain the qualifications in the following passages. Are they effective? Why or why not?

A. Now while it must be admitted that highly educated people are sometimes cruel, I think there can be no doubt that they are less often so than people whose minds have lain fallow. The bully in a school is seldom a boy whose proficiency in learning is up to the average. When a lynching takes place, the ringleaders are almost invariably very ignorant men.

Bertrand Russell

B. To my mind King James's Bible has been a very harmful influence on English prose. I am not so stupid as to deny its great beauty. It is majestical. But the Bible is an oriental book. Its alien imagery has nothing to do with us. Those hyperboles, those luscious metaphors, are foreign to our genius.

Somerset Maugham

C. Today they [political liberals] are apt to echo the common charge that "scientific philosophy" is the root of our evils. Science has indisputably inspired much narrow, harsh philosophy, and much pseudo-science; its disciples have often been inhuman. As inhuman, however, is the fashion of branding all the efforts of intelligence as sinful pride, and all the works of science as mere materialism.

Herbert J. Muller

D. "When the belly is full," runs the Arab proverb, "it says to the head, 'Sing, fellow!' " That is not always so; the belly may get overfull. Such a proverb clearly comes from a race familiar with bellies painfully empty. Yet it remains true, I think, that when the body is in radiant health, it becomes extremely difficult for it not to infect the mind with its own sense of well-being.

F. L. Lucas

3. Listed below are five pairs of sentences, all examples of poor qualification. Revise each pair twice so as to make a proper qualification contained in a single sentence. In the first revision take idea (1) as the main point and (2) as the qualification; in the second reverse that relationship. Keep generally to the wording as it is given, but you may change the order of the clauses and add qualifying words and connectives as needed.

A. (1) College is difficult.
   (2) Some aspects of college are easy.

    B. (1) Baseball is the great American game.
       (2) Its claim to supremacy is being challenged by other professional sports.
    C. (1) The Romans are regarded as culturally inferior to the Greeks.
       (2) They created a great and long-lasting empire.
    D. (1) Exercise is necessary to health.
       (2) Too much exercise or the wrong kind can be harmful.
    E. (1) The "absent-minded professor" is a myth.
       (2) Some professors are forgetful.

4. Using a topic of your own, compose a longer qualification in three sentences. In the first sentence state your primary idea; in the second the qualification; and in the third a reassertion of the main point.

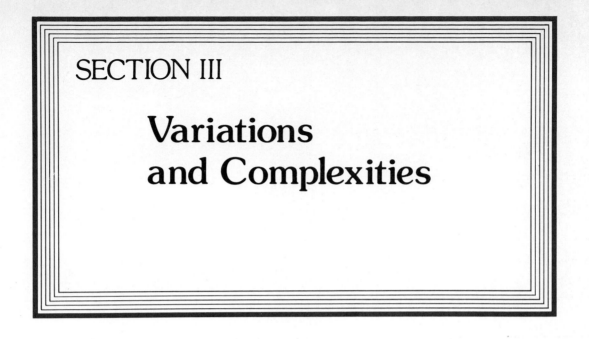

SECTION III

Variations
and Complexities

CHAPTER 24

Variations in the Topic Sentence and in Paragraph Unity

DELAYING THE TOPIC SENTENCE

As we saw in Chapter 14, the topic statement of an expository paragraph usually comes at or very near the beginning. Occasionally, however, it may be delayed until the middle of the paragraph or even until the very end. It appears a bit after the middle of the following paragraph by the novelist William Golding. Golding, who worked as a teacher, is talking about making out progress reports (sentences have been numbered for convenience):

(1) It is possible when writing a boy's report to admit that he is not perfect. (2) This has always been possible; but today there is a subtle change and the emphasis is different. (3) You can remark on his carelessness; you can note regretfully his tendency to bully anyone smaller. (4) You can even suggest that the occasions when he removed odd coins from coats hung in the changing room are pointers to a deep unhappiness. (5) Was he perhaps neglected round about the change of puberty? (6) Does he not need a different father-figure; should he not, therefore, try a change of psychiatrist? (7) You can say all this; because we all live in the faith that there is some machine, some expertise that will make an artificial silk purse out of a sow's ear. (8) But there is one thing you must not say, because it will be taken as an irremediable insult to the boy and to his parents. (9) *You must not say he is unintelligent.* (10) Say that and the parents will be after you like a guided missile. (11) They know that intelligence cannot be bought or created. (12) They know, too, it is the way to the good life, the shaming thing, that we pursue without admitting it, the naked power, the prestige, the two cars and the air travel. (13) Education, pointing still,[1] is nevertheless moving their way; to the world where it is better to be envied than ignored, better to be well-paid than happy, better to be successful than good—better to be vile, than vile-esteemed.

1. A reference to a picture described earlier in which an idealized woman, personifying education, points two beautiful children toward a bright future.

The topic idea does not come until the ninth sentence (the italics are Golding's), although the second sentence ("today there is a subtle change and the emphasis is different") foreshadows it. The first portion of the paragraph, however, is not wasted. It contributes to Golding's larger point, that our culture overvalues sheer intelligence. By illustrating the relative indifference of parents to the moral behavior of their children, this opening section of the paragraph prepares for the startling contrast to the sensitivity of parents concerning their children's intelligence.

Now and then a writer pushes a topic sentence to the very end of a paragraph. Such postponement is most likely when the paragraph opens a piece of writing—an extreme case of delaying the announcement of the subject. Here is a paragraph beginning a history of Spain:

> A dry, barren, impoverished land: 10 per cent of its soil bare rock; 35 per cent poor and unproductive; 45 per cent moderately fertile; 10 per cent rich. A peninsula separated from the continent of Europe by the mountain barrier of the Pyrenees—isolated and remote. A country divided within itself, broken by a high central tableland that stretches from the Pyrenees to the southern coast. No natural centre, no easy routes. Fragmented, disparate, a complex of different races, languages, and civilizations—this was, and is, Spain.
>
> J. H. Elliott

IMPLYING THE TOPIC SENTENCE

It is even possible to compose an expository paragraph without a topic sentence (though it must be clearly implied). In this paragraph, for instance, the historian Carl Becker discusses how technology has affected modern life:

> The first and most obvious result of the technological revolution has been to increase the amount of wealth in the form of material things which can be produced in a given time by a given population. For example, in 1913 there was produced in Great Britain seven billion yards of cotton cloth for export alone. In 1750 the total population of Great Britain, working with the mechanical appliances then available, could have produced only a small fraction of that amount. A second result of the technological revolution is that, as machines are perfected and become more automatic, man power plays a relatively less important part in the production of a given amount of wealth in a given time. Fifty years ago, when all type was set by hand, the labor of several men was required to print, fold, and arrange in piles the signatures of a book. Today machines can do it all, and far more rapidly; little man power is required, except that a mechanic, who may pass the time sitting in a chair, must be present in case anything goes wrong with the machine. And finally, a third result of the technological revolution is that, under the system of private property in the means of production and the price system as a method of distributing wealth, the

greater part of the wealth produced, since it is produced by the machines, goes to those who own or control the machines, while those who work the machines receive that part only which can be exacted by selling their services in a market where wages are impersonally adjusted to the necessities of the machine process.

Becker does not open his paragraph with a broad topic statement, something like this:

The technological revolution has had three important results.

But one can easily understand that such a general idea constrains and shapes the paragraph.

Leaving the topic idea for readers to infer has the advantage of novelty; it varies the standard pattern of topic statement-support. More importantly it can draw readers into the writing, forcing them to supply the idea for themselves and perhaps emphasizing it (on the principle that what we grasp for ourselves we are more likely to remember).

On the other hand, the implicit topic must necessarily be rare. It works only where sufficient context exists to make the controlling idea clear. In shorter compositions, such as those you generally write for English class, the relative lack of context means less chance of, or advantage in, implying the topic rather than stating it.

FIGURATIVE UNITY

In Chapter 15 we saw that you can unify a paragraph first by keeping the ideas within the limits defined by the topic statement and arranging them in a significant order, and second by tying together sentences with repetitions, connectives, or other similar syntactic patterns.

Now and again you can unify a paragraph, or a portion of one, by developing a figure of speech, usually a metaphor or a simile. For example, in the following passage about a Japanese village called Shinohata, the writer discusses social harmony, using a metaphor from seismology:

The "harmony of the village" has its cost. Underneath the placid landscape there are geological faults—a personal incompatibility, a clash of economic interest, a belief that one has been cheated—along which tensions build up which require occasional release. Shinohata, in that sense, is a relatively quake-free zone and has become more so: the tensions rarely become unmanageable, a good deal more rarely than they used to.

Ronald P. Dore

Introduced by "Underneath the placid landscape there are geological faults," the metaphor is continued by "tensions," "release," and "quake-free zone."

A more extensive example, which works through an entire paragraph, appears in the following comic, mock-heroic account of how pasta destroyed Italian painting and culture, the final disaster occurring at Venice. The speaker is an imaginary gourmand named Mr. Vortigern; the unifying figure, a simile likening spaghetti to a monstrous jellyfish or octopus:

> "Then *pasta asciutta* came. It must have taken a century or two to conquer Italy, spreading from the south like a clammy and many-tentacled monster, smothering Italy's genius on its northward journey, and strangling its artists like the serpents of Laocoön. Thousands of seething and dripping tongues of macaroni squirming and coiling up the Apennines, gathering volume every mile, engulfing towns and provinces and slowly subduing the whole peninsula. The North held out heroically for a while—there is no pasta in those banquets of Veronese—and the last stand was at Venice. The rest of the country lay inert under its warm and slippery bonds— slippery, perhaps, but still unbroken today—while Tiepolo held out, and Longhi and Guardi and Canaletto, the last lonely frontiersmen of a fallen Empire.
>
> "But one day some treacherous dauber must have swallowed a streaming green yard or two of *tagliatelle*—the end of the enemy's foremost tentacle, you might say—and, hotfoot, the rest of the huge, victorious monster came coiling into the Veneto and reared itself for the kill. And then——" the end of Mr. Vortigern's malacca cane, which had been describing faster and faster loops in the air, stopped in mid-sweep—"and then, wallop! A billion boiling tons of pasta fell on the town, and the proud city, the sea's bride, with her towers and domes and bridges and monuments and canals, went under. The piazzas were a tragic squirming tangle of spaghetti and lasagne, the lagoon ran red with tomato sauce. Italy's genius was dead, laid low by her own gluttony . . . and not only Italy's painting, but Italian thought and poetry and literature and rhetoric and even Italian architecture. Everything was turned into macaroni."
>
> Patrick Leigh-Fermor

Figurative unity is relatively rare, and it must develop naturally out of subject and tone, never be imposed simply as a way of unifying a paragraph. When an extended figure is appropriate and well-handled—as in the preceding example—it creates an extremely tight-knit paragraph.

EXERCISE

1. Compose three paragraphs, each about six or seven sentences and each on a different topic. In the first, set the topic sentence toward the middle, in the second at the end, and in the third leave it implied.

2. From one of your themes select a paragraph which has the topic sentence in its usual position at or near the beginning. Rewrite the paragraph twice. In the first, put the topic sentence toward the middle, in the second, at the end. Do the revisions make the thought clearer or less clear? More emphatic or less?

3. Compose a short paragraph unified by a simile or metaphor that works all through it. The topic and figure should grow out of your own experience. Here are a few possibilities, but do not feel limited to them:

A. For me English class (or mathematics or German, what you will) is a shipwreck.
B. Shopping with my mother is like going to war.
C. To be in love is to be drunk.
D. College, as John Dos Passos said, is four years under the ethercone.

CHAPTER 25

Paragraph Patterns

INTRODUCTION

We have seen how to develop expository paragraphs by examples, comparisons, reasons, and so on. Here we consider how these methods of development work together to give a paragraph a distinctive shape, which we shall call the *conceptual pattern*. The pattern is, in effect, the outward form of the writer's thought. It is largely an unconscious creation. Writers do not literally plot out the shape of a paragraph in advance, and trying to do so would probably be a mistake. It is better to concentrate on what you want to accomplish; then the pattern takes care of itself.

At the same time an awareness, in the back of your mind, that a well-written paragraph possesses a pattern, a contour determined by subject and purpose, will help. It will also help if you understand the four basic paragraph patterns (though these are not hard and fast and can be combined): the lineal, the ramifying, the circular, and the loose. Now we shall look at one or two examples of each of these conceptual patterns of the expository paragraph.

THE LINEAL PARAGRAPH

The lineal paragraph moves generally in a straight line through a series of closely linked ideas. It carries the reader from an initial assertion to a new but related idea, so its end is not the same as its beginning. An example is the paragraph below from one of *The Federalist* papers by John Madison, written in 1787–1788 to persuade the voters of New York State to accept the Constitution proposed for the newly independent United States. Madison's main point is that the republican (or representative) democracy proposed in the Constitution is better than a pure (or direct) democracy. Here he argues that factions (special-interest groups or political parties as we might call them today) will not be controlled in a pure democracy. (The sentences here and in subsequent examples have been numbered for convenience.)

(1) From this view of the subject it may be concluded that a pure democracy, by which I mean a society consisting of a small number of citizens, who assemble and administer the government in person can admit of no cure for the mischiefs of faction. (2) A common passion or interest will, in almost every case, be felt by a majority of the whole; a communication and concert result from the form of the government itself; and there is nothing to check the inducements to sacrifice the weaker party or an obnoxious individual. (3) Hence it is that such democracies have ever been spectacles of turbulence and contention; have ever been found incompatible with personal security or the rights of property; and have in general been as short in their lives as they have been violent in their deaths. (4) Theoretic politicians, who have patronized this species of government, have erroneously supposed that by reducing mankind to a perfect equality in their political rights, they would, at the same time, be perfectly equalized and assimilated in their possessions, their opinions, and their passions.

Madison's paragraph consists of (1) an assertion, (2) a reason why the assertion is a fact, (3) a result of that fact, and (4) a general conclusion. He puts each of these ideas in a separate sentence, a strategy illustrating the clear correspondence between sentence and idea which, as we saw in Chapter 8, marks a well-written paragraph. The paragraph moves in a straight line from the opening assertion (that pure democracies cannot control faction) to the final point (that theorists who advocate pure democracy are wrong). It does not branch off in other directions or circle back upon itself.

That lineal pattern may be diagrammed like this:

Sentence 1: Assertion   Pure democracy cannot control faction.

Sentence 2: Reason   In pure democracies a tyranny of the majority is inevitable.

Sentence 3: Result   ("Hence") Pure democracies are turbulent and short-lived.

Sentence 4: Conclusion   Political theorists advocating such democracy are wrong.

Note: A word about how these diagrams work. Successive sentences are numbered, and the numbers are followed by a term, printed in red, naming the function of the idea in the paragraph, that is, whether it is a basic assertion, a reason for or a consequence of some preceding point, a restatement or an illustration of that point, and so on. An arrow connects each sentence to the one or ones it relates to in the manner indicated by the red label. Thus in Madison's paragraph the second sentence stands to the first as a reason supporting an assertion, and the first to the second as an assertion supported by a reason. Sometimes a sentence relates conceptually to more than one other statement; in such cases a dashed arrow shows the secondary relationship. A synopsis of the essential idea expressed in the sentence appears below each of the terms in red. Finally, if the writer uses a connecting word or phrase ("hence," "furthermore," "however") it is placed in the diagram in parentheses.

Such diagrams enable us to see how the thought develops through a paragraph, what techniques of paragraph support the writer has used, and how these relate to the sentence structure. Of course, any particular diagram may be simplified. A sentence may contain subordinate ideas not listed in the diagram. For example, the first sentence of Madison's paragraph not only makes an assertion about pure democracy but also defines what pure democracy is. But while they do not reveal everything, the diagrams establish the main conceptual outline of a paragraph.

THE RAMIFYING PARAGRAPH

Not all paragraphs are lineal. Some ramify—or branch—in several directions. While such a paragraph may divide into three or four branches, the simplest and most common case is a single shift of direction, resulting in a dual structure. This paragraph about the life of the medieval peasant provides a clear example.

> (1) Was the peasant's life, in general, tolerable? (2) Or was he a wretched victim of oppression? (3) Modern writers disagree. (4) Some say that his lot was worse than that of the most downtrodden industrial wage slave, the least privileged sharecropper. (5) Others judge him to have been better off than the typical English farm laborer of the nineteenth century. (6) And some even assert that in many regions the peasant's general level of well-being was higher during the Middle Ages than it was to be again until our own enlightened times.
>
> Morris Bishop

Bishop's assertion, spread through the opening three sentences, sets up a twofold analysis. Sentence 4 develops the first part; sentences 5 and 6, the second (see figure on p. 187).

Notice that the paragraph depends for its unity more upon repetition of words and sentence patterns than upon connectives: "Modern writers disagree. . . . Some say. . . . Others judge. . . . some even assert." There are, however, two connective words: "or" in the second sentence,

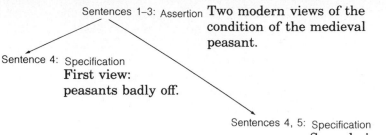

Sentences 1–3: Assertion Two modern views of the condition of the medieval peasant.

Sentence 4: Specification
First view: peasants badly off.

Sentences 4, 5: Specification
Second view: peasants relatively well off.

introducing an alternative; and "and" in the last, signaling a strong restatement.

This more complicated example of ramifying structure also has two parts. Here the writer answers the charge that women are not reliable workers because they quit their jobs to get married:

(1) Now, it is frequently asserted that, with women, the job does not come first. (2) What (people cry) are women doing with this liberty of theirs? (3) What woman really prefers a job to a home and family? (4) Very few, I admit. (5) It is unfortunate that they should so often have to make the choice. (6) A man does not, as a rule, have to choose. (7) He gets both. (8) In fact, if he wants the home and family, he usually has to take the job as well, if he can get it. (9) Nevertheless, there have been women, such as Queen Elizabeth and Florence Nightingale, who had the choice, and chose the job and made a success of it. (10) And there have been and are many men who have sacrificed their careers for women—sometimes, like Antony or Parnell, very disastrously. (11) When it comes to a *choice,* then every man or woman has to choose as an individual human being, and, like a human being, take the consequences.

Dorothy L. Sayers

Sayers's strategy is twofold, which accounts for the pattern of her paragraph. In the first phase she concedes that the complaint is true—at least in part—but discusses the reason for its truth in such a way as to neutralize the fault. If it is the case, she says, that most women prefer marriage to careers it is because the "choice" is forced upon them; it is pointless to blame them for accepting a culturally dictated role. The second phase of Sayers's strategy is to contradict the claim with a counterassertion buttressed by examples both of women who did put the job first and of men who did not.

The paragraph is complicated, using simple restatement, emphatic restatement, restatement in different terms, reason, contrast, contradiction, examples, and conclusion. Yet everything comes together in a coherent, balanced paragraph of two parts:

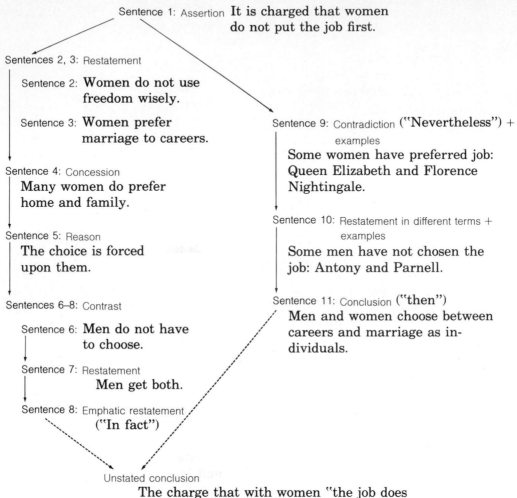

Sentence 1: Assertion It is charged that women
do not put the job first.

Sentences 2, 3: Restatement

Sentence 2: Women do not use
freedom wisely.

Sentence 3: Women prefer
marriage to careers.

Sentence 9: Contradiction ("Nevertheless") +
examples
Some women have preferred job:
Queen Elizabeth and Florence
Nightingale.

Sentence 4: Concession
Many women do prefer
home and family.

Sentence 5: Reason
The choice is forced
upon them.

Sentence 10: Restatement in different terms +
examples
Some men have not chosen the
job: Antony and Parnell.

Sentences 6–8: Contrast

Sentence 6: Men do not have
to choose.

Sentence 7: Restatement
Men get both.

Sentence 11: Conclusion ("then")
Men and women choose between
careers and marriage as individuals.

Sentence 8: Emphatic restatement
("In fact")

Unstated conclusion
The charge that with women "the job does
not come first" is invalid.

THE CIRCULAR PARAGRAPH

The circular paragraph ends where it begins, by repeating its opening assertion. It may move through other, closely related ideas, but it comes back finally to repeat the topic. The following paragraph by the linguist Mario Pei on the difficulties of English spelling is an instance:

(1) English spelling is the world's most awesome mess. (2) The Chinese system of ideographs is quite logical, once you accept the premise that writing is to be divorced from sound and made to coincide with thought-concepts. (3) The other languages of the West have, in varying degrees, coincidence between spoken sounds and written symbols. (4) But the spelling of English reminds one of the crazy-quilt of ancient, narrow, winding streets

in some of the world's major cities, through which modern automobile traffic must nevertheless in some way circulate.

Mario Pei

The paragraph is like a circle: sentences 2 and 3 take us away from the immediate topic of English spelling to that of Chinese and modern western languages. Then sentence 4, leading us with "but," swings back to English, repeating the initial assertion in a metaphor and ending the paragraph where it began:

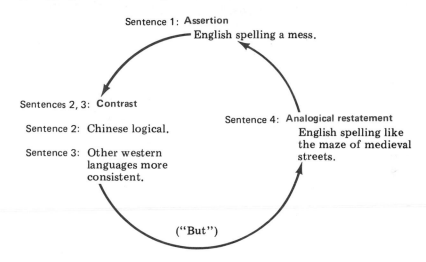

Notice that the fourth sentence fills two functions, contrasting with the idea expressed in sentences 2 and 3 and restating the topic asserted in sentence 1.

Circular paragraphs are neither better nor worse than those which progress to new ideas. "Circular" and "progressive," as applied to paragraph structure, do not imply value. They simply designate different patterns. Circular paragraphs are emphatic, making their point both at beginning and end. They may also function, in effect, as miniature essays with beginnings, middles, and ends. On the other hand, they may prove too complete, especially when used as part of a longer composition, making it difficult for a writer to get out and move on.

THE LOOSE PARAGRAPH

The loose paragraph is usually a special case of the ramifying paragraph, returning to the topic one or more times to take up new aspects. It differs from the kinds of paragraphs we have seen so far chiefly in the nature of the topic. The topic is more diffuse, less sharply defined, consisting simply of a subject rather than of a subject plus a predication. In Madison's paragraph, for instance, the *subject* is "pure democracy," but the *topic* is actually that subject plus an assertion about it, namely, that pure democracy is unable to control faction. The topics of most expository paragraphs consist of a full proposition, that is, of a subject about which something has been predicated: our spelling is a mess; modern historians disagree about the life of the medieval peasant. Such paragraphs are relatively tight in the sense that everything must relate not simply to the subject but also to what has been claimed about it.

The loose paragraph, on the other hand, develops out of a subject alone, without any controlling predication. Rather it consists of stringing together several predications, as the writer returns to the subject to assert new truths about it. Consider this paragraph by the historian G. M. Trevelyan:

> (1) The importance of the Roman roads after their makers had gone, lay in this: no one made any more hard roads in the island until the turnpike movement of the Eighteenth Century. (2) Throughout the Dark Ages and in early medieval times, these stone highways still traversed an island otherwise relapsed to disunion and barbarism. (3) The Roman roads greatly increased the speed of the Saxon, Danish and Norman Conquests, and aided, both in peace and in war, the slow work of Saxon and Norman Kings in uniting England as one State, and making the English nation. (4) Thanks to the Roman legacy, Britain had better national highways under the Saxon heptarchy than in Stuart times, though in the latter period there were more by-roads. (5) The imperial stone causeways, often elevated some feet above the ground, ran from sea to sea, generally keeping the higher land, but where needful marching majestically over bog and through forest. (6) If the bridges soon fell in from neglect, the paved fords remained. (7) For centuries wild tribes who only knew the name of Caesar as a myth, trod his gigantic highways and gave them the fantastic names of Watling Street, Ermine Street, and the Foss Way. (8) Gradually the stones subsided and men were too careless and ignorant to replace them. (9) Next, the road was used as a quarry, when the medieval Englishman, having somewhat exhausted his timber, began to build for himself dwelling-houses of stone. (10) From driving roads they declined into pack-horse tracks, finally disappearing for the most part in moor or plough-land. (11) Stretches of them have been repaired and modernized, and the motor car now shoots along the path of the legions. (12) But other stretches—and those the best beloved—are reserved for the Briton or Saxon who still fares on foot; they are to be traced as green lanes, starting up out of nowhere and ending in nothing, going for miles straight as a die through the magical old English countryside.

The topic of Trevelyan's paragraph is the Roman roads in Britain—the roads, not this or that about them. As the paragraph develops, Trevelyan asserts particular facts about the roads—for example, that they were important—but no single proposition unifies and organizes the entire paragraph. Instead the paragraph moves easily from one assertion about the subject to another in a loose pattern. The roads, it states, were important, and it tells why. But then the paragraph slides into a description of the highways and from that into the history of their gradual decay.

It would be possible to divide the passage into at least two paragraphs: sentences 1–4 and sentences 5–12. But such a division would be more a matter of taste or custom than of necessity. For though it runs a bit longer than the usual expository paragraph today, this one works: everything relates to one central point and furthers our understanding. The paragraph differs only in conceiving its topic as a broad and inclusive subject, not as a specific assertion about that subject.

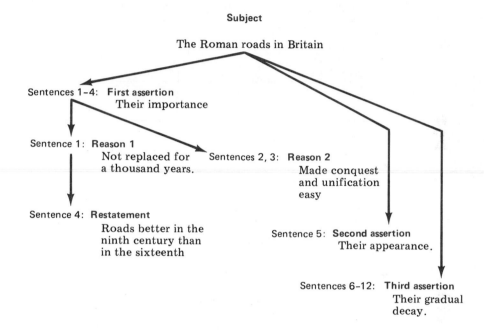

Even looser is this paragraph by the humorist James Thurber from a personal reminiscence:

(1) I left the University in June, 1918, but I couldn't get into the army on account of my sight, just as grandfather couldn't get in on account of his age. (2) He applied several times and each time he took off his coat and threatened to whip the men who said he was too old. (3) The disappointment of not getting to Germany (he saw no sense in everybody going to France) and the strain of running around town seeing influential officials finally got him down in bed. (4) He had wanted to lead a division and his chagrin at not even being able to enlist as a private was too much for him. (5) His brother Jake, some fifteen years younger than he was, sat up at night with him after he took to bed, because we were afraid he might leave the house without even putting on his clothes. (6) Grandfather was against the idea of Jake watching over him—he thought it was a lot

of tomfoolery—but Jake hadn't been able to sleep at night for twenty-eight years, so he was the perfect person for such a vigil.

A kind of conventional expository structure exists here: Thurber's failure to get into the army is compared to his grandfather's, and the grandfather's disappointment leads to his illness and to Jake's all-night vigils, and finally to an explanation of why Jake is the ideal watchman. But the paragraph does not develop in a tight pattern of cause and

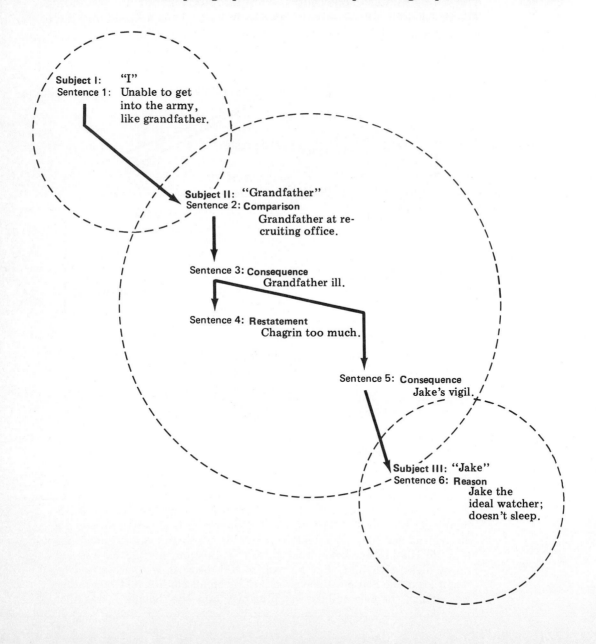

effect. There is no topic in the sense of a single unifying assertion. Even in the looser sense of one encompassing subject, the topic is less clearly defined than in the paragraph by Trevelyan. Thurber's paragraph has, in effect, three subjects: "I," "grandfather," and "Jake." One leads into the next by association. Grandfather and his eccentricities loom larger than the other two subjects and give the paragraph a focus of sorts. (Grandfather figures in all six sentences.) But the focus is soft, allowing the writer to slip easily from subject to subject. The pattern of the paragraph is best suggested by overlapping circles (see figure on page 192).

Loose paragraphs have a less formal tone than those constrained by an opening assertion to which everything must relate. When your purpose is to convey information or ideas you want a tight paragraph. It suits the impersonal tone of relatively formal occasions when the writer's presence is felt primarily as a controlling intelligence, analyzing ideas and organizing them into a coherent whole. The writing you do in term papers, critical essays, discussion answers on exams requires such paragraphs.

In that kind of formal, impersonal writing the associational pattern of a paragraph like Thurber's would be out of place. But it suits the colloquial, personal style of a familiar essay. It fits with the contractions ("couldn't," "hadn't"), slang words like "tomfoolery," the frequent use of coordination in the sentences, and the intrusive statements (explanations set within parentheses or dashes)—all reminiscent of speech.

Thurber's example shows that the deep pattern of a paragraph not only depends on the subject and the writer's habits of mind; it also depends on the tone the writer wishes to achieve.

EXERCISE

1. Diagram each of the following paragraphs to reveal its basic pattern. Summarize each sentence and show how it contributes to the conceptual structure of the paragraph. Follow the general directions for such diagramming as given on pages 185–86 and be guided by the examples of the diagrams used in this chapter.

A. (1) Ants are more like the parts of an animal than entities on their own. (2) They are mobile cells, circulating through a dense connective tissue of other ants in a matrix of twigs. (3) The circuits are so intimately woven that the anthill meets all the essential criteria of an organism.

Lewis Thomas

B. (1) Style, I repeat, is a means by which a human being gains contact with others; it is personality clothed in words, character embodied in speech. (2) If handwriting reveals character, style reveals it still more—unless it is so colourless and lifeless as not really to be a style at all. (3) The fundamental thing, therefore, is *not* technique, useful though that may be; if a writer's personality repels, it will not avail him to eschew split infinitives, to master the difference between "that" and "which," to have Fowler's *Modern English Usage* by heart. (4) Soul is more than syntax. (5)

If your readers dislike you, they will dislike what you say. (6) Indeed, such is human nature, unless they like you they will mostly deny you even justice.

F. L. Lucas

C. (1) In principle this method of manufacture [Henry Ford's assembly line] was far from new. (2) It depended upon Eli Whitney's great discovery of the principle of interchangeable parts. (3) It owed much to the refinement of that principle by such men as Henry M. Leland, who had shown what close machining could do to make these interchangeable parts fit with absolute precision. (4) Moreover, many a manufacturer had used the assembly-line principle to some extent. (5) Cyrus McCormick, for instance, had done so in his reaper works as far back as the eighteen-fifties; and in particular the packers had used an overhead conveyer to carry slaughtered animals past a series of workers. (6) Ford was indebted, too, to Frederick Winslow Taylor for his studies in "scientific management," the careful planning of manufacturing processes so as to save steps and motions. (7) And Ransom Olds had already put a single type of automobile into quantity production—until his financial backers forced him back into the luxury market. (8) Nevertheless the Ford assembly line, with its subassemblies, was unique as a remorselessly complete application of all these ideas.

Frederick Lewis Allen

2. Compose a paragraph according to each of the following plans. Use the topic as your opening assertion and adhere to the given conceptual pattern, supplying verbal links so that the thought flows smoothly from sentence to sentence. This is not, of course, a practical way of composing paragraphs in an actual writing situation; its value here is simply to accustom you to thinking of a paragraph as a structure of related ideas expressed in individual sentences.

A. *Topic:* There were several things I especially liked (disliked) about my high school.

> Sentence 1: Assertion
>
> ↓
>
> Sentences 2, 3, 4, 5: Specification
>
> ↓
>
> Sentence 6: Conclusion

B. *Topic:* Communication between parents and children can be extraordinarily difficult.

> Sentence 1: Assertion
>
> ↓
>
> Sentence 2: Restatement
>
> ↓
>
> Sentence 3: Restatement
>
> ↓
>
> Sentence 4: Reason

C. *Topic:* Learning to write well is not easy.

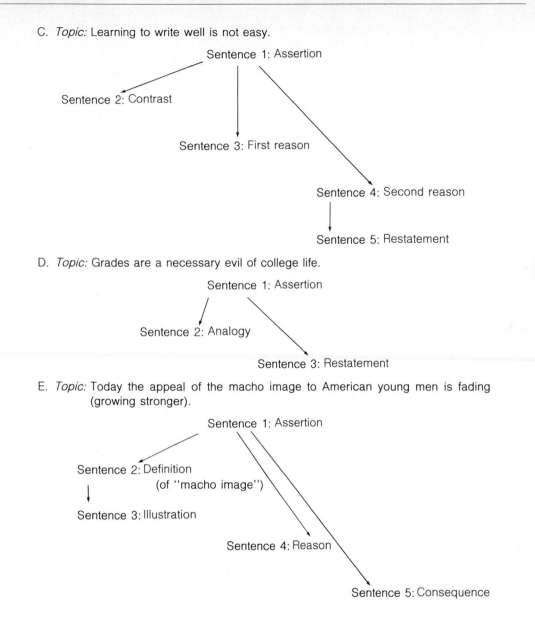

Sentence 1: Assertion

Sentence 2: Contrast

Sentence 3: First reason

Sentence 4: Second reason

Sentence 5: Restatement

D. *Topic:* Grades are a necessary evil of college life.

Sentence 1: Assertion

Sentence 2: Analogy

Sentence 3: Restatement

E. *Topic:* Today the appeal of the macho image to American young men is fading (growing stronger).

Sentence 1: Assertion

Sentence 2: Definition
(of "macho image")

Sentence 3: Illustration

Sentence 4: Reason

Sentence 5: Consequence

F. *Topic:* The present system of tenure in universities should (should not) be abolished.

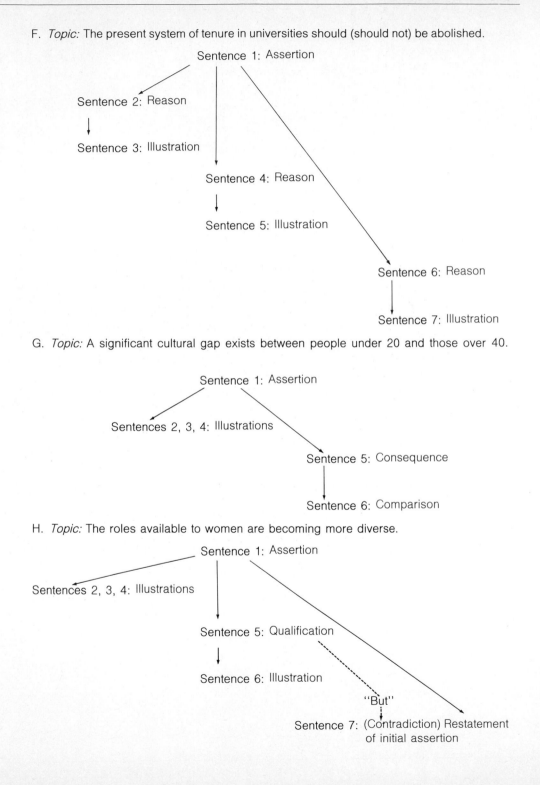

G. *Topic:* A significant cultural gap exists between people under 20 and those over 40.

H. *Topic:* The roles available to women are becoming more diverse.

CHAPTER 26

Sentence Patterns in the Paragraph

INTRODUCTION

The paragraph structures we looked at in the last chapter exist below the surface of the words, in the ideas or feelings which the words express. Another kind of structure lies closer to the surface in the sentence pattern of a paragraph. Roughly speaking, the sentences of a well-written paragraph will be enough alike to seem all of a piece, yet different enough to avoid monotony and to establish subtle shades of emphasis and meaning.

These complementary principles of similarity and variation in sentence length and style always operate together in well-composed paragraphs. However, they rarely operate in perfect balance. In some paragraphs similarity predominates; in others variation is the more important principle. And in many cases the sentences in one portion of a paragraph show considerable similarity, while those in another are more varied.

A paragraph is not simply a pattern of thought. It is also a pattern of sentences, and these must strike the ear as parts of an integrated whole. Moreover, their likenesses and variations ought to reflect the writer's purpose. Where, for instance, he or she is discussing logically parallel ideas—two examples of the same generalization, say—similar patterns in the sentences should reinforce the sameness of thought. On the other hand, a sentence expressing a climactic idea, an idea more important than what has gone before or will follow, should differ in pattern from the other sentences, perhaps shorter and simpler, perhaps longer and more complicated.

The possibilities of creating sentence structure within paragraphs are endless. Here we shall discuss and illustrate only a basic few. In the simplest case sentences of very similar form give unity to a short paragraph. When more variation is wanted, the most common patterns are symmetry, progression, and alternation.

SIMILARITY IN SENTENCE PATTERN

When the ideas of a paragraph are logically identical or parallel, a closer than usual similarity of sentence style often reflects the fact. Sometimes, for instance, a paragraph develops several causes of the same effect (or several effects of the same cause), halves of a contrast, a series of examples supporting the same generalization, several restatements of one basic idea, or a number of particulars making up the same scene. The last is the case in this paragraph describing the haul brought in by commercial fishermen:

> (1) Again the sixty fathoms of ropes, arm over arm. (2) This time the net was well filled. (3) It was too heavy to lift on board without the block and tackle. (4) Once on board, Michael untied the codstring and the deck became alive with struggling fish. (5) They slithered over one another. (6) They flapped, they gasped. (7) The rays lashed their long tails. (8) Dark crabs crept away and hid in the scuppers. (9) A cod gaped and gaped and gaped. (10) A starfish lay motionless. (11) Plaice, brill, whiting, gurnard, haddock everywhere.
>
> Robert Gibbings

Gibbings develops this paragraph by listing the details of the scene. Obviously the sameness of the sentences suits the writer's purpose. Yet close as the sentences are in their basic pattern, they have enough variety to avoid monotony. They range from fourteen words to four. The final seven sentences, the focus of the description, are most alike. But even these show variation. One consists of two clauses ("they flapped, they gasped"); another doubles the verbs ("crabs crept . . . and hid"); still another triples them ("gaped and gaped and gaped").

Moreover the author subtly links some of the variations to the progression of thought. The longest sentence and the one least like the others is the third. This sentence mediates between the two parts of the scene: the hauling in of the trawl and the spilling of the catch onto the deck. The first and last sentences are fragments (the only ones in the paragraph). This repetition serves to round off the paragraph, completing it by returning, in a subtle way, to the pattern with which it began.

This next example of similar sentence patterning comes from a more expository piece, an essay about teaching creative writing:

> (1) Now you'll see that this kind of approach eliminates many things from the discussion. (2) It eliminates any concern with the motivation of the writer except as this finds its place inside the work. (3) It also eliminates any concern with the reader in his market sense. (4) It also eliminates that tedious controversy that always rages between people who declare that they write to express themselves and those who declare that they write to fill their pocketbooks, if possible.
>
> Flannery O'Connor

Notice in O'Connor's paragraph how the second, third, and fourth sentences follow the same pattern. But notice also how she varies that pattern in the fourth sentence, which is much longer and ends with a brief elliptical clause ("if possible") marked off from the preceding words by a comma.

VARIETY IN SENTENCE STRUCTURE

Paragraphs like the two quoted above are uncommon. One cannot maintain a close similarity of sentence style very long without becoming monotonous. Usually, well-written paragraphs possess greater variety in their sentences, consisting of both long, relatively complicated statements and short, relatively simple ones. The variations, however, ought not to be random. They should reflect differences of emphasis, and they should create a form, a shape within which the thought is developed and contained.

SYMMETRICAL SHAPE

Symmetry means a shape in which one half mirrors or somehow repeats the other. Writers sometimes give short paragraphs a symmetrical sentence pattern of short-long-short:

(1) Always, from the very first sight of it, she hated the cottage. (2) She loathed the plain square red-brick box, its blue slate roof, the squalid confusion of currant bushes, black hen coops, falling fences and apple trees in sprawling decay that passed for a garden, the muddy pond at the foot of it and the three withered willows sticking nakedly up from the water, like grey arms caught and fossilized in the act of drowning. (3) Above all she hated the quiet clenching cold.

H. E. Bates

The rough symmetry of its sentence structure not only shapes that paragraph, it also establishes emphasis on the chief reason for the wife's dislike of her husband's vacation retreat—the "quiet clenching cold."

Here is a little longer example of symmetry, a paragraph describing the crash of the stock market in 1929:

(1) The Big Bull Market was dead. (2) Billions of dollars' worth of profits and paper profits—had disappeared. (3) The grocer, the window-cleaner, and the seamstress had lost their capital. (4) In every town there were families which had suddenly dropped from showy affluence into debt. (5) Investors who had dreamed of retiring to live on their fortunes now found themselves back once more at the very beginning of the long road to riches. (6) Day by day newspapers printed the grim reports of suicides.

Frederick Lewis Allen

While these sentences show considerable similarity, the movement from the short opening statement (six words) to the long sixth one (twenty-seven) establishes a controlled variety. Then the final sentence returns us to the first, both in its diction ("suicides" echoing "dead") and in its pattern (a short, simple sentence of ten words), completing a symmetry of sorts and closing the paragraph.

PROGRESSIVE SHAPE

Sometimes the key sentence of a paragraph will be the longest rather than the shortest. More often than not such a sentence appears at or near the end of the paragraph, and the sentence structure builds progressively toward it. The following passage, for example, builds toward the long third sentence, from which it falls in a final sentence comparable in length, though not in style, with the opening topic statement:

(1) The mechanization of modern war greatly reduces the power of the human conscience to keep its abuses in check. (2) It would be hard to induce a youth of ordinary good nature to take a woman with a baby in her arms and tear the two to pieces with a Mills bomb in full view of the explosion. (3) But the same youth, thousands of feet up in a war plane, preoccupied with the management of his machine and the accuracy of his aim, will release a bomb that will blow a whole street of family homes into smithereens, burning, blinding, mutilating scores of mothers and babies, without seeing anything of his handiwork except the glow of a conflagration which is as pretty as a display of fireworks. (4) The hospital surgeon sees what the pilot has done; but it is the pilot and not the surgeon who releases the bomb.

Bernard Shaw

The next example shows an even more climactic sentence structure, the longest statement coming at the very end:

(1) By adding the third character to tragedy, Sophocles changed the nature of drama. (2) By exalting the chorus and diminishing the actors, television has changed entirely the nature of our continuing history. (3) Watching things as they happen, the viewer is a part of events in a way new to man. (4) And never is he so much a part of the whole as when things do not happen, for, as Andy Warhol so wisely observed, people will always prefer to look at something rather than nothing; between plain wall and flickering commercial, the eyes will have the second. (5) As hearth and fire were once center to the home or lair so now the television set is the center of modern man's being, all points of the room coverge upon its presence and the eye watches even as the mind dozes, much as our ancestors narcotized themselves with fire.

Gore Vidal

ALTERNATING SHAPE

We have seen paragraphs in which the sentence structure follows either a roughly symmetrical plan (short-long-short) or a more climactic one (short-longer-longest). A writer can also alternate between longer and shorter statements. The following paragraph closes a book about the inequities of world economics. Notice how the writer effectively uses short statements for emphasis:

(1) If we want to spread the revolution of liberty round the world to complete and reconcile the other great revolutions of our day, we have to re-examine its moral content and ask ourselves whether we are not leaving liberty as a wasted talent and allowing other forces, not friendly to liberty, to monopolize the great vision of men working in brotherhood to create a world in which all can live. (2) But God is not mocked. (3) We reap what we sow and if freedom for us is no more than the right to pursue our own self-interest—personal or national—then we can make no claim to the greatest vision of our society: "the glorious liberty of the sons of God." (4) Without vision we, like other peoples, will perish. (5) But if it is restored, it can be as it always has been the profoundest inspiration of our society, and can give our way of life its continuing strength.

Barbara Ward

CONCLUSION

Well-written paragraphs exhibit both a solid structure of idea and a subtle pattern of sentences. The sentence structure, closer to the surface, evolves from the deeper organization of thought and creates a sense of form which pleases in itself and focuses the writers' ideas and feelings.

The several basic patterns we have looked at—similarity, symmetry, progression, alternation—are, of course, often combined. There is no reason why the sentences of a paragraph must fall into one and only one of these patterns. To see how a paragraph may combine at least two kinds of structure, study this passage:

(1) New York has changed in tempo and in temper during the years I have known it. (2) There is greater tension, increased irritability. (3) You encounter it in many places, in many faces. (4) The normal frustrations of modern life are here multiplied and amplified—a single run of a crosstown bus contains, for the driver, enough frustration and annoyance to carry him over the edge of sanity: the light that changes always an instant too soon, the passenger that bangs on the shut door, the truck that blocks the only opening, the coin that slips to the floor, the question asked at the wrong moment. (5) There is greater tension and there is greater speed. (6) Taxis roll faster than they rolled ten years ago—and they were rolling faster then. (7) Hackmen used to drive with verve; now they sometimes seem to drive with desperation, toward the ultimate tip. (8) On the West Side Highway, approaching the city, the motorist is swept along in a trance—a sort of fever of inescapable motion, goaded from behind, hemmed in on either side, a mere chip in a millrace.

E. B. White

In its ideas the paragraph is organized into two parts, both established by the opening topic statement. The first part—set up by sentence 2 and including 3 and 4—stresses the increased irritability of New Yorkers. The second—set up by sentence 5 and including 6, 7, and 8—stresses the city's greater speed. Within each of these parts the sentences vary considerably as they build progressively toward greater length and complexity. Thus the discussion of growing irritability climaxes in the long, complicated fourth sentence. But while variation occurs among the three or four sentences comprising each half of the paragraph, the halves themselves repeat, in a general way, the same progression from shorter, simpler statements to longer, more complicated ones. Sentence 5, for example, is like sentence 2, and 8 is like 4.

Of course, the similarity between the two groups of sentences is far from exact. Nor should it be exact. Subtlety is here a virtue. Without making the organization obvious, the writer has used sentence structure to create a formal unity closely related to the deeper structure of idea. The increasing complexity of the sentence patterns within each part reflects the accelerating tempo and frustrations of city life. The similarity between the two parts suggests the close connection between a faster mode of life and a constantly more irritating one.

EXERCISE

1. Study the following paragraphs and be able to discuss how each exemplifies the various kinds of sentence patterns discussed in this chapter.

A. (1) I do not yet have the discipline to sit at my desk at an appointed hour each day. (2) I do not know how to seek out my ideas. (3) I must wait till my thoughts come to me. (4) My mind is an uncharted sea. (5) I am like a beachcomber waiting for the waves to wash up shells of ideas. (6) The creativity works independently of what I would like it to do. (7) I cannot force it. (8) I must be patient.

Anne Lasoff

B. (1) The romances of chivalry were tales of love and adventure, in verse or prose. (2) Directed toward an audience of nobles, they glorified the aristocratic way of life and contained long descriptions of luxury, furniture, accessories. (3) They exalted women, who were the poets' patrons and their most receptive public. (4) They exalted also the institution of courtesy, a code of morals and ideals for gentlemen and ladies. (5) Most of all they exalted love, "the origin and foundation of all that is good."

Morris Bishop

C. (1) So the darkness comes on, covering the graves and the withered cedar and the nameless dead. (2) Lights wink on around the Sound. (3) Rats stir in the weeds, among the graves. (4) The smoke still ascends in the night, clean and without guilt, borne like passion with the last dust of the nameless and the unremembered, upward and upward, toward the stars.

William Styron

D. (1) The homesteader got most of his outside items through mail-order catalogues, including, sometimes, his wife, if one could call the matrimonial papers, the heart-and-hand publications, catalogues. (2) They did describe the offerings rather fully but with, perhaps, a little less honesty than Montgomery Ward or Sears Roebuck.

(3) Unmarried women were always scarce in new regions. (4) Many bachelor settlers had a sweetheart back east or in the Old Country, or someone who began to look a little like a sweetheart from the distance of a government claim that got more and more lonesome as the holes in the socks got bigger. (5) Some of these girls never came. (6) Others found themselves in an unexpectedly good bargaining position and began to make all kinds of demands in that period of feminine uprising. (7) They wanted the husband to promise abstinence from profanity, liquor, and tobacco and perhaps even commanded allegiance to the rising cause of woman suffrage. (8) Giving up the cud of tobacco in the cheek was often very difficult. (9) A desperate neighbor of ours chewed grass, bitter willow and cottonwood leaves, coffee grounds, and finally sent away for a tobacco cure. (10) It made him sick, so sick, at least in appearance, that his new wife begged him to take up chewing again. (11) Others backslid on the sly, sneaking a chew of Battle Axe or Horseshoe in the face of certain anger and tears.

Mari Sandoz

2. Compose a paragraph on each of the following models:

   A. A paragraph of four or five short sentences similar in style on a topic connected with your school work. Aim for a clear similarity but with sufficient variation to avoid monotony.
   B. A paragraph of three or four sentences following a symmetrical pattern: short sentence-long sentence(s)-short final sentence similar to the first.
   C. A paragraph in which you work progressively toward a long final statement.
   D. A paragraph of five or six sentences alternating between long and short statements.

PART FOUR

THE SENTENCE

INTRODUCTION

The first of the two sections which make up Part Four discusses the rudiments of the sentence, beginning with a definition and then distinguishing the grammatical kinds of sentences. The second section concerns sentence style, such matters as the rhetorical kinds of sentences, concision, emphasis, and rhythm.

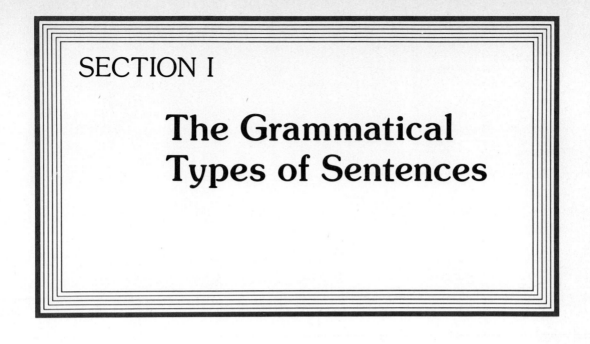

SECTION I

The Grammatical Types of Sentences

CHAPTER 27

The Simple Sentence

INTRODUCTION

Defining a sentence is difficult, and definitions vary according to the use to which they will be put. In composition we understand a sentence to be a group of words which contains a subject and a finite verb and which is grammatically complete.

In addition to subject and verb most sentences also contain other elements—complements and modifiers of various kinds and, less often, appositives and absolutes. We survey the kinds of words and constructions which can serve as subjects, complements, modifiers, appositives, and absolutes in the Reference Grammar beginning on page 720. Here you need only understand a bit more fully what is meant by the description of the compositional sentence given in the first paragraph.

GRAMMATICALLY COMPLETE

Grammatically complete means that a sentence stands by itself without depending upon another statement for its grammatical function. In other words, it cannot serve as an adjective or adverb or as the subject or complement of anything else.

If we were to see the word *perfectly* printed by itself, we should be puzzled. We know what the word means, but all alone it makes no sense; it is not grammatically meaningful. Of course, we do not often encounter a word in complete isolation. Even if it were printed as a separate statement, it would probably occur in the context of other words:

Do you understand?
Perfectly.

In such cases we automatically supply the elements necessary to turn the word into a formal sentence:

[I understand] Perfectly.

When we do come upon a single term isolated from other words, it will make sense only in a situation that gives it grammatical significance, like STOP painted on a red or yellow octagonal sign at a street corner:

[You must] STOP [your car here.]

But while in speech and on signs we often use individual words as though they were complete sentences, relying upon the context to supply their grammatical meaning, the general rule in composition is that a sentence should not depend for its grammatical value upon anything outside itself. A statement punctuated as a sentence but failing to satisfy the rule of grammatical completeness is called a *fragment*. In Chapter 30 we glance at the common kinds of fragments.

FINITE VERB

The second requirement of a sentence is that it contain a finite verb. A *finite verb* is one limited with reference to number, person, or tense. For example, we say "I walk" but not "he walk," and we say "he walks" but not "I walks": the forms *walk* and *walks* are finite, or limited, verbs. The English verb does have nonfinite forms called infinitives and participles (see Reference Grammar, pages 680–82). The present participle *walking* is an example: we may use it with any person, in either singular or plural, and with reference to today, yesterday, or tomorrow.

By convention, however, a formally complete sentence is expected to have a finite verb. "John walks" is a sentence; "John walking" is not. A participle or an infinitive may occur as part of a verb phrase, but in a true sentence such a phrase will itself be a finite form because it contains auxiliaries like *is, has, does, will,* and so on, as in "John is walking," "John has walked," "John will walk."

GRAMMATICAL SIMPLICITY

Sentences are classified according to the number and type of clauses they contain, that is, whether the clauses are independent or dependent. We can distinguish three basic kinds: simple, compound, and complex. In this chapter we study the first of these.

A *simple sentence* consists of one independent clause. The sentence has one subject-verb connection, as in:

Everybody in the room *stood up.*

The definition, notice, does not say "one subject and one verb," but rather "one subject-verb connection." That connection may involve two or three subjects and two or three verbs (in theory any number of either), but as long as the connection itself is single, the sentence is simple. "All the men and women in the room stood up and cheered" is a simple sentence. It has two subjects ("men" and "women") and two verbs ("stood up" and "cheered"), but there is only one connection, or "nexus" as grammarians call it, a fact we diagram like this:

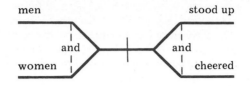

Most simple sentences, however, are short and uncomplicated and do not pile up subjects and verbs.

Simp. s.
awk.
THE AWKWARD SIMPLE SENTENCE

You can make two mistakes with the simple sentence: you can use one where it doesn't work very well, and you can fail to use one where it does work well. The first is the more common error. In a third-grade reader a simple sentence style is necessary. But when writing deals with more complicated subjects than "See Jane run," a series of simple sentences may prove too simple. Primer style not only sounds immature; it fails to establish the subtle relationships of thought and feeling that more complicated sentences make clear. Consider this passage:

> My uncle teaches physics. He is short and stout and about forty-five. He has no personality at all.

Unless very skillfully handled, writing like this has no focus. It comes across as a series of assertions, none of which has primary importance. In a longer, single sentence the writer can make one of the points paramount and gather the others around it. This example permits various possibilities:

1. A teacher of physics, short, stout, about forty-five, my uncle has no personality at all.
2. A teacher of physics with no personality at all, my uncle is short, stout, and about forty-five.
3. My uncle is a teacher of physics—short, stout, about forty-five, and with no personality at all.

Simp. s.
THE EFFECTIVE SIMPLE SENTENCE

But while a sustained simple style does not work very well in college composition, individual simple sentences have many uses. The short sentence provides special clarity, or emphasis, or variety to a series of longer, more complicated statements.

The simple sentence is effective in setting up a topic at the beginning of a paragraph, as in this passage about the invasion of Anzio Beach in World War II:

> *The German reaction, as always, was brisk.* Without withdrawing any troops from the Cassino front, General Kesselring dispatched everything available to contain the beachhead while a major force could be sent to the scene. This force soon began to materialize, thanks to the defensive procedure that had been prepared for just such a contingency.
>
> Fred Majdalany (italics added)

Not only does Majdalany's simple sentence announce the topic clearly; it also reinforces the idea. Like the military response it describes, the sentence is brisk and strong. In good writing the *form* of a sentence ought as much as possible to imitate the *sense*. (This principle applies, of course, to all kinds of sentences, not merely to those which are brief and simple.)

Another advantage of the simple sentence is that it creates emphasis and variety, as it does in the following discussion of the English poet William Cowper:

> Cowper's madness finds its origin far deeper in the sufferings of childhood, it may be in inherent physical defect. *All his life it hung over him.* And religion, so far from being the cause, was the most considerable of the remedies by which he tried to get rid of it. *It failed.*
>
> Lord David Cecil (italics added)

Cecil's short sentences give added significance to what they express, especially by virtue of being set in the context of longer ones. The fact that Cowper's efforts to avert madness by means of religious devotion failed is the final and tragic fact of his life. It is important. See how that importance is muted, almost thrown away, if the final two words are combined with the preceding statement:

> And religion, so far from being the cause, was the most considerable of the remedies by which he tried to get rid of it, but it failed.

EXERCISE

1. Indicate the subject(s) and verb(s) in the following sentences. One of them is not a simple sentence; identify it, and explain why.

   A. Harry felt disappointed.
   B. The father, his two sons, and several cousins left this morning on a hunting trip.
   C. She went to town, visited her mother in the hospital, and did the shopping on her way home.
   D. The town was run-down and everything looked seedy.

E. Men, women, and children traveled together, shared the work, and helped one another in difficult times.

2. The following passages have been rewritten simply to make a single statement out of what are actually two or more. No words of the originals have been changed, deleted, or rearranged. See if you can revise each passage so as to free the short sentence, restoring it to its original form and effectiveness.

A. It was assumed that he would face the issue without alarm or weakening and he did.

Samuel Hopkins Adams

B. As Thompson and the *Transcript* man had said, Vanzetti was naturally and quietly eloquent and disturbing so he was electrocuted.

Phil Strong

C. And that, by every known precedent, should have been that, but it wasn't.

John Lardner

D. But as the years go by they will realize that they are workers, that these are their jobs, that this is their one and only life and something must be done about it.

Barbara Garson

E. For various reasons, Americans *do* take to machines, perhaps more to machines than to tools, more easily than do Europeans, much more easily than do Asians or Africans. This, combined with the national impatience with precedent, the readiness to do anything at least once, was one of the secrets of American wealth and power and it still is.

D. W. Brogan

F. Again, it's an incontrovertible fact that, in the past, when contraceptive methods were unknown, women spent a much larger proportion of a much shorter life pregnant, or nursing infants whom they had borne with little or no medical help, and don't believe that that's a natural, healthy thing for human beings to do, just because animals do it because it isn't.

Elizabeth Janeway

G. The glorification of one race and the subsequent debasement of another—or others—always has been and always will be a recipe for murder and there is no way around this.

James Baldwin

H. It was not: *The Crown Prince is dead—long live the Crown Prince!* It was: *The Crown Prince is dead* . . . and then silence and emptiness as the grandees scatter, maggots twitch, and the cold wind blows.

Frederic Morton

3. In your next writing assignment compose at least one topic statement in the form of a short simple sentence. Elsewhere in the theme use two or three simple sentences to emphasize important points.

CHAPTER 28

The Compound Sentence

A compound sentence consists of two or more independent clauses. An independent clause contains a subject and finite verb. While in theory a compound sentence could have any number of such clauses, most have only two or three. These may be joined in either of two ways: by coordination or by parataxis. *Coordination* means that the clauses are linked by one of the coordinating conjunctions: *and, but, for, or, nor, yet,* or the correlatives *either . . . or, neither . . . nor, not only . . . but (also), both . . . and:*

> The car had broken down, and the driver had pushed it to the side of the road.

Parataxis means that the clauses are simply pushed together without a conjunction. To avoid confusion a strong pause must be marked between the clauses, conventionally by a semicolon, less commonly by a comma or a colon:

> The car had broken down; the driver had pushed it to the side of the road.

Although one may use either coordination or parataxis in any compound sentence, each results in a different tone and emphasis. In any given case, one or the other probably proves more effective.

People commonly make several elementary mistakes with compound sentences.

AWKWARD COORDINATION

Coordinated ideas should be compatible and equally important. *Awkward coordination* means that they are not. There is no ready way of judging when ideas are unequal or incompatible; much depends upon context. When thoughts are properly coordinated the reader senses an underlying unity, as in this sentence:

> The building had been broken into, and a number of things had been stolen.

We understand the stealing to be a consequence of the breaking in. Or in the sentence

> Harry likes jazz, but his sister hates it,

we realize that brother and sister are being compared with reference to the same thing.

On the other hand, the incongruity of thought puzzles us in statements like these:

> The building had been broken into, and its floors were made of marble.

> Harry likes jazz, but his sister is a Girl Scout.

The ideas simply do not go together. However, in judging incongruity we must be guided by context. We could imagine a situation, for instance, in which the second of these seemingly illogical sentences makes sense: the writer might be slyly implying that Girl Scouts do not appreciate jazz.

Overcoord. OVERCOORDINATION

Overcoordination means that a sentence contains too many coordinated clauses. Two independent clauses in a single sentence will serve most purposes; three are generally a practical limit. More than three may confuse a reader by throwing out too many ideas too fast, by obscuring the connections among thoughts, or by failing to establish appropriate emphasis. For example:

> Some of us make a real chore of studying for examinations, and the last night before a test we sit down and cram for ten or twelve hours and probably we learn nothing and we fail the test.

Overloading causes this passage to be less effective than it could be. It needs to be pruned and rephrased as two or even three sentences. For example, we can take out "sit down" (who "crams" standing up?) as well as "test" (it is clearly implied):

> Some of us make a real chore of studying for examinations. The last night before a test we cram for ten or twelve hours. Probably we learn nothing, and we fail.

Dividing the passage into three sentences helps to clarify and emphasize the writer's point: the expense of effort and the poverty of result.

USE PARATAXIS

Two clauses are sometimes more effective when joined paratactically, that is, with a semicolon and without the conjunction. Usually coordination is preferable, but now and then parataxis is clearer and more emphatic. One case is when the second clause merely repeats the first. Here a semicolon is more appropropriate than *and* (which might mislead by suggesting movement to a new idea). The following sentence momentarily confuses by misusing *and* in this way:

> The Coast Guard has withdrawn the keeper from another of Maine's lighthouses, and after more than a century of manual operation the light on Mt. Desert Rock is now fully automated.

A semicolon would signal more clearly that the oncoming clause will restate the point of the first. In the sentence below, Mark Twain uses the semicolon for that purpose:

> The humorous story is told gravely; the teller does his best to conceal the fact that he even dimly suspects that there is anything funny about it. . . .

Another place where a semicolon may be better than a conjunction is between contrasted or contradictory ideas. Parataxis stresses the difference:

> Harry likes jazz; his sister hates it.

> Groups are capable of being as moral and intelligent as the individuals who form them; a crowd is inchoate, has no purpose of its own and is capable of anything except intelligent action and realistic thinking.
>
> Aldous Huxley

In either of these examples the addition of *but* might make a smoother sentence, but it would be a less dramatic one.

Sometimes parataxis also works effectively between clauses expressing a cause-effect relationship, emphasizing the logic by forcing the reader to fill it in for him or herself:

> The men slowly filed from the room; they had no reason to remain.

In the following passage the pause induced by the semicolon and the absence of a word such as *for* or *because* strengthen the tone of cynical wit:

> I never read a book before reviewing it; it prejudices a man so.
>
> Sidney Smith

Finally, a series of short, similar clauses expressing particular instances of a general idea are often better handled by parataxis than by coordination, as in this sentence by the essayist and novelist Virginia Woolf (she is mocking an English tourist's reaction to Paris):

> The habits of the natives are disgusting: the women hawk on the floor; the forks are dirty; the trees are poor; the Pont Neuf is not a patch on London Bridge; the cows are skinny; morals are licentious. . . .

In all these cases parataxis makes for a stronger, more forceful linking of independent clauses than coordination would. This fact does not mean that a semicolon is always better than *and* or *but* or *for*. On the contrary, coordination is generally preferable. It is only occasionally that you need to be emphatic. But when that necessity arises, remember that parataxis exists as an option.

EXERCISE

1. Which of these sentences are compound? Identify those which are coordinated and those which are paratactic. If you think any sentence is not compound, be able to explain why.

    A. Many women and quite a few men voted for him.
    B. Many women voted for him, and quite a few men did too.
    C. Not everyone was convinced; I wasn't.
    D. They did not like the idea, but they agreed.
    E. The books are not at the bookstore yet; we'll just have to get along without them.

2. In which of these sentences does the coordination seem awkward?

    A. It was late and everybody looked tired.
    B. It was late but everybody looked tired.
    C. Jane likes to dance but her sister dances beautifully.
    D. The building had burned down and the day was sunny.

3. This sentence is overcoordinated. Revise it in two or three shorter statements:

> The city was large and inhospitable, and crowds of people hurried by on the streets, and they didn't look at me or even see me, and I felt lonely and frightened.

4. Each of the following sentences would be slightly improved if a semicolon were used in place of the *and* or *but*. Explain why.

    A. The game ended in a tie, and athletic contests often seem inconclusive.
    B. Coordinated ideas should be compatible, and these are not.
    C. People are frightened, and they are terrified.
    D. Not everyone likes football, and my brother is an example.

5. Compose a paragraph on a topic of your choice. Use a short simple sentence for the topic. Elsewhere in the paragraph include three or more compound sentences, at least one of which should be paratactic.

CHAPTER 29

The Complex Sentence

The complex sentence has one independent and at least one dependent clause. (A dependent clause contains a subject and finite verb but cannot stand alone.) The independent construction is called the main clause; the dependent one, the subordinate clause. A complex sentence may have two, three, or more subordinate clauses, although by definition it can have only one main clause. A subordinate clause always fulfills some grammatical function within a complex sentence: as a subject, complement, object of a preposition, or modifier.

       sub. adverbial clause   main clause
1. When the car broke down, the driver pushed it off the road.

      main clause   sub. noun clause, direct obj. of "told"
2. Sarah told everyone that she would go.

       main clause   sub. adj. clause modifying "boy"
3. Mike lent his bike to the boy who lived next door.

    sub. clause, acting as sentence adverb   main clause
4. Although it wasn't true,   Mike told everyone

    sub. noun clause, direct object of "told"
that he had lent his bike to the boy

    sub. adj. clause modifying "boy"
who lived next door.

The complex sentence develops more complicated patterns of thought, feeling, or experience than simple or compound sentences. A good example is this sentence by the novelist Emily Brontë:

Being unable to remove the chain, I jumped over, and, running up the flagged causeway bordered with straggling gooseberry

> bushes, knocked vainly for admittance, till my knuckles tingled and the dogs howled.

We could imagine this in a number of shorter statements:

> I was unable to remove the chain. So I jumped over. Then I ran up the flagged causeway. It was bordered by straggling gooseberry bushes. I knocked vainly for admittance. My knuckles tingled and the dogs howled.

This series of simple sentences (the last is compound) lacks two important qualities of the original: synthesis and focus. Brontë's sentence creates a single experience out of several distinct elements: struggling with the chain, jumping, running, seeing the flagstones and the bushes, feeling the tingle in the knuckles, hearing the dogs. But while the elements remain distinguishable, Brontë synthesizes them into a complex whole. These individual actions and perceptions have been ordered and focused. The jumping and, even more, the knocking are the centers of the experience to which she subordinates the other elements. When that subordination is broken up, the sense of the passage changes. Brontë's sentence, for example, casts only a brief glimpse at the gooseberry bushes as the narrator hurries by; the revision stops and looks at them quite unnecessarily.

A complex sentence is not always better than several simple ones. The point here is simply than a complex sentence conveys shades of meaning that elude a simpler style. The opposite is also true: some subtleties of meaning and emphasis are better conveyed by simple sentences. We shall return to this matter when we study stylistic varieties of the sentence.

For now it is enough to understand what a complex sentence is and some of its pitfalls. Precisely because the complex sentence makes greater use of the different kinds of phrases and clauses and thus offers wider opportunities for expression, it is more difficult to handle. A few basic principles will help.

SUBORDINATE IDEAS OF LESSER IMPORTANCE

Expressing something in a subordinate construction usually implies that it has less significance than what appears in the main clause. It is a mistake, then, to use a main clause or a separate sentence for an idea to which the context assigns minor importance. That is what happens in this passage:

> The first two or three weeks of college are the hardest. Some students do not feel the difficulty. But during these weeks everything seems strange—the professors, the lectures, the textbooks, even your classmates.

The problem appears in the second sentence—not in its grammar but in the fact that it raises a minor point to primary importance. The writer's topic, as the initial sentence makes clear, is the initial strangeness of college. To say in a separate sentence that some students do not experience this strangeness misleads the reader. The qualification is worth making, but it should be subordinated:

> Although some students do not feel the difficulty, the first two or three weeks of college are the hardest.

No sub.

DO NOT SUBORDINATE IDEAS OF PRIMARY IMPORTANCE

Subordinating a major idea is the opposite error to not subordinating a minor one. In the following case, for instance, the writer develops the argument that young people cannot afford to buy homes:

> Most young couples still dream of owning their own home, although for many that dream is becoming increasingly difficult to realize. The effect of inflation upon the costs of labor and building materials and mortgages means that fewer and fewer people can afford to become home owners.

The primary topic is buried in the subordinate *although*-clause. It ought to be in a main clause or a separate sentence. Any of these revisions focuses upon the topic more clearly:

1. Most young couples still dream of owning their own home. For many, however, that dream is becoming increasingly difficult to realize.
2. Most young couples still dream of owning their own home, but for many that dream is becoming increasingly difficult to realize.
3. Although most young couples still dream of owning their own home, that dream is becoming increasingly difficult for many to realize.

Reduce sub.

REDUCE SUBORDINATION TO THE BRIEFEST FORM THAT CLARITY REQUIRES

So far we have spoken of subordination only in terms of clauses. But phrases and even single-word modifiers act as subordinate elements and express secondary ideas. Independent clauses, dependent clauses, phrases, and individual words form a rough scale of diminishing importance. These versions of the same idea illustrate the point:

> *The settlers began.* They had great hopes.
>
> *When they began,* the settlers had great hopes.
>
> *At first,* the settlers had great hopes.
>
> *Initially,* the settlers had great hopes.

How much emphasis you want will determine how much weight you give to a secondary idea. In general, use as brief a subordinate construction as clarity and emphasis allow. Often, for example, an adjectival or adverbial clause can be reduced to a participle or a participial phrase with no loss of clarity. Instead of the full clause "that is needed" in the following sentence, the writer could go directly to the participle:

The material ~~that is~~ needed for the course is free.

You can often use reduction with adverbial clauses:

If a student wishes to change his course, he must see his adviser.

More concise A student wishing to change his course must see his adviser.

Using a word or a phrase in place of a full clause to reduce a subordinate idea to a lower degree does not always improve a sentence. But the possibility always exists, and it may enable you to avoid messy sentences with too many subordinate clauses that crowd each other and confuse the reader.

ARRANGE SUBORDINATE CONSTRUCTIONS IN NATURAL ORDER IF POSSIBLE

Order of Sub.

When you combine clauses and phrases to make a complex sentence, you must pay attention to several different principles of arrangement: (1) the conventions of English grammar, (2) the subject's natural sequence of events or order of thought, (3) the shade of emphasis you want, and (4) considerations of rhythm. Emphasis and rhythm we shall discuss in Chapter 39. Here we want to look at sequence of ideas or events and how this relates to grammatical order.

To indicate grammatical function, our language relies to a considerable degree upon word arrangement. For instance, the five words in "The bear killed the man" mean something very different if we rearrange the subject and complement to say "The man killed the bear." Because of the importance of word order, English has many rules restricting the placement of various kinds of words, phrases, and clauses. Thus we put adjectival clauses and prepositional phrases after the nouns they modify: we write "the boy who lives in Chicago" or "the house on the hill," not *"the who lives in Chicago boy" or *"the on the hill house."

However, English word order is far from rigid, and a writer often has options for placing subordinate constructions. While adjectival prepositional phrases and clauses are relatively immovable, participial adjectives and adverbial constructions shift more freely:

Running down the street, the boy fell and hurt himself.
The boy, running down the street, fell and hurt himself.
The boy fell and hurt himself running down the street.

When the sun came up, the men went home.
The men, when the sun came up, went home.
The men went home when the sun came up.

When you can, arrange a complex sentence in a sequence that reflects the actual order of events or ideas. The sentence by Emily Brontë quoted on page 218 is worth studying in this respect. So is this one, a bit of satire about a fictional medieval pope who tried a stroll upon the water:

> Having wanted to walk on the sea like St. Peter, he had taken an involuntary bath, losing his mitre and the better part of his reputation.
>
> Lawrence Durrell

Technically Durrell's sentence is simple since it has only one subject-finite verb connection ("he had taken") and uses participial phrases as subordinating constructions. Still it nicely illustrates the strategy of following natural order: first the desire, then the action, finally the result.

It is not always easy to arrange a sentence so that it reflects the pattern of thought or of history and at the same time follows the rules determining word order. Sometimes the two types of order clash, and when they do, grammatical structure takes precedence. As we shall see, emphasis or rhythm may require that natural order give way. But when none of these considerations prevents it, this is a good rule to apply: Follow the actual sequence of ideas or events.

THE COMPOUND-COMPLEX SENTENCE

In addition to these three basic types of sentences, a fourth kind results when you combine two of the others. This is the compound-complex sentence. It contains at least two independent clauses and at least one dependent clause:

independent clause   independent clause
The car had stalled, and the driver had pushed it to the side of

dependent adj. clause modifying "road"
the road where it would not block traffic.

As with a compound sentence, be sure that the main clauses of a compound-complex construction fit together coherently and are equally important. As with the complex sentence, be sure that the assignment of ideas to independent and dependent status reflects their true significance and that the clauses appear in a reasonable order.

EXERCISE

1. Identify the complex sentences in the following group and explain in each case why the sentence is or is not complex:

    A. If you see him, say hello.
    B. She saw him, and she said hello.
    C. I like Chicago, although I haven't been there often.
    D. I like Chicago, but I haven't been there often.
    E. Whoever said that doesn't know what he's talking about.
    F. All those who live on the north side will be excused.

2. Revise the following passages so as to turn each into a single complex sentence. You may select any construction as the main clause and arrange the others around it.

    A. It was summer. The beaches were crowded. Many people had less money to spend. This was because of inflation and unemployment.
    B. The books lay in a heap. They were in front of an open window. There was a danger of their being damaged by rain. No one closed the window. No one moved the books.

3. In the light of the sentences that follow it, what is awkward about the subordination of the first sentence of the following passage? Revise it to correct the awkwardness.

    Although college is different, it is not overwhelming. The professors cover more ground than high school teachers, and the textbooks are harder. But the most striking difference is that you are given responsibility for learning; no one nags you to read the assignment or to turn in homework.

CHAPTER 30

The Fragment

Frag.

When a word, a phrase, or a dependent clause stands alone as a sentence—that is, when it begins with a capital letter and ends with a period, a question mark, or an exclamation point—it is called a *fragment*. Fragments do not meet the criterion of grammatical independence which is required of a compositional sentence. We shall later see how fragments may be used effectively, though their value depends upon their relative rarity. More often, however, fragments result from uncertainty about what a sentence is and constitute a serious fault. It is with such pointless fragments that we are concerned here.

You can correct a fragment in either of two ways. You may keep it as a separate statement and change or remove whatever prevents it from being a true sentence. Or you may retain the wording of the fragment but, instead of punctuating it as a sentence, attach it to what it grammatically depends upon—usually something in the preceding or the following sentence.

Here are the common kinds of fragments. (In the examples the fragment appears in italics.)

Frag.
adv. cl.

THE DETACHED ADVERBIAL CLAUSE

It was very late. *When the party broke up.*

The fragment contains a finite verb ("broke up"), but the subordinator "when" keeps the construction from being independent, turning it instead into an adverbial modifying the clause that precedes it.

A detached adverbial clause may be corrected either by attaching it to what it modifies:

It was very late when the party broke up,

or by removing the subordinator and allowing the clause to stand alone:

It was very late. The party broke up.

The alternative solutions, while each results in a true sentence, have slight differences of meaning. Sometimes one, sometimes the other is preferable, depending upon what you wish to say. Here, for instance, the second choice implies a pattern of cause and effect—the party broke up because it was late. "It was late when the party broke up" implies nothing of the sort.

THE DETACHED PARTICIPLE

Frag.
part.

A few people stayed. *The rest of us going home.*

The trouble with "The rest of us going home" is that the verb "going" is not finite; it is a participle. The fragment could be tacked on to the first clause (it would then be what is technically called a nominative absolute):

A few people stayed, the rest of us going home.

Or its separateness could be justified by changing the participle to the appropriate finite verb:

A few people stayed. The rest of us went home.

Here again the options are not exact equivalents. The first focuses more on those who stayed; the second lays more stress on the fact that "the rest of us" left.

THE DETACHED ADJECTIVAL CLAUSE

Frag.
adj. cl.

Everyone left except John. *Who decided to stay longer.*

"Who decided to stay longer" is not independent. Rather it is an adjectival clause modifying "John." One correction is:

Everyone left except John, who decided to stay longer.

The other:

Everyone left except John. He decided to stay longer.

How do these alternatives differ in emphasis?

THE VERBLESS STATEMENT

Frag.
no verb

All people whether they live in the city or in the country.

Here modifiers surround a noun ("people"). But this noun, presumably the would-be subject of a sentence, has no verb; the writer never predicates anything about "people." In cases like this the correction may require extensive revision, although sometimes, if the noun is followed by a modifying clause, the verb of the clause may be adapted as the main verb:

All people live in the city or in the country.

In this instance, although this correction results in a true sentence, it sounds simple-minded. It would be better for the writer to think out what he or she wants to say about "people" and to add the appropriate predication, something like this perhaps:

All people, whether they live in the city or in the country, want the conveniences of modern life.

EXERCISE

Which of the following statements are fragments? Explain why they are fragments and revise them so that they become true sentences.

    A. In the morning when the sun came up. The party broke camp.
    B. Most people are honest. Making an effort, for example, to find the owner of a wallet they pick up on a busy street.
    C. That girl is very nice. The one you introduced me to.
    D. College is not so difficult. If you don't let your work pile up.
    E. Not everyone likes football. My brother, for instance.
    F. Older people who lived through the depression and the second world war and experienced great changes in our society.
    G. The boy climbing the tree. That's my cousin.
    H. Although he wasn't at fault. Everybody blamed him.
    I. Everybody blamed him. But he wasn't at fault.
    J. That man running down the street. He stole this lady's purse.

SECTION II

Sentence Style

CHAPTER **31**

The Serial Sentence

INTRODUCTION

In effective sentences the component words, phrases, and clauses are put together not only according to the grammatical principles of coordination and subordination, but also according to certain stylistic principles. There are three basic ones: serial structure, parallel and balanced structure, and hierarchic structure.

We shall look at these kinds of sentences in the next several chapters, but first, a cautionary note. While different, none of these sentence types is better or worse than the others in any inherent, absolute sense. Each works well for some purposes, poorly for others. And none represents an ideal style. A skillful writer uses them all well, knowing when one works better than another.

A second warning: with regard to style, classifications are rarely adequate. The reality is too subtle. The discussion that follows, like all attempts to classify sentences, cannot avoid oversimplification and arbitrariness. The English sentence, like Shakespeare's Cleopatra, possesses infinite variety. It cannot be reduced to rules. And that is a good thing.

THE SEGREGATING STYLE

Serial structure means that a sentence is built by the simple addition of more or less equal units, one after the other. It exists in three basic varieties: the segregating, the freight-train, and the cumulative sentence.

In its purest form the segregating style consists of a series of short sentences, each consisting of a single idea.[1] "Idea" is difficult to define;

---

1. To be exact, it is inaccurate to speak of the segregating style as a type of serial structure. Although it consists of a series, the units are not parts of a longer sentence; rather each is a simple sentence, complete in itself. However, we shall stretch the category "serial structure" a little and include the segregating style here.

229

here it will mean simply one subject + one predicate: "The night was dark" thus contains one idea. Most sentences, of course, consist of several ideas: "The night was dark, and it was lonely"; or, with the ideas more closely combined, "The night was dark and lonely."

In sophisticated writing the segregating style rarely maintains an absolute ratio of one idea to one sentence. Practically speaking, the style consists of short sentences which are mostly grammatically simple and contain one or two ideas. We can view such a style as a variety of serial structure because the elements are added, though as separate statements, one after another.

Skillfully handled, the segregating style can be very effective. For example, an English essayist once interviewed the novelist Philip Guedalla and described Guedalla's method of writing in this passage:

> He writes, at most 750 words a day. He writes and rewrites. He polishes and re-polishes. He works in solitude. He works with agony. He works with sweat. And that is the only way to work at all.
>
> Beverly Nichols

These short repetitive sentences are as strong and seemingly as monotonous as hammer strokes. The monotony suits the point, for what Nichols says is that writing is often tedious, wearying toil. In part the repeated words create the sense of monotony, but fully as important is the iteration of the short, simple sentence pattern. It is dangerous to imitate tedium in your own sentence style, but Nichols gets away with it. The skill with which he suggests monotony makes his own sentences interesting, though obviously he could not continue in this manner for very long. The passage provides a good example of how a sentence style may be used to reflect and reinforce an idea.

Another advantage of the segregating style is its potential for dramatic description. By isolating individual details of a scene or an action, the short sentence enables—even forces—us to look very closely. This capacity for close-up makes the style useful on occasion in narrative writing. In the following passage an historian describes the scheme of political allies of a French claimant to the Spanish throne to manipulate food prices in early eighteenth-century Madrid. They would then seize control of the government, made more unstable than usual by the illness of the Hapsburg king, Charles II:

> The next object was to get rid of the ministers. Madrid was supplied with provisions by a monopoly. The Government looked after this most delicate concern as it looked after everything else. The partisans of the House of Bourbon took advantage of the negligence of the administration. On a sudden the supply of food failed. Exorbitant prices were demanded. The people rose. The royal residence was surrounded by an immense multitude. The Queen harangued them. The priests exhibited the host. All was in vain. It was necessary to awaken the King from his uneasy sleep, and

to carry him to the balcony. There a solemn promise was given that the unpopular advisers of the Crown should be forthwith dismissed. The mob left the palace and proceeded to pull down the houses of the ministers. The adherents of the Austrian line were thus driven from power, and the government was intrusted to the creatures of Porto Carrero [the Archbishop of Toledo, who had inspired the uprising as a way of forcing his enemies out of office].

Thomas Babington Macaulay

In such narrative the writer has the problem of analyzing a complex action into its separate episodes and then arranging these in a revealing pattern. Using a segregating style throughout the middle of this paragraph, Macaulay shows each phase of the action with dramatic clarity. After four opening sentences in which he establishes the aims of the politicians and the situation they set out to exploit, Macaulay shifts to seven shorter statements which detail the specific events of the story: the failure of food supplies, the inflation, the rising of the populace, their descent upon the palace, the attempts of the priests to control the mob, and their failure.

Not only do these segregating sentences analyze the action, their staccato pace suggests the rush and violence of events. With the climax of the King's being carried out onto the balcony, the sentences grow longer and less breathless, indicating that the conflict has been resolved. The drama is over and the pace slows.[2]

Here are two more examples of the short sentence used for narration. The first concerns the aftermath of a shipwreck: four survivors, barely afloat in an overloaded 10-foot dinghy, are threatened by the attempt of a half-crazed stoker to leap into their boat from a raft. The second describes a criminal in China being led to the place of his execution (the British consul has to be present since the victim of the crime was English):

He had turned into a demon. He was wild—wild as a tiger. He was crouched on this raft and ready to spring. Every muscle of him seemed to be turned into an elastic spring. His eyes were almost white. His face was the face of a lost man reaching upward, and we knew that the weight of his hand on our gunwale doomed us.

Stephen Crane

The judge gave an order and the vice-consul rose and walked to the gateway, where their chairs awaited them. Here stood the criminal with his guard. Notwithstanding his tied hands he smoked a cigarette. A squad of little soldiers had been sheltering themselves under the overhanging roof, and on the appearance of the judge the officer in charge made them form up.

2. Of the fifteen sentences in this paragraph, twelve are grammatically simple, while only two are complex and one compound. Not all the twelve simple sentences are truly segregating. Some combine several ideas by using phrases ("It was necessary to awaken the King from his uneasy sleep, and to carry him to the balcony"). The segregating style is concentrated in the seven sentences from the fifth to the eleventh.

The judge and the vice-consul settled themselves in their chairs. The officer gave an order and the squad stepped out. A couple of yards behind them walked the criminal. Then came the judge in his chair and the vice-consul.

W. Somerset Maugham

Neither of these paragraphs provides as pure an example of the segregating style as do the passages by Nichols and Macaulay. In both, the majority of the sentences are technically simple (six of seven in Crane, five of eight in Maugham), but none of these really treats only a single idea like Macaulay's: "The people rose," "The Queen harangued them," or "All was in vain." Even so, the sentences are relatively short and uncomplicated, and we may fairly regard both passages as variations of the segregating style.

In each case the style analyzes a visual experience, making us see the components clearly. But the writers want us to understand different things. Crane's short sentences are as dramatic as Macaulay's, though they analyze a frozen image rather than a series of events. First we receive an emotional impression, conveyed by such words as "demon," "wild," and "tiger"; then the general image of the figure "crouched . . . ready to spring"; and finally, one by one, the details of his muscles, his eyes, his face.

If we think about what the sentences enable us to see so clearly we realize that Crane suggests a great deal more than he actually states. For he sees—and wants us to see—the stoker under two simultaneous aspects. He is a pitiable human being reduced to the animal instinct for survival, a poor creature deserving our pity and help. At the same time the stoker is to the men in the boat both a danger and a portent of their own deaths (here the image of the tiger is revealing). The final sentence reveals this double vision most clearly. The simple "and" points to a truth which Crane nowhere states but everywhere implies: that we can respond to the humanity in "the face of a lost man reaching upward" and yet leave that face to destruction in the sure and certain knowledge that nothing can be done and that we must each of us save ourselves.[3] Certainly the sentence style does not convey this truth. But the style is an effective vehicle for what Crane wishes us to understand. It makes us *see* the physical details and enables us to discover, if we can, the spiritual significance below their surface.

Maugham also uses the segregating style to render a scene vividly. But his purpose is different. He distances us from what is happening, unlike Macaulay and Crane, who put us "there." Maugham avoids hot words like "demon," "wild," and "tiger," which draw us emotionally into the action. His paragraph is careful reporting, objective and cool.

---

3. This, however, is not the "truth" which informs the great short story Crane developed out of his experience in the shipwreck, "The Open Boat." The theme of that story is more complex.

The sentences break the scene into a series of tableaux, each scrupulously presented without overt comment. The technique is not that of a callous man but that of an artist willing to let the event speak for itself. The style has a truth of its own. Crane reveals something of what it means to face death. Maugham's truth is that in the society he is describing life and death do not matter very much.

The segregating style, then, can be effective in narrative and descriptive writing. It is less useful in exposition, where you must be able to combine ideas in complicated patterns and distinguish subtle gradations of importance among them. Such subtleties are difficult to express in a series of simple, independent statements, which by their very nature treat every idea as having much the same importance as every other.

Individually used, however, segregating sentences can be effective in exposition, especially if set alongside longer statements:

> Before election day he [Huey Long] predicted that he would win if rain didn't keep the mud farmers away from the polls. It rained.
>
> Hodding Carter

Such a short sentence makes a fine way to achieve both emphasis and variety. Later we shall look at more examples of ways to employ segregating sentences for emphasis and variety. Every good writer should be able to handle the short sentence and even, when occasion warrants, to compose brief passages in the segregating style. On the whole, however, the style is too limited for general use in exposition.

EXERCISE

Write a brief paragraph of about 100 words describing a football game or a party (or some other sporting or social event). Use short segregating sentences (eight to ten of them) to analyze the scene or occasion.

THE FREIGHT-TRAIN SENTENCE

The freight-train sentence consists of several (three or more) independent clauses. It may take the form of multiple coordination, in which the clauses are joined by *and* (less often by *but, or, nor*). Or the sentences may be paratactic—that is, the clauses are butted together without conjunctions and separated by stops, usually semicolons, sometimes commas.

As a style in English literature multiple coordination goes back a thousand years or more to Anglo-Saxon narrative. It was frequent in the beginnings of modern English prose: in Thomas Malory's account of King Arthur (*c.* 1470), and in the King James version of the Bible (1611):

And then was he [Sir Tristram] naked and waxed lean and poor of flesh; and so he fell in the fellowship of herdmen and shepherds, and daily they would give him some of their meat and drink.

Sir Thomas Malory

And the rain descended and the floods came, and the winds blew, and beat upon the house; and it fell: and great was the fall of it.

Matthew, 7:27

The freight-train sentence has several virtues. It is useful when you wish to join a series of events, ideas, impressions, feelings, or perceptions as immediately as possible, without judging their relative value or imposing a carefully ordered logical structure upon them. Children often experience reality in this immediate, uncritical accepting way, and authors writing for children or trying to recreate the vision of childhood may employ multiple coordination:

And I'll look out for you, and you'll sing out as soon as you see me. And we'll go down the street arm in arm, and into all the shops, and then I'll choose my house, and you'll choose your house, and we'll live there like princes and good fellows.

Kenneth Grahame

Even in writing that expresses more adult attitudes, the same desire to convey mental experience directly may lead a writer to select the freight-train style. In the following passage by Ernest Hemingway, the hero of *A Farewell to Arms* forgets the dull routine of army life by fantasizing a romantic adventure with the heroine, Catherine Barkley:

Maybe she would pretend that I was her boy that was killed and we would go in the front door [of a small hotel] and the porter would take off his cap and I would stop at the concierge's desk and ask for the key and she would stand by the elevator and then we would get in the elevator and it would go up very slowly clicking at all the floors and then our floor and the boy would open the door and stand there and she would step out and I would step out and we would walk down the hall and I would put the key in the door and open it and go in and then take down the telephone and ask them to send a bottle of capri bianca in a silver bucket full of ice and you would hear the ice against the pail coming down the corridor and the boy would knock and I would say leave it outside the door please.

Both Grahame and Hemingway use the freight-train sentence to describe an experience taking place in the mind of a character. The style enables them to suggest the continuous flow of the dreaming mind; for we fantasize in a stream of loosely connected feelings and ideas and images, not in neatly packaged sentences of intricately related clauses and phrases all tied together with *if, but, therefore, consequently, on the other hand.* Indeed, we may not fantasize in words at all but in imagined perceptions, as Hemingway implies ("and you would hear the ice against the pail"). Hemingway also goes further than Grahame in imitating the stream of fantasy: his one sentence is much longer than Grahame's two and its flow is unimpeded by commas, semicolons, or dashes. This technique is a variety of what novelists and critics call "stream of consciousness," a way of writing so as to suggest the mind feeling, dreaming, and even thinking in a loose, associational manner.

The next example shows how the freight-train sentence can also convey the immediacy of the world as it impinges upon the mind. A seaman is describing what it was like aboard a freighter under attack by a German dive bomber during World War II:

> . . . you're standing there, and you can feel the sweat running down the back of your knees, and he gets closer and closer and you can see those horrible splashes of the bullets as they come across the water and go whipping across the deck. . . .

There is a good deal in the diction of this passage to communicate fear: the word "horrible," the image of the "sweat running down the back of your knees." But quite apart from the words, the sentence structure enhances the feeling of terror. The simple repetition of "and . . . and . . . and" implies a mind that, like a camera, can do nothing but record images; fear has made it incapable of imposing order upon experience. It cannot analyze what is happening into separate sentences, or logically articulate the individual perceptions with *since* and *although* and *consequently*. It can only register raw experience. This—*this*—is happening. The style suggests paralysis. The speaker is too caught up in events to analyze them, too terrified to run; and there is in any case no place to run to.

For Grahame, Hemingway, and the British seaman, the freight-train style best suggests the flow of pure experience, whether it be a fantasy within the mind or an overwhelming reality without. In the following passage, on the other hand, Virginia Woolf employs the style for irony, mocking the insensitivity of an English tourist who visited Paris in 1765 and recorded his impressions in a diary:

> That is what he writes about, and, of course, about the habits of the natives. The habits of the natives are disgusting; the women hawk on the floor, the forks are dirty; the trees are poor; the Pont Neuf is not a patch on London Bridge; the cows are skinny; morals are licentious; polish is good; cabbages cost so much; bread is made of coarse flour.

By hooking detail onto detail in no particular order, Woolf ridicules the diarist, who comes across as a dull, closed-minded bigot.

Woolf uses semicolons instead of conjunctions between her clauses because they carry essentially the same idea. They are repetitive rather than progressive; each new image simply implies again the same underlying theme: that the diarist was dull and insensitive. The fact points to another virtue of the freight-train sentence, one it shares with the segregating style: the repetition of the same sentence pattern implies a common thread of thought or feeling. Of course, you may repeat any sentence pattern for this purpose, but the repetition is more striking and effective with relatively simple constructions like the short sentence or a series of linked independent clauses.

Here is another instance of the freight-train sentence in which the clauses are joined without conjunctions. Mark Twain describes the arrival of a steamboat at a sleepy river town, conveying the excitement of the event in this series of short clauses which follow one another at a rapid pace:

> The town drunkard stirs, the clerks wake up, a furious clatter of drays follows, every house and store pours out a human contribution, and all in a twinkling the dead town is alive and moving.

Again the clauses express a common theme. While there is movement from one detail to another, the details are all specific examples of the generalization Twain expresses in his final clause. Before this clause he signals the change from particular to general by "and." Otherwise he relies upon stops rather than conjunctions, like Woolf (though Twain punctuates with commas instead of the more conventional semicolons).

Most often, however, the freight-train sentence uses conjunctions. These are called for when the clauses represent different phases of an experience and do not repeat an underlying concept. Consider this sentence by Hemingway:

> It was a hot day and the sky was bright and the road was white and dusty.

There is movement here, not repetition. The change involves both the senses by which we perceive the scene and the scene itself. First we feel the heat, and then we see the sky, and finally we lower our eyes and gaze down the road.

A sentence like Hemingway's analyzes an experience much like a series of segregating sentences. But it brings the parts more closely together so that the fluidity of the experience is paramount. This capacity to clarify the parts of an event yet make them coalesce into a complex unity is another virtue of the freight-train sentence. Here is another example, from the newspaper account quoted on page 231, in which Stephen Crane described his experiences in the shipwreck. The four survivors have struggled ashore after their dinghy broaches in the surf, and a man on the beach rushes to help them:

> Then he went after the captain, and the captain sent him to me, and then it was that he saw Billy Higgins lying with his forehead on sand that was clear of the water, and he was dead.

The sentence style directs our senses much as a camera directs them in a film, distinguishing the parts of the scene, moving us from one to the next, yet creating a continuous experience.[4]

To look at the same point from another side, imagine that the final three of the segregating sentences in the paragraph on pages 231–32 by Somerset Maugham were put together in freight-train style:

> The officer gave an order and the squad stepped out and a couple of yards behind them walked the criminal and then came the judge in his chair and finally the vice-consul.

What was a series of snapshots has become a motion picture. The scene is fluid; the images stream into one another. This does not mean that the revision improves what Maugham wrote; it is, in fact, poorer—given Maugham's purpose. The point is simply that a slight adjustment in sentence structure alters how we perceive the event and in a subtle way how we evaluate it.

Along with its advantages, the freight-train sentence has limitations. Like the segregating style, it does not treat ideas subtly. The freight-train sentence implies that the thoughts it links together with grammatical equality are equally significant. But ideas usually are not of the same order of importance; some are major, others secondary. Moreover, this type of construction cannot show very precise logical relationships of cause and effect, condition, concession, and so on. It joins ideas only with such general conjunctions as *and, but, or, nor,* or even less exactly with semicolons and commas.

THE TRIADIC SENTENCE

A second deficiency of the freight-train sentence is that because it is open-ended it lacks definite shape. It has no necessary stopping place; one could go on adding clauses indefinitely. As a way of providing it with a more exact structural principle, the freight-train sentence is sometimes composed as a triad—that is, in three units:[5]

> They loomed, they bulged, they impended.
>
> H. G. Wells

> Her showmanship was superb; her timing sensational; her dramatic instinct uncanny.
>
> Robert Coughlan

4. The sentence is also a fine example of understatement. The short final clause, flatly announcing "he was dead," implies profound feeling, more easily understood if one has read the entire story. Notice, too, the power of the simple "and"; in both this passage and that quoted on page 231 Crane plucks great resonance from "and," usually an empty word.

5. Such a sentence is also called a "tricolon." Loosely, "colon" designated in ancient Greek rhetoric an independent clause which was part of a longer sentence. A "tricolon" is a sentence consisting of three such clauses.

Business executives, economists, and the public alike knew little of the industrial system they were operating; they were unable to diagnose the malady; they were unaware of the great forces operating beneath the surface.

Thurman Arnold

Semicolons or commas, not conjunctions, separate these clauses. This is often the case because the triadic sentence is usually repetitive. But it can use conjunctions where a sufficient change of idea occurs:

Then the first star came out and the great day was over and in the vestibule I saw my grandmother reverently saluted by her sons who wished her a happy holiday.

Ludwig Lewisohn

These new listeners cheered the florid Coughlin rhetoric and the roll of the great metaphors, and they followed with angry approval when the voice rose in jeremiads, and they grew solemn when it fell in heavy warning.

Wallace Stegner

In Lewisohn's sentence the final clause is substantially longer than the two which precede it. This movement to a longer, more complicated final construction is a refinement of the triadic sentence:

The canisters were almost out of reach; I made a motion to aid her; she turned upon me as a miser might turn if anyone attempted to assist him in counting his money.

Emily Brontë

The muezzins are waiting on the minarets; they lean over their balconies; then voice after voice, in the wailing Eastern tone, cries the salutation, one voice striking through another.

Arthur Symons

Occasionally the shift may work in the opposite direction, from long to short:

Calvin Coolidge believed that the least government was the best government; he aspired to become the least President the country had ever had; he attained his desire.

Irving Stone

EXERCISE

1. Compose four freight-train sentences, each consisting of at least four clauses. In two link the clauses with conjunctions; in two join them paratactically, that is, without conjunctions and separated by semicolons.

2. Write four triadic sentences. In two keep the clauses of about equal length; in one make the final clause substantially longer and more complicated; and in the third make it shorter.

THE CUMULATIVE SENTENCE

The last of our types of serial structure is the cumulative sentence. In its usual form a main clause precedes a series—often quite long—of appositive, modifying, or absolute constructions which accumulate details about the scene, person, or event being described. The elements that come after the main clause are technically subordinate, but in fact they carry the main load of the sentence. They are at least equal in significance to the main clause and frequently more important.

Cumulative sentences most often appear in description. The writer begins with a general picture, like an artist's charcoal sketch; then he or she fills in the picture with details added in the grammatical form of appositives, adjectives, participles, prepositional phrases, relative or adverbial clauses, and nominative absolutes. We shall briefly analyze the grammatical structure in the cumulative sentences below to give you an idea of how the writers employ the various building blocks out of which a sentence is constructed. (The building blocks themselves are reviewed in the Reference Grammar on pages 720ff.)

A creek ran through the meadow, winding and turning, clear water running between steep banks of black earth, with shallow places where you could build a dam.

Mark Schorer

The initial clause establishes the general scene: "A creek ran through the meadow." Details follow in (1) a pair of participles ("winding and turning"); (2) a modified appositive + a participle modified by two prepositional phrases ("clear water running between steep banks of black earth"); and (3) a prepositional phrase + a relative clause ("with shallow places where you could build a dam").

It was a very poor quarter, a labyrinth of squalid bamboo huts, patched with palm-leaf, winding all over a steep hillside.

George Orwell

The main clause is followed by (1) an appositive + a prepositional phrase ("a labyrinth of squalid bamboo huts"); (2) a participle + a prepositional phrase ("patched with palm-leaf"); and (3) a second participle + an adverb + a prepositional phrase ("winding all over a steep hillside").

Back of the raised stage is the Throne, two carved chairs under crimson hangings, in an elaborate alcove with columns on either side and indirect but effective lighting.

Joy Packer

The writer has inverted the main clause, placing the subject ("Throne") at the end so that it is next to the constructions which modify it. These take the form of (1) a modified appositive + a prepositional phrase ("two carved chairs under crimson hangings"); and (2) a series of prepositional phrases ("in an elaborate alcove with columns on either side and indirect but effective lighting").

7000 Romaine St. looks itself like a faded movie exterior, a pastel building with chipped *art moderne* detailing, the windows now either boarded up or paned with chicken-wire glass and, at the entrance, among the dusty oleander, a rubber mat that reads WELCOME.

Joan Didion

Main clause + (1) an appositive with a prepositional phrase ("a pastel building with chipped *art moderne* detailing"); (2) a nominative absolute ("the windows now either boarded up or paned with chicken-wire glass"); and (3) a second absolute ("and, at the entrance, among the dusty oleander, a rubber mat that reads WELCOME ").

The cumulative sentence is very useful in character description:

Mrs. Faithful came in—a thin woman with what we used to call the touch of good breeding upon her, wearing a nervous smile and offering hospitality in words that tumbled over each other.

Morris Markey

She [Anne Morrow Lindbergh] was then twenty-one, a year out of Smith College, a dark, shy, quiet girl with a fine mind and a small but pure and valuable gift for putting her thoughts and fancies, about the earth, sky, and sea, on paper.

John Lardner

Though it is less often used for narrative than for descriptive writing, the cumulative sentence can also handle a series of events, as in this passage about an English military expedition into France in 1359:

The unwieldy provision carts, draught horses, and heavily armed knights kept the advance down to nine miles a day, the huge horde moving in three parallel columns, cutting broad highways of litter and devastation through an already abandoned countryside, many of the adventurerers now traveling on foot, having sold their horses for bread or having slaughtered them for meat.

John Gardner

In the next two descriptive examples the cumulative sentence approaches narration in its emphasis upon movement and action:

The candle stood on the counter, its flame solemnly wagging in a draught; and by that inconsiderable movement, the whole room was filled with noiseless bustle and kept heaving like a sea: the tall shadows nodding, the gross blots of darkness swelling and dwindling as with respiration, the faces of the portraits and the china gods changing and wavering like images in water.

Robert Louis Stevenson

Then he would talk to the boy, the two of them sitting beneath the close fierce stars on a summer hilltop while they waited for the hounds to bring the fox back within hearing, or beside a fire in the November or December woods while the dogs worked out a coon's trail along the creek, or fireless in the pitch dark and heavy dew of April mornings while they squatted beneath a turkey-roost.

William Faulkner

Like all serial construction the cumulative sentence has the problem of being open-ended, with no natural stopping place. The deficiency is sometimes made good by not only beginning with one general statement but closing the sentence with another. In the following example, William Gibson, describing a photograph of his parents, opens with the clause "When they sat for a photograph together," follows this with accumulated details, and ends with an assessment of the meaning of the picture:

> When they sat for a photograph together—two neat slim bodies, the girl unsmiling and her eyes astare, elbows and knees tight, hands clenched in her lap, immaculate to the throat in lacy white, and the young man with grin and straw hat both aslant, jaunty on the bench arm, one leg crossed, natty in his suit and tie complete with stickpin, his arm around her with fingers outspread possessively upon her shoulder—it was a portrait not only of contrasts, but of a nation's lower middle class coming out of its cocoon.

What happens in that sentence is that the accumulation, instead of following an introductory statement, is gathered between dashes and intruded into the middle of a complex sentence. That sentence—"When they sat for a photograph together . . . it was a portrait not only of contrasts, but of a nation's lower middle class coming out of its cocoon"—becomes a frame enclosing the details, a pattern nicely suited to the content the sentence describes.

The series of details is also neatly closed off in these two examples. In the first Robert Louis Stevenson is writing about the guilty reaction of a man who has just committed murder and imagines that all the neighbors have heard the sounds of the crime:

> It was impossible, he thought, but that some murmur of the struggle must have reached their ears and set on edge their curiosity; and now, in all the neighbouring houses, he divined them sitting motionless and with uplifted ear—solitary people, condemned to spend Christmas dwelling alone on memories of the past, and now startingly recalled from that tender exercise; happy family parties, struck into silence round the table, the mother still with raised finger: every degree and age and humour, but all, by their own hearts, prying and hearkening and weaving the rope that was to hang him.

Stevenson pulls the details together in the final clause with the image of all these people "weaving the rope that was to hang him."

In the second example Ernest Hemingway opens with the generalization "He sat on the logs," describing the hero relaxing after fishing for trout. He closes the sentence, following a semicolon, with the clause "slowly the feeling of disappointment left him," and restates this point in two progressively shorter sentences which round off the passage:

> He sat on the logs, smoking, drying in the sun, the sun warm on his back, the river shallow ahead entering the woods, curving into the woods, shallows, light glittering, big water-smooth rocks, cedars along the bank and white birches, the logs warm in the sun, smooth to sit on, gray to the touch; slowly the feeling of disappointment left him. It went away slowly, the feeling of disappointment that came sharply after the thrill that made his shoulders ache. It was alright now.

Before we leave it entirely we should look at another variety of the cumulative sentence. In this type the order of elements is reversed: the numerous details precede the main clause instead of following it. The first of the two examples, about Mary Queen of Scots, opens with a series of adjectivals which bear upon the generalization closing the sentence. In the second a novelist, discussing her art, begins by listing the essential of a story.

> Frivolous, extravagant, careless, emotional, utterly self-centered, lacking in judgment and temper, unmindful of the interests of her country, she looked upon the world largely as it advanced or retarded her personal aspirations.
>
> D. Harris Willson

> Conflicts or rivalries and their resolution, pride and its fate, estrangement and reconciliation, revenge or forgiveness, quests and searches rewarded or unrewarded; abidingness versus change, love and its proof—these are among the constants, the themes of the story.
>
> Elizabeth Bowen

Notice in Bowen's sentence how she uses the pronoun *these* to sum up all the preceding nouns and to act as the subject of the sentence. Employing a pronoun in such a way (*this, that, those, such* are others) is an easy way of drawing together a string of initial concepts and focusing them as a subject.

EXERCISE

1. Compose three cumulative sentences on the patterns of examples used in the preceding pages. In two of them place the accumulated details after the main clause; in the third, before.

2. In each of the following passages identify whether the basic sentence style is segregating, freight-train, or cumulative. Be able to explain why you think so:

A. The fight was now over; night had closed in, and those among the English host who had not fallen around their King had left the field under the cover of darkness.

Edward Freeman

B. The trouble was that war was already on us as the result of the underlying causes of the great stock market crash. The political and economic institutions of the nineteenth century were dying. The cement was falling out of social structures. Peoples and goods were hemmed in by obsolete national boundaries. When goods fail to move, armies march.

Thurman Arnold

C. Markheim gave ear to it [a church hymn] smilingly, as he sorted out the keys; and his mind was thronged with answerable ideas and images: churchgoing children and the pealing of the high organ; children afield, bathers by the brookside, ramblers on the brambly common, kite-fliers in the windy and cloud-navigated sky; and then, at another cadence of the hymn, back again to church, and the somnolence of summer Sundays, and the high genteel voice of the parson (which he smiled a little to recall) and the painted Jacobean tombs, and the dim lettering of the Ten Commandments in the chancel.

Robert Louis Stevenson

D. The first intimation of a break in the program came by letter from Europe. Colonel Charles R. Forbes, traveling for his health, resigned. It was an ominous note. Harding failed to recognize its import. He urged his old crony to reconsider; but Forbes knew now that he could never stand up to the threatened Senatorial investigation. He was through.

Samuel Hopkins Adams

E. Such was the political Louisiana in which Huey Pierce Long was born in 1893, in impoverished Winn Parish in north Louisiana, a breeding ground of economic and political dissenters.

Hodding Carter

F. He [Joe Louis] had fought often since then, and I had seen his two bouts with Jersey Joe Walcott on television, but there hadn't been any fun in it.

A. J. Liebling

G. Well that was yesterday and to-day is the landing and we heard Eisenhower tell us he was here they were here and just yesterday a man sold us ten packages of Camel cigarettes, glory be, and we are singing glory hallelujah, and feeling very nicely, and everybody has been telephoning to us congratulatory messages upon my birthday which it isn't but we know what they mean.

Gertrude Stein

H. Shouldering one another to get next at the sharpening stone, were men stripped to the waist, with the stain all over their limbs and bodies; men in all sorts of rags, with the stain upon those rags; men devilishly set off with the spoils of women's lace and silk and ribbon, with the stain dyeing those trifles through and through.

Charles Dickens

I. Sustained, then, by an outlook on life which helped them in the daily needs of economic existence; conscious of a bond of unity with others who shared their convictions; aware of themselves as an aristocracy of the spirit against which aristocracies of this world were as naught; fortified by the earthly victories which this morale helped to bring about: how should the hard core of convinced Puritans not have believed that God was with them and they with him?

Christopher Hill

J. He [Karl Marx] went to Manchester with Engels. He went to Paris. He went to Holland to borrow money from rich relatives. He dispatched Engels to Paris.

Fritz J. Raddatz

K. And they brought their cattle unto Joseph: and Joseph gave them bread in exchange for horses, and for the flocks, and for the cattle of the herds, and for the asses: and he fed them with Cattle for that year.

Genesis 47:17

L. It was not long before the wreck-train arrived, a thing of flat cars, box cars and cabooses, with hospital cots made ready en route, and a number of doctors and nurses who scrambled out with the air and authority of those used to scenes of this kind.

Theodore Dreiser

M. Meanwhile the news from Anzio was worse. The red light was very red. The German counter-offensive was due in a day or two. Anzio was in danger. Anzio had to be saved.

Fred Majdalany

N. The next day all the shops were open and hundreds of fatigued assistants were pouring out their exhaustless patience on thousands of urgent and bright women; and flags waved on high, and the gutters were banked with yellow and white flowers, and the air was brisk and the roadways were clean.

Arnold Bennett

O. So ended our occupancy of the middle-class house I had always hated, the lodestar of our morning together, and ever after most bitterly loved; I would see it again only from the sidewalk opposite, time gone by and the trees grown up in a streetful of cheapwork dwellings run down, all needy of paint, our lawn neglected and trodden back to dirt, and the house so poor and narrow with the two mute windows upstairs behind which my father had died, I believed by an effort that he had died, and our lives had happened.

William Gibson

P. The battery in the next garden woke me in the morning and I saw the sun coming through the window and got out of bed.

Ernest Hemingway

Q. These one or two hundred and fifty thousand native people constituted twenty-one known nationalities, or small nations, which were in turn further separated into sub-nationalities, and these again into tribes or tribelets to a total number of more than two hundred and fifty—exactly how many can never be known because of the obliteration in modern times of whole peoples and cultures by the Spaniard and Anglo-Saxon alike without record of tribal name and affiliation.

Theodora Kroeber

R. After the door shut behind me, I stood still, afraid to sit down or to lean against the wall. Then I groped for the door. Gradually the darkness paled. I caught a faint sound approaching; I heard a key turn in the lock. A matron appeared. I recognized Miss Johnson, the one who had frightened me out of my sleep on my first night in the penitentiary.

Emma Goldman

S. But man, who knows good and evil, is cruel for cruelty's sake; he who has a moral law is more brutal than the brutes, who have none; he alone inflicts suffering on his fellows, with malice aforethought.

Susanne K. Langer

T. He [James II] felt an assurance that every obstacle would give way before his power and his resolution. His Parliament withstood him. He tried the effect of frowns and menaces. Frowns and menaces failed. He tried the effect of prorogation. From the day of prorogation the opposition to his designs had been growing stronger and stronger.

Thomas Babington Macaulay

U. And you go into the shops, and take anything you want—chocolates and magic-lanterns and injirubber balls—and there's nothing to pay; and you choose your

own house and live there and do just as you like, and never go to bed unless
you want to!

Kenneth Grahame

3. Describe a day in school, first in a passage of five or six segregating sentences. Then cover the same material in two freight-train sentences, and again in one long cumulative sentence.

4. Do the same thing on the subject of a Saturday night date or a visit to a relative's home.

CHAPTER 32

Parallel and Balanced Sentences

INTRODUCTION

Parallelism and balance are closely related; in fact, some people use the terms almost interchangeably. Here we shall make a distinction. *Parallelism* means that two or more words, phrases, or clauses have the same grammatical form and an identical grammatical relationship to the same thing. "Jack and Jill went up the hill" offers a simple example. "Jack" and "Jill" are parallel because both act as subjects of the one verb and are proper nouns.

Balance means that two or more words or constructions have essentially the same form and length and have similar functions. "Jack went up the hill, and Jill went with him" consists of two balanced clauses, both independent and both of the same pattern and of similar length.

Constructions may be parallel without being balanced, balanced without being parallel, or both parallel and balanced. As an example of parallelism without balance, consider this sentence:

> He said that he would go, and that, considering the situation, we ought to go too, even though we don't approve of the meeting.

The two *that*-clauses are parallel because each is a complement of "said." However, they are not balanced: the second is much longer than the first and more complicated, containing an interrupting participial phrase ("considering the occasion") and an adverbial clause ("even though we don't approve of the meeting").

In the following case the clauses are balanced but not parallel:

> He said that he would go early and added that we should come later.

Since both *that*-clauses act as complements of verbs and are almost identical in length and form, we may fairly describe them as balanced.

But because they are objects of different verbs, they are not parallel. (The verbs themselves, however, are parallel, sharing the same subject.)

When the sentence is revised as follows, the clauses become parallel as well as balanced:

He said that he would go early and that we should come later.

The essential feature of parallelism, then, is that constructions share a grammatical role. The essential feature of balance is that constructions close to one another be of the same structure, roughly of the same length, and fulfill similar grammatical functions (if the functions are identical, then, of course, the constructions are also parallel). In the next few pages we shall look at examples first of parallelism, second of balance, and finally of several passages that combine the two.

THE PARALLEL SENTENCE

The freight-train style treats ideas as separate units and couples them in a loose series. One can handle ideas more simultaneously by paralleling them, making several perform the same grammatical role within the sentence.[1] A sentence in which such constructions play a major role is described as a "parallel sentence." Here are a few examples (the parallel constructions here and in subsequent examples are italicized):

In *its energy, its lyrics, its advocacy of frustrated joys,* rock is one long symphony of protest.

Time magazine

Three parallel objects of the preposition "in."

The Department of Justice began a vigorous campaign *to break up the corporate empires, to restore the free and open market, and to plant the feet of industry firmly on the road to competition.*

Thurman Arnold

Three parallel infinitive phrases, all modifying "campaign."

Here is *where the hot wind blows and the old ways do not seem relevant, where the divorce rate is double the national average and where one person in every thirty-eight lives in a trailer.*

Joan Didion

Although they follow the verb, the three *where*-clauses are parallel subjects of "is"; the first of these constructions actually consists of two clauses in parallel.

---

1. You can never achieve literal simultaneousness in writing. Writing, by definition, arranges words one after another in space, and these must be read one after another in time. But parallelism—while it cannot escape the inevitable serial order of language— does impose a nonsequential pattern in which the elements function simultaneously in a common grammatical relationship.

I am interested *in the folklore of Howard Hughes, in the way people react to him, in the terms they use when they talk about him.*

Joan Didion

Three parallel prepositional phrases all modifying "am interested."

They [Voltaire's writings] reflect the society in which he lived, *complacent, limited, content to enjoy its luxuries and its elegance, proud of its theatre and Poussin's pictures, its silks and its sparkling wines, light in its loves, troubled by no mysteries, sure of itself because it knew the steps in the formal dance that was its life.*

H. N. Brailsford

Seven adjectives—two of them participles—modify "society"; some of these introduce adjective phrases which contain parallel elements within themselves. Note that this passage is a good example of how parallelism may be used to build a cumulative sentence.

There never did, there never will, and there never can exist a Parliament, or any description of men, or any generation of men, in any country, possessed of the right or the power of binding and controlling posterity to the *"end of time,"* or of commanding for ever how the world shall be governed, or who shall govern it; and, therefore, all such clauses, acts or declarations by which the makers of them attempt to do what they have neither the right nor the power to do, nor the power to execute, are, in themselves, null and void.

Thomas Paine

This sentence is so replete with parallelism that a page would be necessary to explain it fully. One example: the first three constructions consist of the expletive "there" + "never" + an auxiliary verb, all being hooked to the same base verb, "exist."

To complain of the age we live in, to murmur at the present possessors of power, to lament the past, to conceive extravagant hopes of the future, are the common dispositions of the greatest part of mankind. . . .

Edmund Burke

Four infinitive phrases function as common subjects of "are."

Paral.

Shift.

All these sentences illustrate an important stylistic rule about parallelism: *parallel constructions must be identical in their grammatical form.* Thus in Burke's sentence the four subjects are expressed as infinitive phrases. It would violate the rule to shift from an infinitive to a gerundive construction or to an abstract noun modified by a prepositional phrase. Such mistakes are called *shifted constructions*:

To complain of the age we live in, murmuring against the present possessors of power, lamentations of the past, conceiving extravagant hopes of the future. . . .

Of course, the writer could have started with a gerund or with an abstract noun, but the same criticism would apply: the constructions that follow ought to be the same, and they are not.

Observing this rule of parallelism has the practical value of helping readers sort out a sentence. For example, by repeating form words such as prepositions, conjunctions, or auxiliary verbs, you can avoid ambiguities like that in this sentence:

They will work hard to defeat the bill and help you in any way they can.

Is "help" parallel to "work" or to "defeat"? The appropriate form word would tell us:

They will work hard to defeat the bill and will help you in any way they can.

They will work hard to defeat the bill and to help you in any way they can.

Aside from showing us that good writers observe the rule of placing parallel constructions in the same form, these examples indicate that parallel sentences are far from old-fashioned: four of the six are by contemporary writers. The final two come from the eighteenth century and reveal something about the development of the parallel style in English. Both political writers, Paine and Burke composed in a tradition of oratory that goes back to Cicero, the greatest of the Roman orators, and beyond him to teachers of rhetoric in Athens and other Greek city-states. The parallel style suits the needs of speakers well. It is impressive and pleasing to hear—elaborate yet rhythmic and ordered, following a master plan with a place for everything and everything placed. Listeners, moreover, can easily follow the plan, no small advantage to speakers, who risk losing their audiences if their sentences are too complicated. But if they can satisfy their listeners' sense of form by neatly fulfilling the expectations raised by the pattern, then they gain their hearers' favor and, perhaps, their support.

Parallelism has other advantages. It is economical, using one element of a sentence to serve three or four others. In the following examples the writers use a single subject to introduce several predicates, saving themselves the bother of repeating the subject each time:

We must somehow *take a wider view, look at the whole landscape, really see it, and describe what's going on out there.*

Annie Dillard

The station-keeper *lights his lamps, kindles a fire of twigs, prepares our beds.*

Lafcadio Hearn

He *shuffled into the room, dumped his notes on the desk, and began his usual dull lecture.*

Student

Piling up several predicates probably constitutes the most common form of parallel construction. Paralleling verbs is particularly effective in organizing a short description of a complex process or event. Each verb establishes a phase of the event; the sequence of verbs indicates its progress; and the concentration upon verbs without the constant intervention of the subject focuses the sentence properly—upon the action:

As the danger drew near they [prairie dogs] would *wheel about, toss their heels into the air, and dive in a twinkling into their burrows.*

Francis Parkman

It [a giant water bug] *seizes a victim with these legs, hugs it tight, and paralyzes it with enzymes injected during a vicious bite.*

Annie Dillard

In the following case parallel verbs organize a landscape rather than an event or a process. Yet their progression suggests the enlargement of the observer's view and creates an illusion of movement in the scene itself:

Hill and valley followed valley and hill; the little green and stony cattle-tracks *wandered in and out of one another, split into three or four, died away in marshy hollows, and began again sporadically on hillsides or at the borders of a wood.*

Robert Louis Stevenson

The same parallel pattern also helps to summarize action that extends over longer periods of time:

From 1925 to 1928 Huey [Long] *mended political fences, kept himself in the headlines, and built up a lucrative practice as attorney for some of the vested interests against which he ranted.*

Hodding Carter

Charles *borrowed his way through Savoy, disappeared into the Alps, and emerged, early in September, at Asti, where his ally met him and escorted him to the suburbs of Milan.*

Ralph Roeder

The sentence is about an invasion of Italy by Charles VIII of France in 1494.

This last example suggests another advantage of parallel verbs (and of parallelism in general). Parallelism enhances meaning by increasing the possible connections of words within the sentence. For instance, Roeder's sentence brings "borrowed," "disappeared," and "emerged" into a special relationship. Taken together they imply an ironic assessment of the French king; one could guess from this sentence Roeder's view that the expedition was a hare-brained, fly-by-night affair, ill planned and ill executed. Of course, it is the words, not parallelism in the abstract, that implies all this. But the parallel style, by widening the range of relationships into which words can enter, enables the added meaning to emerge.

In this sentence, a series of parallel verbs leading to an unlooked-for conclusion implies a sly amusement:

She *laid two fingers on my shoulder, cast another look into my face under her candle, turned the key in the lock, gently thrust me beyond the door, shut it; and left me to my own devices.*

Walter de la Mare

And Bernard Shaw, writing about Joan of Arc, insinuates a sardonic view of humanity under the surface of this prosaic summary of Joan's life:

> Joan of Arc, a village girl from the Vosges, was *born about 1412, burnt for heresy, witchcraft, and sorcery in 1431; rehabilitated after a fashion in 1456; designated venerable in 1904; declared Blessed in 1908; and finally canonized in 1920.*

The feeling focused by parallel verbs does not have to be irony or amusement. It may be anger, as in Thomas Jefferson's indictment of George III:

> He has *plundered our seas, ravaged our Coasts, burnt our towns, and destroyed the lives of our people.*

Or eloquence:

> Let every nation know, whether it wishes us well or ill, that we shall *pay any price, bear any burden, meet any hardship, support any friend, oppose any foe to assure the survival and success of liberty.*
>
> John F. Kennedy

In its fullest development the parallel sentence has something of the grandeur of a great building, and this architectural impressiveness gives weight and dignity to what the sentence says. Sentences like the two that follow from Samuel Johnson's "Preface to the Dictionary" provide good examples:

> That it [the Dictionary] will immediately become popular I have not promised myself: a few wild blunders, and risible absurdities, from which no work of much multiplicity was ever free, may for a time furnish folly with laughter, and harden ignorance in contempt; but useful diligence will at last prevail, and there can never be wanting some who distinguish desert; who will consider that no dictionary of a living tongue can ever be perfect, since while it is hastening to publication, some words are budding, and some falling away; that a whole life cannot be spent upon syntax and etymology, and that even a whole life would not be sufficient; that he, whose design includes whatever language can express, must often speak of what he does not understand; that a writer will sometimes be hurried by eagerness to the end, and sometimes faint with weariness under a task, which *Scaliger* compares to the labours of the anvil and the mine; that what is obvious is not always known, and that what is known is not always present; that sudden fits of inadvertency will surprize vigilance, slight avocations will seduce attention, and casual eclipses of the mind will darken learning; and that a writer shall often in vain trace his memory at the moment of need, for that which yesterday he knew with intuitive readiness, and which will come uncalled into his thoughts tomorrow.
>
> I have protracted this work till most of those whom I wished to please have sunk into the grave, and success and miscarriage are empty sounds: I therefore dismiss it with frigid tranquillity, having little to fear or hope from censure or from praise.

The first of these sentences shows just how complicated the parallel style can become. It strings together eight *that*-clauses as complements of "who will consider," and each of these, with one exception, contains parallelism within itself. So elaborate a sentence pattern seems artificial in that it results from conscious art, not from everyday speech. But artificiality does not here imply phoniness or coldness. As you can see from the second of Johnson's sentences, the parallel style is able to convey—even to intensify—deep emotion. Like many people, Johnson felt the emptiness of success delayed and the sadness of losing those whom one wishes most to please. But instead of crying out his feelings directly, Johnson gives them greater dignity and force by shaping them in a parallel sentence.

The parallel style has its limitations, of course. While it can handle ideas better than the freight-train sentence, primarily it suits ideas that are logically parallel: several effects of the same cause, for instance, or three or four conditions of a single effect. The parallel style, especially in its more extended forms, is a bit formal for modern taste. And it can become wordy if a writer allows the style to dominate him, padding out ideas to make a parallel sentence, instead of making a parallel sentence to organize ideas. Even so, we have seen that parallelism still appears frequently in contemporary prose. It remains an effective way of ordering ideas or perceptions, of shaping a sentence, and of attaining economy and emphasis.

EXERCISE

1. Identify the parallel constructions in the following sentences:

   A. The men dozed, waked, sighed, groaned.

   Joseph Conrad

   B. It was certainly true that Starr had played with his children, that the two families had seen a good deal of each other, and that he had been alone with Starr on many occasions.

   Morris Markey

   C. Before me lies a bundle of these sermons, rescued from six-score years of dust, scrawled on their title-pages with names of owners dead long ago, worm-eaten, dingy, stained with the damps of time, and uttering in quaint old letter-press the emotions of a buried and forgotten past.

   Francis Parkman

   D. They are scurrying around, collecting consensus, gathering as wide an acceptance as possible.

   Barbara W. Tuchman

   E. He ground the powders, mixed the pills, rode with the doctor on his rounds, held the basin when the patient was bled, helped to adjust plasters and to sew wounds, and ran with vials of medicine from one end of the town to the other.

   John Bach McMaster

F. If the movies had never gone beyond photographing plays on stage, if they had not created their own rhythm of presentation through cutting and camera movements, they would have remained a small and shabby thing.

Gilbert Seldes

G. Sometimes somebody has to start trying to change things, start to say something, do something, be politically expendable.

Shirley Chisholm

H. Perhaps her fading mind called up once again the shadows of the past to float before it, and retraced, for the last time, the vanished vision of that long history—passing back and back, through the cloud of years, to older and ever older memories—to the spring woods at Osborne, so full of primroses for Lord Beaconsfield—to Lord Palmerston's queer clothes and high demeanour, and Albert's face under the green lamp, and Albert's first stag at Balmoral, and Albert in his blue and silver uniform, and the Baron coming in through a doorway, and Lord M. dreaming at Windsor with the rooks cawing in the elm-trees, and the Archbishop of Canterbury on his knees in the dawn, and the old King's turkey-cock ejaculations, and Uncle Leopold's soft voice at Claremont, and Lehzen with the globes, and her mother's feathers sweeping down towards her, and a great old repeater-watch of her father's in its tortoise-shell case, and a yellow rug, and some friendly flounces of sprigged muslin, and the trees and the grass at Kensington.

Lytton Strachey [of the dying Queen Victoria]

I. We see an old woman tending the fire outside a house, a man spearing fish beside a stream, a half-grown boy paddling down stream in a dugout canoe.

Theodora Kroeber

2. Following the pattern of the sentence by Edmund Burke (page 248), construct parallel sentences on two of these topics:

   A. The duties of a policeman or other official.
   B. The complaints of a college student.
   C. The mistakes you make in composition.

3. Using the sentence by Bernard Shaw about Saint Joan (page 251) as a model, write a parallel sentence synopsizing the life of someone you know well, yourself or a relative, for instance.

THE BALANCED SENTENCE

Balanced constructions have similar form and function and approximately equal length, and they usually occur in the same sentence. (Occasionally balanced phrases, clauses, and so on come in successive sentences.) But unlike parallel constructions, elements in balanced constructions do not necessarily relate grammatically to the same thing. Balanced elements may be played against one another, sometimes repeating the same idea, sometimes expressing contrasting ideas. When the contrast is sharply pointed it is called *antithesis*: antithetical constructions are simply balanced phrases or clauses expressing opposed ideas.

While balance can involve any sentence element, it is most common with independent clauses. A sentence comprising two such clauses of roughly the same length and separated by a comma, semicolon, or colon is called a *balanced sentence*:

Visit either you like; they're both mad.

Lewis Carroll

The road lay white in the sun, and the railway ran just beyond.

Hilaire Belloc

Children played about her; and she sang as she worked.

Rupert Brooke

In a few moments everything grew black, and the rain poured down like a cataract.

Francis Parkman

The measures taken did not lack merit, but they did not rise above the plane of expediency.

Thurman Arnold

These are all compound sentences; most of them coordinated, one without a conjunction. However, not all compound sentences are balanced, nor are all balanced sentences necessarily compound. Balance requires that the sentence divide into roughly equal halves on either side of a central pause. This may occur even in sentences that are not technically compound:

They read hardly at all, preferring to listen.

George Gissing

We live in an ascending scale when we live happily, one thing leading to another in an endless series.

Robert Louis Stevenson

He seemed like a walking blasphemy, a blend of the angel and the ape.

G. K. Chesterton

To the fundamental question of human existence, astronomy has little of value to offer.

Sir James Jeans[2]

The examples we have seen thus far exhibit elementary balance between two equal and unbroken units (——/——). But writers vary the pattern in many ways. Sometimes they further divided one half into two (——/—— or ——/——); sometimes into three (——/—— — or — — —/——). They may split both halves in two (— —/— —), and so on. Here are a few examples:

2. Not everyone would agree that such sentences may fairly be described as "balanced." Some argue that balanced constructions must be of the same grammatical order and therefore that a balanced sentence cannot exist unless its halves are independent clauses. However, to the degree that a sentence sounds to the ear as if it consists of two parts more or less equal in length and importance, it has balance. The balance is more exact when the parts are independent clauses cut to the same pattern.

For being logical they strictly separate poetry from prose; and as in prose they are strictly prosaic, so in poetry, they are purely poetical.

(——/——)   G. K. Chesterton

Yet she stepped deliberately, to be surefooted in a dusky room; she touched along the wall and came to the door, where a footstool nearly tripped her.

(——/——)   George Meredith

But called by whatever name, it is a most pleasant fruitful region; kind to the native, interesting to the visitor.

(——/——)   Thomas Carlyle

I stood like one thunderstruck, or as if I had see an apparition: I listened, I looked round me, but I could hear nothing, nor see anything.

(——/————)   Daniel Defoe

Balance involving smaller elements, such as words, phrases, or subordinate clauses, may occur within clauses as well as between them:

In Plato's opinion man was made for philosophy; in Bacon's opinion philosophy was made for man; it was a means to an end; and that end was to increase the pleasures and to mitigate the pains of millions who are not and cannot be philosophers.

Thomas Babington Macaulay

The opening prepositional phrases of Macaulay's first two clauses are balanced against one another. He achieves more antithetical balance by repeating "man" and "philosophy" in reversed order.[3] Further on in the sentence, the infinitive phrases "to increase the pleasure and to mitigate the pains" are balanced; note that while the infinitives themselves are also parallel (both are subjective complements of "was"), the key terms "pleasure" and "pains," which focus the antithesis, are balanced but not parallel (they follow different infinitives). Finally, "are not and cannot be" is both a balanced and a parallel construction.

As Macaulay's sentence shows, parallelism and balance often go hand in hand. Here are several other examples:

As for me, I frankly cleave to the Greeks and not to the Indians, and I aspire to be a rational animal rather than a pure spirit.

George Santayana

The sentence balances two coordinated clauses of similar structure and length. Within the first clause, the prepositional phrases "to the Greeks and not to the Indians" are parallel and antithetical. In the second clause "a rational animal rather than a pure spirit" is a parallel, balanced construction, and an antithesis is provided by playing "rational animal" against "pure spirit."

The notice which you have been pleased to take of my labours, had it been early, had been kind; but it has been delayed till I am indifferent,

---

3. The device of switching the order of a pair of key terms was called *antimetabole* or *chiasmus* by Greek rhetoricians, who were fond of it.

and cannot enjoy it; till I am solitary, and cannot impart it; till I am known, and do not want it.

Samuel Johnson

In this passage from a famous letter to Lord Chesterfield, Johnson refuses a belated offer to help pay the cost of the great dictionary he had just finished editing. While coordinated, the sentence does not really balance because the second part is much longer and more complicated than the first. However, considerable internal balance, as well as parallelism, can be found in the sentence. "Early" balances "kind" (the two *had*-constructions are not parallel, however). The second part of the sentence—from the semicolon on—consists of a main clause followed by three parallel and balanced adverbial clauses, each introduced by "till" and each consisting of two parallel and balanced predicates. The pattern sets up an elaborate network of balances and antitheses: "delayed"/"indifferent"; "indifferent"/"cannot enjoy it"; "solitary"/"cannot impart it"; "known"/"do not want it"; "indifferent"/"solitary"/"known"; "cannot enjoy it"/"cannot impart it"/"do not want it."

Most people, of course, made no distinction between a Communist—who believed in nothing but government—and such philosophical anarchists as Vanzetti—who believed in no government at all.

Phil Strong

"A Communist" and "such philosophical anarchists as Vanzetti" are parallel objects of "between," though the second is too much longer than the first to constitute a balance. However, balance does occur in the two *who*-clauses, though these are not parallel, modifying, as they do, different nouns.

Balanced construction has several advantages. It displays key terms, not only emphasizing them but encouraging the reader to explore the implications suggested by playing one word against another. For example, the following sentence by Charles Dickens forces us to consider the difference between having money and having ability as well as the plight of those who lack the cash to turn their ideas to account:

Talent, Mr. Micawber has; capital, Mr. Micawber has not.

Anthony Hope implies a skeptical assessment of politicians and bureaucrats in this balanced sentence:

Ability we don't expect in a government office, but honesty one might hope for.

Here are a few more examples of how balance enables a writer to focus on important words and to enhance similarities and contrasts:

In itself, a love idyll like this may seem harmless, but it won't be by itself very long.

Pauline Kael

Kael complains that one consequence of movies like *Love Story* is that they usher in a host of shoddy imitations. Notice how the sentence swings and advances on the two phrases "in itself" and "by itself."

Thinking out loud was his pastime; exchanging ideas was his passion.

Roscoe Drummond

Written about one-time presidential candidate Wendell Willkie, this sentence balances key terms at either end of its two clauses; one pair—"pastime" and "passion"—are reinforced by alliteration.

> We talk of the inimitable grandeur of the old cathedrals; but indeed it is rather their gaiety that we do not dare to imitate.
>
> G. K. Chesterton

The balanced clauses focus on "grandeur" and "gaiety" (again stressed by alliteration), two contrasting, if not antithetical, terms, and compel us to think about our conventional response to religion and about older responses our culture no longer comprehends.

> Our heritage of Greek literature and art is priceless; the example of Greek life possesses for us not the slightest value.
>
> George Gissing

The structure of the sentence underscores the contrast between "art" and "life," as well as between "priceless" and "without the slightest value."

Beyond highlighting important ideas and thereby enhancing meaning, balanced construction is pleasing in itself. It gives shape to a sentence, one of the essentials of good writing, and pleases the eye and ear of the reader. At its best, balance expresses the way a writer looks at the world, just as freight-train or cumulative sentences express different angles of vision. Implicit in the balanced style is a sense of objectivity, control, and proportion that reinforce a writer's argument or ideas. In the following passage about Lord Chesterfield,[4] the English critic F. L. Lucas half convinces us of his argument by the reasonableness of his balanced sentences. The very style seems to confirm the fairness and lack of dogmatism suggested by such phrases as "seem to me" and "I think."

> In fine, there are things about Chesterfield that seem to me rather repellant; things that it is an offense in critics to defend. He is typical of one side of the eighteenth century—of what still seems to many its most typical side. But it does not seem to me the really good side of that century; and Chesterfield remains, I think, less an example of things to pursue in life than of things to avoid.

Because the balanced sentence tends to keep a distance between writer and subject, it works well for irony and comedy. The novelist Anthony Trollope, for instance, expresses disapproval of a domineering female character like this:

> It is not my intention to breathe a word against Mrs. Proudie, but still I cannot think that with all her virtue she adds much to her husband's happiness.

The balanced construction implies the objectivity of the author and increases the credibility of his criticism, while at the same time the second clause comically reveals him indulging in the very gossip he forswears in the first.

Comic, too, is the effect of this sentence from the autobiography of

4. An eighteenth-century British nobleman, he is famous for his letters to his son. To modern eyes the advice he gives the young man on how to succeed seems cold-blooded and calculating, even cynical.

Edward Gibbon, the famous historian of Rome's decline, describing an unhappy love affair of his youth broken off at his father's insistence:

> After a painful struggle I yielded to my fate: I sighed as a lover, I obeyed as a son; my wound was insensibly healed by time, absence, and the habits of a new life.

Writing from the calmer waters of age, when the tempests of twenty seem less catastrophic, Gibbon is smiling at himself. The very parallelism and balance of this sentence, as nimbly formal as an old dance, make an ironic comment upon the passions of youth.

Finally, you can use the balanced style effectively by violating it. The French satirist Voltaire sometimes adopts the trick of deliberate imbalance. Here, for instance, is the first part of a sentence from his satire *Candide*:

> It was reported that there had been at Constantinople, a kind of freethinker, who had insinuated, that it was proper to inquire into the truth of the Alcoran's [the Koran] having been actually written with a quill taken from the angel Gabriel. . . .

Against this long, elaborate first clause, with its hesitations and careful qualifications, Voltaire sets this simple construction to complete the sentence:

> but he was stoned.

The very imbalance makes us smile—somewhat cynically—at the contrast between the scholarly intellectual—timid, learned, diffidently choosing his words—and the man in the street, rock in hand.

SUMMARY

Balance and parallelism do not communicate meaning by themselves. The primary units of meaning, of course, are words. But balanced and parallel constructions do reinforce and enrich meaning. Or, to be more exact, certain kinds of meaning. Not every sentence can be, or should be, cast in this mold. Like every style, parallelism and balance have limitations as well as potentialities. Their very sanity, reasonableness, and control make them unsuitable devices for conveying the immediacy of raw experience or the intensity of strong emotions. Such formal architecture is likely to seem too elaborate to modern readers, a less "natural" way of writing than the segregating style or the freight-train or cumulative sentences.

However, we ought not to equate formality with artificiality or to suppose that naturalness is the only ideal. All well-constructed sentences result from art, even those—perhaps especially those—like Hemingway's which create the illusion of naturalness. Moreover, "natural" is a tricky word. To men and women of the eighteenth century, parallelism and balance reflected nature, which they understood as a vast but comprehensible structure of ordered parts.

Perhaps the most fundamental lesson a modern writer can learn from the parallel and balanced style is the necessity of giving form to what he or she thinks and feels. The form congenial to the eighteenth century comes less naturally to us; less elegant in language (as in manners and dress), less inclined to think in parallel and balanced patterns, we subscribe more to the cultural goal of living and feeling deeply and sincerely than to that of thinking reasonably and urbanely. But while we no longer wish to write like Jefferson or Johnson, we can still use their techniques as ways of organizing at least some of our experience and knowledge, and as a means of attaining economy, emphasis, and variety in our sentences.

EXERCISE

1. The following sentences all exhibit balanced construction. Some exhibit a simple one-to-one balance; others are more complicated. Identify the general pattern of each, whether ——/——; ——/— —; — —/——; and so on.

A. I was enjoying the privilege of studying at the world's finest universities; Negroes at home were revolting against their miserable condition.

Stanley Sanders

B. As for me, I am no more yours, nor you mine, Death hath cut us asunder; and God hath divided me from the world and you from me.

Sir Walter Raleigh

C. For aristocrats and adventurers France meant big money; for most Englishmen it came to seem a costly extravagance.

Geoffrey Hindley

D. Then she shrieked shrilly, and fell down in a swoon; and then women bare her into her chamber, and there she made overmuch sorrow.

Sir Thomas Malory

E. Heaven had now declared itself in favour of France, and had laid bare its outstretched arm to take vengeance on her invaders.

David Hume

F. The more we saw in the Irishman a sort of warm and weak fidelity, the more he regarded us with a sort of icy anger.

G. K. Chesterton

G. Building ceases, births diminish, deaths multiply; the nights lengthen, and days grow shorter.

Maurice Maeterlinck

H. In a few moments everything grew black, and the rain poured down like a cataract.

Francis Parkman

I. He could not keep the masses from calling him Lindy, but he convinced them that he was not the Lindy type.

John Lardner

J. In literature there is no such thing as pure thought; in literature, thought is always the handmaid of emotion.

J. Middleton Murry

2. Choosing different subjects from those in the text, compose five balanced sentences modeled upon examples in the preceding question.

CHAPTER 33

Hierarchic Structure

INTRODUCTION

Serial structure hooks together short, relatively self-contained, nonparallel units, one after the other. In parallel and balanced sentences, similar ideas are expressed in similar constructions organized in proportioned, counterpoised patterns. In both types of structure, the units are treated more or less as equally significant.

In hierarchic structure, however, inequality replaces equality. One idea becomes paramount and the others are gathered around it. Subordination, rather than coordination, parallelism, or balance, is the principle of hierarchic structure.

The relative positions of the main clause and the subordinate constructions in hierarchic sentences take four possible arrangements:

1. Main clause + subordinate constructions: called *loose* structure.
2. Subordinate constructions + main clause: called *periodic* structure.
3. Main clause interrupted + subordinate constructions + main clause completed: called *convoluted* structure.
4. Subordinate constructions + main clause + additional subordinate constructions: called *centered* structure.

THE LOOSE SENTENCE

At its simplest the loose sentence consists of a main clause followed by a subordinate construction:

> We must always be wary of conclusions drawn from the ways of the social insect, since their evolutionary track lies so far from ours.
>
> Robert Ardrey

One can easily increase the length and complexity of the loose sentence by adding further phrases and clauses, related either to the main clause or to something in the preceding subordinate constructions:

> We arrived at Odiham about half after eleven, at the end of a beautiful ride of about seventeen miles, in a very fine and pleasant day.
>
> William Cobbet

> Ten minutes later the steamer is under way again, with no flag on the jackstaff and no black smoke issuing from the chimney.
>
> Mark Twain

> I found a large hall, obviously a former garage, dimly lit, and packed with cots.
>
> Eric Hoffer

> I knew I had found a friend in the woman, who herself was a lonely soul, never having known the love of man or child.
>
> Emma Goldman

As the number of subordinate constructions increases, the loose sentence approaches the cumulative style, discussed earlier as a type of serial structure. It is impossible to draw a clear line between loose and cumulative sentences. The difference lies in relative length and weight of the subordinate constructions. In the cumulative sentence these take over the sentence, even though they are inferior in a technical grammatical sense, becoming more significant than the main clause which introduces them. The three following passages fall about halfway between loose, hierarchic structure and the cumulative sentence:

> The walls of his great hall are covered with the horns of the several kinds of deer that he has killed in the chase, which he thinks the most valuable furniture of his house, as they afford him frequent topics of discourse, and show that he has not been idle.
>
> Joseph Addison

> Llanblethian hangs pleasantly, with its white cottages, and orchard and other trees, on the western slope of a green hill; looking far and wide over green meadows and little or bigger hills, in the pleasant plain of Glamorgan; a short mile to the south of Cowbridge, to which smart little town it is properly a kind of suburb.
>
> Thomas Carlyle

> It will frequently be found convenient for a sailing vessel to keep tolerably close to the coast of Borneo, especially when working to windward against the northeast monsoon, as favourable tidal streams will be found near the shore when a strong current is running to the southward some distance from it.
>
> *China Sea Pilot*

The loose sentence is ideal for writing that aims at being colloquial, informal, relaxed. It puts first things first, as most of us do when we talk. Even so, the loose sentence lacks emphasis, and it can easily become formless. Its unity derives not so much from a structural principle as from the coherence of what it expresses. A loose sentence is well formed to the degree that it mirrors the writer's sense that he or she has completed a complicated idea or perception. In describing his great-grandmother's Brooklyn home, this writer begins his paragraph with the following loose sentence:

> Her house was a narrow brownstone, two windows to every floor except the ground, where the place of one window was taken by a double door of solid walnut plated with layers of dust-pocked cheap enamel. Its shallow stoop. . . .
>
> William Alfred

The perception that Alfred's sentence encloses—the view of the facade of the house—unifies it. When that perception ends and our eyes focus upon the stoop, the writer wisely begins a new sentence. Of course, this question of when to stop, of knowing when one statement should end and a new one should begin, applies to all kinds of sentences. But it causes special problems when you use loose structure, where the absence of a clear stopping place may tempt you to run on and on.

THE PERIODIC SENTENCE

The term *periodic* comes from ancient rhetoricians who used it to describe a sentence organized into at least two parts and expressing a complex thought not brought to completion until the close. English scholars adapted the term to describe elaborate balanced and antithetical sentences which also came together only at the end. Today this quality of not completing the essential idea until the close of the sentence—rather than elaborate organization into parts—is what we mean by *periodic structure*. Here we use the term to designate that kind of complex sentence which places the main clause after subordinate constructions (with the warning that periodic structure did not mean exactly this to the Greeks and Romans or even to the seventeenth- or eighteenth-century English). Here are a few examples of simple periodic sentences:

> If there is no future for the black ghetto, the future of all Negroes is diminished.
>
> Stanley Sanders

> As we started to arrange the pieces on the board, I was startled by the sight of his crippled right hand.
>
> Eric Hoffer

> In all the years since the Fetterman fight at Fort Phil Kearny, he had studied the soldiers and their way of fighting.
>
> Dee Brown

Such sentences reverse the pattern of loose structure, following instead a climactic order with the major point last. To the degree that more and more subordinate constructions piled up at the beginning of the sentence postpone the main clause, the sense of climax increases:

Whoever wishes to be made well acquainted with the morbid anatomy of governments, whoever wishes to know how great states may be made feeble and wretched, should study the history of Spain.

Thomas Babington Macaulay

Given a moist planet with methane, formaldehyde, ammonia, and some usable minerals, all of which abound, exposed to lightning or ultraviolet radiation at the right temperature, life might start almost anywhere.

Lewis Thomas

When you and I are in Heaven with the angels, the troubled people of Ivorytown, Rinsoville, Anacinburg, and Crisco Corners, forever young or forever middle-aged, will be up to their ears in inner struggle, soul searching, and everlasting frustration.

James Thurber

As these examples show, there is no single formula for the periodic sentence. Macaulay delays his main idea by paralleling two lengthy *whoever*-clauses, which together make up the subject of his sentence. Thomas opens with two relatively long participial phrases before coming to the main clause. And Thurber begins with a subordinate adverbial clause and further delays the main clause by listing four imaginary towns of radio soap-opera and by the interrupting modifiers "forever young or forever middle-aged."

The periodic sentence is emphatic. By delaying and preparing the way for the principal thought, you alert readers to its importance. The style is also formal and literary, suggesting not the flow of familiar talk, but the writer at his desk.

THE CONVOLUTED SENTENCE

In this special type of periodic structure the subordinate elements, instead of preceding the main clause, split it apart from the inside:

None of this, naturally, was true.

Wallace Stegner

White men, at the bottom of their hearts, know this.

James Baldwin

John Quinn, a heavy-set, blond college student, was always happy.

Student

And once in a spasm of reflex chauvinism, she called Queen Victoria, whom she rather admired, "a goddamned old water dog."

William Alfred

> Such clothes as they wear, a skirt of shredded bark, a buckskin breechclout, an occasional fur or feather cape, also blend into the natural background.
>
> Theodora Kroeber

Convoluted structure has certain virtues. As an occasional rather than an habitual style, it is a good way of achieving variety in sentence movement. It also establishes strong emphasis by throwing weight upon the words preceding the commas or dashes which set off the intruding constructions:

> Now demons, whatever else they may be, are full of interest.
>
> Lytton Strachey

Here both "demons" and "full of interest" draw our attention, expressing the principal idea more strongly than either loose or straightforward periodic structure would:

> Now demons are full of interest, whatever else they may be.

> Now whatever else they may be, demons are full of interest.

This does not suggest anything inherently better about the convoluted pattern: it is simply a convenient way to establish certain kinds of emphasis.

Convoluted structure, however, is formal, and it can tax the reader's attention, especially as the interrupting elements grow longer and more complicated, a problem clear in these examples:

> Even the humble ambition, which I long cherished, of making sketches of those places which interested me, from a defect of eye or of hand was totally ineffectual.
>
> Sir Walter Scott

> Over this little territory, thus bounded and designated, the great dome of St. Paul's, swelling above the intervening houses of Paternoster Row, Amen Corner, and Ave Maria Lane, looks down with an air of motherly protection.
>
> Washington Irving

> The life story to be told of any creative worker is therefore by its very nature, by its diversion of purpose and its qualified success, by its grotesque transitions from sublimation to base necessity and its pervasive stress towards flight, a comedy.
>
> H. G. Wells

These are good sentences, carefully articulated and precise; but they are not easy to read. They require an attentive mind capable of remembering a suspended construction, of recognizing when that construction is resumed, and of putting the pieces together. Used sparingly, the long, intricate convoluted sentence has the advantage of the unusual: it draws attention to itself and, more important, to what it says, and it can be a pleasant and stimulating challenge to the reader. A steady diet of such challenges, however, is indigestible.

THE CENTERED SENTENCE

The type of hierarchic sentence which places the main clause more or less in the middle, with subordinate elements on either side, has no common label. It has been called "circuitous" and "round composition"; we shall use the term *centered*. Whatever we call it, we see it often. (The main clauses have been italicized in the examples that follow.)

> But when Custer reported that the hills were filled with gold "from the grass roots down," *parties of white men began forming like summer locusts,* crazy to begin panning and digging.
>
> Dee Brown

> In every crack that offered a foothold, *a species of wild begonia grew,* spreading its dark-green plate-shaped leaves against grey rock, the delicate sprays of waxy yellow flowers drooping down towards the rushing waters beneath.
>
> Gerald M. Durrell

> Looking out from the bushes, *I saw her trotting towards an open space of lawn the other side the pond,* chattering to herself in an accustomed fashion, a doll tucked under either arm, and her brow knit with care.
>
> Kenneth Grahame

Centered construction, while not as emphatic as periodic or as relaxed and informal as loose construction, has several advantages, especially in longer sentences with a number of subordinate elements. It enables a writer to sort out and place those elements more clearly. If half a dozen phrases or clauses all come before the main elements (as in the usual periodic sentence) or after them (as in the loose), some of these may seem to float free, so far from what they connect with that the connection becomes obscure. But if some of the subordinate constructions can come before and others after the principal clause, this reduces the chances of missed connections.

The centered sentence also gives the writer a better chance to arrange elements to reflect the natural order of events or of ideas. Jonathan Swift does this in this passage criticizing England's participation in the War of the Spanish Succession (1701–1714):

> After ten years' fighting to little purpose, after the loss of above a hundred thousand men, and a debt remaining of twenty millions, we at length hearkened to the terms of peace, which was concluded with great advantages to the empire and to Holland, but none at all to us, and clogged soon after with the famous treaty of partition.

Granted a broad and uncritical meaning of "idea," we may say that the sentence contains nine of them: (1) the "ten years' fighting"; (2) the "little purpose," or lack of result; (3) the "loss" of the men; (4) the "debt remaining"; (5) the "hearkening" to peace; (6) the conclusion of the peace; (7) the "advantages" that followed for England's allies;

(8) the absence of such advantages for England herself; and (9) the "clogging" of the peace. Here syntax mirrors events. In the sentence, as in history, the fighting comes first; then the absence of results, the loss of life, and the debt; then follows the hearkening to peace, its conclusion, and so on. Effecting a workable compromise between the natural order of thought or reality on the one hand, and the order of sentence elements on the other, is one of the most difficult tasks a writer faces. The centered sentence makes the task a little easier.

EXERCISE

1. Which of these sentences are loose, periodic, convoluted, or centered?

   A. California's first people, so far as is presently known, were American Indians, ancestors of today's Indians and in no significant way different from them.

      Theodora Kroeber

   B. And yet, in spite of it all, the silence persists.

      Aldous Huxley

   C. When the white men in the East heard of the Long Hair's defeat, they called it a massacre and went crazy with anger.

      Dee Brown

   D. These camps were originally established by Governor Rolph in the early days of the Depression to care for the single homeless unemployed of the state.

      Eric Hoffer

   E. To Bishop Whipple's listeners, this seemed a strange way indeed to save the Indian nations, taking away their Black Hills and hunting grounds, and moving them far away to the Missouri River.

      Dee Brown

   F. The Protestants—for that was what she called native-born Americans—moved out as the immigrants moved in.

      William Alfred

   G. Born in a region long poor and defeated, into a family itself humble and moneyless, often at the mercy of capricious economic, social or political forces, the boy at home faced those first insecurities, those early rivalries, hates and struggles which often set the pattern for later ones.

      Robert Coles

   H. Having wanted to walk on the sea like St. Peter he had taken an involuntary bath, losing his mitre and the better part of his reputation.

      Lawrence Durrell

   I. One inmate had been kept there for twenty-eight days on bread and water, although the regulations prohibited a longer stay than forty-eight hours.

      Emma Goldman

   J. A fat woman lounges, her arm on a cheap yellow chest of drawers, behind him.

      Virginia Woolf

K. As the sun went down and the evening chill came on, we made preparation for bed.

Mark Twain

L. Standing on the summit of the tower that crowned his church, wings upspread, sword uplifted, the devil crawling beneath, and the cock, symbol of eternal vigilance, perched on his mailed foot, Saint Michael held a place of his own in heaven and on earth which seems, in the eleventh century, to leave hardly room for the Virgin of the Crypt at Chartres, still less for the Beau Christ of the thirteenth century at Amiens.

Henry Adams

M. In spite of all the silly talk about his vulgarity, he really had, in the strict and serious sense, good taste.

G. K. Chesterton

N. Paralyzed by the neurotic lassitude engendered by meeting one's past at every turn, around every corner, inside every cupboard, I go aimlessly from room to room.

Joan Didion

O. Success in almost any calling in England, unless the aspirant happens to be endowed with energies and talents almost superhuman, depends in a great measure on the possession of money.

W. H. Hudson

P. The mountain paths stoop to these glens in forky zigzags, leading to some grey and narrow arch, all fringed under its shuddering curve with the ferns that fear the light; a cross of rough-hewn pine, iron-bound to its parapet, standing dark against the lurid fury of the foam.

John Ruskin

Q. And in one corner, book-piled like the rest of the furniture, stood a piano.

Kenneth Grahame

R. He is a dapper little fellow, with bandy legs and pot belly, a red face, with a moist merry eye, and a little shock of red hair behind.

Washington Irving

S. Japanese families—and the villagers were conscious of the fact—were enterprises and not just groups of people bound by blood and sentiment.

Ronald P. Dore

2. Revise each of these loose sentences so as to make it periodic.

   A. The men returned to camp in the morning after the sun had come up, burning the dew off the ground.
   B. They lost control when the boat buried her bow in a heavy sea, which broke on the foredeck, washing away the cabin and flooding the cockpit.
   C. The books filled the room, piled anywhere, on tables and on chairs and heaped on the floor.

3. Turn these periodic sentences into loose ones:

   A. To the best of my recollection, which is far from perfect, I never said any such thing.

    B. Although few people really know what happened that night, many have expressed opinions.

    C. Boisterous, drinking heavily, ill-mannered and badly dressed, he annoyed everyone.

4. Using the sentence about Custer by Dee Brown on page 267(E) and the sentence by Swift on page 266 as models, compose two centered sentences on topics of your own choice.

5. Compose a convoluted sentence describing someone getting up in the morning, a student preparing to study, or a teacher getting ready to lecture. Follow the sentence below carefully, beginning with some adverbial word such as *then* (*but, thus, finally*), following this with a participial construction, then the main subject and the first of two verbs, a second interrupting phrase, and finally the second predicate.

> Then, tucking his paper under his pillow, he popped out the guttering candle, and turning around upon his side with a smile of exceeding sweetness, settled himself to sleep.
>
>   Austin Dobson

CHAPTER 34

Sentence Patterns: Summary

The kinds of sentences we have so far discussed differ essentially in how they organize ideas, perceptions, feelings. Segregating sentences analyze thought into separate units. The freight-train sentence, still maintaining a clear separation, brings ideas more closely together by hooking them one after another within a single statement. Parallel and balanced sentences express logically comparable ideas in the same form and assign them identical or similar grammatical roles. The style is suited to underscoring similarities or sharp differences of thought. Hierarchic sentences select one idea as focal and subordinate others to it, placing them after the main thought (the loose style), before (the periodic), inside of (the convoluted), or around (the centered).

These various styles are capable of combination. One can use parallelism and balance, for example, to organize some of the subordinate elements within essentially hierarchic sentences. Or one can employ subordination within parallel sentences. Sometimes freight-train sentences contain parallel or subordinate elements. Of course, certain combinations are impossible by definition. You cannot have a sentence both loose and periodic, or a hierarchic segregating sentence. But allowing for such exceptions, numerous combinations are possible. Here, for instance, is a case of parallelism within a loose sentence:

> It was a big, squarish frame house that had once been white, decorated with cupolas and spires and scrolled balconies in the heavily lightsome style of the seventies, set on what had once been our most select street.
>
> William Faulkner

The main clause comes first ("It was a big, squarish frame house"). Parallelism partially organizes the modifiers that trail after it in loose style: two parallel participial phrases ("decorated . . . set"), and within the first of these, three objects of the same preposition ("with cupolas . . . spires . . . balconies").

In the next example parallelism combines with a periodic style:

When they learn that in France young fellows, notorious for their debauchery, raised to their bishoprics by women's intrigues, make love in public, amuse themselves by composing tender lyrics, spread daily and at length a luxurious dinner, and go from it to pray for the guidance of the Holy Ghost, while they boldly claim to be the successors of the apostle, Englishmen thank God that they are Protestants.

Voltaire

Voltaire's sentence contains considerable parallelism, all centered on the subject "fellows": two adjectival phrases ("notorious . . . raised") plus four predications ("make . . . amuse . . . spread . . . and go"). Yet the general frame is periodic, the main clause reserved for the end.

Finally, this basically parallel sentence about Indian tribes in California also has an interruption between subject and verb in convoluted fashion:

After two or three thousand years, "barbarians" from outside, Wintun and others, who were by then stronger and more numerous than the older population, engaged in one of their thrusts of history making, invaded Yana country, occupied the richer parts of it, and pushed the smaller older population back into the hills.

Theodora Kroeber

The interrupter "Wintun . . . older population" comes between the subject " 'barbarians' " and the four parallel verbs ("engaged . . . invaded . . . occupied . . . and pushed").

The test for mixing styles is whether or not the mix works. These styles exist, and each has capacities and limitations making it useful on some occasions, awkward on others. An individual writer may find one style especially congenial, but a good writer is able to use them all.

CHAPTER 35

Concision in the Sentence

INTRODUCTION

Concision is brevity relative to purpose, not to be confused with absolute brevity. A sentence of seven words is brief; but if you could express the idea with equal clarity in five, your sentence is not concise. On the other hand, a sentence of fifty-three words is in no sense brief; but it is concise if you cannot make your point in fewer words.

Words that do not contribute in any way to a writer's purpose are called *deadwood*. Most deadwood results from poor word choice. In Chapter 38, which deals with diction problems, we shall look at this kind of wordiness. Some words, however, die when you construct inefficient sentences, and we shall study that problem here.

Dead. s. The abbreviation "dead. s." serves as a blanket term to cover all cases of unneeded words. A second abbreviation indicates the particular kind of error. For example, "dead. s./paral." means "deadwood because of a failure to use parallelism." For convenience the eight specific errors appear alphabetically.

Dead. s./
adj.-adv. ## USE SINGLE ADVERB OR ADJECTIVE

WORDY	He acted *in an unnatural way.*
CONCISE	He acted unnaturally.
WORDY	The organization of a small business can be *described in a brief statement.*
CONCISE	The organization of a small business can be briefly described.
WORDY	He prefers wines *having a French origin.*
CONCISE	He prefers French wines.

In good writing, adverbs and adjectives should link as directly and concisely as possible to what they really modify. The writers of the

272

first two examples above are afraid of adverbs. "Unnatural" really describes "acted," but instead of directly connecting it to that verb, the writer attaches it to "way" in an unnecessary prepositional phrase. And "brief" is one step removed from "described"; turned into an adverb it is a clearer and more efficient modifier. In the last example the writer used a whole phrase for a job a single adjective could do with more clarity and less effort: "French wine" is by definition "wine having a French origin."

Now and then you may boil down an entire clause to a phrase or even a single word:

WORDY	American exploration was rapid considering the means *which the pioneers had available to them.*
CONCISE	American exploration was rapid considering the means available to the pioneers.
WORDY	The targets *that are supplied in skeet shooting are discs made of clay.*
CONCISE	Skeet targets are clay discs.

Dead. s./ **AVOID AWKWARD ANTICIPATORY CONSTRUCTIONS**
Antic.

WORDY	*This is the kind of* golfer *that is* called a hacker.
CONCISE	This kind of golfer is called a hacker.
	Such golfers are called hackers.

In an anticipatory sentence the notional subject—that is, what the sentence is really about—is not the grammatical subject. Instead, it is introduced (or "anticipated") by a pronoun such as *it, this, that, these, those, there* which functions as the grammatical subject. (The *there*-construction is actually different grammatically but works the same way for all practical purposes.) A verb like *is, are,* or *seems* links the notional subject to the pronoun, and an adjectival clause or phrase that tells us what is being predicated about the subject usually follows:

This is the man who witnessed the accident.

There are many property owners who object to new schools.

Those are the people from Chicago.

Anticipatory constructions use more words than comparable direct statements. Sometimes the extra words serve a valid purpose, such as emphasis or idiom. When they do not, they are simply deadwood, and you should recast the sentence to make better use of its main elements. *Seems* and its close relative *appears* are especially frequent in wordy anticipatory sentences. From an excess of caution or politeness some writers habitually hedge their bets, preferring a hesitant claim like

It seems that this professor did not prepare his lectures very well

to the bolder assertion:

This professor did not prepare his lectures very well.

Now and then such qualification is necessary, but more often it is not.

About any anticipatory construction, then, ask yourself whether idiom or emphasis justifies it. To revise "It is true that we did not like the idea at first" to "That we did not like the idea at first is true" prunes one word but results in a stiff, highly formal sentence, less effective than the original. To change "This is the man who witnessed the accident" to "This man witnessed the accident" deemphasizes the point, hardly an improvement if the writer wants to make a strong statement. But when such justifications do not exist, the anticipatory construction is wordy, and you ought to replace it by a more direct sentence.

 ## USE COLON OR DASH

WORDY   There were many reasons for the Civil War, *which include* slavery, economic expansion, the issue of states' rights, cultural differences, and sectional jealousies.

CONCISE   There were many reasons for the Civil War: slavery, economic expansion, the issue of states' rights, cultural differences, and sectional jealousies.

WORDY   Pitchers are divided into two classes. *These classes are* starters and relievers.

CONCISE   Pitchers are divided into two classes—starters and relievers.

In sentences like these the colon and dash say: "Here follows a series of particulars." If you let the punctuation mark do the talking, you won't need deadwood like "which include" or "these classes are." The only difference between the colon and the dash in this function is that the colon is more formal. (However, each mark has other, very different tasks.)

The conciseness of the colon and dash also sets up a key idea delayed for emphasis:

WORDY   But a counterforce has been established within the weapons platoon. *This counterforce is* the anti-tank squad.

CONCISE	But a counterforce has been established within the weapons platoon—the anti-tank squad.

USE ELLIPSIS

Dead. s./
ellip.

WORDY	He is taller than his brother *is*.
CONCISE	He is taller than his brother.
WORDY	When *you are* late, you must sign yourself in.
CONCISE	When late, you must sign yourself in.
WORDY	He lost his wallet; she *lost* her pocketbook.
CONCISE	He lost his wallet; she, her pocketbook.

Ellipsis (plural: *ellipses*) means omitting words from a sentence which are necessary to complete the grammar but not the sense. The writer using ellipsis assumes that the reader can supply the missing terms for him- or herself. Ellipses are common in sophisticated writing. In fact, one may regard some of the other matters in this chapter as types of ellipsis: parallelism, for instance, or certain kinds of direct modification.

The omission of words often secures concision with no cost in clarity or emphasis. It may even enhance those qualities. In the first example above the sense does not require the second "is," and the revision allows the statement to end on the key term "brother." In the second, the concise version lays more stress upon "late" and avoids repeating "you." In the third, dropping "lost" from the second clause makes for an unusual, striking statement.

The unusual quality of some ellipses sets a limit to their usefulness, however. "He lost his wallet; she, her pocketbook," for example, has a literary flavor that might seem odd in a matter-of-fact, colloquial passage.

USE PARALLELISM

Dead. s./
paral.

WORDY	These books are not primarily for reading, but *they are used* for reference.
CONCISE	These books are not primarily for reading but for reference.
WORDY	The beginner must work more slowly, *and he must work* more consciously.
CONCISE	The beginner must work more slowly and more consciously.

Parallelism means that two or more words, phrases, or clauses are grammatically related in the same way to the same thing. In "The man and the boy came in together," "man" and "boy" are parallel because each acts as a subject of the same verb ("came in"). Or in "She stood and raised her hand," "stood" and "raised" are parallel because each is a verb of the same subject.

Parallelism resembles factoring in mathematics; instead of repeating "a" in $2ax + 3ay + az$ the mathematician writes $a(2x + 3y + z)$. In a parallel grammatical construction the governing term need not be stated two or three times. In our first example the phrase "for reference," by being made parallel to "for reading," does duty for the whole second clause, while in the other example "more consciously" replaces a clause.

At times, however, parallelism improves nothing. Emphasis or rhythm justifies a certain amount of repetition. For instance, to stress "work" the writer might well stick with the original of the second example.

USE PARTICIPLES

Dead s./ part.

WORDY	It leaves us *with the thought* that we were hasty.
CONCISE	It leaves us thinking that we were hasty.
WORDY	This is the idea *that was* suggested last week.
CONCISE	This is the idea suggested last week.

Participles are economical. In cases like that in the first example an abstract noun ("thought"), which requires a preposition and an article, can be replaced by the single participle "thinking," which needs neither of those form words.

The second example shows how to prune an adjective clause consisting of a relative word ("that") + a linking verb ("was") + a participle, predicate noun, or predicate adjective ("suggested"). By dropping the relative word and the linking verb, you can move directly from the noun to the participle (or predicative) which modifies it.

However, if the revision sacrifices emphasis or causes the sentence to move so rapidly that it confuses the reader, then you should retain the full clause.

You can sometimes cut back an entire adverbial clause to the operative participle:

| WORDY | *Because they were* tired, the men returned to camp. |
| CONCISE | Tired, the men returned to camp. |

Even an independent clause or a separate sentence may work more efficiently as a participial construction:

WORDY	These football plays are very effective, *and they* often prove to be long gainers.
CONCISE	These football plays are very effective, often proving to be long gainers.
WORDY	The women of the settlement would gather together at one home to work on the quilt. *They would* bring their children *with them* and spend the entire day and chatter gaily as they worked.
CONCISE	The women of the settlement would gather together at one home to work on the quilt, bringing their children and spending the entire day, chattering gaily as they worked.

The last two examples illustrate again that the virtue of concision lies less in saving words than in improving the clarity and focus of a sentence. In the first case the revision makes clearer that certain football plays are effective because they result in long gains, while the more concise version of the second corrects an awkward shift in subject from the quilting bee to the children.

USE PREDICATE ADJECTIVES

Dead s./ pred.

WORDY	Riots became frequent *affairs.*
CONCISE	Riots became frequent.
WORDY	Mr. Martin is *a* quiet, patient, and cautious *person.*
CONCISE	Mr. Martin is quiet, patient, and cautious.
WORDY	The day was *a* perfect *one.*
CONCISE	The day was perfect.

A *predicate adjective* stands after its noun, connected to it by a linking verb (*is, are, was, were, seems, becomes,* and so on):

The house is *large.*

An *attributive adjective* stands before the noun it modifies:

a *large* house.

Predicate adjectives are not inherently better. But it *is* better not to repeat a word or idea just to use an adjective attributively, as the examples above do. "Affairs," "person," and "one" are empty words here. Each merely repeats the subject, serving as a peg upon which to hang an attributive adjective. When the adjective appears as a predicative, the need for the empty word disappears. And more important, the adjectives get the attention they deserve.

DO NOT WASTE THE SUBJECT, VERB, AND OBJECT

WORDY   *The fact of* the war *had the effect of* causing many changes.

CONCISE   The war caused many changes.

The subject, verb, and object form the main core of a sentence. They ought to convey the core of the thought, too. Isolating the subjects, verbs, and objects from these examples, we get:

fact had effect
war caused changes

Clearly the revision—less than half as long as the original—makes better use of the main elements: one can quickly grasp the nub of the idea. But who could guess the writer's point from "fact had effect"?

Not employing your main sentence elements efficiently sometimes results from not knowing what your topic really is. If you let the wrong idea become the subject, the sentence will stagger under a load of excess wordage as you struggle to say what you mean:

The first baseman wears a special leather glove that is designed for easy scooping and long-range catching, while the catcher wears a large glove that is heavily padded to protect him from fast pitches.

The grammatical subject of the first clause is "the first baseman," and of the second, "the catcher"; but the players are not being compared—their gloves are. Because they begin wrong, both clauses require an otherwise unnecessary verb ("wears") to get from the grammatical subject to the real one. If the appropriate subject ("glove") appears in the right place to begin with, the sentence gets off on the right foot and steps along easily:

The first baseman's glove is designed for easy scooping and long-range catching, while the catcher's is large and heavily padded to protect him from fast pitches.

EXERCISE

1. Remove the deadwood from the following sentences. Stick as closely as possible to the original wording, concentrating upon removing what is unnecessary rather than drastically revising the statement. Be able to justify your changes.

    A. She worried about the cooking of the dinner.
    B. The book is an interesting one.
    C. It seems that nobody liked the decision of the committee.
    D. John said that he was ready to go and said that you were ready to go too.
    E. With the hope that all would turn out well he left for the meeting.
    F. The medical profession is indeed a worthwhile one.

G. Parties became more and more boisterous affairs.

H. There are many aspects that make up civilization.

 I. The repeating of words with the same meaning is a source of deadwood.

J. This is a gibberish type statement.

K. There were many reasons for the depression. These reasons include credit buying, widespread speculation in stocks, and overoptimism among businessmen.

L. With a knowledge of why he is reacting in such a fashion a neurotic husband may be able to overcome his problems.

M. This is done by executing a climbing turn. A climbing turn is a maneuver in which the nose of the plane is held up to gain altitude and the plane is banked just enough to counter the centrifugal force.

N. Such a display of confidence had a relaxing effect upon the patient.

O. When he is at a party, he thinks first of himself.

2. Using a red pencil, go over your last theme, marking examples of the kinds of deadwood discussed in this chapter.

36

The Emphatic Sentence

INTRODUCTION

Emphasis means strength or force, and we use it all the time in communication. In speech we achieve emphasis in a variety of ways: by talking loudly (or sometimes very softly); by speaking slowly, carefully separating words that ordinarily we run together; by altering our tone of voice or by changing its timbre. We also stress what we say nonvocally. A rigid, uncompromising posture; a clenched fist; a pointing finger; any of many other body attitudes, gestures, and facial expressions, which vary considerably from culture to culture—all signal degrees and kinds of emphasis.

Writers can rely upon none of these signals. Yet they too need to be emphatic. What they must do, in effect, is to translate loudness, intonation, gesture, and so on into writing, and equivalents are available. Some are merely visual symbols for things we do in speech: much punctuation, for example, stands for pauses we would make if we were talking. Other devices, while not unknown in speech, belong primarily to composition, and we shall look at these in this chapter and the next.

Before examining the specific ways in which writers draw attention to certain words, we must distinguish between the emphatic sentence and emphasis within the sentence. The emphatic sentence is one that stands out in its totality. Consider these variants:

An old man sat in the corner.
In the corner sat an old man.

The first is matter-of-fact, attaching no special importance to what it tells us. The second, by merely rearranging the same words, does more than state a fact; like a close-up in a film, it hints that the fact is significant. Now this distinction does *not* mean that the second version is superior to the first: simply that it is more emphatic. Whether or not the emphasis makes it better depends upon what the writer wants to say.

By their nature, emphatic sentences must occur rarely; their effectiveness hinges upon rare appearance. Writing in which every sentence is emphatic—or even every other—has the effect of somebody shouting in your ear.

Emphasis within the sentence does not involve the total statement, only a part of it.

It suddenly began to rain.
Suddenly, it began to rain.

The second version lays more stress upon "suddenly" by moving it to the opening position and isolating it with a comma. Again we have no question of better and worse. Both sentences are clear and concise. Which is preferable depends upon the writer's intentions.

Most well-written sentences make one word or phrase or clause the most important. The problem is to ensure that the reader understands correctly *which* is the most important. It is easy to make mistakes in emphasis, and they seriously disrupt communication. The writer must construct his or her sentences to establish clear and appropriate emphasis. This is what we are about to study.

Emph. s. The abbreviation "Emph. s." means that a passage deserves stronger expression than you have given it. If the construction is already a separate sentence, you need to revise it in a more emphatic form. If it is a clause within a longer sentence, you should recast it as a strong separate statement. The additional eight abbreviations suggest particular types of emphatic construction that would solve the problem (though these suggestions do not rule out other solutions).

Emph. s. ## ANNOUNCEMENT
Announce.

Announcement means introducing an idea of special importance by a preliminary statement which says to the reader "watch out; here comes something important." Often that something important follows in the latter part of the same sentence:

Finally, last point about the man: he is in trouble.

Benjamin DeMott

It has, of course, the quality common to all special and unbalanced types of virtue, that you never know when it will stop.

G. K. Chesterton

Whether the stressed idea is an independent statement, as in the first of these examples, or a dependent clause, as in the second, it must be separated from the introduction by appropriate punctuation: a colon, dash, semicolon, or comma.

Occasionally the key thought may appear in a new sentence:

> For the true rationale of humanistic study is now what it has always been,
> even though now it is not only in decay, but dead. I allude to the arts of
> rhetoric.
>
> Allen Tate

But however it relates to the clause announcing it, the stressed idea
should be phrased as concisely and vigorously as possible, gaining further strength by contrast to the comparatively looser statement that
introduces it.

Anticipatory constructions are also a variety of announcement.[1] They
give less attention to the introduction, making do with little more than
a pronoun or "there" + the verb:

> This was the consequence we feared.
>
> Student

> It's tragic—this inability of human beings to understand each other.
>
> Joy Packer

Like all types of emphatic sentences, announcement works only when
you use it now and then. Indeed, misusing it can seem worse than
misusing some of the other varieties of emphasis. No one wants to
read something trivial after great fanfare. Moreover, announcement
requires words, and if emphasis does not really justify them, these extra
words become so much deadwood.

BALANCE

Emph. s.
bal.

The balanced sentence divides into two approximately equal parts on
either side of a central pause. Usually a comma or other stop marks
the pause, though now and then it may go unpunctuated. The halves
of the balanced sentence are generally independent clauses, but sometimes one side may be a dependent clause or even a phrase. In any
case, the two parts must be roughly the same in length and of comparable significance, although they need not be of the same grammatical
order.

Balanced sentences achieve emphasis by playing key terms against
one another:

> Man has no property in man; neither has any generation property in the
> generations which are to follow.
>
> Thomas Paine

---

1. In anticipatory construction the notional subject—that is, the thing the predication
is about—is introduced by an expletive, such as *there, this, it,* followed by a linking
verb, usually *is* or some other form of *to be.* For example, in the sentence "There are
some people who do not agree," the notional subject is "people." But instead of being
in the usual subject position at the beginning of the sentence, it follows "there are."

It is a sort of cold extravagance; and it has made him all his enemies.

G. K. Chesterton

Till he had a wife he could nothing; and when he had a wife he did whatever she chose.

Thomas Babington Macaulay

The balanced ideas here are similar in nature. Chesterton's sentence clarifies the connection between "cold extravagance" and "making enemies." Paine stresses the essential similarity between one "man" and a "generation" of men. And Macaulay, playing "do nothing" against doing "whatever she chose," comments wryly on the freedom of the married man.

Often, however, the key terms in a balanced sentence are in sharp contrast, or antithesis:

Cattiness is a cold war staged by women; macho is a hotter war fought by men.

Student

The modest man mentions skyscrapers he saw in Chicago; the braggart saw taller ones in New York.

Student

Sometimes the positions of the key terms are reversed, a rhetorical device called *chiasmus* or *antimetabale*:

If there had never been a danger to our constitution there never would have been a constitution to be in danger.

Herbert Butterfield

Books are among the best of things, well used; abused, among the worst.

Ralph Waldo Emerson

THE FRAGMENT

Emph. s.
frag.

A *fragment* is a construction that begins with a capital letter and ends with full-stop punctuation, but which does not satisfy the traditional definition of the compositional sentence.[2] While often serious grammatical faults, fragments can on occasion add effective emphasis, if you know how to handle them. Clear and direct like short sentences, they draw attention by their difference:

And that's why there's really a very simple answer to our original question.
What do baseball *managers* really do?
Worry.
Constantly.
For a living.

Leonard Koppett

2. See pages 209–10 for that definition.

"Many a man," said Speer, "has been haunted by the nightmare that one day nations might be dominated by technical means. That nightmare was almost realized in Hitler's totalitarian system." Almost, but not quite.

Aldous Huxley

Sweeping criticism of this type—like much other criticism—throws less light on the subject than on the critic himself. A light not always impressive.

F. L. Lucas

Going off her diet, she gained back all the weight she had lost. Also the friends.

Student

You must not overuse fragments, or they will lose their advantage and become an awkward mannerism. Some teachers, in fact, prefer that you not use them at all. Others may ask you to identify a deliberate fragment by writing "frag." in the margin, which tells them that you know what you're doing. Before you employ a fragment for emphasis, you should learn your instructor's policy.

Emph. a imp.

THE IMPERATIVE SENTENCE

At its simplest the imperative sentence is a command:

Come here!
Listen to me!

Its distinguishing feature—usually—is that it drops the subject and begins with the verb, although some commands use a noun of address or an actual subject:

John, come here!
You listen to me!

Such commands are rare in composition, where issuing orders to readers is not necessary or advisable. But imperative sentences are useful for emphasis, catching our attention by opening with the verb:

Insist on yourself; never imitate.

Ralph Waldo Emerson

Let us spend one day as deliberately as Nature, and not be thrown off the track by every nutshell and mosquito's wing that falls on the rails.

Henry David Thoreau

Consider, for example, those skulls on the monuments.

Aldous Huxley

Give me that critic who rushed from my play to declare furiously that Sir George Crofts [one of the characters] ought to be kicked.

George Bernard Shaw

Aside from providing emphasis, the imperative sentence has the additional virtue of establishing a direct link between writer and reader.

Emerson does not say "men and women must insist on themselves"; he addresses *you*. Thoreau urges *you* to participate in a new way of life, and Huxley invites *you* to accompany him as he inspects sixteenth- and seventeenth-century monuments.

Given their emphatic nature, imperative sentences serve several functions. Some are simply strong statements: Shaw's "Give me" is an emphatic way of saying "I prefer." Other imperatives are more inspirational than self-assertive: the passages by Emerson and Thoreau, for instance. And still others provide the reader with ways to move quickly and clearly from one point to another: Huxley's sentence.

Emph. s.
int.

THE INTERRUPTED SENTENCE

A sentence or clause moves naturally from subject to verb to complement. Interruption breaks up that movement by inserting various kinds of constructions between these main sentence elements. In the next chapter we shall see that interrupted movement helps to attain emphasis within the sentence by isolating important words or phrases. In the following cases, however, interruption makes the entire statement emphatic:

> White men, at the bottom of their hearts, know this.
>
> > James Baldwin
>
> And finally, stammering a crude farewell, he departed.
>
> > Thomas Wolfe
>
> When I first saw her I was very impressed, not only by her appearance (she was, indeed, a beautiful girl), but by her smile.
>
> > Student
>
> The lax mouth, the veiled, weary, wistfully cynical eyes, betrayed, through the slackening flesh, the softening brain. . . .
>
> > Ralph Roeder [a description of Pope Alexander VI]

Any of those sentences could be expressed more conventionally in straightforward order:

> White men know this at the bottom of their hearts.
> And he finally departed, stammering a crude farewell.

But while more natural, these are weaker statements (again, the revisions are not therefore "poorer"; value depends upon purpose).

Interrupted movement does make demands upon the reader, as Ralph Roeder's sentence shows. Pushed too far or used too often, interrupted construction is likely to tire and repel readers. But kept reasonably simple—as in Baldwin's sentence—interrupted movement is a valuable means of emphasis.

Emph. s.
inv.

THE INVERTED SENTENCE

Inversion primarily means rearranging the main elements of a sentence in some order other than subject-verb-object, as in this clause from Milton's *Paradise Lost*:

| | Him the Almight Power hurl'd. . . . |
| RATHER THAN | The Almight Power hurl'd him. . . . |

A secondary type of inversion consists of moving an adverbial construction from after the verb to before the subject:

| | In the morning the men left. |
| RATHER THAN | The men left in the morning. |

Inversion is not always emphatic. It may signal a question or a condition contrary to fact:

Are you going into town today?
Had I only been there.

But because words appear in odd positions relative to one another—a verb before a subject, say—inversions draw attention. When employed for emphasis, inversion usually takes the form of adverb-verb-subject:

On the floor lay the weapon, a heavy iron poker.

Student

And in one corner, book-piled like the rest of the furniture, stood a piano.

Kenneth Grahame

Here, in addition to the adverbial prepositional phrase, a nonrestrictive adjective construction precedes the verb.

Less commonly, emphatic inversion is of the object-subject-verb type:

Wrangles he avoided, and disagreeable persons he usually treated with a cold and freezing contempt.

Douglas Southall Freeman

Emphatic inversion occurs less commonly in contemporary prose than it did a hundred or more years ago. Some examples from nineteenth-century prose strike a modern ear as stilted:

Him Nature solicits with all her placid, all her monitory picture; him the past instructs; him the future invites.

Ralph Waldo Emerson

Remarkable it is truly . . .

Thomas Carlyle

Our time is lower-key, and such formal, literary sentences sound high-faluting (though in fairness to Emerson and Carlyle, we must remember that they wrote for other ears).

Finally, inversions are tricky, subject to various and subtle conventions of idiom, too numerous and complex to bother with here. When you don't know whether or not a particular inverted sentence will work, the best thing is to read it out loud and trust your ear. If the inversion sounds peculiar, take it out. If you do not listen to what you write, you will find it all too easy to cobble up inversions that simply are not English:

    $\times$ From the sun his complexion never gets tan.
    $\times$ The house was out of the question; mere plasterboard was it.
    $\times$ Dimly-lit, smelling of must, is the common peasant's home.
    $\times$ Twenty-six, handsome, well-built was he.

NEGATIVE-POSITIVE RESTATEMENT

The negative-positive sentence combines restatement, balance, and announcement. It stresses an idea by telling us first what is *not* the fact, then what is:

Ethan Frome is not a tragic fool; he is a tragic hero.

Student

Color is not a human or personal reality; it is a political reality.

James Baldwin

This is more than poetic insight, it is hallucination.

J. C. Furnas

The poor are not like everyone else. They are a different kind of people. They think and feel differently; they look upon a different America than the middle class looks upon.

Michael Harrington

Generally the same sentence contains the negative and positive halves (as in the first three examples above). In an extended passage, however, negative and positive may go in separate sentences (as in the fourth example). In either case, repeating the same type of clause or sentence strengthens the contrast between what something is not and what it is.

Less commonly, the progression goes from positive to negative, as in this passage by G. K. Chesterton about social conventions:

Conventions may be cruel, they may be unsuitable, they may even be grossly superstitious or obscene, but there is one thing they never are. Conventions are never dead.

All this could be put more briefly:

Although conventions may be cruel, unsuitable, or even grossly superstitious or obscene, they are never dead.

But not so strongly.

Emph. s.
per.

THE PERIODIC SENTENCE

A periodic sentence (sometimes called a "suspended" sentence) does not complete its main thought until the end:

> That John Chaucer was only an assistant seems certain.
>
> John Gardner

> That the author of Everyman was no mere artist, but an artist-philosopher, and that the artist-philosophers are the only sort of artists I take seriously will be no news to you.
>
> George Bernard Shaw

> If you really want to be original, to develop your own ideas in your own time, then maybe you shouldn't go to college.
>
> Student

Periodic structure differs from "loose," which places the main clause at the beginning and adds subordinate ideas. The last example above would become loose if revised like this:

> Maybe you shouldn't go to college, if you really want to be original, to develop your ideas in your own time.

A loose sentence can be broken off somewhere before the end and still make its essential point: the revision above, for example, could end after "original." Periodic structure, on the other hand, forces us to keep reading until the very end if we are to grasp what the writer is saying.

Periodic sentences have various constructions.[3] Some develop a long subject like the noun clause in the Gardner and Shaw examples. Others, like the student sentence, begin with subordinate constructions and place the main clause last. But whatever means they use to postpone completion, periodic sentences achieve emphasis by requiring that we pay close attention and suspend our understanding until the final words pull everything together.

This type of sentence has limitations. Too many grow tiresome, for the alertness such a style demands is not easy to sustain. Furthermore, periodic sentences have a formal, literary tone, less suitable than that of loose sentences for informal, conversational occasions. Despite these limitations, however, a periodic sentence now and then supplies valuable emphasis and has the further advantage of varying your sentence patterns.

Emph. s.
rhet. q.

THE RHETORICAL QUESTION

Some rhetorical questions are sometimes devices of organization: a writer asks them so that he or she can generate a paragraph or an

---

3. See page 263 for a fuller discussion of the periodic sentence.

essay by supplying the answer. Others—the kind we are concerned with here—are a type of emphatic statement. In either use, rhetorical questions exist to serve the writer's purpose, not to elicit information like an ordinary query. Formally, however, they are genuine questions and must be closed by a question mark.

Many emphatic rhetorical questions are, in effect, disguised assertions. In the following example "How could I?" is a way of saying "I could not possibly":

> I took the exam on a cold, rainy January morning and I tried hard, but I didn't pass. How could I?
>
> > Student

And here the writer means that the young man was "*not* desirable":

> A desirable young man? Dust and ashes! What was there desirable in such a thing as that?
>
> > Lytton Strachey

Some emphatic questions are more open and complicated in meaning, combining an implicit avowal with an actual query:

> Yet this need not be. The means are at hand to fulfill the age-old dream: poverty can be abolished. How long shall we ignore this under-developed nation in our midst? How long shall we look the other way while our fellow human beings suffer? How long?
>
> > Michael Harrington

Even here, however, Harrington tries not so much to elicit a definite answer as to force upon us the view that allowing poverty to continue is indefensible.

In addition to being a rhetorical question, each of the examples above also illustrates at least one other means of emphasis. Harrington uses a short sentence to open his paragraph, and asks his question not once but three times, ending with a two-word sentence. In the student passage the question is also a short statement and contrasts with the longer one that precedes it. And Strachey writes two brief fragments (one closed by an exclamation point) and poses his question twice. This multiplication of effort points to a significant fact: emphasis need not rely on a single device. In fact, when a writer feels strongly about something, emphasis will probably show itself in several simultaneous ways.

Emph. s.
rhy.

RHYTHM AND RHYME

We think of rhythm and rhyme as properties of poetry, not of prose. Some people even condemn them in prose, though they are more common than these critics suppose. But rhythm and rhyme are not invariably awkward in prose: rather they become awkward when they are obvious. If readers immediately recognize the meter of a sentence or the repetition of certain sounds, then the writer has failed. Rhythm and rhyme should be so unobtrusive that they enter into the reader's experience of a passage subconsciously. A reader may be struck by a sentence and remember it, but be only dimly aware, if aware at all, that he or she has responded in part to how it sounds. In Chapter 39 we shall discuss how to control the rhythm and sound patterns of sentences and some of the special effects you can achieve by such control. For the moment we shall only observe that when well handled, rhythm and rhyme give a sentence extra strength.

In a piece about the collapse of the stock market in 1929, Frederick Lewis Allen wrote:

> The Big Bull Market was dead.

Read aloud, the sentence sounds as if it were written "The BIG BULL MARket was DEAD." Several factors help to create this impression. One, certainly, is the shortness and directness of the statement. Of more telling effect are the monosyllabic words and the alliteration of *Big Bull*. Moreover, three successive stresses[4] occur in the first four

words ("The Big Bull Market"), slowing down the sentence and, in a

subtle way, inducing pauses between key words. These pauses are reinforced by the fact that in each case the closing sound of the preceding word differs from the opening sound of the following word, and therefore one needs an instant's hesitation to rearrange the lips and tongue. Even when reading silently most of us have enough "speech memory" to respond to difficult sound combinations. All these things make "The Big Bull Market was dead" an exceptionally strong statement. It conveys a sense of unalterable finality, which is as much a part of the writer's intended meaning as the simple fact that the market collapsed.

See how an assertion gains force by the clustering of stressed syllables here:

He speaks and thinks plain, broad, downright, English.

> William Hazlitt

The next case has a different rhythmic effect: a metrical run (a series of regularly spaced stressed and unstressed syllables).

For one brief moment the world was nothing but sea—the sight,

the sound, the smell, the touch, the taste of sea.

> Sheila Kaye-Smith

---

4. A stressed syllable is spoken heavily, an unstressed syllable softly. In part, rhythm consists of establishing patterns of stressed and unstressed syllables. For purposes of discussion stresses are marked ′, nonstresses ˣ.

In addition to controlling the stress patterns of words, a writer may also use rhyme. Most often this takes the form of alliteration, beginning successive or near-successive words or syllables with the same sound, which draws attention to them:

Reason will be replaced by Revelation.

W. H. Auden

She sighed a sigh of ineffable satisfaction.

Charlotte Brontë

Emph. s.
Short
THE SHORT SENTENCE

The short sentence is inherently emphatic. If well constructed, it is clear, concise, and convincing. It will seem especially strong in the context of longer, more complicated statements. Often the contrast in length reinforces the contrast in thought:

The first premise of the college elective system is that the subjects and courses of the curriculum are of approximately equal value. Well, they are not.

Brand Blanshard

As Thompson and the *Transcript* man had said, Vanzetti was naturally and quietly eloquent—and disturbing. So he was electrocuted.

Phil Strong

In other cases a brief emphatic sentence repeats a preceding idea:

The glorification of one race and the subsequent debasement of another—or others—always has been and always will be a recipe for murder. There is no way around this.

James Baldwin

I didn't have a very nice time that summer on the Cape and I couldn't forgive my family for making me go. I still can't.

Student

But however it relates to its context, the short sentence draws attention to itself by virtue of its difference. We can see how much is lost by not exploiting the value of the short sentence if we revise any of the examples above to combine the shorter statement with the longer. Examine this revision of James Baldwin's sentence:

The glorification of one race and the subsequent debasement of another—or others—always has been and always will be a recipe for murder, and there is no way around this.

EXERCISE

1. Explain how emphasis is achieved in the following examples of strong sentences.

Out rushes the husband.

Student

They don't build houses to last in Belgrade because they know that in ten years or so there will be another war, and the whole thing will be blown to pieces again. That is the sort of spirit one met the whole time. Nothing permanent. No trust. No faith. I looked into a photographer's shop and saw a photograph of the Parliament in session. So pompous, so threadbare, so utterly, damnably sad.

Beverley Nichols

But envy? Why envy? And hatred? Why hatred?

Norman Podhoretz

Caesar flies to his hunting lodge pursued by ennui; in the faubourgs of the Capital, Society grows savage, corrupted by silks and scents, softened by sugar and hot water, made insolent by theaters and attractive slaves; and everywhere, including this province, new prophets spring up everyday to sound the old barbaric note.

W. H. Auden

The profit system is oppressive not because relatively trivial luxuries are available, but because basic necessities are not.

Ellen Willis

Close beside the bier crowded the thieves and sycophants of the inner circle. Beyond wept the poor.

Hodding Carter

Americans, I believe, will never forget Wendell Willkie.

Roscoe Drummond

There is a fight, they win, and we retreat, half whimpering, half with bravado. My first nauseating experience of cowardice.

Norman Podhoretz

But if thought corrupts language, language can also corrupt thought.

George Orwell

Prudery in this matter is less pardonable than youthful arrogance, but it is of a piece with it.

Richard Poirier

In my childhood we were taught to sing nursery rhymes and, in pious households, hymns. Today the little ones warble the Singing Commercials. Which is better—"Rheingold is my beer, the dry beer," or "Hey diddle-diddle, the cat and the fiddle"? "Abide with me" or "You'll wonder where the yellow went, when you brush your teeth with Pepsodent"? Who knows?

Aldous Huxley

Dickens's descriptions, of Fagin's pupils or of any others, are not exaggerations; they are photographic.

Raymond Postgate

In March, 1876 I broke loose.

George Bernard Shaw

And, if the need came, she meant to be as good as her word.

Lytton Strachey

"Every man a king, but no man wears a crown," Huey's campaign banners had proclaimed. No man, that is, but Huey himself.

Hodding Carter

Today the inns are established for the sake of the travelers but during the Middle Ages many a monk turned traveler for the sake of the inns.

Lawrence Durrell

Instead, the audience must have taken away some vague impression that the poor jungle villagers of South-East Asia all have perfect complexions, and fly elaborately lovely kites. They don't.

Ian Watt

There is a tendency which, being human, is like most human things not unpardonable, but like many human things is rather irritating—to try to make out that everything is something else.

George Saintsbury

Unfortunate, ignorant creatures they were, but still with hope in their hearts because they had not yet been convicted.

Emma Goldman

Besides Sidney Herbert, she had other friends who, in a more restricted sphere, were hardly less essential to her.

Lytton Strachey

The Renaissance created Antiquity at least as much as Antiquity created the Renaissance.

André Malraux

It began to seem that the machinery of the organization I worked for was turning over, day and night, with but one aim: to eject me.

James Baldwin

But then, Yogi was a remarkably garrulous, gregarious, gossipy sort who just couldn't keep still on a ball field no matter where he was.

Leonard Koppett

At dead of night in the house which he had bought from the President, Forbes' right-hand man, Cramer, shot himself.

Samuel Hopkins Adams

Again, it's an incontrovertible fact that, in the past, when contraceptive methods were unknown, women spent a much larger proportion of a much shorter life pregnant, or nursing infants whom they had borne with little or no medical help. And don't believe that that's a natural, healthy thing for human beings to do, just because animals do it. It isn't.

Elizabeth Janeway

And crippling poor is what the family was.

William Alfred

Trust thyself; every heart must vibrate to that iron string.

Ralph Waldo Emerson

It is Ethan who must suffer; it is Ethan who must watch his dream die.

Student

If we had lived from childhood with a boa constrictor, we should think it no more a monster than a canary-bird.

W. S. Landor

Honest the man might be, and ingenuous, but nothing was so mischievous as misguided virtue. . . .

Ralph Roeder

2. Develop each of these topics in a sentence (or pair of sentences) using negative-positive restatement:

A. The strangeness of college (or its lack of strangeness, if you prefer)
B. The difficulties of the child-parent relationship
C. The pleasures of sailing, skiing, or any other recreation.

3. Compose a paragraph of about 250 words on a topic of your own choice. In that paragraph use a fragment, a rhetorical question, and a short sentence for emphasis.

4. Write three different sentences on the topics in question 2, this time using announcement:

CHAPTER 37

Emphasis within the Sentence

In the last chapter we looked only at emphatic sentences. The necessity of making such strong statements, however, does not occur as often as the need to stress a portion of a sentence by directing the reader's eye and mind to a particular word or phrase. A good writer does this subtly. Rather than scattering exclamation points, underlinings, and capital letters throughout a piece of prose, he or she relies upon the selection and positioning of words. The following pages list, alphabetically, the more common techniques of achieving inner emphasis.

Emph.

The abbreviation "Emph." means that the stress does not fall where your context suggests it should. The more specific abbreviations that come after it suggest what you might do to put the stress where it belongs.

Emph. adj. **ADJECTIVES**

Adjectives are an important source of emphasis. A special class of modifiers called intensives (it includes adverbs as well as adjectives) exists simply to stress the term the modifier is attached to: *great, greatly, much, very, extremely, terribly, awfully,* and many, many more. On the whole, intensives are not very satisfactory. Used too often and too indiscriminately, these terms have become devalued, a process that results in a never-ending search for fresh intensives. Writers can certainly find unusual and effective ones, as in this description of the modern superstate:

> These moloch gods, these monstrous states. . . .
>
> Susanne K. Langer

Still, you should not rely primarily upon intensives as a way of establishing inner emphasis.

Better than calling upon adjectives to intensify other words, use them as emphatic terms in their own right. One way is to put an adjective in an unusual place. For example, revising "The heartbroken family gave up the search" in either of the following ways puts greater weight upon "heartbroken":

Heartbroken, the family gave up the search.
The family gave up the search, heartbroken.

And in this passage Susanne K. Langer emphasizes the adjectives by placing them after the term they modify and at the end of the sentence:

The national states are not physical groups; they are social symbols, profound and terrible.

PAIRED ADJECTIVES

Adjectives may also be made emphatic by being paired, as in the sentence just quoted. Here are several others:

They [a man's children] are his for a brief and passing season.

Margaret Mead

And as a war should be undertaken upon a just and prudent motive. . . .

Jonathan Swift

This antiquated and indefensible notion that young people have no rights until they are twenty-one. . . .

Student

A dilapidated macaw with a hard, piercing laugh, mirthless and joyless, with a few unimaginative phrases, with a parrot's powers of observation and a parrot's hard and poisonous bite.

Edith Sitwell

[Describing an eighteenth-century intellectual named Lady Mary Wortley Montague.]

Working as a team, such adjectives impress themselves upon the reader. And they can do more, reinforcing a point by restatement ("a brief and passing season") or suggesting subtle contrasts and amplifications of meaning. For instance, Sitwell's sentence leads us to think about the relationship between "mirth" and "joy" and about how a laugh can be both "hard" and "piercing."

PILING UP ADJECTIVES

Another way to use adjectives for emphasis is to string together three, four, or more, as in this passage about the writer's family:

. . . a wilful, clannish, hard-drinking fornicating tribe.

William Gibson

In the following case the writer is complaining about a neighbor's singing:

> A vile beastly rottenheaded foolbegotten brazenthroated pernicious piggish screaming, tearing, roaring, perplexing, splitmecrackle crashmegiggle insane ass of a woman is practicing howling below-stairs with a brute of a singingmaster so horribly, that my head is nearly off.
>
> Edward Lear

Passages like these, especially the second, are virtuoso performances in which exaggeration becomes an end in itself. Of course, exposition cannot indulge itself like this very often. But sobriety needs occasional relief, and such exuberance of language delights and dazzles. Whatever may be the objective truth of such passages, they bring us into electrifying contact with the mind and feelings of the writer—and that is the essence of communication.

On the other hand, adjectives can easily lead us astray. Some writers seem unwilling ever to let a noun stand alone. Use adjectives only when they say something necessary, and use them emphatically only when emphasis is called for.

ELLIPSIS

Emph.
ellip.

Ellipsis means omitting words necessary to grammar but not to sense. Certain commonplace elliptical constructions cause no surprise: in "He is taller than his brother," for instance, the omission of "is tall" after "brother" is conventional. Sometimes, however, an unusual ellipsis requires us to pause to figure out the grammar, thereby making us feel the force of the idea:

> Other things are necessary too, but memory is fundamental: without memory no knowledge.
>
> Carl Becker

> The nurses in the hospitals were especially notorious for immoral conduct; sobriety almost unknown among them.
>
> Lytton Strachey

ISOLATION

Emph.
isol.

Isolation means cutting off a word or phrase from the rest of the sentence. The isolated term can occur anywhere, but it appears most frequently at the beginning or end, positions naturally emphatic in themselves:

> Leibnitz, it has sometimes been said, was the last man to know everything.
>
> Colin Cheery

> Children, curled in little balls, slept on straw scattered on wagon beds.
>
> Sherwood Anderson

If the King notified his pleasure that a briefless lawyer should be made a judge or that a libertine baronet should be made a peer, the gravest counsellors, after a little murmuring, submitted.

                                                                      Thomas Babington Macaulay

Ruben said something in a hurried whisper, made rather an impressive gesture over his head with one arm, and, to say it as gently as possible, died.

                                                                      Katherine Anne Porter

And then, you will recall, he [Henry Thoreau] told of being present at the auction of a deacon's effects and of noticing, among the innumerable odds and ends representing the accumulation of a lifetime, a dried tapeworm.

                                                                      E. B. White

You may use both ends of the sentence. The first of these next two examples neatly isolates the two key words "position" and "difficult." The second (about the famous economist Thorstein Veblen) forces a contrast between the adjectives at the beginning and those at the end, a contrast which implies a good deal about the disparity between communal values and the values of the individual:

The position—if poets must have positions, other than upright—of the poet born in Wales or of Welsh parentage and writing his poems in English is today made by many people unnecessarily, and trivially, difficult.

                                                                      Dylan Thomas

In the bustling, boosting, gregarious community in which he lived, he stood apart: uninvolved, unentangled, remote, aloof, disinterested, a stranger.

                                                                      Robert L. Heilbroner

Placing the isolated term in the middle of the sentence is less common, but it can be done:

I was late for class—inexcusably so—and had even forgotten my homework.

                                                                      Student

Whether the isolated term comes at the opening of the sentence, at the close, or in the middle, you must set it off by commas, dashes, or a colon. The mark you choose roughly controls the degree of emphasis. Dashes, for example, convey more emphasis than commas: "Suddenly— it began to rain" draws our eyes to the adverb even more than "Suddenly, it began to rain." A colon before a final word is stronger than a comma, though about the same as a dash. (As isolating marks, colons never go around words in the middle of a sentence, though a colon may sometimes follow an initial word. Commas and dashes, on the other hand, may be employed for isolation—though they are not necessarily equal in regard to other uses—anyplace in the sentence.)

Successful isolation involves more, however, than simply sticking commas or dashes before, after, or around words you want to emphasize. You must construct the sentence so that the pause signaled by the punctuation occurs at the appropriate point in a way that sounds natural. Suppose, for instance, that you wanted to revise the following sentence to throw more weight upon "Harry":

Clearly Harry was not the man for the job.

You could not merely separate the noun from the adverb and the verb; the result would not be English:

Clearly, Harry, was not the man for the job.

But you could move the noun to the opening position and set "clearly" after it as an interrupter:

Harry, clearly, was not the man for the job.

Or you might prefer to expand "clearly" into a clause:

Harry—it was clear—was not the man for the job.

There are other possibilities: for example, opening with "He" and restating "Harry" in an isolated final position:

He was not the man for the job—not Harry.

All these revisions stress "Harry" by means of isolation, and all achieve the isolation naturally.

The emphasis gained by isolation—like emphasis in general—does more than add strength. It enables you to convey overtones of meaning, nuances that depend less upon individual words than upon their arrangement. To see how these nuances change as word positions are altered, imagine that the end of the sentence by Macaulay on page 298 went like this:

. . . the gravest counselors submitted, after a little murmuring.

The words are the same, and the syntax and logic—but not the implications. Macaulay, while admitting that the counsellors of Charles II occasionally protested, stresses their submissiveness; the revision, while acknowledging that finally they submitted, lays a bit more stress upon their protest. In short, the two versions evaluate the King's ministers slightly differently.

Or suppose that E. B. White's sentence were changed to this:

And then, you will recall, he told of being present at the auction of a deacon's effects and of noticing a dried tapeworm among the innumerable odds and ends representing the accumulation of a lifetime of endeavor.

Again no change occurs in diction or syntax, only in arrangement. But the revision, like a poor comedian who throws away a punch line by delivering it too rapidly, obscures the point: that a lifetime spent accumulating property is a lifetime misspent.

Finally, notice how isolation amplifies the irony of "friends" in the following passage:

> There was a quarter-page advertisement in *The London Observer* for a computer service that will enmesh your name in an electronic network of fifty thousand other names, sort out your tastes, preferences, habits, and deepest desires and match them up with opposite numbers, and retrieve for you, within a matter of seconds, friends.
>
> Lewis Thomas

Emph.
mech.

MECHANICAL EMPHASIS

Mechanical emphasis consists of printing or writing words in an unusual way. In exposition the most common mechanical devices are the exclamation mark and italics (in handwriting or typing the equivalent of italicization is a single underline):

> It is so simple a fact and one that is so hard, apparently, to grasp: *Whoever debases others is debasing himself.*
>
> James Baldwin

> Yet this government never of itself furthered any enterprise, but by the alacrity with which it got out of its way. *It* does not keep the country free. *It* does not settle the west. *It* does not educate.
>
> Henry David Thoreau

> Worse yet, he must accept—how often!—poverty and solitude.
>
> Ralph Waldo Emerson

Other devices of mechanical emphasis include quotation marks, capital letters, boldface and other changes in the style or size of type, different colored inks, wider spacing of words or letters, and lineation—placing key words or phrases on separate lines. Advertisements reveal how well all these techniques work.

In composition, however, they work less effectively. Experienced writers do not use even exclamation points or underlining very often. Overindulged in, they quickly lose their value. And they reveal that a writer does not know how to create emphasis so, in effect, he or she shouts.

Certainly in the examples above the italics and the exclamation point are effective. But in each case the device merely strengthens an emphasis already attained by more compositional means. Baldwin's sentence puts the key idea last and carefully prepares its way with a colon. Thoreau draws our attention to "it" not only by using italics but by repeating the word at the beginning of three successive brief, emphatic sentences. And Emerson stresses "how often" more by isolating it between dashes than he does by the exclamation mark.

In summary, then, it is best, not to avoid underlining and exclamation points altogether, but to save them for occasions when they really count.

Emph.
list.

POLYSYNDETON AND ASYNDETON

Despite the formidable sound of the words, *polysyndeton* and *asyndeton* are, quite simply, ways of handling lists or series. Polysyndeton places a conjunction (*and, or*) after every term except the last. Asyndeton uses no conjunctions and separates terms of the series with commas. Both differ from the conventional treatment of lists and series, which places only commas (under rare conditions semicolons) between all items except the last two, where a conjunction is used (the conjunction may or may not be accompanied by a comma—it is optional).

> We stopped on the way to the camp and bought essential supplies: bread, butter, cheese, hamburger, hot dogs, and beer.
>
> Student

With polysyndeton this becomes:

> . . . bread and butter and cheese and hamburger and hot dogs and beer.

With asyndeton:

> . . . bread, butter, cheese, hamburger, hot dogs, beer.

The conventional series emphasizes no particular item, though the last usually seems a bit more significant. In polysyndeton the emphasis falls more evenly upon each member of the list, and also more heavily:

> It was bright and clean and polished.
>
> Alfred Kazin

> It is the season of suicide and divorce and prickly dread, whenever the wind blows.
>
> Joan Didion

In asyndeton too the series takes on more significance as a whole than it does in the conventional pattern. But the stress on each individual item is lighter than in polysyndeton, and the passage moves more quickly:

> His care, his food, his shelter, his education—all of these were by-products of the parents' position.
>
> Margaret Mead

Polysyndeton or asyndeton does not always improve lists and series. Most of the time the usual treatment will be more appropriate, as it is in the student example where the writer probably wants a little extra stress on "beer." However, when you need a different sort of emphasis on a series than the common method permits, remember polysyndeton and asyndeton.

Emph.
pos.

POSITION

Two positions in a clause or sentence convey inherent emphasis: the opening and the closing. Elsewhere, emphasis must depend upon inversion, isolation, modification, restatement, and so on. Of course you can also use these techniques with initial and final words, where they will seem especially effective because they are working upon terms already prominent by virtue of their position.

Opening with the key word has much to recommend it. Immediately the reader sees what matters. E. M. Forster, for example, begins a paragraph on curiosity with the following sentence, identifying his topic at once:

Curiosity is one of the lowest human faculties.

Putting the essential idea first is natural, suited to a style aiming at the simplicity and straightforwardness of forceful speech, as in this case:

Great blobs of rain fall. Rumble of thunder. Lightning streaking blue on the building.

J. P. Donleavy

Donleavy's sentences suggest the immediacy of the experience he describes by going at once to what dominates his perception: the heavy feel of rain, thunder, lightning. (The two fragments also enhance the forcefulness of the sentences.)

Beginning with the principal idea has still another advantage. It becomes easy to underscore a contrast by starting the next sentence or clause with an opposing term:

Science was traditionally aristocratic, speculative, intellectual in intent; technology was lower-class, empirical, action-oriented.

Lynn White, Jr.

Postponing a major point to the end of the sentence is more formal and literary. The writer must have the entire sentence in mind from the first word. On the other hand, the final position is a bit more emphatic than the first, perhaps because we remember best what we have read last:

One thing is so certain that it seems stupid to verbalize it: both modern technology and modern science are distinctively Occidental.

Lynn White, Jr.

So the great gift of symbolism, which is the gift of reason, is at the same time the seat of man's peculiar weakness—the danger of lunacy.

Susanne K. Langer

Here the emphasis is also a function of announcement and isolation.

In topic sentences the final position is often reserved for the idea the paragraph will develop (if it can be done without awkwardness). This enables the writer to step off neatly into the development.

This sentence, for instance, begins a paragraph discussing Welsh Christianity:

> The third legacy of the Romans was Welsh Christianity.
>
> George Macaulay Trevelyan

And this one, a paragraph about the development of the black press in the United States:

> The Negro found another means of expression in the press.
>
> Oscar Handlin

Like the opening position, the closing is also useful for reinforcing comparisons and iterations:

> We can never forget that everything Hitler did in Germany was "legal" and everything the Hungarian freedom fighters did in Hungary was "illegal."
>
> Martin Luther King, Jr.

The quotes around "legal" and "illegal" also have an emphatic effect, though their primary purpose is to signal the words are ironic.

> But Marx was not only a social scientist; he was a reformer.
>
> W. T. Jones

Inexperienced writers often waste the final position:

> ✕ As the military power of Kafiristan increases, so too does the pride that Dravot has.

The sentence fails to give enough weight to the important words "pride" and "Dravot" (Daniel Dravot is the hero of Rudyard Kipling's story "The Man Who Would Be King" and makes himself ruler of an Asian country called Kafiristan). The writer spends the final space foolishly on the empty word "has." Depending upon the shade of emphasis wanted, either of these arrangements would be preferable:

> As the military power of Kafiristan increases, so too does Dravot's pride.

> As the military power of Kafiristan increases, so too does the pride of Dravot.

Finally, a writer may exploit both ends of a sentence for special emphasis, as Martin Luther King, Jr., does in the following instance, playing off "lukewarm acceptance" against "outright rejection":

> Lukewarm acceptance is much more bewildering than outright rejection.

Emph.
rep.

REPETITION

In a strict sense, repetition is more a question of diction than of sentence structure. But because the simplest way to emphasize an idea is to repeat it, we shall consider it here.

Repetition is sometimes a virtue, sometimes a fault; it is not easy to draw the line. Generally, however, when a writer concentrates on an essential point and when he or she controls what is going on, repetition will probably work. The reader's attention goes where the writer wants it. On the other hand, using repetition for a less important point seems awkward and misleads the reader, who supposes the ideas are more significant than they are. Don't be annoyed, then, if a teacher advises you at one place in an essay to reiterate something for emphasis, yet warns at another place that something you did repeat is awkward. This is not inconsistency, just asking that you use repetition wisely.

Repetition may take two basic forms: restating the same idea in a different word (called *tautologia* by the Greeks), or repeating the same exact word (or sometimes a variant form of the same word).

TAUTOLOGIA: REPEATING AN IDEA IN DIFFERENT WORDS

In tautologia the synonyms are frequently stronger than the original term:

> That's camouflage, that's trickery, that's treachery, window-dressing.
>
> Malcolm X

"Treacherous" is more damning than "trickery"; "window-dressing," a false and showy appearance to hide an empty or shabby reality, is worse than "camouflage," a military term not implying disapproval.

A second term need not be strictly synonymous with the first, and often it is not. Rather than simply restating the initial idea, the second or third term may carry it a step further:

> October 7 began as a commonplace enough day, one of those days that sets the teeth on edge with its tedium, its small frustrations.
>
> Joan Didion

> One clings to chimeras, by which one can only be betrayed, and the entire hope—the entire possibility—of freedom disappears.
>
> James Baldwin

In Didion's sentence "frustrations" signifies a different and worse condition than "tedium," but the ideas relate to the extent that tedium may contribute to frustration. Baldwin neatly exploits the added power of a second word: "possibility" includes more and here suggests more despair than "hope." Now and then a writer uses an expression just to replace it with another:

> That consistent stance, repeatedly adopted, must mean one of two—no, three—things.
>
> John Gardner

Finally, repetition of idea may take the form of simile or metaphor:[1]

> It follows that any struggle against the abuse of language is a sentimental archaism, like preferring candles to electric light or hansom cabs to aeroplanes.
>
> George Orwell

> In [Henry] James nothing is forestalled, nothing is obvious; one is forever turning the curve of the unexpected.
>
> James Huneker

Similes and metaphors certainly have uses other than emphasis, as we shall see in Chapter 43. Here we need only observe that one way to repeat an important idea is to employ a simile or metaphor. The specific image contained in the figure adds value by expressing an abstraction in concrete terms or translating an unfamiliar idea into a more common one. In Orwell's sentence (incidentally, he is paraphrasing a view he does *not* agree with; he believes that abuses of language should be struggled against) we cannot *see* a "sentimental archaism" and we may not even know what one is. But we do recognize candles and electric light, and we understand what a perverse preference for the former signifies. Figurative repetition may also have a vivid, surprising quality. Practicing the very quality Huneker praises in the novelist Henry James, he startles us by shifting suddenly from the man to a road curving to reveal an unexpected vista.

REPEATING THE SAME WORD

Repeating words is a frequent means of emphasis. Ancient writers liked it so much that Greek and Roman rhetoricians developed an elaborate set of terms (about two dozen) to distinguish various types. We have used some of these terms, followed by an explanation, but if you find the Greek labels hard to remember, forget them.

The point is to remember that the patterns of repetition are still very much alive. Contemporary writers still employ them—not writers consciously imitating the classics, but men and women who belong to our world and have something important to say about it. The patterns remain vital because they appeal to a fundamental quality of the mind: the delight in combining things, or in discerning what others have combined. A decorator arranging vases on a table is doing what writers do when they repeat words in particular patterns: creating configura-

---

1. A simile is a literal comparison commonly introduced by "like" or "as": Robert Burns's famous line "my luv is like a red, red rose" is a simile. A metaphor is a literal identification—as if Burns had written "my luv is a red, red rose"—though we are to understand that the two things are still being compared. Some metaphors carry identification another step, fusing the two concepts by substituting the second term for the first, as in "my red, red rose" to mean "my love."

tions that please. But for writers, of course, pleasing readers by word arrangement is less an end in itself than a means of helping them recognize what is most important.

ANADIPLOSIS: THE SAME TERM ENDS ONE CLAUSE, BEGINS THE NEXT.

To philosophize is to understand; to understand is to explain oneself; to explain is to relate.

Brand Blanshard

ANAPHORA: THE SAME TERM BEGINS SUCCESSIVE CLAUSES.

I wanted to go to Austria without war. I wanted to go to the Black Forest. I wanted to go to the Harz Mountains.

Ernest Hemingway

I didn't like the swimming pool, I didn't like swimming, and I didn't like the swimming instructor, and after all these years I still don't.

James Thurber

ANTISTROPHE: THE SAME TERM ENDS SUCCESSIVE CLAUSES.

When that son leaves home, he throws himself with an intensity which his children will not know into the American way of life; he eats American, talks American, he will be American or nothing.

Margaret Mead

He [Aldous Huxley] gives one a sense, in his writings, of a little group of intelligentsia clinging unhappily together in a grossly hostile world. Not merely unsympathetic or lacking in understanding, but grossly, actively hostile.

Beverley Nichols

DIACOPE: THE SAME TERM IS SEPARATED BY ONE OR TWO OTHERS.

A middle-aged lady, frail, very frail; exceedingly pale from long ill health, prematurely white haired, with beautiful gray eyes, gentle but wonderfully bright.

W. H. Hudson

Everyone is a little weary now, weary and resigned, everyone except Sandy Slagle, whose bitterness is still raw.

Joan Didion

It is nonsense, and costly nonsense, none the less.

Brand Blanshard

I am neat, scrupulously neat, in regard to the things I care about; but a book, as a book, is not one of these things.

Max Beerbohm

EPANALEPSIS: THE SAME WORD APPEARS AT THE BEGINNING AND THE END.

Problem gives rise to problem.

Robert Louis Stevenson

EPIZEUXIS: A WORD IS IMMEDIATELY REPEATED.

Life is tragic simply because the earth turns and the sun inexorably rises and sets, and one day, for each of us, the sun will go down for the last, last time.

James Baldwin

But why, *why* should it have been so different between the Negroes and us?

Norman Podhoretz

A horse is galloping, galloping up from Sutton.

Amy Lowell

POLYPTOTON: A WORD REPEATED IN A DIFFERENT FORM.

She smiled a little smile and bowed a little bow.

Anthony Trollope

Dead darkness lay on all the landscape, dead darkness added its own hush to the hushing dust on all the roads.

Charles Dickens

PLOCHE: A WORD REPEATED WITH A DIFFERENT SENSE.

Visitors whom he [Ludovico Sforza, a Renaissance duke] desired to impress were invariably ushered into the Sala del Tesoro, they rubbed their eyes, he rubbed his hands, they returned home blinded, he remained at home blind.

Ralph Roeder

While the literal meanings of the two uses of "rubbed" are the same, their implications are different. Sforza's guests rubbed their eyes dazzled and amazed by his riches; the Duke rubbed his hands in self-satisfaction. Their blindness was a literal blurring of vision, his a blindness of spirit.

SYMPLOCHE: ONE TERM REPEATED AT THE BEGINNING OF SUCCESSIVE CLAUSES, AN-OTHER TERM REPEATED AT THEIR END.

The average autochthonous Irishman is close to patriotism because he is close to the earth; he is close to domesticity because he is close to the earth; he is close to doctrinal theology and elaborate ritual because he is close to the earth.

G. K. Chesterton

TAUTOTES: A WORD REPEATED TWO OR MORE TIMES.

Mr. and Mrs. Veneering were bran-new people in a bran-new house in a bran-new quarter of London. Everything about the Veneerings was spick and span new.

Charles Dickens

Music is mere beauty; it is beauty in the abstract, beauty in solution.

G. K. Chesterton

Notices how Chesterton advances the position of "beauty" in the three clauses

Joe Median was an average man—average build, average weight, average temper, average everything.

Student

EXERCISE

1. What word or words are especially emphatic in the following sentences? Explain how they are made so.

   A. However lofty the grounds for Hoover's inhibitions, they added up to one thing in the minds of the people: impotence.

   Arthur M. Schlesinger, Jr.

   B. Dark and somber and ominous was that face, solid and stolid and expressionless, with eyes that smouldered and looked savage.

   Jack London

   C. The cause of pornography is *not* the same as the cause of free speech. There *is* a difference.

   Barbara W. Tuchman

   D. We watch the movements of his body, the waving of his arms; we see him bend down, stand up, hesitate, begin again.

   Joseph Conrad

   E. His twisted leer was smoothed into a pleasant smile.

   James Thurber

   F. The crotchety, cynical, spiteful, clever, worldly, witty son of a great father, Horace Walpole, twenty years younger than she, became the object of this unnatural and celebrated passion.

   Nancy Mitford

   G. Opponents should not be argued with; they should be attacked, shouted down, or, if they become too much of a nuisance, liquidated.

   Aldous Huxley

   H. The creek is the mediator, benevolent, impartial, subsuming my shabbiest evils and dissolving them, transforming them into live moles, and shiners, and sycamore leaves.

   Annie Dillard

   I. The man whose mastery emerged most clearly out of the excitement and anxiety of 1933 was, of course, the President.

   Arthur M. Schlesinger, Jr.

   J. Our saddles and all our equipments were worn and battered, and our weapons had become dull and rusty.

   Francis Parkman

   K. He bubbled and sparkled and glittered.

   Beverley Nichols

   L. Going to sea became the last resource of the dregs of the waterfront, the vicious, the improvident, the incompetent, and the irresponsible.

   William McFee

2. Identify the particular kind of repetition used in these sentences:

   A. Bernard Shaw is a Puritan and his work is a Puritan work. He has all the essentials of the old, virile and extinct Protestant type. In his work he is as ugly as a Puritan. He is as indecent as a Puritan.

   G. K. Chesterton

B. . . . the old, old words worn thin, defaced by ages of careless usage.

Joseph Conrad

C. Who has ever poured out at the grave of his loved one, half, a hundredth, even a thousandth part of the tears he shed daily for her cruelty in life?

Lawrence Durrell

D. Incredibly, this amused me, and, incredibly, it amuses me still.

Annie Dillard

E. But Puritanism defies human nature, and human nature, repressed, emerges in disguise.

Robert Coughlan

F. No man of note was ever further separated from life and fact than Lindbergh. No man could be more reluctant to admit it.

John Lardner

G. Their pointless unrest was suddenly pointed.

Wallace Stegner

H. Capital would not invest in it [the U.S. merchant marine], the average citizen would not sail in it, and the working, native-born American would not accept employment in it.

William McFee

I. They were brave days—days of desperation and of hope, of drama and of triumph.

Arthur M. Schlesinger, Jr.

J. He took things easy, and his fellow freebooters took almost everything easily.

Hodding Carter

K. Such plants to operate successfully had to run at capacity. To run at capacity they needed outlets for their whole output.

Thurman Arnold

L. He was wild—wild as a tiger.

Stephen Crane

M. Mr. Hoover, who was then President, was an engineer with an engineering mind.

Thurman Arnold

N. It is the fact, the irrefragable fact.

Jack London

O. We rowed around to see if we could not get a line from the chief engineer, and all this time, mind you, there were no shrieks, no groans, but silence, silence, and silence and then the Commodore sank.

Stephen Crane

P. Whether the system made the criminal or the criminal made the system it is here irrelevant to enquire.

'W. H. Lewis

Q. Engineers will become leaders, leaders will become engineers, and the world will have a better chance of avoiding disaster.

Samuel C. Florman

R. If Periclean imperialism caused the war, says Thucydides, war in turn produced violence, and violence political chaos.

C. A. Robinson, Jr.

S. She left the village running, she entered the wood running, looking at nothing, hearing nothing.

Victor Hugo

T. For if ever there was a group dedicated to—obsessed with—morality, conscience, and social responsibility, it has been the engineering profession.

Samuel C. Florman

3. Revise each of the following sentences so that the underlined word, phrase, or clause receives stronger stress. Use any of the devices of emphasis we have discussed (except underlining, capitalization, or other mechanical means, which offer too easy a solution). You may break any of the longer passages into two sentences if you feel that a clause requires separate statement for emphasis.

A. I never cared for *tapioca pudding*.
B. The *leaderless* mob rioted down the street.
C. Dancing seems quite hard but *actually it is easy*.
D. The day was *clear, sunny, and cold*.
E. The trip was long and tiresome and when we got there we found the store closed and *all in all the day was wasted*.

4. Revise each of the following sentences three times so that in each revision a different one of the underlined words or constructions receives maximum stress.

A. The sloop was *clawing* her way off a *lee shore* with *her sails close-hauled*.
B. The achievement of *Athens* is all the *greater* when we consider how *relatively brief* was the period of her political supremacy.
C. *Chaucer's characters* in *The Canterbury Tales* are *so lifelike* that it is difficult to think they never existed.
D. The *brilliant* scheme *slowly* took shape *in her mind*.

5. In the following cases the stress on the underlined expressions is muted by awkward construction. Revise each sentence so that the key terms are played off against one another more emphatically.

A. The novelist creates a *purely imaginary world*, but we can see the *dramatist's world* right before our eyes.
B. Sailors are generally *cheerful* fellows, while *somberness and sobriety* are the chief characteristics of lawyers.
C. *Geography* is the science of the earth, while the nature of animals is what *biologists* study.
D. College is more *interesting* than high school, but college is less *fun* than high school.
E. He likes *chocolate* ice cream but has never cared for *strawberry* very much.

38

Variety in Sentences

Var.

INTRODUCTION

The Art Cinema is a movie theater in Hartford. Its speciality is showing uncensored films. The theater is rated quite high as to the movies it shows. The movies are considered to be good art.

Passage from a theme

The Smith disclosures shocked [President] Harding not into political housecleaning but into personal reform. The White House poker parties were abandoned. He told his intimates that he was "off liquor." Nan Britton [Harding's mistress] had already been banished to Europe. His nerve was shaken. He lost his taste for revelry. The plans for the Alaska trip were radically revised. Instead of an itinerant whoopee, it was now to be a serious political mission.

Samuel Hopkins Adams

Both of these passages consist chiefly of short, simple sentences. The first uses them poorly, the second effectively. Where does the difference lie? The first writer has failed to understand the twin principles of recurrence and variety which govern sentence style. Adams, a professional author, understands them very well.

Recurrence means repeating a basic sentence pattern; *variety* means changing the pattern. Paradoxical though it sounds, good sentence style must use both. Enough sameness must appear in the sentences to make the writing all of a piece, enough difference to avoid monotony.

The proportions of recurrence and variety change with subject and purpose. For instance, when you repeat the same essential point or state a series of parallel ideas, the similarity of subject justifies—and is even enhanced by—a close similarity in sentence structure. Thus Adams repeats the same pattern in his second through seventh sen-

311

tences because they have much the same content, detailing the specific steps President Harding took to offset a scandal that threatened his administration. Here the recurrent style evolves from the topic.

In the other passage, however, the writer makes no such connection between style and subject, and so the recurrence seems awkward and monotonous. The ideas expressed in the separate sentences are not really of the same order of value: for example, the fact that the theater is in Hartford is less important than that it shows "uncensored films." The sentence style, in short, does not reinforce the writer's ideas: it obscures them.

Nor has the writer understood the need for variety. He offers no relief from his short, straightforward sentences. Adams does use variety. Moreover he uses it effectively to structure his paragraph, opening with a relatively long sentence, which, though technically simple, is complicated by the correlative "not . . . but" construction. And he ends the paragraph by offering a new sentence opening, for the first time beginning with something other than the subject.

Adams' brief sentences work because the subject matter justifies them and because they have sufficient variety. Lacking similar justification or relief, the four sentences of the first passage are ineffective. They could be improved like this:

> The Art Cinema, a movie theater in Hartford, specializes in uncensored films. It is noted for the high quality of its films; in fact, many people consider them good art.

There is still recurrence: in effect the passage consists of three similar short clauses plus an appositive. But now there is more variety. The first sentence has an appositive interrupting subject and verb; the second contains two clauses rather than one, and opens the second of these with the phrase "in fact." Moreover subordinating the information about Hartford brings the writing more sharply into focus upon the nature of the films, the essential point.

Of course, a writer rarely strives for variety alone. In composing a sentence that differs from most of his or her other sentences, the goal may be to emphasize a key idea; the variety is simply a by-product. Whatever the writer's intention, however, good prose must show some degree of variation in sentence style.

VARY LENGTH AND PATTERN

Var. length

The simplest kind of variation is changing sentence length and pattern:

> We took a hair-raising taxi ride into the city. The rush-hour traffic of Bombay is a nightmare—not from dementia, as in Tokyo, nor from exuberance, as in Rome; not from malice, as in Paris; it is a chaos rooted in years of

practiced confusion, absent-mindedness, selfishness, inertia, and an incomplete understanding of mechanics. There are no discernible rules.

James Cameron

From the beginning she had known what she wanted, and proceeded single-minded, with the force of a steam engine towards her goal. There was never a moment's doubt or regret. She wanted the East; and from the moment she set eyes on Richard Burton, with his dark Arabic face, his "questing panther eyes," he was, for her, that lodestar East, the embodiment of all her thoughts. Man and land were identified.

Lesley Blanch

It is not necessary to maintain a strict alternation of long and short, nor even desirable. You need only an occasional brief sentence to change the pace of predominantly long ones, or a long sentence now and then in a passage composed chiefly of short ones:

One of the great arts of modern living is to go through a whole week without spending more money than you earn. It is practically a lost art. Nearly everybody we know has a shylock [loan shark] next to his bed each morning, testing his breath with a hand mirror to make sure he is alive.

Jimmy Breslin

Dave Beck was hurt. Dave Beck was indignant. He took the fifth amendment when he was questioned and was forced off the executive board of the AFL-CIO, but he retained enough control of his own union treasury to hire a stockade of lawyers to protect him. Prosecution dragged in the courts. Convictions were appealed. Delay.

John Dos Passos

Now and then a writer uses variation in length to lead up to a key idea. In the following passage the historian Herbert Butterfield moves through two long sentences (the second a bit shorter than the first) to a strong short sentence:

The whig historian is interested in discovering agency in history, even where in this way he must avow it only implicit. It is characteristic of his method that he should be interested in the agency rather than in the process. And this is how he achieves his simplification.

And here a writer prepares for the three short sentences which close his passage by two longer statements (note that these are varied by a short sentence placed between them):

In the alleys the snow packed down on the mounds of garbage and provided us with hills for sliding. It leveled the uneven sidewalks. It even painted the buildings and filled the holes in the streets and in the yards, and laid lawns, for once, all over the neighborhood. It was clean. It was pure. It was good.

Ronald L. Fair

FRAGMENTS

Var.
frag.

Fragments, usually a special kind of short sentence, make for effective variation—easy to see and easy to use:

> Sam steals like this because he is a thief. *Not a big thief.* He tried to be a big thief once and everybody got mad at him and made him go away to jail. He is strictly a small thief, and he only steals for his restaurant.
>
> Jimmy Breslin [italics added]

> Examinations tend to make me merry, often seeming to me to be some kind of private game, some secret ritual compulsively played by professors and the institution. I invariably became facetious in all the critical hours. *All that solemnity for a few facts!* I couldn't believe they were serious. But they were. I never quite understood it.
>
> Mary Caroline Richards [italics added]

> This is not a personal memoir, except as I am the eyes that saw, the ears that heard, the fingers that felt or touched. Millions of us have the same story, but different ways of telling it. *Or of keeping silent.*
>
> R. L. Duffus [italics added]

Used with restraint, fragments like these offer a simple and effective way to change your sentence pattern. But they are more at home in a colloquial style than in a formal one.

RHETORICAL QUESTIONS

Var. rhet.
q.

Rhetorical questions are also a source of variety, though we should repeat that, like fragments or any other kind of unusual sentence, they should not be used for variety alone. Their primary purpose should be to emphasize a point or to set up a topic for discussion; when occasionally employed for such reasons, they achieve variety naturally:

> But Toronto—Toronto is the subject. One must say something—*what* must one say about Toronto? What can one? What has anybody ever said? It is impossible to give it anything but commendation. It is not squalid like Birmingham, or cramped like Canton, or scattered like Edmonton, or sham like Berlin, or hellish like New York, or tiresome like Nice. It is all right. The only depressing thing is that it will always be what it is, only larger, and that no Canadian city can ever be anything better or different. If they are good they may become Toronto.
>
> Rupert Brooke

VARIED OPENINGS

Var. open.

Beyond altering length, complexity, and type of sentences, you can also change how they begin. Most sentences open with subject and verb, but other openings are easily possible: an initial prepositional phrase or adverbial clause, a connective word like *therefore* or an attitudinal adverb like *naturally,* or a nonrestrictive adjective or a participial phrase modifying the subject:

> In the first decade of the new century, the South remained primarily rural; the beginnings of change, in those years, hardly affected the lot of the Negro. The agricultural system had never recovered fully from the destruction of the old plantation economy. Bound to the production of staples—tobacco, cotton, rice, sugar—the soil suffered from erosion and neglect. Those who cultivated it depended at best upon the uncertain returns of fluctuating world markets. But the circumstances under which labor was organized, particularly Negro labor, added to those difficulties further hardships of human creation.
>
> Oscar Handlin

The five sentences composing this paragraph show considerable variation in their openings: a prepositional phrase, the subject, a participial phrase, the subject, and a connective word.

Var. int. ## INTERRUPTED MOVEMENT

Interrupted movement—placing a modifying or absolute element within a sentence so that it requires pauses on either side—serves to vary several straightforward sentences, as in these examples:

> I had halted on the road. As soon as I saw the elephant I knew with perfect certainty that I ought not to shoot him. It is a serious matter to shoot a working elephant—it is comparable to destroying a huge and costly piece of machinery—and obviously one ought not to do it if it can possibly be avoided.
>
> George Orwell

> I do not wish to quarrel with any man or nation. I do not wish to split hairs, to make fine distinctions, or set myself up as better than my neighbors. I seek rather, I may say, even an excuse for conforming to the laws of the land. I am but too ready to conform to them.
>
> Henry David Thoreau

Notice how Thoreau uses parallelism in the second sentence as a way of varying the pattern of his simple statements.

EXERCISE

1. Study these paragraphs and explain how the authors vary their sentences. Notice differences in sentence opening, length, and principle of construction.

> 1. It all comes back. Even that recipe for sauerkraut: even that brings it back. I was on Fire Island when I first made that sauerkraut, and it was raining, and we drank a lot of bourbon and ate the sauerkraut and went to bed at ten, and I listened to the rain and the Atlantic and felt safe. I made the sauerkraut again last night and it did not make me feel any safer, but that is, as they say, another story.
>
> Joan Didion

2. Part of Andrew's charisma was his appearance. Erect in posture and slight of build—indeed cadaverously thin—he exuded rawboned toughness and strength. A ramrod, he stood six feet one inch in height and had a long, thin face accentuated by a strong shaft of a jaw. His nose was slender and flared a little at the tip. But there was considerable dramatic play about the eyes, those remarkable, deep blue eyes that could shower sparks when excited by passion. Anyone could tell his mood by watching his eyes; and when they started to blaze it was a signal to get out of the way *quickly*. But they could also register tenderness and sympathy, especially around children, when they generated a warmth and kindness that was most appealing.

Robert V. Remini [about Andrew Jackson]

2. What similarities, or recurrences, do you notice about the sentences in each of the foregoing paragraphs?

3. Compose a paragraph of about 200 words on a topic of your choice. Try consciously to vary your sentences and yet to maintain enough recurrence to keep your writing unified. Be able to explain what you did and why.

CHAPTER **39**

Rhythm in the Sentence

INTRODUCTION

Scholars do not completely agree about how prose rhythm works. But anyone who has thought at all about this complicated subject acknowledges that rhythm is important. In fact, effective rhythm is absolutely essential to good writing. An essay or book may have profound significance, it may develop fresh ideas or reveal important new information, but it cannot be described as well written if it falls flat upon the ear.

"Upon the ear"—rhythm is something we hear. Since it exists in the mind's ear, so to speak, an element of subjectivity inevitably colors any talk about prose rhythm. We do not all hear exactly the same thing. Just as eye-witnesses to an accident see it in different ways, so readers—even experienced, sensitive readers—"hear" the same sentence differently.

We cannot say, however, that rhythm is purely a question of the reader's perception. To a considerable degree it is controlled by the arrangement of the sentence. Writers can—and good writers do—regulate what their readers hear, not completely, but within fairly clear limits. To understand how they do this you need to have a rudimentary idea of what rhythm is.

Roughly, rhythm consists of repeating similar patterns. At least three kinds of patterns can be found in prose rhythm. The most obvious involves the loudness and softness of syllables; heavy syllables are said to be *stressed* and are indicated by ′, while light syllables are called *unstressed* or *nonstressed* and marked by ˣ.[1] Writers create syllabic

1. Distinguishing only two degrees of loudness and softness is arbitrary. In actual speech, innumerable gradations exist. However, limiting the number to two makes it convenient to discuss syllabic stress. Sometimes an intermediate stage, called *secondary stress,* is distinguished and marked ˋ. The process of analyzing and marking syllabic rhythm is known as *scansion* or *scanning.*

rhythm by maintaining a more or less regular alternation of stressed and unstressed syllables:

    x ′ x   ′   x ′   x   ′
A lucky few escaped the fire.

A second type of sound pattern is intonation, the rise and fall in the pitch of the voice, a kind of melody we use in speaking. Think, for example, of the many shades of meaning you can give to the word *yes* merely by altering how you say it. Or speak the following sentences out loud and you will hear a difference in intonation which we use to differentiate certain kinds of questions from statements:

Are you going downtown?

I am going downtown.

A kind of rhythm based on intonation is created by repeating phrases or clauses of similar construction so that the same melody plays over several times, as in these famous lines of poetry by Tennyson:

The long day wanes: the slow moon climbs: the deep
Moans round with many voices.

We hear this as a three-part structure with an identical pattern of intonation in the first two clauses, repeated in the opening four words of the third but varied by the concluding phrase. This kind of rhythm coexists with the rhythm of syllables. Thus in addition to similar intonations, Tennyson's lines also have an almost perfect alternation of stressed and lightly stressed syllables:

     x ′   ˋ   ′   x ′   ˋ   ′   x ′
The long day wanes: the slow moon climbs: the deep
   ˋ   ′   x ′x   ′x
Moans round with many voices.

The third kind of repeated pattern is that of similar or identical sounds: for example, the *b* in "a big beautiful blond." While such repetitions are important and affect the rhythm we hear in sentences, they are more a matter of rhyme. We touch upon rhyme briefly at the end of this chapter, but for the moment let us return to more central matters of rhythm.

Var. rhy.   **EFFECTIVE RHYTHM**

Effective rhythm means that a sentence pleases the ear and even, in the best cases, moves to a beat that reinforces the idea or feeling the sentence conveys. Prose has effective rhythm when the sentence or passage is laid out in clear syntactic units (phrases, clauses, whole sentences) which have something in common (length, intonation, grammatical structure) and when, within these units, a loose but discernible pattern of stressed and unstressed syllables is faintly to be heard. Generally the repetition of the units making up the sentence or passage is not exact. Nor is the alternation of syllables. That is the essential difference between prose and traditional accented poetry (not free verse): poetry has a much more regular pattern of stressed and unstressed syllables than prose. You can hear clear, effective prose rhythm in the following passage:

```
 x    x  x  / x   x  x  /     x  x   /    x   x  / x   x   x
There was a magic, and a spell, and a curse; but the magic has been
   /    x /   x   x   /    / x   x   x   /  x   x  /   x   /
waved away, and the spell broken, and the curse was a curse of sleep
 x   /  x   /
and not of pain.
```

R. L. Duffus

The sentence proceeds in carefully arranged units: two primary ones separated by the semicolon, and within each of these, three secondary units marked by commas. Each of the six units has a pattern of stressed and unstressed syllables, a pattern regular enough to be sensed, yet not so steady that it dominates the sentence, turning it into a kind of sing-song.

Here are several other examples:

```
 x   x   / x  / x   /    x   / x    /   x   x   x    / x x
We have little logic here, and simple faith, but we have energy.
```

Henry Adams

```
 x  x   x    / x   x   /   x   /   /   /  x   /   x   x   / x
So she came holding her dress with one fair rounded arm, and her taper
 x /  x   / x   /    x  /  x  /    / x
before her, tripping down the stair to greet Esmond.
```

William Makepeace Thackeray

```
 x  /   /    / x   x  /   x / x   x /   x  /   x  x   /
He sat bolt upright, his back supported against the wall, and his legs
 / x x   x x   /   / x  x x   /   / x  /  /  x   / x
pendulous, within three inches of the ground, being too short to reach it.
```

George Borrow

```
 x    /    /   /  x   /    x  x   / x   /   x   /   x
The poor man died that night, and we gathered stones and piled them
 x  x  / x  x  x   x  / x  x   x / x   x   /   x /   x
on his body so that the foxes and caranchos should not devour him.
```

W. H. Hudson

```
 x   x   /  x  x   / x   x / x x /   x  x / x  x    /     x
We came up on the railway beyond the canal. It went straight toward
 x   /   x /  x   /   /   x   x  /   x  /  x x   / x   / x
the town across the low fields. We could see the line of the other railway
 x /  x  /
ahead of us.
```

Ernest Hemingway

AWKWARD RHYTHM

Poor rhythm usually results from one or a combination of two causes. You may have failed to organize a sentence into parts of similar length or intonation. Or you may have failed to group syllables in patterns related to the sense and pleasing to the ear: stresses and nonstresses may be so irregular that no pattern at all emerges or so regular that a steady and obtrusive beat overrides everything else.

Look at this example of poor rhythm:

> Each party promises to make the city bigger and better before the election, but what happens after the election?

The sentence has two difficulties. The initial clause does not separate into well-defined groups. This fault can be corrected by changing the position of the adverbial phrase:

> Before the election, each party promises to make the city bigger and better . . .

Or it could be positioned as an interrupter:

> Each party promises, before the election, to make the city bigger and better . . .

Either strategy improves the rhythm.

The writer has also mixed a statement and a question in the same sentence. The clash of different intonations that results leaves the ear vaguely dissatisfied. It would be wiser to express the ideas as separate sentences:

> Before the election, each party promises to make the city bigger and better. But what happens after the election?

Other minor improvements might be made. For instance, it sounds better to shorten the second sentence to "But what happens afterwards?" which makes it less repetitious and more emphatic. But our revision, just as it stands, adds no words and takes none away, showing that you can often improve the rhythm of a sentence simply by rearranging the words.

Sometimes, on the other hand, you need changes beyond mere rearrangement to make an awkward sentence better. Consider this case:

> x ′ x ′ x x x ′ x ′ x ′ x ′ x
> The man was standing on the stairs and far below we saw the
> ′ x ′ x ′ x ′ x ′ x ′
> boy, who wore an old, unpressed, and ragged suit.

This sentence has partly the same difficulty as the first example: it needs to be divided more clearly (or at least its first two clauses do).

But it also has a different problem: its syllabic rhythm is too regular. With one exception the sentence scans as a series of unvaried iambs.[2] This kind of regularity dominates the sentence, obscuring shadings of emphasis. The passage is much improved like this:

 x   / /   x x / / x / x / x / /
The man stood on the stairs; far below we saw the boy, dressed

 x x /   x /   / x /
in an old, unpressed, ragged suit.

The specific changes—substituting "stood" for "was standing" and "dressed" for "who wore," and replacing two "ands" with a semicolon and a comma—break up the excessive sameness of the syllabic beat, yet leave pattern enough to please the ear. Furthermore, the rhythm, by clustering stresses, now focuses the reader's attention upon key

    /   /   /   /   /   x /   / x   /
points: "man stood," "boy, dressed," and "old, unpressed, ragged suit."

METRICAL RUNS

Despite the last example, do not suppose that regular syllabic rhythm is always bad in prose. Sometimes relatively unvaried patterns of stressed and unstressed syllables work quite effectively; these are called metrical runs:[3]

  / x   x / x   x / x x / x
Hamlet is quite the reverse of the skeptic.

                 G. K. Chesterton

 x / x / x /   x / x / x x / x /
I love to lie in bed and read the lives of the Popes of Rome.

                Logan Pearsall Smith

  /   x x /x x x   /   x /   x x / x /   x x /   x
This is a story about love and death in the golden land, and begins with

 x   / x
the country.

                 Joan Didion

Smith and Didion achieve their metrical runs in part by the skillful use of prepositional phrases. This is an easy way to control rhythm since the prepositional phrase has a built-in meter. Typically the phrase consists of a one- or two-syllable preposition, a noun marker (*a, an,*

2. An *iamb* is a group of two syllables, a nonstress followed by a stress, as in the word

x /   x x x
above. The one exception in the sentence above is in the four syllables "-ing on the

 /
stairs."
3. One sense of the word *meter* is a regularly maintained grouping of stressed and unstressed syllables; hence the term *metrical run.*

the, this, that, and so on), and an object of one or two syllables. Neither the preposition nor the marker is stressed, while the object is, so that

one of these meters is likely: x ′ ("at home"), x x ′ ("in the house"),

x x ′ x ("in the morning"), or x x x ′ ("above the hill"). In such *rising meters* the strong syllable comes after one or more weak ones. By adding modifiers or doubling the objects of a preposition or stringing together several phrases, a writer can sustain these meters over a portion of a sentence:

about love and death in the golden land.

In the following examples the phrases creating the metrical run come at the end of the sentence, bringing it neatly to a close:

Smoke lowering from chimney pots, making a soft black drizzle, with flakes of soot in it as big as full grown snow-flakes—gone into mourning, one might imagine, for the death of the sun.

Charles Dickens

Beyond the blue hills, within riding distance, there is a country of parks and beeches with views of the far-off sea.

Logan Pearsall Smith

There was the sea, sheer under me, and it looked grey and grim, and streaked with the white of our smother.

John Masefield

Like fragments and rhetorical questions, metrical runs work because they are uncommon, and cease to be effective when they are overused. Their effect is subtly to draw our notice. We feel that a sentence is somehow memorable. We don't know why exactly, but it sticks in our minds, partly because the rhythm pleases us. Certainly good rhythm will not dignify a silly idea, but it does help us remember something worth remembering.

RHYTHMIC BREAKS

One of the advantages of maintaining a fairly regular rhythm is that you can suddenly alter it for special effect, as Amy Lowell does in this sentence:

The roses have faded at Malmaison, nipped by the frost.

The abrupt break at "nipped" throws great weight upon that word, making it the center of the sentence. And a sentence by Logan Pearsall Smith quoted on page 322 replaces a series of rising meters with a cluster of stresses to end on his key image, the sea:

```
x   /   x   /   /     x x    / x    / x     x   x x  / x   x   /
Beyond the blue hills, within riding distance, there is a country of parks
x     / x   x    /   x x   / /   /
and beeches with views of the far-off sea.
```

MIMETIC RHYTHM

Mimetic means imitative. Mimetic rhythm enhances the meaning of a sentence by imitating the reality the sentence describes or the feeling or idea it expresses. It might be fanciful to suggest that in Amy Lowell's sentence the abrupt stress upon "nipped" suggests the action the verb names. But take a look at the mimetic rhythm of this passage by the naturalist Rachel Carson discussing the movement of the tides:

```
x   /   / x   /   /   / x   / x x   x   x / x   /
The tide reaches flood stage, slackens, hesitates, and begins to ebb.
```

It needs no great imagination to hear the sentence moving to a rhythm like that of the tidal flow: running smoothly and uninterruptedly to a midpoint, slowing down, pausing (the commas), then picking up and running faster until the end.

The following passage describes Niagara Falls, and it, too, imitates what it describes:

```
x   x   /   x   x / x   x   / x   /   x   / x   x /   x /     x
On the edge of disaster the river seems to gather herself, to pause, to
/  x   /    / x x   / x   x   /   x x   /   / x   x   /   x x
lift a head noble in ruin, and then, with a slow grandeur, to plunge into
x   x / x   / x   x   /   / x   x /
the eternal thunder and white chaos below.
```

Rupert Brooke

Mimetic rhythm may also imitate subjects more abstract than physical movement, as in this sentence describing the life of European peasants:

```
/     /     /     /     /     /   x / x x   /     / x   /   x   / x
Black bread, rude roof, dark night, laborious day, weary arm at sunset;
x   /   /   x /
and life ebbs away.
```

John Ruskin

The six unrelieved stresses at the beginning suggest the dreary monotony of existence for the peasant. The nonstressed syllables become more numerous, and the sentence picks up speed and runs to a close, just as life slips away (in Ruskin's view) from the peasant before he has held and savored it.

Finally let us look again at a sentence opening a description of the collapse of the stock market in 1929 (quoted earlier on p. 290):

x   ′   ′   ′ x   x   ′
The Big Bull Market was dead.

It is a classic topic sentence: clear, concise, emphatic, the key term placed at the end. But apart from that, its rhythm neatly fits its subject, amplifying the idea of finality. The relatively large number of stresses (four syllables in seven), the clustering of three stresses near the beginning, and the strong ending make the statement uncompromising: the boom is over, over absolutely.

RHYME

Rhyme is the repetition of sounds in positions close enough to be noticed. We associate this aspect of language with poetry, usually in the form of end rhyme—the closing of successive or alternate lines with the same sound:

The grave's a fine and private place,
But none I think do there embrace.

Andrew Marvell

Poetry also has inner rhyme—the repetition of sounds within the line, as the *a* and *i* vowels and the *p*'s of the first line of Marvell's couplet.

Despite its association with poetry, rhyme also occurs in prose, more often than people think. It is usually a kind of inner rhyme—prose writers rarely structure sentences or clauses by ending them with the same sound. Like rhythm, rhyme can affect the ear both pleasantly and unpleasantly, and it can enhance meaning.

Probably sounds do not have "natural," culture-free meanings in themselves. Psychologists who have investigated the question of sound symbolism, as it is called, have found no evidence to suggest that meanings inhere in sounds. We may think that the long *e* in *teeny, weeny, eeny* is a "small" sound, but, chances are, this feeling results from cultural conditioning and in other languages different conditioning leads people to think other vowel sounds "tiny." Even onomatopoetic words (those which, like *slurp,* directly imitate a sound) differ greatly from culture to culture. *Coquerico,* the French word for a rooster's crowing, sounds different from *cockadoodle-doo,* though presumably roosters crow the same in Brittany and New England.

But even if sounds contain no universal inherent meanings, within a specific culture particular sounds do evoke particular attitudes. Even here, however, one must be very careful in talking about the "meaning" of sounds. Such meanings are very broad and resist precise interpretation. In the following description by Mark Twain of a Mississippi River town, the frequent *l* sounds, the *s*'s, the *m*'s, and the *n*'s probably contribute to the sense of peace and quiet the passage describes. Words like *lull, lullaby, loll, slow, silent, ssh, shush, hush* have conditioned us to associate these sounds with quietness; but that is about all we can say.

> After all these years I can picture that old time to myself now, just as it was then: the white town drowsing in the sunshine of a summer's morning; the streets empty or pretty nearly so; one or two clerks sitting in front of the Water Street stores, with their splint-bottomed chairs tilted back against the walls, chins on breasts, hats slouched over their faces, asleep— with shingle shavings enough around to show what broke them down; a sow and a litter of pigs loafing along the sidewalk, doing a good business in watermelon rinds and seeds; two or three lonely little freight piles scattered about the "levee"; a pile of "skids" on the slope of the stone-paved wharf, and the fragrant town drunkard asleep in the shadow of them; two or three wood flats at the head of the wharf, but nobody to listen to the peaceful lapping of the wavelets against them; the great Mississippi, the majestic, the magnificent Mississippi, rolling its mile-wide tide along, shining in the sun; the "point" above the town, and the "point" below, bounding the river-glimpse and turning it into a sort of sea, and withal a very still and brilliant and lonely one.

If we do not insist upon interpreting their "meaning" too closely, then, it is fair to say that sounds can convey or reinforce certain moods. Sounds may also contribute to meaning in another, less direct way. By rhyming key words, a writer draws attention to them. In prose such rhyme often takes the form of alliteration, the repetition of initial sounds in successive or near-successive words. In the following sentence the writer emphasizes "wilderness" by repeating *w* and "decay" by repeating *d*:

> Otherwise the place is bleakly uninteresting: a wilderness of windswept grasses and sinewy weeds waving away from a thin beach ever speckled with drift and decaying things,—worm-ridden timbers, dead porpoises.
>
> Lafcadio Hearn

Alliteration can be risky. Hearn succeeds, but G. K. Chesterton rides alliteration too hard and too long in the *k* sounds of this sentence:

> Thus a creed which set out to create conquerors would only corrupt soldiers; corrupt them with a craven and unsoldierly worship of success: and that which began as a philosophy of courage ends as a philosophy of cowardice.

Excesses like this have led some people to damn and blast all alliteration—and, in fact, all other varieties of rhyme—in prose. There is no doubt that in prose a little rhyme goes a long way. The trick is to keep the rhyme unobtrusive, subordinate to the sense. Composition suffers when rhyme comes to the surface, a fault to which poets are sometimes given when they write prose:

> Her eyes were full of proud and passionless lust after gold and blood; her hair, close and curled, seems ready to shudder in sunder and divide into snakes.
>
> Algernon Charles Swinburne

> His boots are tight, the sun is hot, and he may be shot. . . .
>
> Amy Lowell

These contain too much rhyme for most tastes. Lowell's sentence, an example of what she called polyphonic prose, seems especially awkward, employing in "hot/shot" the kind of vowel-consonant end rhyme common in poetry. The unrelieved meter of the sentence also contributes to its awkwardness:

>   x   ′   x   ′   x ′ x ′ x x ′ x ′
> His boots are tight, the sun is hot, and he may be shot. . . .

Yet despite such abuses, it is extreme to say that rhyme has no place in prose. It is more reasonable to acknowledge that the sounds of words play an inevitable part in their effect upon a reader. Negatively, certain things should be avoided: obvious and jingling rhyme or combinations of awkwardly dissimilar sounds. Positively, sounds can create a tonal harmony which pleases the ear and makes us more receptive to what the sentence says, as in this passage by John Donne (a seventeenth-century poet who also wrote great prose):

> One dieth at his full strength, being wholly at ease, and in quiet, and another dies in the bitterness of his soul, and never eats with pleasure; but they lie down alike in the dust, and the worm covers them.

Or sounds can enhance the effect of an image:

> Dust swirls down the avenue, hisses and hurries like erected cobras round the corners.
>
> Virginia Woolf

Thus rhyme is—or can be—a positive element in composition. It is less significant than rhythm, but far from negligible. Too great a concern with sound, too much "tone painting," is a fault in prose (and in poetry, too). Controlled by a sensitive ear, however, the sounds of a sentence enrich and widen its meaning.

SUMMARY

Rhythm and rhyme are not a writer's first order of business. The principal thing is to get your ideas and feelings clearly expressed. If you succeed in doing that, rhythm may well take care of itself. At the same time, remember that how a sentence sounds helps to determine its effect. A good idea in a dull sentence is not a formula for effective communication.

As to the practical question of how to achieve good rhythm, keep in mind two basic principles: arrange your sentences (or longer passages) in clearly grasped syntactic units, and maintain within these units an unobtrusive pattern of syllabic stress which will please the reader and, if possible, underscore your meaning.

These things happen best when they happen naturally. Good writers rarely concentrate consciously upon rhythm and rhyme—counting syllables, looking up words that will alliterate with each other. They trust their ears, and when a sentence grates they revise it until it sounds right. The revision may be based upon a clear understanding of how rhythm works, or it may be trial-and-error tinkering. But in the final event good writers listen to what they write. That is what you must do: listen to your sentences and work on them until you are pleased by what you hear.

EXERCISE

1. The following sentences are all examples of the skillful use of rhythm. Scan each one to indicate the stressed and nonstressed syllables. Do any words seem to be thrown into special prominence by the rhythm?

A. Perhaps she too would see the sunset and pause for a moment, turning, remembering, before he faded with her sleep into the past.

F. Scott Fitzgerald

B. Sad is the day and worse must follow, when we hear the blackbird in the garden, and do not throb with joy.

W. S. Landor

C. For, though I speak to you, I think the king is but a man, as I am: the violet smells to him as it doth to me; all his senses have but human conditions: his ceremonies laid by, in his nakedness he appears but a man; and though his affections are higher mounted than ours, yet, when they stoop, they stoop with the like wing.

William Shakespeare

D. When all is done, human life is at the greatest and the best but like a froward child, that must be played with and humoured a little to keep it quiet till it falls asleep, and then the care is over.

Sir William Temple

E. The world of Goldsmith's poetry is, of course, a flat and eyeless world; swains sport with nymphs, and the deep is finny.

Virginia Woolf

F. There was a stiff, dry west wind blowing, and a blue haze in the air.

Herbert Quick

G. Bitter are the waters of old age, and tears fall inward on the heart.

D. H. Lawrence

2. These sentences have rhythms especially appropriate to sense. Explain why.

A. [Of a city suddenly attacked at night.] The city so sound asleep one minute past, was now awake and alive in every fibre. Bugles sounded there; arms and armour rang, and fierce voices in a strange tongue shouted passionate commands. Dogs bayed, horses neighed, women shrieked, and children wept; and all the time the noise of trampling feet sounded like low thunder, a bass accompaniment to all that treble.

Standish O'Grady

B. [Of a beech tree seen from a fast-moving automobile.] It grew bigger and bigger with blinding rapidity. It charged me like a tilting knight, seemed to hack at my head, and pass by.

G. K. Chesterton

C. [Of a trip down the Thames in a speedboat.] Everything was alive about them, flashing, splashing, and passing, ships moving, tugs panting, hawsers taut, barges going down with men toiling at the sweeps, the water all a-swirl with the wash of the shipping, scaling into millions of little wavelets, curling and frothing under the whip of the unceasing wind.

H. G. Wells

D. [Of whaleboats approaching a whale] Like noiseless nautilus shells, their light prows sped through the sea; but only slowly they neared the foe.

Herman Melville

E. The traffic jam reaches its highest peak, tapers off, and slowly disappears, all within forty-five minutes.

Student

F. Suddenly, as if the movement of his hand had released it, the load of her accumulated impressions of him tilted up, and down poured in a ponderous avalanche all she felt about him.

Virginia Woolf

3. Modeling your work upon the sentences in exercise 2, compose six passages on similar topics, using the movement of the sentence to reinforce the sense, as Melville, for instance, slows down his final clause to suggest the caution with which the boats approach the whale.

4. Indicate the rhyme in the following sentences. Do you think it is effective? Why (or why not)?

A. And I see a bay, a wide bay, smooth as glass and polished like ice, shimmering in the dark.

Joseph Conrad

B. The day is fresh and fair, and there is a smell of narcissus in the air.

Amy Lowell

C. So we lurch and lumber through the most famous novels in the world.

Virginia Woolf

D. Green field, and glowing rock, and glancing streamlet, all slope together in the sunshine towards the brows of ravines, where the pines take up their own dominion of saddened shade; and with everlasting roar in the twilight, the stronger torrents thunder down, pale from the glaciers, filling all their chasms with enchanted cold, beating themselves to pieces against the great rocks that they have themselves cast down, and forcing fierce way beneath their ghastly poise.

John Ruskin

E. Ruskin's sentence branches into brackets and relative clauses as a straight strong tree branches into boughs and bifurcations. . . .

G. K. Chesterton

F. Walking one day into a field that I had watched yellowing beyond the trees, I found myself dazzled by the glow and great expanse of gold.

Logan Pearsall Smith

PART FIVE

DICTION

INTRODUCTION

Diction refers to words. The study of diction concerns choosing words so that they carry the meanings you want them to carry. In the next four sections we shall consider these questions: (1) What does "meaning" mean? What is involved in saying that a word "means" something? (2) What are the qualities that distinguish poor diction and good? (3) How may words be used in bold and unusual ways? And (4) how can dictionaries and thesauri help to improve your knowledge and handling of words?

The first section is more or less theoretical, though on an elementary level. The others are practical. Theory and practice, however, go hand in hand. Using words well requires at least a rudimentary understanding of what words do.

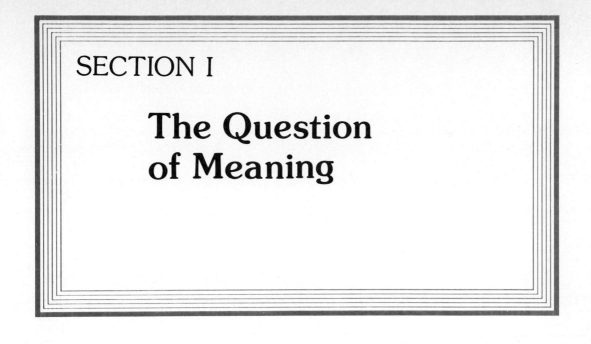

SECTION I

The Question of Meaning

40

Meaning

To say that a word has meaning is to say that it has purpose. The purpose may be to signify something—that is, to refer to an object or person other than the writer, to an abstract conception such as "democracy," or to a thought or feeling in the writer's mind. On the other hand, the purpose may be to induce a particular response in the readers' minds or to establish an appropriate relationship between the writer and those readers. We shall consider each of these three uses of words— modes of meaning, we shall call them.

Before we do that, however, we need to glance at several misconceptions about words and also at two aspects of meaning fundamental to all the purposes for which words may be used. These aspects concern denotative and connotative meaning and the various levels of usage.

First the misconceptions.

WORDS ARE NOT ENDOWED
WITH FIXED AND "PROPER" MEANINGS

When people object to how someone else uses a word, they often say, "That isn't its proper meaning." The word *disinterested,* for example, is frequently employed in the sense of "uninterested," and those who dislike this usage argue that the proper meaning of *disinterested* is "objective, unbiased."

In such arguments "proper meaning" generally signifies a meaning sanctioned by past usage or even by the original, etymological sense of the word. But the dogma that words come to us out of the past with proper meanings—fixed and immutable—is a fallacy. The only meanings a word has are those that the speakers of the language choose to give it. If enough speakers of English use *disinterested* to mean "uninterested," then by definition they have given that meaning to the word.

Those who take a conservative attitude toward language have the right, even the duty, to resist changes which they feel lessen the efficiency of English. They should, however, base their resistance upon demonstrating why the change does make for inefficiency, not upon an authoritarian claim that it violates proper meaning.

As a user of words you should be guided by consensus, that is, the meanings agreed upon by your fellow speakers of English, the meanings recorded in dictionaries. We shall look at what dictionaries do in Chapter 45. For now, simply understand that dictionary definitions are not "proper meanings" but succinct statements of consensual meanings.

In most cases the consensus emerges from an activity in which individual language users participate without knowing that they are, in effect, defining words. The person who says "I was disinterested in the lecture" does not intend to alter the meaning of *disinterested*. He or she has simply heard the word used this way before. In a few cases people do act deliberately to establish a consensual meaning, as when mathematicians agree that the word *googol* will mean "10 raised to the 100th power." In any case, meaning is what the group consents to. This is the only "proper meaning" words have, and any subsequent generation may consent to alter a consensus.

But while the unconscious agreement which establishes the meaning of a word is a group activity, it originates with individuals. Particular speakers began using *disinterested* in the sense of "uninterested" or *square* in the sense of "extremely conventional and unsophisticated." From the usage of individual people the change spreads through the group—for better or worse.

By such process word meanings change, sometimes rapidly, sometimes glacially. Often the change occurs as a response to historical events. When the eighteenth-century historian Edward Gibbon writes of "the *constitution* of a Roman legion" he means how it was organized, not, as a modern reader might suppose, a written document defining that organization. The latter sense became common only after the late eighteenth century, with the spread of democratic revolutions and the formal writing down of a new government's principles.

Because words must constantly be adapted to a changing world, no neat one-to-one correspondence exists between words and meanings. On the contrary, the relationship is messy: a single word may have half a dozen meanings or more, while several words may designate the same concept or entity. Thus *depression* means one thing to a psychologist, another to an economist, and another still to a geologist. But psychological "depression" may also be conveyed by *melancholia, the blues,* or *the dismals, in the dumps, low,* and so on.

One-to-one correspondences do in fact exist in the highly specialized languages of science and technology and mathematics. To a chemist *sodium chloride* means only the compound NaCl, and that compound is always designated in words by *sodium chloride*. The common term *salt,* in contrast, has a number of meanings, and we must depend upon the context (that is, the words around it) to clarify which sense the writer intends:

Pass the *salt.*
She's the *salt* of the earth.
They're not worth their *salt.*
He's a typical old *salt.*
Her wit has considerable *salt.*
The crooks intended to *salt* the mine.
They are going to *salt* away all the cash they can.

But while one-to-one correspondences might seem desirable, having a distinct word for every conceivable object and idea and feeling would not be practical. The vocabulary would swell to unmanageable proportions. And probably we would like it less than we suppose. The inexact correspondence of words and meanings opens up possibilities of conveying subtleties of thought and feeling which an exactly defined vocabulary would exclude. The fact that *sodium chloride* means one thing and only one thing is both a virtue and a limitation. The fact that *salt* means many things is both a problem and an opportunity.

Words, then, are far from being tokens of fixed and permanent value. They are like living things, complex, many-sided, and responsive to pressures from their environment. They must be handled with care.

DENOTATION AND CONNOTATION

Denotation and connotation are aspects of a word's meaning, related but distinct. *Denotation* is a word's primary, specific sense, as the denotation of *red* is the color (or, from the viewpoint of physics, light of a certain wavelength). *Connotation* is the secondary meaning (or meanings), associated with but different from the denotation. *Red,* for instance, has several connotations: "socialist," "anger," and "danger," among others.[1]

Using a circle to represent a word, we may show the denotation as the core meaning and the connotation as fringe meanings gathered about that core. The line enclosing the denotation (D in the diagram) is solid to signify that this meaning is relatively fixed. The line around the connotation (C) is broken to suggest that the connotative meanings of a word are less firm, more open to change and addition.

Connotations may evolve naturally from the denotation of a word, or they may develop by chance associations. *Rose* connotes "fragrant," "beautiful," "short-lived" because the qualities natural to the flower have been incorporated into the word. On the other hand, that *red* connotes "socialist" is accidental, the chance result of early European socialists' using a red flag as their banner.

1. In logic *denotation* and *connotation* are used in somewhat different senses.

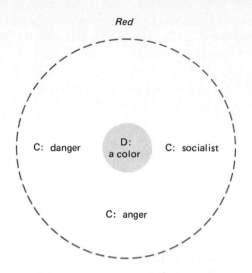

Sometimes a connotative meaning splits off and becomes a second denotation, the nucleus, in effect, of another word configuration. Thus "socialist" has become a new primary meaning of *red* when used as a political term. Around this second nucleus other connotations have gathered, such as (for most Americans) "subversive," "un-American," "traitorous," and so on:

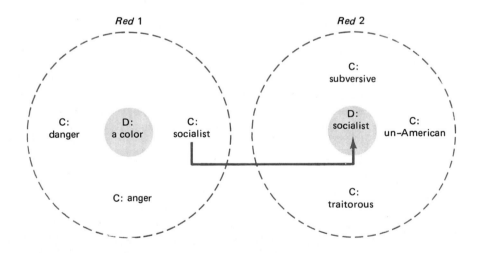

Often, though not inevitably, connotative meanings imply degrees of approval or disapproval and may arouse emotions such as affection, admiration, pity, disgust, hatred. Like positive and negative electrical charges, emotive connotations attract or repel readers with regard to the thing or concept the word designates (though the exact degree of attraction or repulsion depends upon how particular readers are them-

selves charged concerned the thing or concept). These positive and negative charges are extremely important to a word's connotation, and in later diagrams we indicate them by + and − signs.

Individual words vary considerably in the relative weight of their denotative and connotative meanings. Most technical terms, for example, have very little connotation. That is their virtue: they denote an entity or concept precisely and unambiguously without the possible confusion engendered by fringe meanings: *diode, spinnaker, cosine.* We may think of such words as small and compact—all nucleus, so to speak. They have no circle of connotations around them.

Connotation looms larger than denotation in other cases. Some words have large and diffuse meanings. What matters is their secondary or suggestive meanings, not their relatively unimportant denotations. The expression *old-fashioned,* for instance, hauls a heavy load of connotations. It denotes "belonging to, or characteristic of, the past." But far more important than that central meaning is the connotation, or rather two quite different connotations, that have gathered about the nucleus: (1) "valuable, worthy of honor and emulation" and (2) "foolish, ridiculous, out-of-date; to be avoided." With such words the large outer, or connotative, circle is significant; the nucleus small and insignificant.

For many words denotation and connotation are both important aspects of meaning. *Rose* (in the sense of the flower) has a precise botanical denotation: "any of a genus (*Rosa* of the family Rosaceae, the rose family) of usu.[ally] prickly shrubs with pinnate leaves and showy flowers having five petals in the wild state but being often double or semidouble under cultivation."[2] At the same time *rose* also has strong connotations: "beautiful," "fragrant," "short-lived," and so on.

CONTEXT

The denotation of any word is easy to learn: you need only look in a suitable dictionary. Understanding connotations, however, is more difficult. Dictionaries cannot afford the space to treat them, except in a very few cases. You can gain practical knowledge of a word's range of connotation only by becoming familiar with the contexts in which the word is used.

Context means the surroundings of a word. In a narrow sense, context is the other terms in the phrase, clause, sentence—a word's immediate linguistic environment. More broadly, context comprises all the other words in the passage, even the entire essay or book. It widens further to include a composition's relation to other works, why it was written, and so on. In speech, context in this inclusive sense involves the occasion of a conversation, the relationship between the talkers, who else may be listening.

2. *Webster's Seventh New Collegiate Dictionary* (Springfield, Mass.: G. & C. Merriam Company, 1963).

But one does not have to explore all the ramifications of context to get at a word's connotation. Usually the terms immediately around it supply the vital clue. *Real old-fashioned flavor* printed on an ice cream carton tells us that here *old-fashioned* connotes "valuable, rich in taste, worthy of admiration (and of purchase)." *Don't be old-fashioned—dare a new experience* in an ad for men's cologne evokes the opposite connotation: "foolish, ridiculous, out-of-date."

Linguistic context acts as a selective screen lying over a word, revealing certain of its connotations, concealing others. Thus "real" and "flavor" mask the unfavorable connotation of *old-fashioned*, leaving us aware only of the positive one. Here is a diagram of *old-fashioned* in the "real/flavor" context:

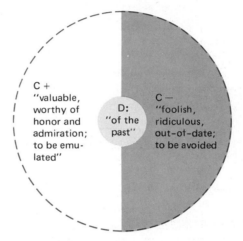

In the context of "don't/dare a new experience," the screening effect is just the opposite:

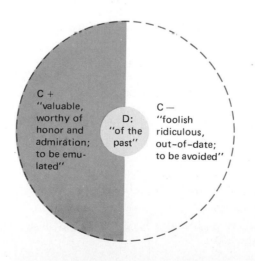

Not only does the linguistic context serve both to reveal and to hide certain of a word's connotations. It may also activate latent implications that ordinarily are not associated with a word. The meaning "rich in taste," for instance, is not one we customarily associate with *old-fashioned*. Yet in *real old-fashioned flavor* it comes to the surface.

Context also helps you determine whether a word is functioning primarily in its denotative or the connotative sense. With words like *rose* that carry both kinds of meaning, only context reveals which is operating, or if both are in varying degrees. Clearly this sentence calls upon only the denotation of *rose*:

> Our native wild roses have, in spite of their great variety, contributed little to the development of our garden roses.

But when the poet Robert Burns tells of his feelings for a young lady, while still denoting the flower, he uses the word primarily for its connotations:

> O, my luv is like a red, red rose
> That's newly sprung in June.

In choosing words, then, you must pay attention both to denotative and to connotative meaning. With a purely denotative word like *cosine*, say, the problem is simple. If you make a mistake with such a word, it is simply because you do not know what it means and had better consult a dictionary (or textbook). But when words must be chosen with an eye to their connotations, the problem is more difficult. Connotative meaning is more diffuse, less readily looked up in a reference book, more subtly dependent upon context. Here mistakes are easier to make. For instance, if you want readers to like a character you are describing, it would be unwise to write "a fat man with a red face," even though the words are literally accurate. *Fat* and *red* are negatively charged in such a context. More positive would be "a stout [*or* plump] man with rosy cheeks."

LEVELS OF USAGE

Level of usage refers to the kind of situation in which a word is normally used. Most words suit all occasions. Some, however, are restricted to formal, literary contexts, and others to informal, colloquial ones. Consider three verbs which roughly mean the same thing: *exacerbate, annoy, bug*. Talking with your friends, you would not be likely to say, "That teacher really exacerbated me." On the other hand, answering a question on a history examination you wouldn't (or shouldn't) write, "The Spartan demands bugged the Athenians." But you could use *annoy* on both occasions, without arousing either your friends' derision or your instructor's red pencil.

The three words differ considerably in their levels of usage. *Exacerbate* is a literary word, appropriate to formal occasions. *Bug* (in this sense) is a colloquial, even slang, term appropriate to speech and very informal writing. *Annoy* is an all-purpose word, suitable for any occasion. When in the next chapter we discuss the practical problem of appropriateness, we shall use the labels *formal, informal,* and *general* to distinguish these broad levels of usage.

From the more theoretical viewpoint we are taking here, we may think of level of usage as a peripheral part of a word's connotation. While not an aspect of a word's designation, the situation in which it is customarily used is part of its meaning in a broad sense. As with connotation in general, it is not easy to look up the level of usage of any particular word. Dictionaries label an occasional term "colloquial" or "slang," but not in every case; and they do not label formal words like *exacerbate* at all. You have to depend upon your own knowledge as a guide.

In recent years the line between formal and informal usage has blurred considerably (though not enough for Spartans to bug Athenians). The distinction still exists, however, and careful writers pay attention to it.

TELIC MODES OF MEANING

Finally we shall discuss the point with which we began—the purpose a word is chosen to serve. This aspect we shall call the "telic mode" of meaning, from the Greek word *telos,* meaning "end," and the Latin *modus,* meaning "manner." Though the phrase sounds forbidding, it is a useful brief label for an obvious but important fact: that part of a word's meaning is the purpose it is expected to fulfill, and that words may serve different purposes.

To get a bit further into this matter it will help to look at a well-known diagram called the "communication triangle":

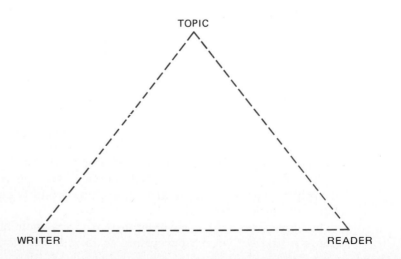

TOPIC

WRITER   READER

The diagram simply clarifies the fact that any act of communication involves three things: someone who communicates (for our purposes, a writer); something the communication is about (the topic); and someone to whom the communication is made (the reader). The broken lines joining these elements indicate an indirect relationship between them.

It is indirect because it must be mediated by words. Directly, each corner of the triangle connects only to words. The writer selects them, the reader interprets them, and the topic is expressed by them. Words thus occupy a central, essential, mediating position in the triangle:

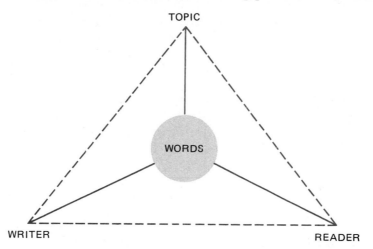

In selecting his or her words, a writer may be concerned primarily with any of the three areas of the triangle: writer-topic, writer-reader, or reader-topic. These areas correspond to the three "telic modes" of meaning. We shall call them respectively: "referential," "interpersonal," and "directive."

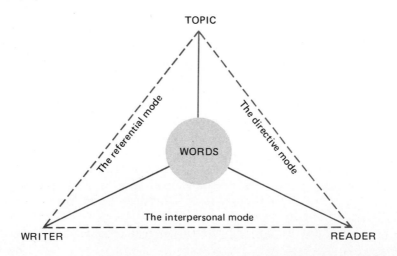

THE REFERENTIAL MODE

Referential meaning connects writer and topic. In this mode the writer chooses words for the exactness and economy with which they signify, or refer to, what he or she observes, knows, thinks, feels—in short, what is in his or her mind. Your compositions involve chiefly this mode of meaning. Here are three examples:

Mary [Queen of Scots] had returned to Scotland in 1561, a young widow of nineteen, after an absence of thirteen years in France. . . .

D. Harris Willson

The principle of verification is supposed to furnish a criterion by which it can be determined whether or not a sentence is literally meaningful.

Alfred Jules Ayer

Calculus is a lousy subject.

Student

In all these cases the writers select words for their referential value, to make clear what is in their minds. The historian, aiming to be factually accurate, and the philosopher, aiming to be conceptually exact, chose diction on the basis of denotation: "in 1561," "a young widow of nineteen," "verification," "criterion." The student, expressing how he feels, selects "lousy" for its connotation; and while it would be more difficult to ravel out all the implications of "lousy" than to explain the meanings of "widow" or "criterion," the word is exactly right.

In each case, of course, the diction will affect readers' attitudes toward both subject and writer, and to that degree the words will operate in the interpretive and interpersonal modes. Ayer's abstract diction may well bore people uninterested in philosophy, for instance. A mathematician, depending upon his sense of humor, might be amused or annoyed by the student's characterization of calculus. But although such spillover effects are very real, the fact remains that in all these examples the diction aims at referential accuracy and operates primarily in that mode of meaning.

THE INTERPERSONAL MODE

In your writing assignments you will choose words chiefly for their referential meanings. Those words, however, will also affect the link between readers and you. It follows that you should select even referential diction with an eye on the reader. You must consider what readers know and do not know, how they resemble you and how they differ, what degree of formality or informality you wish to establish with them. Such considerations may lead you, for example, to look for an easier word even though it is a bit less exact than a technical term they probably do not know.

But beyond showing a general concern for readers in choosing the words with which you discuss your topic, you may also wish occasionally to include words that will directly affect the readers' attitude toward you. Now you are in the interpersonal mode of meaning.

First, certain expressions create a favorable image of yourself. Inevitably you exist in your words—whether you wish to or not—as an unseen presence, a hidden voice of which readers are aware, sometimes dimly, sometimes with acute consciousness. The word *persona,* from the Latin for "mask," signifies this presence of a writer in his or her work. Despite its etymology, a persona does not have to be a false face, though probably it will be only one—even if a true one—of the many faces each of us possesses.

Since a persona is inevitable, you had better strive for an attractive one. Modesty, for instance, is generally a virtue in a writer. An occasional expression like *I think, it seems to me, to my mind* suggests to readers that here is a modest writer, undogmatic, aware of his or her fallibility. The following passages illustrate such interpersonal diction (the italics are added):

> What, then, can one learn from [Samuel] Johnson in general? First, *I think,* the inestimable value of individuality.
>
> F. L. Lucas

> Whether this slowing-down of traffic will cause a great or a small loss of national income is, *I am told,* a point on which expert economists are not agreed.
>
> Max Beerbohm

> That this is so can hardly be proved, but it is, *I should claim,* a fact.
>
> J. L. Austin

Such personal disclaimers are not always a virtue. At times modesty may strike a note that is weak or false. At times a subject may demand an impersonal point of view, making the use of *I, my, me* impossible. Even when modesty is called for and a personal point of view is possible, a few *I think*s and *in my opinion*s go a long way. Used in every second or third sentence they may well draw too much attention and annoy the reader. Still, occasionally acknowledging your limitations is one way of creating a favorable impression upon readers.

Beyond suggesting a diffident, nonassertive persona, you can also use words in the interpersonal mode which graciously acknowledge your readers' presence. Without being insincere or obsequious you can draw readers into your exposition so that they seem to share more directly in your ideas and feelings. The judicious use of *we, our, us,* for instance, implies a common ground of knowledge and values, subtly flattering to readers. (Again, in the examples that follow italics are added):

> Let *us* define a plot.
>
> E. M. Forster

No doubt, if one has more than one self (like most of *us*), it had better
be one's better self that one tries to become.

F. L. Lucas

When *we* look more closely at this craft of philosophic expression, *we*
find to *our* relief that it is less exacting than the art of the true man of
letters.

Brand Blanshard

Any words, then, that refer to the writer in the role of writer or to
the reader in the role of reader operate in the interpersonal mode of
meaning. To the degree that such words create an attractive image
of the former and graciously acknowledge the latter, they will add to
the effectiveness of any piece of writing. In exposition, however, such
diction, while important, necessarily remains infrequent.

THE DIRECTIVE MODE OF MEANING

The last of the three modes of meaning relates to the reader-topic
side of the communication triangle. Here you select words primarily
for their value in assisting readers to understand or feel about the
topic. Understanding and feeling are quite different responses: the first
a function of intelligence, the other of emotion. Words concerned with
facilitating understanding we shall call *constructive diction;* words intended to evoke emotion, *emotive diction.*

Constructive diction includes the various connectives and signposts
which clarify the organization of a composition and the flow of its ideas:
however, even so, on the other hand, for example, in the next chapter,
and so on. While such words and phrases indicate real connections
within the topic, their essential function is to help readers follow the
construction of thought.

How much constructive diction you include in a composition depends
both upon the amount of help you think readers need and upon your
own preferences for spelling out logical relationships or leaving them
implicit. You can overuse such diction, boring or even annoying readers
with too many *however*s and *therefore*s. Most people, however, are more
likely to err on the other side, giving readers too little help.

The other kind of interpretive diction aims at feeling. In emotive
diction, connotations play a major role, especially those carrying strong
negative or positive charges. Examples abound in advertising copy. The
word *Brut* on a man's cologne tells us nothing referential, nothing
about the product. *Brut* aims at our emotions. Cleverly combining
strong macho connotations with others of sophistication and elegance,
the name is intended to overcome masculine resistance to toiletries

as "sissy" (or perhaps to appeal to women, who buy most of these products for their men).[3]

Emotionally loaded diction is also the stock-in-trade of the political propagandist. The Marxist who writes of "the *bourgeois lust* for personal liberty" uses *bourgeois* (a leftist sneer word for all things pertaining to capitalism) and *lust* for their capacity to arouse disapproval in a socialist audience. Similarly the conservative who complains of "*pinko liberals* in Washington" employs rightist sneer words. Diction may also be loaded positively, calling forth feelings of affection and approval: "*grass-roots* Americanism," "*old-fashioned* flavor," "an *ancient* and *glorious* tradition."

There is nothing wrong in trying to arouse the emotions of readers. It is the purpose for which the emotion is evoked that may be reprehensible, or admirable. The devil's advocate uses loaded diction, and so do the angels.

Many words operate in both the referential and directive modes simultaneously. In fact, it is not always easy to know which mode is paramount in particular cases. Both Marxist and conservative, for example, may believe that *bourgeois* and *pinko* really denote facts. Still, most of us feel that such words are largely empty of reference and have their meaning chiefly in their emotive force. On the other hand, some words work effectively in both modes, like those italicized in the following passage (the author is describing some fellow passengers on a bus tour of Sicily):

> Immediately next to me was an aggrieved French couple with a small child who looked around with a *rat-like* malevolence. He had the same face as his father. They looked like *very cheap microscopes.*
>
> Lawrence Durrell

Rat-like and *cheap microscopes* have genuine reference; they would help an illustrator drawing a picture of this father and son. At the same time the words arouse the emotional response that Durrell wants in the reader.

CONCLUSION

The relative importance of the three modes of meaning varies considerably from one kind of writing to another. Scholarly and scientific papers, for example, make the writer-topic axis paramount; advertising and political propaganda use that of reader-topic; applications for jobs and letters of appeal, for example, lie along the writer-reader axis. We can

---

3. The sophistication and elegance derive from the French word *brut*—meaning "dry, unsweet"—which appears on fine champagne labels. The macho connotation follows from the fact that *brut* is pronounced "brute."

suggest such differences in emphasis in our triangular diagram by moving the circles representing words from the center of the triangle toward one or another of its sides. Some of the examples we have used might be visualized like this:

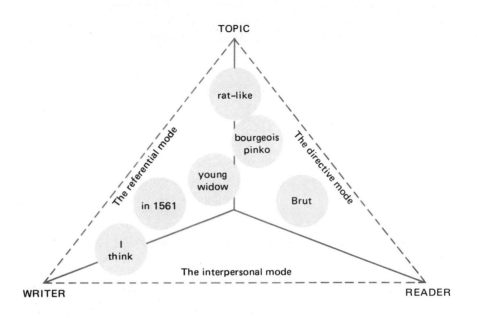

Some expressions (*in 1561*, for instance) are chosen solely for reference, that is, to explain the topic; a few solely to influence readers' feelings about the topic (*Brut*). Other words function in two areas of meaning: either primarily within one but extending partially into another (*pinko, bourgeois, I think, young widow*), or more evenly balanced (*rat-like*).

But whether designed to serve a single end or several, diction succeeds only to the degree that it does in fact serve an end—enabling readers to comprehend your observations, ideas, feelings, and affecting their responses both to the topic and to you in ways that you wish. To the degree that it fails to achieve your purpose, your diction fails entirely.[4]

You must, finally, realize that words inherently have meaning in some or in all of the modes we have enumerated. If you do not choose words wisely, words will, in effect, choose you, saying things about the topic you do not intend and affecting readers in ways you do not want.

---

4. A purpose itself may be silly or stupid, of course, but then the fault lies in the writer's conception—what he or she wants to say—not in the diction—how it is said. Writers may use words well by a happy chance, that is, without really understanding their effect, and thus achieve a purpose they are blind to. But lucky prose is rare. The general truth holds: good diction is diction chosen to achieve a conscious purpose.

EXERCISE

1. In the following sentences decide whether the italicized words are effective because of their denotation or their connotation (or perhaps equally effective on both counts):

 A. A *Democrat* denouncing a budget deficit is like *a hog with wings*.

Russell Baker

 B. An *arrogant jackass* who *brays* from Louisiana. . . .

A Texan on Huey Long

 C. The use of *metaphor* aims deliberately at *imprecision*.

Oscar Handlin

 D. . . . these *dusky, wolf-eyed warriors* [Mongols], who teemed in the *wide, arid plain land of Asia* like *rats on an old threshing-floor*.

H. H. Munro

 E. I am always entertained—and also *irritated*—by the *newsmongers* who inform us, with *a bright air of discovery,* that they have questioned a number of *female workers* and been told by one and all that they are "sick and tired of the office and would love to get out of it."

Dorothy L. Sayers

2. In these sentences discuss whether the italicized words primarily refer to the topic, direct the reader's feelings about or understanding of the topic, or establish a particular relationship between writer and reader:

 A. *Peaceful. It's a quiet country interlude. A bright blue sky laced with clouds and filled with sunshine.* . . .

Advertisement for Puerto Rico

 B. The *Joint Chiefs of Staff* have reported that, over the last fifteen years, *less than ten percent of the United States defense budget* has been directed to strategic nuclear forces.

Laurence W. Beilenson and Samuel T. Cohen

 C. By now *you have rightly inferred* that I find something important in *Emerson*.

Howard Mumford Jones

 D. Watching a fight on television has always seemed *to me* a *poor substitute* for being there.

A. J. Liebling

 E. He liked getting up early in the morning, to the tune of *cocks crowing on the dunghill.*

H. L. Mencken

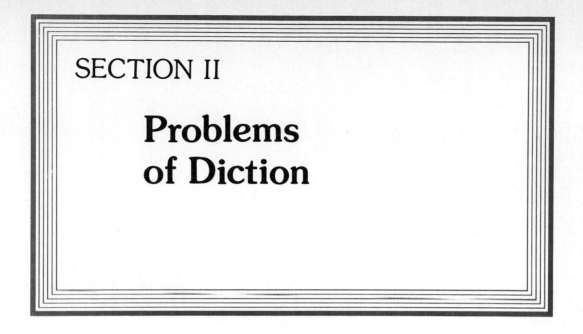

SECTION II

Problems
of Diction

CHAPTER 41

Wrong Words

INTRODUCTION

To be effective in the sense we have just discussed, words must be precise. In workaday terms *precision* means that your words are adequate to your purpose, and in exposition that purpose is largely to express exactly what you see, hear, think, know, feel. Precision comes into play, of course, in establishing an appropriate relationship with your readers and of guiding their responses. But mostly the problem lies in expressing your topic clearly.

This is more complicated than it sounds. It is not simply a question of deciding what in fact you do see, think, feel, or know, and then of choosing the right words. The distinction between what goes on in our minds and how we put it into language is not nearly so clear-cut. Words condition and limit what we are able to perceive and feel and think. We do not so much "choose" words to fit our perceptions and ideas as we perceive and think in terms of the words we know. More exactly, the two processes—thinking, feeling, seeing, knowing, on the one hand; using words, on the other—feed upon and reinforce one another. As we learn new words, our capacity to respond to and understand ourselves and the world around us increases. As our sensitivity to self and world expands, our alertness quickens to new words that express the subtler, more complicated self we are becoming.

Diction, or word choice, then, is the heart of composition. Sentences count; paragraphs count; a clear organization counts. But in the beginning, most important of all, are words.

We may proceed positively or negatively in this study of how to use words well, stressing either the qualities of good diction, or the faults of bad. In Section III we study diction from a more positive approach—for example, how to be provocative and original by using images, figures, allusions, clever words and phrases, and so on. But in this chapter and the next we shall concentrate on why words go wrong—the sins of diction, so to speak.

When one sins one sins against the cardinal virtues of diction: precision, of course, but also simplicity, brevity, emphasis, originality, variety. These virtues are not absolute. Simplicity, for instance, which relates to purpose, means that diction should be no more difficult than a topic requires; on occasion diction may be simple yet the thought hard to understand. Nor are these virtues always compatible. Precision or variety may justify a difficult term over a simple one. Simplicity may sanction a common word that is not quite as precise as a more learned one. Emphasis may warrant several words instead of only one.

In general, however, precision, simplicity, forcefulness, originality, and variety are what you should strive for. When you fail to achieve any of these qualities you make an error in diction. We treat these errors in two broad categories: wrong words (in this chapter) and dead words (in the following).

D. "D"—meaning "diction"—is the general symbol for wrong words. "D" is not a lot of help by itself. Words may be ineffective for many different reasons, and to make the symbol more informative we add a more precise abbreviation of the error. Some marks indicate that a word is totally wrong for your purpose. But most signify that a word, while not completely inaccurate, is fuzzy or awkward. In all cases you should replace the term with a better choice.

It is important that you try to make that choice. Your instructor may indicate one or two possibilities; look them up, and if you still don't understand the suggestion, ask. Or an instructor may prefer that you search out a better word for yourself; use your dictionary and again do not hesitate to ask advice. Education, someone has said, is simply enlarging your vocabulary. But this means more than acquiring new words; it also means using more effectively the ones you already know.

For convenience the various kinds of errors in diction that follow appear in alphabetical order, and negative examples are marked ×.

D. abst. ## TOO ABSTRACT

Make your words as concrete and specific as the topic allows. *Concrete words* refer to perceptible things: *a rose, the odor of violets, a clap of thunder. Abstract words* signify conceptions that cannot be directly perceived: ideas, feelings, generalizations of various kinds.

No hard and fast distinction exists between abstract and concrete. Often it is a matter of degree. Depending upon its context, the same term may find use now abstractly, now concretely, like *rose* in these sentences:

> On the hall table a single yellow tea rose stood in a blue vase. (CONCRETE)

Roses were growing in the garden. (CONCRETE, BUT LESS SO)

The rose family includes many varieties. (ABSTRACT)

The closer a word comes to naming a single, unique object or person, the more concrete it is. When diction moves from the specific and perceptible toward the general and imperceptible, it becomes abstract.

Extensive or unnecessary abstraction leads to dullness and confusion. Many readers find it hard to enjoy or understand words too removed from the experience of the senses. There is no reason to use abstract diction when writing about what you see and hear. Notice how specific, concrete words improve the following passage:

× The large coves are surrounded by *various building structures.*

The large coves are surrounded by *summer cottages, boat houses, and piers jutting into the water.*

On the other hand, abstractions are not inherently wrong. Often they are unavoidable, in dealing with ideas, for example. All the abstractions in the following sentences are essential to the clarity of the thought:

All too often the debate about the place, purpose, and usefulness of films as a means of instruction and communication is clouded by confusion, defensiveness, and ignorance.

Sol Worth

To define pedantry is not an easy thing to do. The meaning assigned should be neutral and fixed. But pedantry is relative to occasion.

Jacques Barzun

Yet even in cases like these, wise writers do not stay too long on the heights of abstraction. Occasionally, for both variety and clarity, they pull their ideas back to the level of the senses by examples, analogies, metaphors, or similes:

It is often said truly, though perhaps not often understood rightly, that extremes meet. But the strange thing is that extremes meet, not so much in being extraordinary, as in being dull. The country where the East and the West are one, is *a very flat country.*

G. K. Chesterton [italics added]

And in the following description of Japanese trains, notice how the abstract terms "trim" and "dapper" are given substance by concrete details:

Everything about them is trim and dapper; the stylized flourishes of the white-gloved guard, for instance, as he waves the flag for the train to start from Sano station, or the precise unfumbling way the conductor, in equally clean white gloves, clips one's ticket, arms slightly raised, ticket held at correct angle and correct distance from the body, clipper engaged and operated in a sharp single movement.

Ronald P. Dore

Diction like "clean white gloves" and "ticket held at correct angle and correct distance from the body" presents us with images, that is, with things we can perceive. We say more about images in Section III; for the moment, simply notice how much interest and value they add to this passage.

D. ambig. **AMBIGUITY**

Ambiguity means that a word may be read in two or more ways and the context does not make clear which is intended. The remedy is to change either the word or the context.

Often ambiguity results when a word has two different meanings:

It was a *funny* affair. ("Laughable" or "strange"?)

He's *mad.* ("Crazy" or "angry"?)

Large abstractions may be ambiguous, particularly if they involve value judgments. Words like *democracy, romantic, sentimental, Christian* have a wide range of meanings, some of them contradictory. A writer or a reader can easily make mistakes with such words, slipping unconsciously from one sense to another, an error logicians call "equivocation."

Pronouns suffer ambiguity if they appear in situations where they may refer, both grammatically and logically, to more than one antecedent:

Children often anger parents; *they* won't talk to *them.*

We sat near the heater, as *it* was cold. (The "heater" or the unmentioned room?)

Some connectives are prone to ambiguity. *Or,* for instance, can mean (1) a logical disjunction, that is, *a* or *b* but not both; and (2) an alternative name or word for the same thing: "The shag, or cormorant, is a common sea bird." *Because* after a negative statement may be ambiguous:

We didn't go because we were tired. (We did not go and the reason was that we were tired; *or* We did go and emphatically were not tired.)

Sometimes ambiguity exists not in a single word but in the entire statement:

× I liked this story as much as I liked all his others.

× So be it, until Victory is America's and there is no enemy, but Peace.

Like (in the sense of "enjoy") is not normally ambiguous. But here, in the context of the qualifying clause, it is possible to read the word ironically: the writer didn't like any of his books. And "Peace," despite the comma, could be the final "enemy."

Clever stylists now and then use this sort of ironic ambiguity deliberately to say one thing while meaning another, or to imply a more complex meaning than appears on the surface. Describing a bride's gown, Joan Didion makes a wry comment on marriage by exploiting the ambiguity of *illusion*, using it literally in the technical, dressmaker's sense of a bridal veil but also implying its more common meaning of a false hope or dream:

A coronet of seed pearls held her illusion veil.

Similarly the nineteenth-century novelist and statesman Benjamin Disraeli had a standard response to all would-be authors who sent him unsolicited manuscripts:

Many thanks; I shall lose no time in reading it.

D. barb: BARBARISM

A *barbarism* is either a nonexistent word or an existing one used ungrammatically. Inventing new words is not necessarily a fault; imaginative writers create new words for special purposes—"neologisms," they are called. But a genuine neologism fulfills a need. When an "invented" word is merely an ungrammatical form of a term already in the language, it serves no purpose and is a barbarism:

× Americans used to pay more attention to *idealison*. (For *idealism*)

× I've always been a *dutifulled* daughter. (For *dutiful*)

Barbarisms are often spawned by confusion about suffixes, those endings we use to extend the meaning or alter the grammatical form of a word, as *ness* turns the adjective *polite* into the noun *politeness*. Sometimes a barbarism results from adding a second, unnecessary suffix to a word to restore it to what it was in the first place:

× He has great *ambitiousness*. (For *ambition*)

× The story contains a great deal of *satiricalness*. (For *satire*)

Aside from nonexistent words, barbarisms also include legitimate words used ungrammatically. Confusions of sound often cause this:

× The fact is that *are* natural resources are being used up. (For *our*)

× Garbage is also used to fill holes *were* houses are to be built. (For *where*)

× The average man is not *conscience* of his wasteful behavior. (For *conscious*)

× I should *of* gone. (For should*'ve*)

× A *women* stood on the corner. (For *woman*)

The chances of confusion are even greater with true homonyms, that is, different words having the same pronunciation (though they may differ in spelling)—*bear* ("carry"), *bear* ("animal"), and *bare* ("naked"); or *right, rite*. Especially prone to barbarous misuse are the forms *there* (adverb), *their* (possessive pronoun), and *they're* (contraction of *they are*); and *to* (preposition), *too* (adverb), and *two* (adjective).

Legitimate words may also become barbarisms as a result of grammatical shift. Grammatical shift occurs when you use a word which commonly functions as one part of speech to serve as another part of speech, as in "The car *nosed* down the street." In Section III we shall see the value of grammatical shift. But misused, it creates barbarisms:

× Our *strive* for greatness is one of our best qualities. (For *striving*)

× They made their *deciding*. (For *decision*)

× From this *predictable* arises the problem. (For *predictability*)

Awkward shifts are common with adjectives and adverbs. Usually the problem lies in leaving off a necessary *-ly*:

× She dances *beautiful*. (For *beautifully*)

× They did it *satisfactory*. (For *satisfactorily*)

Occasionally the problem may be the other way round, using the adverbial form where the adjectival is called for:

× The very *simply* things we use in daily life. (For *simple*)

As a rough rule, adverbs of three or more syllables end in *-ly*; of those having one or two syllables, some always have *-ly*, a few never do, and some may be used in either form. (See Reference Grammar, pages 711–12.)

On the fringe of barbarism are a number of trendy words like *finalize* and the adverbial forms ending in *-wise* (*weatherwise, economywise, universitywise*). There seems little justification for a word like *finalize*, which says nothing that *complete* or *finish* does not say already. On the other hand, one can argue that *weatherwise* is at least briefer than a phrase like *in regard to the weather*. The safest bet in an English theme, however, is to avoid such expressions. Even if a teacher does not reject them as barbarisms, he or she may properly feel that they are clichés or jargon (which we shall come to shortly).

D. cl.   **CLARITY**

In a broad sense most problems of diction are problems of clarity. The abbreviation *cl.,* however, means specifically that a word clashes with its context, that it does not say what the words around it suggest it should say.

This often results from carelessness. For example, in the haste of writing it is easy to omit a *not*. In the following case the writer meant to say "not recommend":

✕ I thought the characters dull and the plot unbelievable. I would recommend the book to anyone else.

Or lapses in clarity may result from not seeing your words as others will see them and thus they convey a meaning very different from what you intended:

✕ Many people now realize that the word *wastefulness* is dangerous.

Obviously the writer meant that wastefulness itself is dangerous.

Lack of clarity of this kind often occurs with the careless use of the auxiliary *would,* as in:

✕ A good example of false economy would be buying cheap tires for an expensive automobile.

In such passages *would* implies conditionality, that is, that something is the case *if* a condition is fulfilled, as in:

We would do it if you helped.

In the sentence about the tires, however, the writer means that buying cheap ones *is* an example of false economy, and the words should say that.

An extreme failure of clarity leads to *contradiction,* the assertion in the same passage of two things which cannot both be true:

✕ The disaster is Creon's fault, but cannot be helped.

To say that a disaster is someone's fault implies that it could have been helped (that is, avoided); to assert that it cannot be helped denies that it is anyone's fault. The root of contradictions usually lies deeper than poor word choice: it lies in fuzzy thinking. Even so, the fault reveals itself in diction and must be corrected by thinking more clearly about what words imply.

D. cliché   CLICHÉ

A *cliché* is a trite expression, that is, one devalued by too much use:

an agonizing reappraisal	the bottom line
at this point in time	the finer things of life
cool, calm, and collected	the moment of truth
history tells us	the voice of the people

Many clichés are overworked similes and metaphors:

cool as a cucumber	Mother Nature
dead as a doornail	pleased as Punch
gentle as a lamb	sober as a judge
happy as a lark	the patience of Job
in the pink	the pinnacle of success
light as a feather	white as snow

Others are quotations quoted too often:

"fools rush in"
"lend me your ears"
"something is rotten in the state of Denmark"
"to be or not to be"
"to damn with faint praise"

Clichés have two faults: they are dull and unoriginal; worse, they impede clear perception, feeling, and thought. Clichés are verbal molds into which we pour experience. As molds they are convenient and save work, but we lose in the saving. Instead of shaping reality for ourselves, we hand it on pre-cast—and probably miscast—in stale words.

Clichés, however, ought not to be confused with dead metaphors. Expressions like *the key to the problem, the heart of the matter, the mouth of the river* have passed beyond being clichés and are acceptable diction. A cliché attempts to be original and perceptive and fails because the words are not really new or revealing. A dead metaphor, on the other hand, makes no pretense at newness; it has dried and hardened into a useful expression for a common idea.

A special kind of cliché is the *euphemism,* an expression designed to soften or conceal an improper or unpleasant reality. Euphemisms for death include *to pass away, to depart this life, to go to that big* [whatever] *in the sky,* all equally trite. Poverty, sexual relationships, and disease are often described euphemistically. Politicians, diplomats, advertisers are adept in their use: *dedication to public service* = "personal ambition," *a frank exchange of views* = "continued disagreement," *tired blood* = "anemia."

Euphemisms have a time and place, but not generally in exposition. They are language that conceals, not language that reveals.

d. Colloq. COLLOQUIALISM

A *colloquialism* is a word or expression appropriate to a conversational level of usage, but which sounds out of place in a composition:

× The rhyme scheme *goes along with* the meaning of the poem.
(BETTER: *is appropriate to, enhances*)

×We have a *swell* professor of mathematics. (BETTER: *pleasant, interesting, capable*)

Colloquial words are not invariably a fault in writing. They cause problems only when they fit awkwardly with their contexts or when they are vague. And frequently colloquialisms are vague, for many have loose meanings unacceptable in writing even though they may get by in speech. (What, for example, does *swell* mean in the sentence above?) In speech you can compensate for vagueness in a word by gestures, tone of voice, the common ground of knowledge and experience shared with your listener. None of these aids to communication is available in writing, and the vagueness may be fatal.

However, not all colloquialisms are vague. Some indeed are remarkably expressive, and these are more acceptable today than they were a generation ago. Contemporary writers no longer observe restrictions concerning level of usage as carefully as people used to do. They feel free to range widely in their diction, mixing formal words and colloquialisms in the same passage. The mixture, if accomplished with control and taste, represents a gain in precision and variety:

> Joan's voices and visions have *played tricks* with her reputation.
>
> George Bernard Shaw [italics added]

> But because of the way our society is set up it's [passive femininity] what usually happens to women; and so their adaptations to the situation have become "normal" and expected.

> There's another *wrinkle* to this.
>
> Elizabeth Janeway [italics added]

Slang is an extreme form of colloquialism. We all recognize it but find it very difficult to define. Some slang takes the form of ordinary words given a special meaning: *square* in the sense of holding to old-fashioned values, *heavy* in the sense of serious, *cool* in the sense of unperturbed or a little better than all right. Other slang is in the form of special words used only as slang; *hep,* for example. Slang tends to be short-lived: that of one generation sounds silly or incomprehensible to another (though some slang words have been around for years: *okay, dough* = "money," *hooker* = "prostitute"). Slang tends to be richly suggestive in meaning, conveying a wide range of attitudes and values in a brief expression (*square, hep*). But the richness has a certain imprecision: often one feels that while a slang word says exactly what one wants to say, it is impossible to explain what that something is.

Even more than colloquialisms generally, slang carries the atmosphere of informal social situations. That tone can be useful, establishing an appropriate writer-reader relationship or creating an attractive persona. Used intelligently, an occasional slang expression can have the virtue of saying just the right thing and of pleasing us by surprise:

The authors had a reputation for being jealous of each other's fame and losing no opportunity of *putting the boot in* [= "kicking a fallen enemy"]. . . .

Frank Muir [italics added]

I don't mean to suggest that Segal is as *gaga* as this book [*Love Story*]— only that a part of him is.

Pauline Kael [italics added]

D. connot. CONNOTATION

The connotation of a word is its fringe or suggested meanings, including implications of approval or disapproval (see pages 337–39). When a word's connotation pulls awkwardly against its context, even though its denotative meaning fits, you should replace the word by one with more appropriate implications. In the following sentence, *unrealistic* has the wrong connotations:

> ✕ In such stories it is exciting to break away from the predictable world we live in and to enter an unrealistic world where anything can happen. (BETTER: *fantasy, unpredictable, imaginary*)

The writer approves of the story she is describing because it stimulates her imagination. But *unrealistic* generally connotes disapproval ("Don't be so unrealistic"; "His plan is much too unrealistic to work"). Thus while the basic meaning fits the writer's purpose, the connotations of *unrealistic* do not. *Fantasy,* on the other hand, is a more neutral term, and even, in an appropriate context as here, has positive connotations.

In the following passage the connotations of *open spaces* are also inappropriate. Usually the expression connotes desirable freedom and solitude; the context requires more negative overtones:

> ✕ Across this country one sees deep holes in the ground where man has mined, oil rigs working day and night, and open spaces which were once cradles of trees. (BETTER: *wasteland, emptiness, barren ground, deserts*)

Errors in connotation are subtle, often demanding more discussion than an instructor can conveniently supply in a marginal note. If you don't understand an error in connotation, ask for an explanation.

D. denot. DENOTATION

An error in denotation means that you have used a word wrong: the term simply does not mean what you think it does. Unlike a mistake in connotation, which results in a fuzzy word, a mistake in denotation results in an inaccurate word. In the following cases the writers were probably led astray by vague resemblances of sound:

$\times$ If they're paid by the hour some workers *ponder* away their time. (FOR *squander*)

$\times$ Yet all that has been said does not *conclude* inefficiencies in other fields. (FOR *include*)

Errors like these in which a word is misused—especially a big word—because of a phonetic similarity to another one are called "malapropisms." Mrs. Malaprop, a character in *The Rivals,* an English comedy by Richard Brinsley Sheridan (1775), was a silly woman who tried to appear more learned than she was and who constantly gave herself away by misusing words. For example, she describes another character as being as "headstrong as an allegory on the banks of the Nile," confusing the symbolic story and the reptile. Malapropisms are still used for comedy (Archie Bunker uses them all too often), but in exposition they are not funny, or, if they are, the humor does you no credit.

Sometimes a wrong denotation results less from not knowing a word's definition than from not thinking clearly:

$\times$ Another *flaw in* wastefulness is the manufacturing companies. (FOR *cause of*)

$\times$ The space program is a waste of time and money. There are problems to solve in our *land,* such as starvation in Cambodia and overpopulation in South America. (FOR *planet*)

It is fair to suppose that the writers of those sentences knew the meanings of *flaw* and *land.* But they did not think through what they wanted to say and so chose words that did not jibe with their contexts.

d. fig. ## AWKWARD FIGURE OF SPEECH

Figures of speech are words and phrases used less for their literal meaning than for their capacity to clarify or intensify feelings or ideas. Chief among the many kinds of figures for the writer of exposition are the simile and metaphor.

A *simile* is a comparison, generally introduced by *like* or *as.* The essayist Robert Lynd describes the bleak houses of a nineteenth-century city as looking "like seminaries for the production of killjoys." A metaphor is more complicated. For now let us say only that a *metaphor* expresses an implicit comparison, not a literal one as does a simile:

> When I walked to the mailbox, a song sparrow placed his incomparable seal on the outgoing letters.
>
> E. B. White

White does not literally say that the bird's song is like a bright stamp or seal, but the comparison is there.

Section III contains a fuller discussion of how similes and metaphors may be used positively to improve diction (pages 388–99). Here we are concerned with their misuse. A metaphor or simile becomes faulty under three conditions: when it is awkwardly mixed, when it is inappropriate; or when it overwhelms the idea it was intended to serve.

Mixed metaphors ask us to visualize two or more things simultaneously that cannot go together without seeming ridiculous:

> × He put his foot in his mouth and jumped off the deep end.

> × We must feel with the fingertips of our eyeballs.

A good metaphor introduces an image, that is, something we can perceive. But a mixed metaphor turns that virtue into a vice by yoking images that simply cannot go together without making us laugh.

Even when it is not mixed, a simile or metaphor may be *inappropriate*—that is, it has a different effect than the writer wanted:

> × A green lawn spread invitingly from the road to the house, with a driveway winding up to the entrance like a snake in the grass.

Since the writer intended no sinister implications, comparing the driveway to the snake is misleading. Moreover, the simile, aside from being trite, is a bit ridiculous. A snake in the grass is a kinetic image—one involving motion—and a wriggling driveway is silly.

Overwhelming metaphors ride roughshod over the main idea. In the following sentence (about the considerable girth of the comedian Jackie Gleason) the metaphor overwhelms the thought (and is not especially appetizing to begin with):

> × Out of that flesh grew benign tumors of driving energy and unsatisfied appetite that stuck to his psyche and swelled into a galloping disease that at once blights and regenerates him.

D. gen.   ## TOO GENERAL

A good rule in writing is to use diction as specifically as possible. A specific word names a particular one within a class of people, things, concepts, and so on. A general word designates the class. *Fallacy*, for example, is a generic term including all kinds of false arguments. *Argumentum ad populum* (an appeal to the emotions and prejudices of a crowd) is a specific term, naming one of those false arguments.

In the following cases the words are too general:

> × Thrift is not one of our *attributes*. (BETTER: *virtues*)

> × The novel has far too many *people*. (BETTER: *characters*)

> × Hardy's poem *allows* the reader to experience the crashing of the iceberg and the ship. (BETTER: *forces, makes*)

As a complementary rule to using as specific a word as possible, employ the appropriate general term when the context requires it:

The *attributes* that distinguish a social class are often difficult to define.

People differ considerably in their religious beliefs.

The distinction between general and specific is relative rather than absolute. Consider the words *emotion, fear,* and *terror. Emotion* is the most general; and *fear,* with regard to *emotion,* is more specific. At the same time *fear* is general with reference to *terror.* These degrees of generality/specificity can be indicated by inclusive circles:

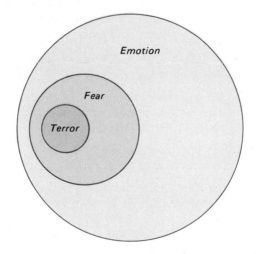

 FALSE HYPERBOLE

Hyperbole is deliberate exaggeration designed to heighten the importance or emotional force of a topic. Though no hyperbole is intended to be taken literally, we can properly call it false only when the exaggeration far outdistances the real value of what is being described:

× Football is the most magnificent sport ever developed by the mind of man. It tests physical skill, stamina, courage, and intelligence more thoroughly than any other human activity.

× One shudders to think what the world would have been like had Shakespeare never written *The Tempest.*

Such claims make us smile. Even those who admire football are likely to be put off by such grandiloquent praise. As for *The Tempest,* it is a great play, but a world without it would be much the world we know.

Even false hyperboles have a use at times—in politics or advertising, for instance—but in exposition they are likely to produce an effect opposite to what the writer hoped for. Rather than impressing us with the importance of the subject, they make us laugh at it.

A distinction deserves to be made, however, between false and legitimate exaggeration. Hyperbole is an old and useful figure of speech, though not as fashionable today as it was a hundred or more years ago. In the nineteenth century politicians dealt in hyperbolical, spread-eagle oratory, and historians embraced a rhetorical style in which hyperbole figured prominently. In the following passage, for example, the American historian William H. Prescott writes about the ill effects of the gold which fifteenth-century Spain garnered from the New World:

> The golden tide, which, permitted a free vent, would have fertilized the region, through which it poured, now buried the land under a deluge which blighted every green and living thing.

Mark Twain was a master of hyperbole. Here he describes a tree after an ice storm:

> . . . it stands there the acme, the climax, the supremest possibility in art or nature, of bewildering, intoxicating, intolerable magnificence. One cannot make the words strong enough.

But Twain is at his best—at least to modern ears—when he uses hyperbole for comic effect:

> [On the New England weather] In the spring I have counted one hundred and thirty-six different kinds of weather inside of four-and-twenty hours.

> [On the music of Richard Wagner] Another time we went to Mannheim and attended a shivaree—otherwise an opera—called "Lohengrin." The banging and slamming and booming and crashing were beyond belief, the racking and pitiless pain of it remains stored up in my memory alongside the memory of the time I had my teeth fixed.

Today most readers find hyperbole—in its serious uses, anyhow—not much to their taste. Still, there is hyperbole and there is hyperbole. Not all of it is false.

D. id.   ## WRONG IDIOM

An *idiom* is a combination of words that commonly function together as a unit of meaning, as in "take the subway [bus, streetcar, etc.] home." Often one or more of the words has a special sense different from its usual meaning and confined to that idiom: thus *take* in the above expression means to "get on and travel in." In such cases the word cannot be replaced by any of its normal equivalents: we cannot "*carry, bring,* or *fetch* the subway home."

Always a difficulty in learning a foreign language, idioms do not easily reduce to rule and have to be memorized. Even native speakers make mistakes with idioms. Sometimes they err in combining words, sometimes in using an idiomatic combination in a wrong context. Errors in idiom, however, are difficult to classify and explain. When you make such a mistake, you simply have to accept the fact that you have just not used the word or expression right.

The most frequent errors involve prepositions. Verb/preposition patterns are highly idiomatic. In the following cases the writers chose wrong prepositions:

×I complained *with* my parents about their attitude toward dating. (IDIOMATIC *to*)

×She concluded *in* saying. . . . (IDIOMATIC *by*)

×That is where we fool ourselves *of* our efficiency. (IDIOMATIC *about*)

×They can't decide what to do *with* their problem. (IDIOMATIC *about*)

Such errors may be the result of confusing two idioms (*argue with/ complain about, cure ourselves of/fool ourselves about*), or of selecting an inappropriate one of several possible verb/preposition idioms (we do *with* physical objects—"What shall we do with this vase?"—but we do *about* problems, arguments, difficulties, abstractions of various sorts—"What shall we do about the crack in the vase?").

While most frequent with prepositions, mistakes in idiom occur with other grammatical patterns. Some verbs, for example, may not combine idiomatically with certain objects. The following sentences use verbs unidiomatically:

×People only *look out* for prestige. (Prestige is *looked for, valued, esteemed.*)

×Robert Frost *gives* the image of a silken tent in a field. (Poets *create, present, develop* images.)

Adjectives and nouns also enter into idiomatic combinations, and failure to observe these leads to mistakes in diction:

×We have a *great* standard of living. (IDIOMATIC *high*)

×The English prefer *dining-room* comedy. (IDIOMATIC *drawing-room*)

D. jarg. **JARGON**

Jargon is technical language misused. Technical language is simply the highly specialized diction of experts in particular fields writing for other experts. So used, it is legitimate and efficient. But extended beyond their proper limits, technical terms become jargon.

Social scientists often use jargon, perhaps because their subjects make more contact with everyday concerns. Writing for a general audience more frequently than does, say, a nuclear physicist, the sociologist or economist more often succumbs to temptation and uses technical terms where simple ones would do:

> Given a stockpile of innovative in-house creativity for the generation of novel words, substituting numbers for the input of letters whenever feasible, and fiscally optimized by computer capacitization for targeting in on core issues relating to aims, goals and priorities, and learned skills, we might at last be freed from our dependence on the past.

This is in fact a parody of "socialese" by Lewis Thomas, a biological scientist who does *not* write jargon. It catches the faults of jargon perfectly: the abstract, general, polysyllabic Latinisms (*capacitization, optimized*); the trendy words (*creativity, in-house, input, core issues*); the pointless redundancy (*aims, goals and priorities*); and the awesome combinations of modifiers and headwords (*innovative in-house creativity, computer capacitization*).

At its worst jargon is incomprehensible. (The word originally meant the twittering of birds.) But even when it can be puzzled out, jargon is nothing more than puffed-up language, a kind of false profundity in which relatively simple ideas are padded out in polysyllabic dress to make them seem larger than they are.

D. mean?   ## MEANING?

The problem here is plain incomprehensibility. We have already spoken about clarity. But when clarity fails, the reader can probably figure out what you intended to say. When meaning fails, he or she simply cannot understand what you have said. The problem may be that the word has no discernible meaning. In such cases it is difficult to suggest a correction.

Or the word may have a meaning but the reader does not know it and cannot easily look it up. Incomprehension of this kind often results from the careless use of technical terms, archaisms, or foreign expressions. A writer mindful of readers does not use diction beyond their capacity. If now and then you must employ an older word, a technical phrase, or a foreign term, define it briefly and unobtrusively. In the following passage, for example, the writer neatly explains an eighteenth-century sense of *ordinary*:

> Travellers on the road could usually eat in their inns. Inns and taverns usually provided an "ordinary", a set meal at a set price, which the traveller could sit down to alongside a mixed group of cut-purses, poets, whores and fellow-travellers, or order something better to be served in his room.
>
> Frank Muir

D. pret.

PRETENTIOUS DICTION

A pretentious word is a fancy, learned one used where a common term will do. Pretentiousness is a sin against simplicity:

> × Upon *receiving an answer in the affirmative,* he *proceeded* to the bulletin board. (BETTER: Told yes, he went to the bulletin board.)

> × Television shows which *demonstrate participation in physical exercise* will improve his muscle tone. (BETTER: Television exercise shows are beneficial.)

Pretentious diction is likely to result in long-winded, wooden sentences filled with deadwood (a problem we take up in the next chapter). Using shorter, simpler words generally means writing shorter, clearer sentences.

In spirit, pretentiousness is closely akin to jargon. In fact, jargon is a special variety of pretentiousness. But jargon is technical language used off limits, but conceivably could be legitimate in technical discourse. The kind of diction we are talking about here cannot be justified on any ground.

However, not all unusual, learned words are a flaw, even when they could be replaced by simpler ones. They can be a virtue. Skillful writers now and then employ such words to draw our attention or to imply a nuance of meaning:

> Among those who distrust the [literary] critic as an intrusive middleman, edging his vast steatopygous bulk between author and audience, it is not uncommon to wish him away, out of the direct line of vision.
>
> Carlos Baker

Baker cleverly uses a learned word to conceal a vulgar insult, which readers have the fun of discovering for themselves by way of a dictionary.

D. rep.

REPETITIOUSNESS

A word may sound awkward when repeated too closely. In such cases you should replace it by a pronoun or synonym:

> × The auto industry used to produce cars that lasted, but they didn't make enough money so planned obsolescence came into use. (BETTER: . . . came into *fashion, existence, being*)

> × This narrative is narrated by a narrator whom we cannot completely trust. (BETTER: This *story* is *told* by a narrator whom we cannot completely trust.)

However, repetitiousness must be distinguished from legitimate restatement, in which words are deliberately repeated either for clarity or, more commonly, for emphasis:

He [a lax governor] took things easy, and his fellow freebooters took almost everything easily.

Hodding Carter

[Oliver Goldsmith's "The Deserted Village" is] a poem written not in ink but in tears, a rich suffusion of emotion rising up in a grubby room in Grub street for a grubby little Irish village.

Sean O'Faolain

The line between awkward repetition and effective restatement is not easy to draw. We may very valuably repeat words of key importance and construct sentences to draw attention to them. But not when the words are unimportant; then repetition diverts attention from the main point.

D. sd.

AWKWARD SOUND

Words may *sound* awkward, even when nothing is wrong with their meaning. They may inadvertently repeat a nearby sound, create an awkward combination of sounds, or establish a dull or unsuitable rhythm.

We choose words primarily because of what they mean, but we must remember that words are also units of sound and rhythm. Even people adept at silent reading respond semiconsciously to the "sound" of what they see, and they will be put off by awkward patterns of sound, though they may not know exactly what disturbs them.

Mostly the awkwardness results from the accidental repetition of sounds:

×There is a grow*ing* awareness of the slow*ing* down of growth affect*ing* our economy. (BETTER: There is a growing awareness *that diminished rates* of growth *are* affecting our economy.)

×Built-in obsole*scence* has become the es*sence* of our society. (BETTER: . . . has become the *basis* of our society.)

×At the top of the hill were three *fine pine* trees standing in a l*ine*. (BETTER: . . . *beautiful* pine trees in a *row*.)

One can easily make the mistake of repeating the same suffix several times in the same group of words or of succumbing to an unconscious rhyme. Probably most common is unconscious alliteration, that is, beginning successive or near-successive words with the same sound:

×We should *th*ank *th*em for a *th*ird *th*ing—*th*e *th*oroughness with which *th*ey *th*ought out the plan. (BETTER: We should thank them for *something else*—the thoroughness with which they *developed* the plan.)

You need not always break up awkward rhymes altogether. Sometimes, as in the last example, you can simply tone them down.

It would be a mistake to conclude that rhyme has no place at all in prose. As we saw in Chapter 36 (pages 289–90), you can create emphasis by repeating sounds. Repetition of this sort does work:

> After engaging in the Great War with all its mud and blood and ravaged ground, its disease, destruction, and death, we allowed ourselves a bare twenty years before going at it all over again.
>
> Barbara W. Tuchman

> . . . those Hairbreadth Harrys of History [who] save the world just when it's slipping into the abyss.
>
> Arthur Herzog

As is so often the case with diction, it is not easy to say why repeated sound is a fault in one case, a virtue in another. Generally rhyme is awkward when it is accidental or when—even if it is deliberate—it is too obvious. In the first case it draws too much attention to relatively unimportant words. In the second it distracts us from the meaning even of important words. Effective rhyme involves key terms and does not obtrude, drawing us to the words but without hitting us on the head.

The best guard against awkward repetitions of sound is to read your work aloud. If something sounds awkward, it probably is. Change it.

CHAPTER 42

Unnecessary Words

INTRODUCTION

"Dead." stands for deadwood, an expression we use in two slightly different senses. Literally *deadwood* means that a word fulfills no function—that is, it conveys no meaning and contributes in no significant way to the relationship between writer and reader or to the rhythm and movement of the sentence. Such a word should be crossed out. More loosely, the term also indicates prolixity, or wordiness—that an idea expressed in six words could be said as clearly in three.

Deadwood violates the virtue of concision, which as we saw in Chapter 35 is brevity relative to purpose. Deadwood is not a fault because it wastes words—words are not gold coins that we must hoard. The problem is that deadwood interferes with communication. Dead words get in the way of vital words.

Most of this chapter treats the ways in which deadwood reveals itself in actual writing. We can better understand the problem, however, if we consider for a moment some mistaken assumptions that produce deadwood. One is the fallacy of verbal profundity, the notion that just because a word or phrase looks profound and sounds profound it *is* profound. The person, for example, who exclaims of a painting that it "exhibits orderly and harmonious juxtapositions of color patternings" seems to be saying a great deal. But if the phrase means anything more than "color harmony," it is difficult to see what.

Closely related to verbal profundity is the desire to endow a commonplace subject with dignity or elegance. Someone who writes

A worker checks the watch's time-keeping performance

is trying to cast a verbal spell over the job of quality control in a watch factory. It would be better to say simply ". . . the watch's accuracy." To describe chemistry as that branch of science that is deeply involved with the nature of matter

is to suggest a false aura that this simpler statement avoids:

Chemistry is that physical science which studies matter.

(It is beside the point that neither version is a particularly enlightening definition of chemistry.) The trouble with phrases like "time-keeping performance" and "deeply involved with the study of" is that not only are they long-winded, they are also pretentious. False and windy diction does not convey the real importance of quality control, of chemistry, or of anything else.

Confusion and ignorance also contribute to deadwood. Uncertainty about your point leads to a great deal of wordiness:

Music is similar to dress fads in that its style changes from time to time. Perhaps the change is subtle, but no one style of music will remain on top for a very long time. I am not talking about classical music, but rather about popular music that appeals to the majority of young people.

As soon as we read the last sentence we realize that this student writer was unsure of his point from the beginning. Without taking a minute or two to think out the problem, he chose too big a topic ("music"), then had to restrict it. Had he known what he wanted to say from the beginning, he could have started,

Popular music is similar to dress fads. . . .

eliminating altogether the last twenty-word sentence.

Sometimes the reason for deadwood is less confusion about what you want to say than simple ignorance of the words you need to say it. That was the problem for the writer of this sentence:

In this novel, part of the theme is stated directly in so many words and part is not so much said in specific words but is more or less hinted at.

Had she known the terms *explicit* and *implicit* she could have made the point more clearly and more concisely:

In this novel, part of the theme is explicit and part of it implicit.

Vocabulary limitations of this sort are no disgrace. In varying degrees we all suffer from them, and education, in part, is the process of removing them. But pardonable or not, ignorance of words often results in deadwood and obscurity. It helps to keep a list of words, as you learn them, which, like *explicit* and *implicit,* enable you to make distinctions quickly and neatly: *extrinsic/intrinsic, concrete/abstract, actual/ideal, absolute/relative* are other examples.

Finally, excessive caution also contributes to deadwood. Some people are afraid to express anything as certain; they will write, "It seems that Columbus discovered the New World in 1492." There are certainly times that call for caution, and no one can lay down a blanket rule about when qualification is necessary and when it is deadwood. We shall look at overqualification in more detail later in the chapter; for the moment remember that in composition extreme caution is more often a vice than a virtue.

A false sense of what is significant, confusion about what you want to say, ignorance of words, and excessive caution are some of the main psychological factors leading to deadwood. It remains to examine deadwood more closely as it reveals itself in diction. To make the general label "Dead." more helpful, it is followed in the discussion below by more precise abbreviations, and these are arranged in alphabetical order.

OVERLONG CONNECTIVE

**KEEP CONNECTIVES—PREPOSITIONS AND
CONJUNCTIONS—AS BRIEF AS POSSIBLE.**

Piling up connectives is a common variety of deadwood (here and throughout the chapter a red line through a word or phrase means that it is dead):

by
More than one game has been decided ~~on the~~ ₍ₐ₎basis of a fumble.

Wordy equivalents for *because, how,* and *so . . . that* are especially common:

because
The bill failed ~~as a result of~~ ₍ₐ₎the fact that the Senate was misin-

formed.

how
She will show us ~~the way~~ ₍ₐ₎in which to do it.

so
He becomes ₍ₐ₎self-conscious ~~to the extent~~ that he withdraws into

himself.

Dead d./
def.

UNNECESSARY DEFINITION

DON'T EXPLAIN WHAT READERS CAN REASONABLY BE EXPECTED TO KNOW.

Unnecessary definitions result from not thinking about the reader. Not only do needless definitions create deadwood; they also interfere with communication in a more serious way, annoying readers by appearing to insult their intelligence. Here is an example:

> Accountants sometimes function as auditors (~~men from outside a company who check the books kept by the company's own accountants)~~.

The writer did not need to define *auditors*; the term is within the general knowledge of the audience.

Granted, it is not easy to decide when a word should be defined. The answer requires a well-developed sense of reader. In the following instance the naturalist Joseph Wood Krutch, writing for general readers, realizes that they will not understand technical terms and neatly explains what they need to know:

> To even the most uninstructed eye a scorpion fossilized during the Silurian or Devonian epoch—say something like three hundred million years ago—is unmistakably a scorpion.

The best plan is to ask yourself if a definition is necessary *for the reader at whom you are aiming*. (And remember that it is not too much to expect readers to consult a dictionary, though you shouldn't ask them to reach for it every sentence or two.)

Dead. d./
d. w. d.

DISTINCTION WITHOUT DIFFERENCE

POINTLESS DISTINCTIONS MISLEAD READERS.

A distinction without a difference is naming several varieties of something when those varieties do not matter for your purpose:

> Under the honor system, teachers do not have to stand guard during
>
> ~~exams,~~ tests, ~~and quizzes.~~

There are legitimate differences between exams, tests, and quizzes, and had the writer been concerned with the various kinds of testing college students must endure, the distinction would have been necessary. But in fact the topic was the honor system, and the distinction is pointless. One word would do, probably *tests,* the most general.

Dead. d/
gen.   **WORD IS TOO GENERAL**

CHOOSE THE MOST SPECIFIC WORD AVAILABLE.

Using a term more general than your point requires creates a need
for wordy modification:

> Freshmen
> ~~People~~⋏~~who enter college for the first time~~ find it difficult to ad-

> just to the teaching.

The problem is with *people*. It is too inclusive a word. In order to specify
the kind of people, the writer must add seven words of modification.
English provides no single word meaning "people who enter college
for the first time," except for *matriculants*, a Latinism too learned
and forbidding for the writer's purpose here. The best choice is *freshmen*,
even though second-semester freshmen are not, strictly speaking, new
entrants.

While it is most common with nouns, the failure to be specific occurs
with verbs as well:

> enraged him
> The sudden change⋏~~motivated him into a rage~~.

> Chemistry is that branch of physical science that ~~deals~~
>   studies
> ~~with~~⋏~~the study of~~ matter.
>   won
> They ~~emerged~~⋏~~victorious~~.

The too-general verb is especially likely to be a form of *be, have,* or
seem. When these are used merely to link a noun or modifier to the
subject, they can often be replaced by a more expressive verb:

> affected
> The war ~~had an~~⋏~~effect upon~~ the economy.

Notice that in the original sentence the words that convey the thought
are *war, effect, economy,* all nouns. The other five words do not carry
important semantic meanings but rather serve grammatical functions:
the and *an* are determiners—that is, they mark nouns; *upon* is a preposition linking *economy* to *effect*; and *had* an empty verb linking *effect*
to *war*.

All this is not to say that such grammatical words are unimportant.
But it is a good rule of style to keep them to the minimum that clarity

requires. Failing to employ verbs, when possible, to convey an essential part of your idea sacrifices a valuable resource of diction. Verbs are able both to connect other words and to carry meaning in their own right. Using them in this double capacity makes for stronger, more concise prose:

> supported
> The people ~~were~~ ~~supportive of~~ conservation.

> know
> Teachers have to ~~have a~~ ~~knowledge of~~ their students.

> oversees
> The president of the company ~~is the~~ ~~overseer of~~ the other executives.

OBVIOUS BY IMPLICATION

DON'T SPELL OUT IDEAS THAT ARE CLEARLY IMPLIED.

A common source of deadwood is literally stating an idea included in a word by definition. *Blue,* for instance, implies "color"; it is unnecessary to say:

Her dress was blue ~~in color~~.

Occasionally it may be desirable, for emphasis or clarity, to state such implicit meanings, but mostly it is not. Not only is the dead term pointless; it is also likely to dampen the main idea, as the phrase *in color* prevents the sentence above from ending on the key term *blue*.

Noun/adjectival combinations often contain deadwood caused by overexplicitness. Sometimes the noun is dead:

They committed ~~an act of~~ burglary.
Staubach is noted for his passing ~~ability~~.
The last major barrier to the westward expansion ~~movement~~ was the Rocky Mountains.
It has existed for a long ~~period of~~ time.
She was an unusual ~~kind of~~ child.
The punt return resulted in a fumble ~~situation~~.

The categorizing words *kind, sort, type, class,* and so on are especially prone to dead use. Certainly it sometimes pays to write "He is the kind of man who. . . ." But unless the advantage of the longer construction is clear, it is better to write more concisely "He is a man who. . . ."

Often in these noun/adjective combinations, the adjective can be used substantively—that is, to stand for the noun. For example, look at this sentence about the development of quilting in colonial America:

On quilts, silk patches replaced ~~the~~ homespun ~~ones.~~

In other noun/adjective combinations containing deadwood, it is the modifier that is unneeded:

There is considerable danger ~~involved.~~
We question the methods ~~employed.~~
The equipment ~~needed~~ is expensive.
The store stocks many products ~~to be sold.~~
Each football play has a special purpose ~~when it is used.~~
This question has two sides ~~to it.~~
Most countries ~~of the world~~ have their own coinage.

Verbs, too, carry implicit meanings, which, whether expressed as a complement or as a modifier, are often better left unstated:

She always procrastinates ~~things.~~
He tends to squint ~~his eyes.~~
I have been told ~~by various people~~ that smoking is sophisticated.
I measured her foot ~~for the correct size.~~

Sometimes an idea is clearly implied by the total context rather than by any single word:

Writing poetry requires experience as well as sensibility. A prerequisite ~~to writing poetry~~ is being able to write prose.

In the context of the first of these two sentences, *prerequisite* clearly implies *to writing poetry*. The context also makes the phrases deleted in the following examples clear:

I dislike television. Most programs ~~on television~~ are unbelievable.
A good personality will help anyone, no matter what profession he or she chooses ~~in life.~~

Finally, a special but frequent form of overexplicitness is the unnecessary connective, especially common with conjunctive adverbs like *however, therefore, furthermore,* and so on. For instance, the following passage does not really need *however*:

People think that stamp collecting requires money; ~~however,~~ it doesn't.

The negative verb establishes the contradiction, and removing *however* even strengthens it.

On the whole, beginning writers tend to use too few conjunctive adverbs rather than too many. And certainly it is better to risk annoying readers by being too explicit than to confuse them by taking too much for granted. Even so, it pays to check connectives like *however, thus, consequently, then, still,* and so on to be sure that you really need them, or rather that your reader really needs them.

Dead. d./
mod.

WORDY MODIFICATION

THE BEST MODIFICATION IS CONCISE AND DIRECT.

In practice this principle often means not using a phrase when a single-word modifier will do. Three- or four-word adverbial phrases, for example, may often be replaced by one term:

                                                                briefly
The organization of a small business can be~described ~~in a brief~~

~~time.~~

                                        irrationally
She conducted herself ~~in an~~ ~irrational manner~~.

Adjectival constructions may be similarly tightened:
offensive
Each~player ~~on the offensive team~~ . . .
                        his doctor's
He didn't take~the advice ~~given to him by his doctor~~.
                            thinking
It leaves us ~~with~the thought~~ that. . . .
    Skeet   clay
~~The~~~targets ~~that are supplied in skeet shooting~~ are~discs ~~of~~

~~clay.~~

A very common kind of adjectival wordiness occurs when you use a full relative clause to introduce a participle or adjective that could be attached to the noun directly:

This is the same idea ~~that was~~ suggested last week.
The family ~~who are~~ living in that house are friends of mine.

In such clauses the relative word (*that, which, who*) acts as the subject and is immediately followed by a form of *be* which is, in turn, followed by a participal or an adjective. The relative word and *be* contribute nothing: they merely hook the adjective or participal to the modified noun. Occasionally clarity, emphasis, or rhythm justify writing out the entire clause. But mostly they don't.

The direct, economic use of participles is a resource of style used too little by inexperienced writers. It sometimes applies also to adverbial clauses, which can be boiled down to one or two operative words:

                                    T
Because they were⌄tired, the men returned to camp.

Although ~~they were~~ tired, the men did not return to camp.

Even independent clauses or separate sentences may now and then be handled more concisely and clearly by being pruned and subordinated:

                                                            proving
These football plays are very effective, ~~and they~~ often⌄~~prove~~ to be

long gainers.

The women of the settlement would gather at one home and
                                            , bringing
work together on the quilt. ~~They⌄brought~~ their children and
    spending   , chattering
~~spent~~⌄the entire day ~~and⌄chattered~~ happily as they worked.

Participles are also more economical than gerunds, and when idiom and meaning permit, they make for a smoother sentence with fewer form words. In the following examples, using the *-ing* verbal as a participal makes the noun markers and prepositions unnecessary:

She worried about ~~the~~ cooking ~~of the~~ dinner.
Basketball players must be skilled in ~~the~~ dribbling and handling ~~of~~ the ball.

Notice, however, that you must think about meaning in such a revision. "She worried about the cooking of the dinner" might make sense if someone else were doing the cooking.

Dead. d./
pass.

WORDY PASSIVE

AVOID UNNECESSARY USE OF THE PASSIVE VOICE.[1]

The passive is not always a poor choice in composition. Sometimes the focus of thought or tact make it preferable to the active voice. Generally, however, you should compose in the active voice. The passive usually lards your sentences with needless auxiliaries (another instance of an undue number of form words):

---

1. The voice of verbs is treated in the Reference Grammar, page 680.

The writer must clearly state his point
~~The writer's point~~ ∧ ~~must be clearly stated by him~~ at the beginning of the paragraph.

He must write the paper
~~The paper~~ ∧ ~~must be written by him~~ by tomorrow.

Dead. d./ qual. ## OVERQUALIFICATION

DON'T BE AFRAID TO STATE FACTS AS FACTS.

Excessive caution causes deadwood. The most obvious example is overqualification, when you hedge what you say with words like *seems, somewhat, maybe, perhaps*:

Theater-in-the-round ~~somewhat~~ resembles an arena.

This is too timid. *Resembles* does not mean "identical with"; it is a cautious word and does not require the extra protection of *somewhat*. Writing like this is like holding up your pants with a belt, suspenders, and several huge safety pins.

In addition to *seems,* the verbs *would, tend,* and the windy phrase *can be said to be* (in place of plain *is*) often signal overqualification:

After a square dance the people are pretty tired, but ~~it seems that~~ when they have tried it once, they want more.

      is
This play ~~tends~~ ∧ ~~to be~~ a comedy.

      is
Ethan ~~would~~ ∧ ~~be~~ an example of a tragic hero.

Certainly qualification does not always suggest weakness or timidity. Sometimes it is vital. The writer who lowers his or her head and plunges straight ahead may plunge into a stone wall. Excessive caution, however, is wordy foolishness. Qualify, but qualify only when you must.

Dead. d./ red. ## REDUNDANCY

**DON'T SAY ANYTHING TWICE UNLESS IT IS IMPORTANT
AND UNLESS LOGIC ALLOWS THE REPETITION.**

Redundancy consists of pointlessly repeating an idea. Often it appears in combinations of headwords and modifiers, the modifier emptily repeating the headword or another adjective or adverb:

bisect ~~in half~~
modern life ~~of today~~

vital essentials
sufficiently satisfied
a positive gain
It is clearly evident that. . . .
He hanged himself, thereby taking his own life.

You may need to restate an idea for clarity or emphasis. But redundancies are awkward and illogical, special instances of not understanding what words mean. For example, "important ideas" implies that some ideas are not important, and that is true enough. Similarly "vital essentials" implies that some essentials are not vital, but this statement is self-contradictory: by definition, essentials are that which is vital. Can you bisect anything without cutting it in half? Can a man hang himself without thereby taking his own life? (Never mind about the rope breaking. *Hang* in such a context means to cause death.)

Dead. d./
scaf.
SCAFFOLDING

DON'T WASTE WORDS EXPLAINING YOUR INTENTIONS IF THEY ARE OBVIOUS.

Scaffolding includes all the words concerned less with developing the topic itself than with explaining to the reader what you intend to do, have done, are doing, or will not do at all. Some writers find a great deal of scaffolding useful in early drafts; it helps them to move their own ideas about. But when they revise, experienced writers dismantle as much of this as they can. Some scaffolding should remain—enough to help readers. But when they can follow your intentions for themselves, scaffolding gets in the way, obscuring your thought as the staging around a new building conceals its shape.

A particularly awkward kind of scaffolding is announcement—that is, saying that you are going to say something. An overworked formula of this sort is "Let me say" (variants: "Let me make it clear," "Let me explain," "Let me tell you something"). Now and then emphasis justifies an announcement, but not often. Occasionally, too, a modest "I suggest" or "it seems to me" creates a favorable impression. More than an occasional one, however, changes that impression for the worse.

Announcements frequently get wordy and unnecessary when you set up topics, at the beginning of an essay or a section of one. Most people react negatively to this sort of opening:

The theme that follows is about baseball. Specifically it will deal with the business organization of a major league team.

Identify your topic clearly, but give your reader some credit:

Supporting every major league baseball team is a complex business organization.

Good writers help their readers, but they do not assume that readers are helpless.

Dead. d./
un. ideas

UNDEVELOPED IDEAS

DON'T OPEN UP TOPICS YOU WILL NOT DEVELOP.

Aside from not focusing clearly upon the topic, much deadwood results from irrelevancies—that is, side issues that are not part of the topic. A common mistake is to bring up ideas interesting in themselves but beside the point:

The people had come to the new world for freedom ~~of several different kinds~~, and had found injustice instead.

There is nothing inherently dead in the phrase "of several different kinds." But the writer does not discuss these kinds of freedom (nor does the purpose of the paper require it). To mention them at all, then, is awkward. The phrase contributes nothing to the main point. Even worse, it mutes the contrast between the key terms *freedom* and *injustice* and misleads readers by pointing to a path of development they will not find.

Dead. d./
vb.

TOO MANY VERBS

AVOID POINTLESS STRINGS OF VERBS AND VERBIDS.

Speakers of English are adept at stringing verbs together, as in:

I *was going to go* tomorrow.

In that example the string conveys a real meaning. But when such a chain of verbs says nothing that cannot be said in fewer words, it is so much deadwood:

The current foreign situation should ~~serve to~~ start many Americans

~~to begin~~ thinking.

Nucleonics investigates the smaller particles that ~~go to~~ make up

the nucleus of the atom.

must
The four infielders ~~have the job of building a barrier to~~ stop any

ball hit to them.

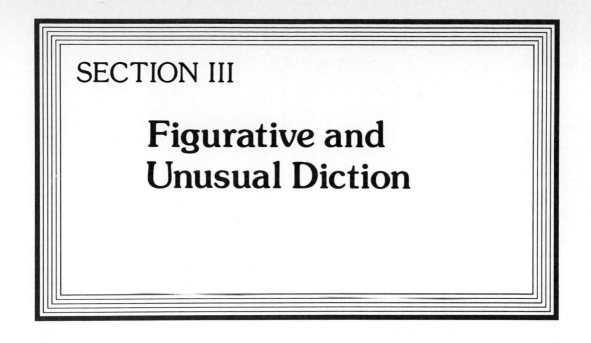

SECTION III

Figurative and Unusual Diction

43

Figurative Language

INTRODUCTION

Whenever language is simple, plain, direct, whenever it employs words to mean what they conventionally mean, we say that it is literal. *Literal* comes from the Latin *litera,* "letter"; what is literal is according to the letter. Literal meanings are those recorded in a dictionary. Consider, for example, this literal statement: "A writer's style should be purposive, not merely decorative." The words are direct and straightforward; they mean exactly what they say.

Figuratively, on the other hand, the same idea has been expressed like this: "Style is the feather in the arrow, not the feather in the cap." *Figurative* indicates that the meaning of a word is stretched to accommodate a larger or even a very different sense than it usually carries. The writer can make the stretch because of some likeness (or other relationship) between two different things that the context makes apparent. Thus the literal meaning of "feather in the arrow" is the stabilizer that keeps the arrow on target; the figurative meaning is that style makes communication fly to the target and hit it.

A writer's task is to provide clues so that attentive readers will understand the special sense carried by a figurative word. In speech we signal such meanings by gestures, facial expressions, pronunciation, or tone of voice (think of the intonation we give to the word *generous* in order to twist its sense to "stingy" when we say sarcastically of a cheap acquaintance, "He's a generous guy!"). In writing, it is primarily the context—the other words in the sentence, paragraph, or even the total composition—that controls a figurative word, pointing its meaning in a special direction.

The effective use of figurative language depends upon total diction, upon all of the words you choose. You cannot improve a composition by sticking in occasional similes and metaphors. You must weave them into the fabric of your composition.

When woven in, figurative language adds enormously to the richness of prose. Consider once again the comparison of style to the feathers on an arrow. For one thing, the figure clarifies a relatively unfamiliar and abstract idea (style) by expressing it in a striking image that readers can visualize and understand. For another, it emphasizes and enlarges the idea of style, giving it the functions we associate with the feather in an arrow (providing stability and guidance, directing the shaft swiftly and surely to its target), and denying it the qualities we associate with a feather in a cap (pretentious, pointless decoration, an expression of vanity). For a third, it implies a personal judgment which influences our feelings about the subject: we approve of style in the functional, directive sense and disapprove of it in the self-glorifying sense. Finally, the figure entertains. We take pleasure in the cleverness and succinctness with which a complicated idea is expressed in the images of the two feathers.

Thus figures clarify, they expand, they express feelings and judgments, and they are a source of delight. We shall observe these virtues over and over as we study some of the more common figures of speech. The most frequent and most useful are similes and metaphors. Similes first.

SIMILES

A simile is a brief comparison, usually introduced by the preposition *like* or the conjunction *as*. *Like* is used when the following construction is less than a clause—that is, when it is a phrase or a word:

> My words swirled around his head like summer flies.
>
> E. B. White

As is used when the construction is a clause—that is, when it contains its own subject and verb:

> The decay of society was praised by artists as the decay of a corpse is praised by worms.
>
> G. K. Chesterton

A simile consists of two parts: tenor and vehicle. The *tenor* is the primary subject—"words" in White's figure, the "decay of society [and] artists" in Chesterton's. The *vehicle* is the thing to which the main subject is compared—"summer flies" and the "decay of a corpse [and] worms."

Usually, though not invariably, the vehicle is or contains an image. An *image* is a word or expression designating something we can perceive with one or another of the senses. "Summer flies," for example, is an image, primarily a visual one, though like many images it has a secondary perceptual appeal: we can hear the flies as well as see them.

Most commonly the vehicle follows the tenor, as in the two examples above. Occasionally, however, the vehicle may be put first:

> Like a crack in a plank of wood which cannot be sealed, the difference between the worker and the intellectual was ineradicable in Socialism.
>
> Barbara Tuchman

Such a pattern is a bit more emphatic. Not only does delay to the final position stress the main point, the initial *like*-phrase arouses our curiosity by putting the cart before the horse.

Although generally brief, now and then similes may be expanded. Most often this is done by analyzing the vehicle into its parts and applying these to the tenor. Thus a historian, writing about the Italian patriot Garibaldi, explains:

> . . . his mind was like a vast sea cave, filled with the murmur of dark waters at flow and the stirring of nature's greatest forces, lit here and there by streaks of glorious sunshine bursting in through crevices hewn at random in its rugged sides.
>
> George Macaulay Trevelyan

SIMILES CLARIFY

Similes have many uses. One is to clarify an unfamiliar idea or perception by expressing it in more familiar terms:

> Cold air is heavy; as polar air plows into a region occupied by tropical air . . . it gets underneath the warm air and lifts it up even as it pushes it back. A cold front acts physically like a cowcatcher.
>
> Wolfgang Langewiesche

Finding a familiar equivalent often involves *concretion,* the translation of an abstraction into an image readers can see or hear. The writer and critic Virginia Woolf says of one of Thomas Hardy's novels that it has a plot

> as complicated as a medieval mousetrap.

Even though few of us have actually seen a medieval mousetrap, the phrase suggests a labyrinthine contrivance of Rube Goldberg design, easy to imagine.

Occasionally the process may be reversed so that a simile "abstracts"—that is, it moves from the concrete to the abstract:

> The taste of that crane soup clung to me all day like the memory of an old sorrow dulled by time.
>
> John G. Neihardt

Then the apse [of a medieval cathedral] is pure and beautiful Gothic of the fourteenth century, with very tall and fluted windows like single prayers.

Hilaire Belloc

Clarifying similes easily lend themselves to emphasis, especially when they are placed last, closing a sentence or passage, like those by Neihardt and Belloc. Such summarizing similes make a strong impression. They are memorable statements of an idea or feeling or perception.

SIMILES EXPAND IDEAS

Most similes—even those whose primary purpose is to explain—do more than provide a perceptible equivalent for an abstract idea. Any vehicle has meanings of its own, and these enter into and enlarge the meaning of the tenor. Belloc's phrase "single prayers" does not help us to see the windows of the cathedral. But it does enlarge our understanding of those windows, endowing them with the connotations we associate with *prayer*: the upward lift of spirit, the urge to transcend mortal limits.

Here are two other similes notable for their implicative richness. In the first a writer describes the "war-is-fun" reminiscences of old soldiers:

The easy phrases covered the crudities of war, like the sand blowing in over the graves of their comrades.

Thomas Pakenham

The image suggests the capacity of the mind to obscure unpleasant reality and to hide, even from those who endured it, the true horror of war.

In the second the novelist Isak Dinesen discusses life on a farm in South Africa:

Sometimes visitors from Europe drifted into the farm like wrecked timbers into still waters, turned and rotated, till in the end they were washed out again, or dissolved and sank.

Her image implies a great deal about such drifters: their lack of will and purpose, the futility with which they float through life, their incapacity to attach themselves to anything solid, their inevitable and unmarked disappearance.

You can begin to see that one of the advantages of similes—and of other figures as well—is economy of meaning. By compressing a spectrum of ideas and feelings into few words, they are one of the chief means that writers use to give denseness to their prose.

SIMILES EXPRESS FEELINGS

Many similes are emotionally charged. Pakenham's image of sand blowing over the graves of fallen soldiers, for example, is heavily freighted with sadness at the futility of human life. In the following figure the naturalist Rachel Carson does more than simply describe the summer sea; she expresses her sense of its beauty:

> Or again the summer sea may glitter with a thousand thousand moving pinpricks of light, like an immense swarm of fireflies moving through a dark wood.

Often emotional connotations involve judgments: the writer likes or dislikes what he or she sees, thinks, feels. The poet Rupert Brooke, writing about a conversation with a salesman, describes how the man's mind works:

> The observer could see thoughts slowly floating into it, like carp in a pond.

The simile works on several levels: it translates an abstraction (the process of thinking) into sharp visual terms; it implies the slowness and ponderousness with which this particular mind works. But also it expresses a judgment, an adverse judgment, even though it is put humorously: this is not a mind the writer admires; admirable minds are not inhabited by slow, carp-like thoughts.

Here are two other examples of judgmental similes. The first describes a character in a story and ensures that we will be repelled by him:

> . . . a bald-headed man whose face looked like a pot of lard that has boiled over and eventually congealed in white, flabby, unhealthy drifts and folds.
>
> H. E. Bates

The second, which is more extended, occurs in a discussion by Barbara Tuchman of the beliefs of certain English Socialists in the years just before World War I.

> What was needed was a strong [Socialist] party with no nonsense and a business-like understanding of national needs which would take hold of the future like a governess, slap it into clean clothes, wash its face, blow its nose, make it sit up straight at table and eat a proper diet.

Tuchman's image of the bossy nanny suggests perfectly the unyielding self-righteousness of many Socialists of the period—their smug self-assurance, their certainty that they alone knew what was best for humanity and that it was their duty to impose their values upon people too childish to understand what was good for them. Tuchman, of course, is passing judgment on these dictatorial do-gooders. Her mocking image reveals the disdain for common people which underlay their drive for social and political reform.

The judgments conveyed by such similes are far more than sober, objective opinions. The images in which they are expressed give them great persuasive force. The disgust we feel at the man in Bates' story is not reasoned, but emotional, conjured up by the loathsomeness of cold, congealed lard. Tuchman's simile plays upon the resentment we carry over from childhood against all those righteous Brobdingnagians who forced us to live by their rules.

Because of their silent power to arouse us, not by appealing to reason but by working almost unfelt upon deep pleasures and disgusts, such images should be used responsibly by writers. And by readers they should be understood with cool sophistication.

SIMILES ENTERTAIN

All good writing is a source of pleasure. We enjoy seeing anyone, at work or at play, who really knows what he or she is doing, who performs with the smooth, relaxed surety of a professional. But figurative language can give us special delight, beyond this general pleasure. It most clearly reveals a writer's wit, his or her sensibility. For example, Tuchman's simile reducing imposing Socialists who would reform the world to bossy nannies pontificating in a nursery is amusing (whether it is fair is something else). Here is another example from the same writer. She is talking about the orders a British general gave to his subordinates at the outbreak of World War I in 1914:

> "The special motive of the Force under your control," he wrote, "is to support and cooperate with the French Army . . . and to assist the French in preventing or repelling the invasion by Germany of French or Belgian territory." With a certain optimism, he added, "And eventually to restore the neutrality of Belgium"—a project comparable to restoring virginity.

H. L. Mencken's witty simile makes fun of grammatical purists:

> There are fanatics who love and venerate spelling as a tomcat loves and venerates catnip. There are grammatomaniacs; schoolmarms who would rather parse than eat; specialists in the objective case that doesn't exist in English; strange beings, otherwise sane and even intelligent and comely, who suffer under a split infinitive as you and I would suffer under gastroenteritis.

SIMILES EXPAND OUR CONSCIOUSNESS

Finally, beyond their capacity to familiarize the strange, to expand ideas economically, to express feelings and impose judgments, and to give us pleasure, similes have an even greater power: they put us more deeply in touch with the world in a more complex way. They push us toward more total awareness by bringing together diverse aspects of experience, joining our perceptions so that we enter into life more fully and richly.

Listen to a writer describing the hands of an old woman:

Their touch had no substance, like a dry wind on a July afternoon.

Sharon R. Curtin

Curtin's simile likens a less familiar experience to a more familiar one, implies something about the loneliness and dryness of age, even passes a kind of judgment—not on the old woman, but on life. But past all this, the simile unifies perceptions that most of us would never have put together.

A simile may also cut across the boundaries that separate the senses:

There was a glamour in the air, a something in the special flavour of that moment that was like the consciousness of Salvation, or the smell of ripe peaches on a sunny wall.

Logan Pearsall Smith

In that image two sense perceptions blend in some inexplicable way into a unified experience—the smell of peaches and the vision of them on a sunlit wall. And this fused perception connects with the writer's inner consciousness, his feelings of religious mystery.

It is this capacity to *connect* that gives similes—and, as we shall see, metaphors—their greatest value: the power to enlarge and harmonize experience.

METAPHOR

A metaphor is also a comparison. The difference is that a simile compares things explicitly—that is, it states literally that X is *like* Y. A metaphor compares things implicitly. Read literally, it does not state that things are alike; it says that they are the *same* thing, that they are identical:

Cape Cod is the bared and bended arm of Massachusetts. . . .

Henry David Thoreau

Thoreau writes "is," not "is like." We understand, however, that he is making a comparison—that he means the Cape resembles an arm, not really is an arm. The metaphor has simply carried the comparison a degree closer and expressed it a bit more economically and forcefully.[1]

A metaphor has the same two parts as a simile: the tenor—or main

---

1. It is sometimes argued that metaphors are more powerful figures than similes and even in some ways essentially different. But for our purposes it is not necessary to assume any greater virtue for metaphors. They are more economical (that is, they make do with fewer words) and generally more emphatic. For those reasons they are sometimes preferable to similes. Sometimes, on the other hand, the more explicit comparison made by a simile is better.

subject—and the vehicle—or image introduced by way of comparison. In Thoreau's case the tenor is "Cape Cod" and the vehicle is "the bared and bended arm." In many metaphors both parts are stated. In some, however, the writer refers only to the vehicle, depanding upon the context to supply the full comparison. Such a figure is called an *implied,* rather than a full, metaphor. Had Thoreau written "the bared and bended arm of Massachusetts" in a context that made clear he meant Cape Cod without actually saying so, that would be implied metaphor.

Fused metaphors sometimes involve metonymy. *Metonymy* means substituting for one concept another that is associated with it—as a quality or a consequence, for example. The novelist Joseph Conrad, discussing the difficulty of saying exactly what one wants to say, speaks of

the old, old words, worn thin, defaced by ages of careless usage.

Conrad does not literally say that words are coins, but he implies the full metaphor by the terms "worn thin" and "defaced"—that is, in terms of the qualities of old coins. The logic of the metaphor goes like this:

Words are (like) old coins.

Old coins can be worn thin by usage and the faces nearly rubbed off.

Therefore words can be "worn thin" and "defaced."

Another closely related figure often found in metaphors is *synecdoche,* which is substituting a part for the whole, as when one refers to a ship as "a sail." In the following passage the religious revivals staged by the evangelist Aimee Semple Macpherson are compared (implicitly) to an amusement park:

With rare ingenuity, Aimee kept the Ferris wheels and merry-go-rounds of religion going night and day.

                                                                    Carey McWilliams

The logic is:

Aimee's revivals were (like) an amusement park.

An amusement park contains Ferris wheels and merry-go-rounds.

Therefore events at the revivals were (like) Ferris wheels and merry-go-rounds.

Many metaphors use synecdoche or metonymy.[2] Usually a writer wants to introduce as precise an image as possible into the vehicle of a metaphor, thus appealing immediately to the reader's eyes or ears. "Ferris wheels" and "merry-go-rounds," for instance, name things eas-

2. Some scholars distinguish metonymy as a figure separate from metaphor. Here it is convenient to regard it—along with synecdoche—as a variety of metaphor.

ier to visualize than does the more abstract phrase "amusement park." And since these images imply the park in its entirety, as well as evoking vivid pictures of revolving vertical and horizontal wheels, they are rich in meaning.

THE USES OF METAPHOR

Metaphors have the same functions as similes. They are valuable in clarifying unfamiliar concepts and in translating abstractions into images that readers can intuit directly, as in this passage about science:

> [Science] pronounces only on whatever, at the time, appears to have been scientifically ascertained, which is a small island in an ocean of nescience.
>
> Bertrand Russell

Russell's image of a small island (scientific knowledge) in a wide and lonely sea (the vastness of all we do not know) is a memorable expression of the relationship between knowledge and ignorance.

Metaphors also enrich meaning by implying a range of ideas and feelings and evaluations. Consider all that is suggested by the term "idol" in this metaphor:

> We squat before television, the idol of our cherished progress.
>
> Student

"Idol" means a false god and thus questions the value of the progress television symbolizes and celebrates. The word implies also the unreason and subservience of those who worship it.

Such a metaphor not only complicates an idea, it also implies judgment. In the next example the judgmental quality of the metaphor is even more pronounced. Speaking of ancient Romans, a writer remarks that

> They were marked by the thumbprint of an unnatural vulgarity, which they never succeeded in surmounting.
>
> Lawrence Durrell

The image of a greasy thumbprint, like one left on a china cup or a white wall, is a graphic signature of crudeness. In the metaphor below, the judgment is ironic; the passage, by Hodding Carter, concerns Huey Long, a powerful Louisiana politician of the 1930s:

> He designated his old benefactor, O. K. Allen of Winnfield, as the apostolic choice for the next full term.

"Apostolic," with its echoes of Christ and his disciples, is a wry comment upon Long: upon the power he wielded, upon the veneration he was accorded by his followers, and even perhaps upon how he regarded himself.

Sometimes, too, as with similes, a metaphor functions to carry an emotional charge, pleasant or, as in the following instance, unpleasant (the writer is describing a spoonful of castor oil forced upon him as a child):

> . . . a bulge of colorless slime on a giant spoon.
>
> William Gibson

Metaphors (and similes too) are often used to clinch a point. Placed at the end of a statement, the figure not only clarifies or concretizes a complex, abstract idea, it also restates it emphatically, leaving a memorable image in the reader's mind:

> What distinguishes a black hole from a planet or an ordinary star is that anything falling into it cannot come out of it again. If light cannot escape, nothing else can and it is a perfect trap: a turnstile to oblivion.
>
> Nigel Calder

Metaphors, finally, are sometimes extended through several sentences or even an entire paragraph. In fact, exploring and expanding the parts of a metaphor is an effective way of generating a piece of prose. Here are two brief examples:

> He assured me that beauty there [in Athens] was in bud at thirteen, in full blossom at fifteen, losing a leaf or two every day at seventeen, trembling on the thorn at nineteen, and under the tree at twenty.
>
> Walter Savage Landor
>
> Man is born broken. He lives by mending. The grace of God is glue.
>
> Eugene O'Neill

And here are two more extended instances. In the first the writer expands a metaphor comparing *Time* magazine to nursery tales:

> *Time* is also a nursery book in which the reader is slapped and tickled alternately. It is full of predigested pap spooned out with confidential nudges. The reader is never on his own for an instant, but, as though at his mother's knee, he is provided with the right emotions for everything he hears or sees as the pages turn.
>
> Marshall McLuhan

Notice how much of the diction the metaphor determines: "slapped and tickled," "predigested pap," "spooned out," "nudges," "never on his own," "mother's knee," "provided with." Even the phrase "as the pages turn" suggests the passivity of a child for whom the baby-sitter turns the pages as she reads.

The second example returns us to the metaphor with which we began this section. Thoreau's comparison of Cape Cod to a bent arm actually begins a paragraph through which the metaphor is developed:

> Cape Cod is the bared and bended arm of Massachusetts: the shoulder is at Buzzard's Bay; the elbow, or crazy-bone, at Cape Mallebarre; the wrist at Truro; and the sandy fist at Provincetown—behind which the State stands on her guard, with her back to the Green Mountains, and her feet planted on the floor of the ocean, like an athlete protecting her Bay—

boxing with northeast storms, and, ever and anon, heaving up her Atlantic adversary from the lap of earth—ready to thrust forward her other fist, which keeps guard the while upon her breast at Cape Ann.

The figure organizes the entire paragraph. The image of "the bared and bended arm" is developed in two directions. Thoreau analyzes it into its parts—"elbow," "wrist," "fist"—and applies each of these to Cape Cod. He also expands the metaphor into the larger inclusive image of the boxer—"back," "feet," "other fist," "breast"—each detail of which is applied to other parts of Massachusetts.

If you wish to develop a metaphor (or simile), remember that you can work in either of these ways or even, like Thoreau, in both: inward, differentiating the elements of the image and connecting these to your main topic; or outward, exploring the larger reality of which the image is a natural part, as the "bared arm" is part of a boxer in a defensive stance.

FINDING METAPHORS

There is no formula for discovering metaphors. Sometimes a word or detail applying literally to a scene or idea lends itself to figurative use, as in the following sentence. (The writer is explaining that she was not allowed into a large office to observe the regimented life of the clerks.)

> I knew those rooms were back there, but I couldn't get past the opaque glass doors any more than I could get past the opaque glass smiles.
>
> Barbara Garson

Another source of metaphors is metonymy, describing something in terms of its associated qualities. In this instance the writer talks about the coming of spring:

> The birds have started singing in the valley. Their February squawks and naked chirps are fully fledged now, and long lyrics fly in the air.
>
> Annie Dillard

Dillard's figure works by applying to the songs of the birds images that are visual rather than auditory ("fully fledged," "fly through the air"). Thus she describes the songs in words more usually restricted to describing the birds themselves.

More often, however, a metaphor or simile comes from outside— that is, it involves a comparison which, while apt and revealing, does not grow naturally out of the subject as do the images by Garson and Dillard. Here, for instance, is a philosopher discussing the style of another philosopher:

> The style is not, as philosophic style should be, so transparent a medium that one looks straight through it at the object, forgetting that it is there;

it is too much like a window of stained glass which, because of its very richness, diverts attention to itself.

Brand Blanshard

But whether a metaphor or simile comes from "inside" the topic or from "outside," its coming depends upon imagination. There is no magic way of spurring the imagination to the discovery of metaphors. Some people are more adept at seeing resemblances than the rest of us. Still, we can all profit from relaxing a little and letting our minds run free of inhibitions. When you are first drafting a composition, don't be frightened of a simile or metaphor, however wild and far-fetched it may sound.

USING METAPHORS EFFECTIVELY

When you revise, however, become more detached and critical about figures of speech (as you should become about all phases of your writing). To use metaphors and similes effectively, remember these principles:

1. *Metaphors and similes should be fresh and original.*

Avoid tired, trite figures: "quiet as a mouse," "white as a sheet," "the game of life," "a tower of strength." Clever humorists can sometimes make such clichés work for them, but most of us are better off leaving them alone. If you cannot think of anything better than "his face was as white as a sheet," you should probably say, "his face was very white."[3]

2. *The vehicle should fit the tenor.*

The vehicle of a simile or metaphor is loaded with many meanings. It is easy to focus only upon the meaning that suits your purpose and to overlook others which do not. Here is an example:

The town hall has been weathered by cold winds and harsh snows like an old mare turned out to pasture.

While an old mare is an image of decrepitude, it has other characteristics which make it unsuitable as a vehicle for a building. Can you imagine a town hall in a pasture, nibbling grass and slapping flies with its tail?

3. *Metaphors and similes should be appropriate to the context.*

Metaphors and similes should contribute to the general atmosphere

---

3. Such trite similes and metaphors may be described as "dying," to distinguish them from figures already dead. These—"mouth of a river" is an example—have passed through the figurative stage and acquired an additional literal meaning. In that meaning they are perfectly legitimate, and you should not feel doubtful about using them. At one time "mouth of a river" was a fresh and vivid image. Now this sense of "mouth" is not really understood as a metaphor at all; it is simply one of the denotations of the word. Many words develop new meanings by such metaphorical extension.

and tone of a piece of writing, not shatter them. Figures have their own levels of formality and informality. Even when it does not possess specifically awkward connotations, a simile or metaphor should not be too literary or learned for your occasion, or too colloquial. It would not do in a paper for a history course to write that "Napoleon went through Russia like a dose of salts."

4. *Metaphors or similes should not be awkwardly mixed.*

When several similes or metaphors appear in the same passage, they should harmonize to produce a consistent and logical sequence of ideas and picture. Avoid awkward mixtures like this:

> The moon, a silver coin hung in the draperies of the enchanted night, let fall her glance, which gilded the roof tops with a joyful phosphorescence.

Perhaps this *sounds* impressive, at least until one begins to think about it and to imagine the scene so lushly described. But if the moon is "a silver coin" how can "she" let fall a "glance"? How can "silver" be used for "gilding," which means to cover with gold? Why mix three elements—gold, silver, and phosphorus—indiscriminately? Can "phosphorescence" be "joyful"? Do "coins" hang in "draperies"?

Even dead metaphors and similes ought not to be mixed. Although such expressions no longer have any real figurative value ("the eye of a needle," for example, or "a bottleneck"), they are likely to come embarrassingly to life if used too close together. The dead metaphors "leg of a journey" and "mouth of a river," for instance, work well enough alone, but it would be clumsy to write: "The last leg of our journey began at the mouth of the river."

You must be careful, too, about the other words you use with a simile or metaphor, even when these words are not actually a part of the figure. Because this writer was careless with his contextual diction, the following metaphor fails:

> The teacher leaves the student to develop that foundation of learning.

Foundations, of course, are "built" or "laid," not "developed."

5. *Metaphors and similes should not be used very often.*

Especially in expository writing, metaphors and similes ought not to be sprinkled about profusely. Even when they do not clash, too many similes and metaphors are likely to cancel out one another. The effectiveness of figures depends upon their being relatively uncommon. If every other sentence contains a simile or a metaphor, the reader soon begins to discount them.

The trouble with overusing metaphors and similes is that they draw too much attention to themselves, thus defeating their chief purpose of enhancing the main idea or perception.

PERSONIFICATION

Personification means referring to inanimate things or abstractions as if they were human. Personification is really a special kind of metaphor. At its simplest it consists of using personal pronouns for objects, as when sailors speak of their ship as "she."

Personification can be more subtle. Thus Washington Irving personifies the social changes in a London neighborhood:

> As London increased, however, rank and fashion rolled off to the west, and trade, creeping on at their heels, took possession of their deserted abodes.

"Rank" and "fashion" are abstractions personifying aristocratic Londoners, "trade" an abstraction signifying the merchant class. These abstractions behave in the manner of the beings they personify, "rolling off" elegantly (in carriages) and "creeping" in with the deference of inferiors.

The purpose of personification—like that of metaphor generally—is to explain, to expand, to vivify. The following figure implies, in only eight words, a great deal about the relationship between ambition and greed:

> Ambition is but Avarice on stilts and masked.
>
> Walter Savage Landor

Even in strictly expository writing, personification renders objects and abstractions with vivid detail. This passage by Bruce Catton, for instance, illustrates how effectively personification can serve the historian:

> There is a rowdy strain in American life, living close to the surface but running very deep. Like an ape behind a mask it can display itself suddenly with terrifying effect. It is slack-jawed, with leering eyes and loose wet lips, with heavy feet and ponderous cunning hands; now and then when something tickles it, it guffaws, and when it is angry it snarls; and it can be aroused more easily than it can be quieted. Mike Fink and Yankee Doodle helped to father it, and Judge Lynch is one of its creations; and when it comes lumbering forth it can make the whole country step in time to its own irregular pulse beat.

By personification (or what in this case might be called "animalification"), Catton evaluates his subject and makes his point with extraordinary emphasis. Personification has the additional advantage of bringing a topic into a human and social relationship to the reader. Mindlessness, Bruce Catton reminds us, is not an abstraction but a real and personal menace. Skillfully employed, personification is one of the most dramatic of all figures of speech.

ALLUSIONS

An allusion is a brief reference to a well-known person or place or happening. Sometimes the reference is explicitly identified:

> As it is, I am like that man in *The Pilgrim's Progress,* by some accounted mad, who the more he cast away the more he had.
>
> W. H. Hudson

More often the reference is indirect, and the writer depends upon his readers' recognizing the source and significance:

> We [Western peoples] tend to have a Micawberish attitude toward life, a feeling that so long as we do not get too excited something is certain to turn up.
>
> Barbara Ward

A writer using an allusion should be reasonably sure that his or her intended audience will be familiar with the reference. Otherwise readers are likely to be not only mystified but annoyed. Barbara Ward, for instance, could fairly refer to Mr. Micawber, confident that her readers were familiar with Charles Dickens' novel *David Copperfield* and would remember the luckless Micawber, burdened by family and debt, yet cheerfully, and foolishly, optimistic that something would turn up to save him from ruin. But had Ward been writing for people who could not be expected to know Dickens, her allusion would have been foolish and pretentious.

Sometimes a literary allusion is not to a character or a title but to a well-known passage—a verse from the Bible, say, or a line from a Shakespeare play. The passage may be repeated literally or paraphrased or echoed in a free adaptation. Unlike a direct quotation, it is not usually enclosed in quote marks. The writer, however, is not plagiarizing; he or she assumes the reader will recognize the lines and know the source. In this sentence, for instance, the allusion is to the Old Testament book of Ecclesiastes (3:1–8):

> I didn't know whether I should appear before you—there is a time to show and a time to hide; there is a time to speak, and also a time to be silent.
>
> Norman O. Brown

Most allusions are drawn from literature or the Bible. But some refer to actual events or people in history, ancient or more recent:

> These moloch gods, these monstrous states, are not natural being. . . .
>
> Susanne K. Langer

(Moloch was an ancient Semitic deity to whom children were sacrificed.)

Others refer to contemporary events or personalities:

And it is not opinions or thoughts that *Time* provides its readers as news comment. Rather, the newsreel is provided with a razzle-dazzle accompaniment of Spike Jones noises.

Marshall McLuhan

(Spike Jones was a popular orchestra leader of the 1940s, famous for his wacky, comic arrangements of light classics and pop tunes. He used automobile horns, cow bells, steam whistles, and so on.)

But whatever the source of an allusion, its purpose is to pack a load of meaning into a few words. To use allusions successfully, remember two things: (1) the allusion must be appropriate to your point, really saying what you want it to say; and (2) it must be within your readers' general knowledge.

IRONY

Irony is using words in a sense very different from their usual meaning, often, in fact, the very reverse of it. The simplest form occurs when a term is given its opposite meaning. Here, for example, an historian describes a party at the court of the English king James I:

> Later the company flocked to the windows to look into the palace courtyard below. Here a vast company had already assembled to watch the King's bears fight with greyhounds, and mastiffs bait a tethered bull. These delights were succeeded by tumblers on tightropes and displays of horsemanship.

C. P. V. Akrigg

By "delights" we are expected to understand "abominations," "detestable acts of cruelty."

In subtler forms irony plays more lightly over words, pervading an entire passage rather than twisting any single term into its opposite. An instance occurs in this sentence (the writer is commenting upon the decline of the medieval Knight at Arms):

> In our end of time the chevalier has become a Knight of Pythias, or Columbus, or the Temple, who solemnly girds on sword and armor to march past his own drugstore.

Morris Bishop

None of Bishop's words means its reverse. Indeed the whole sentence is to be read literally. Still, Bishop intends us to smile at modern men playing at knighthood. The irony is found in the fact that some of the words ought *not* to be taken literally. A twentieth-century businessman ought not to "solemnly gird on sword and armor," unconscious of the disparity between romantic ideals and modern life.

Disparity is the common denominator in both these examples of irony: the difference between the ideal and the actual, between what we profess and what we do, between what we expect and what we get. In stressing such disparities, irony is fundamentally different from simile and metaphor, which build upon similarity. The whole point of irony is that things are *not* what they seem or what they should be or what we want them to be. They are different.

Irony reveals these differences in several ways. One is by using words in a double sense ("delights"), making them signify both the ideal and the actual. Another is by setting side by side contrasting images of what could be (or once was) and what is (the chevalier girding on his sword and the neighborhood druggist). Either way, we are made conscious of the gap between "ought" and "is": people *ought* to treat dumb animals kindly, for instance; they *do* take pleasure in sadism.

Writers using irony must be reasonably sure that readers will understand the special sense in which they use their language. Sometimes the ironist depends upon the general knowledge and attitudes of the audience. Akrigg's ironic use of "delights" is successful because modern readers know that such amusements are not delightful. Here is another example of irony that is played off against the reader's values and expectations:

> Proud as a peacock, NBC announced an uplifting afternoon of sports programming. "In Chicago," the press release read, "six night club strong men compete for the title of 'America's Toughest Bouncer' as they throw a 110-pound stuntman for distance and accuracy, and run an obstacle course by leaping a bar and threading through a maze of chairs and tables to crash through a door."
>
> Supplementing this cultural offering was a chug-a-lug drinking contest, a tug-of-war between Teamsters and Longshoremen and a field-gunnery meet in which teams of the Royal Navy in England race against the clock to dismantle a cannon, reassemble it and fire it three times.
>
> Melvin Durslag

Durslag does not label the words "uplifting" and "cultural offering" as ironic. He depends upon his readers knowing that tossing a human being "for distance and accuracy" is not an uplifting cultural event. And he assumes—properly—that they will recognize the allusion to NBC's promotional slogan ("Proud as a peacock") and understand the irony.

When it is used more subtly, however, irony may need signalling so that readers will not pass over it. Study this passage by the historian Barbara Tuchman (she is discussing the question of the guilt of the German Nazis):

> When it comes to guilt, a respected writer—respected in some circles—has told us, as her considered verdict on the Nazi program, that evil is banal—a word that means something so ordinary that you are not bothered by it; the dictionary definition is "commonplace and hackneyed." Somehow that conclusion does not seem adequate or even apt. *Of course,* evil is commonplace; *of course* we all partake of it. Does that mean that we must

withhold disapproval, and that when evil appears in dangerous degree or vicious form we must not condemn but only understand?

The specifically ironic words are "respected" and "considered verdict." The first is turned around by a signal, the qualification "respected in some circles." Immediately we feel the barbed insinuation: "respected— but not by me or you." "Considered verdict," on the other hand, acquires its ironic value not so much from a cue as from the total context. If "banality" is the only judgment the other writer can make, her judgment is not—Tuchman suggests—worth considering. "Verdict" also has an extra ironic value. The word signifies an official decision of a court of law, and Tuchman implies that her opponent is presumptuous in delivering a "verdict" as if she were a judge or jury.

In other ways, too, Tuchman reveals her feelings and thus contributes to the tone of irony. The repetition of "of course" and the italic emphasis of the phrase imply: "These ideas are so commonplace that we do not need to be told them." And the rhetorical question, by stressing the undeniable truth of Tuchman's point, underscores the folly of the writer she is attacking.

Irony may be used in a variety of tones. Some irony is genial—amused and amusing—like the examples by Bishop and Durslag. These writers disapprove, but their disapproval is good-natured. Sometimes irony is more serious, as in the passage by Akrigg; and sometimes even angry, as in the example by Tuchman.

Whatever its emotional tone, irony contributes importantly to a writer's persona. It is a form of comment—though an oblique form. Thus it represents an intrusion of the writer into the writing. He or she stands forth, moreover, in a special way: as a subtle, complex, witty person, deliberately using intellect to distance emotion from the subject. This does not mean that irony diminishes emotion. On the contrary: irony more often than not acts like a lens, focusing and intensifying the emotions forced through it. But it does mean that the writer using irony constrains his or her emotions rather than letting them gush out.

Irony, finally, may function in prose as a specific figure of speech, a device for expressing a particular judgment. Or it may be more general, used less in momentary and passing figures than as a mode of thought, an encompassing vision of people and events. In this broad sense irony becomes the stance some writers take toward experience. Such writers may be described as ironists. But even the rest of us, though we are not ironists, can profitably employ irony now and then.

OVERSTATEMENT AND UNDERSTATEMENT

Overstatement and understatement are special kinds of irony. Each depends for its effect upon a disparity between the reality the writer is describing and the words he or she uses.

In overstatement the diction exaggerates the subject, deliberately magnifying it far beyond its true dimensions. Understatement takes the opposite tack: the words are intentionally inadequate to the importance or value of the reality.

OVERSTATEMENT

The rhetorical name for overstatement is *hyperbole,* from a Greek word meaning "excess." Loosely speaking, there are two kinds of overstatement: comic and serious. Comic hyperbole ridicules or burlesques. Like caricature, which exaggerates distinctive features of an individual's face, so overstatement overblows the offending or silly features of what the writer is describing.

Comic overstatement has deep roots in American literature. It is a major element in the tall tales told of such culture heroes as Davy Crockett and Mike Fink. Much of Mark Twain's humor depends upon overstatement. Here, for instance, is part of his essay "The Awful German Language" from *A Tramp Abroad.* (If you have ever struggled with German verbs or word order, you will appreciate it especially.)

> An average sentence in a German newspaper, is a sublime and impressive curiosity; it occupies a quarter of a column; it contains all the ten parts of speech—not in regular order, but mixed; it is built mainly of compound words constructed by the writer on the spot, and not to be found in any dictionary—six or seven words compacted into one, without joint or seam—that is, without hyphens; it treats of fourteen or fifteen different subjects, each inclosed in a parenthesis of its own, with here and there extra parentheses which reinclose three or four of the minor parentheses, making pens within pens: finally, all the parentheses and reparentheses are massed together between a couple of king-parentheses, one of which is placed in the first line of the majestic sentence and the other in the middle of the last line of it—*after which comes the* VERB, and you find out for the first time what the man has been talking about; and after the verb—merely by way of ornament, as far as I can make out—the writer shovels in *"haben sind gewesen gehabt haben geworden sein,"* or words to that effect, and the monument is finished.

Serious overstatement differs only in its end, which is not to make us laugh but to persuade us. The writer may wish us to admire and experience something of great value. Or he or she may want to shock us into accepting a harsh truth. Shock is the tactic of H. L. Mencken, who used overstatement to cudgel what he saw as the venality, stupidity, and smugness of American life in the 1920s:

> It is . . . one of my firmest and most sacred beliefs, reached after an enquiry extending over a score of years and supported by incessant prayer and meditation, that the government of the United States, in both its legislative arm and its executive arm, is ignorant, incompetent, corrupt, and disgust-

ing—and from this judgment I except no more than twenty living law-makers and no more than twenty executioners of their laws. It is a belief no less piously cherished that the administration of justice in the Republic is stupid, dishonest, and against all reason and equity—and from this judgment I except no more than thirty judges, including two upon the bench of the Supreme Court of the United States. It is another that the foreign policy of the United States—its habitual manner of dealing with other nations, whether friend or foe, is hypocritical, disingenuous, knavish, and dishonorable—and from this judgment I consent to no exceptions whatever, either recent or long past. And it is my fourth (and, to avoid too depressing a bill, final) conviction that the American people, taking one with another, constitute the most timorous, sniveling, poltroonish, ignominious mob of serfs and goose-steppers ever gathered under one flag in Christendom since the end of the Middle Ages, and that they grow more timorous, more sniveling, more poltroonish, more ignominious every day.

Overstatement, comic or serious, relies upon several devices. It likes the superlative form of adjectives, referring to the longest spans of time, the hugest numbers, to extremes of all sorts. It prefers sweeping generalizations: *every, all, always, never, none*. It offers few qualifications or disclaimers, and if it does qualify, it may turn the concession into another exaggerated claim (like Mencken's "and from this judgment I except no more than twenty living law-makers"). It rides hard upon words with strong emotional connotations like "sniveling," "poltroonish," "ignominious," "knavish." Its sentence structure is likely to be emphatic, with strong rhythms and frequent repetitions. Short statements are stressed by being set beside longer ones. In short, overstatement calls upon all the devices of emphasis the language affords.

Thus overstatement is powerful. It shocks us or infuriates us or makes us double over with laughter. But this very power is also a limitation. Its effectiveness, like that of most figures of speech, depends upon rarity. Overstatement is hard to take for very long; it quickly loses its capacity to shock or amuse. More important, overstatement is easily abused. It is a kind of assertion, not a reasoned argument. At its worst it can degenerate into shrill emotionalism or name-calling.

UNDERSTATEMENT

Understatement deliberately plays down the magnitude or intensity of a subject. It emphasizes by avoiding emphasis. Like overstatement it can be either comic or serious. Twain, for example, is being funny in the following sentence:

I have been strictly reared, but if it had not been so dark and solemn and awful there in that vast, lonely room, I do believe I should have said something then which could not be put into a Sunday-school book without injuring the sale of it.

And note the black humor (if there is any) in this famous remark by the satirist Jonathan Swift:

> Last week I saw a woman flayed, and you will hardly believe how it altered her appearance for the worse.

Understatement works by a paradox: increasing emotional impact by carefully avoiding emotive language. It is a species of irony in that the actual value of the words differs from their surface meaning. Swift's phrase "altered her appearance for the worse" seems woefully inadequate: no streaming blood, no raw, quivering flesh, no screams of the victim—just that "it altered her appearance for the worse." But of course Swift tricks us, shocking us into imagining the scene for ourselves.

In the following paragraph Ernest Hemingway increases our horror by writing as if there is no horror, as if cold-blooded executions are merely routine, which in time of war they are (and *that's* the horror):

> They shot the cabinet ministers at half-past six in the morning against the wall of the hospital. There were pools of water in the courtyard. There were wet dead leaves on the paving of the courtyard. It rained hard. All the shutters of the hospital were nailed shut. One of the ministers was sick with typhoid. Two soldiers carried him downstairs and out into the rain. They tried to hold him up against the wall but he sat down in a puddle of water. The other five stood very quietly against the wall. Finally the officer told the soldiers it was no good trying to make him stand up. When they fired the first volley he was sitting down in the water with his head on his knees.

Sometimes, as Hemingway's style suggests, words are unequal to events. Then a writer's best strategy is understatement, using words as simply and directly as possible to re-create the event:

> In the heart of the city near the buildings of the Prefectural Government and at the intersection of the busiest streets, everybody had stopped and stood in a crowd gazing up at three parachutes floating down through the blue air.
>
> The bomb exploded several hundred feet above their heads.
>
> The people for miles around Hiroshima, in the fields, in the mountains, and on the bay, saw a light that was brilliant even in the sun, and felt heat.
>
> Alexander H. Leighton

To unsophisticated readers understatement may appear callous or insensitive. Some of Swift's contemporaries, for instance, misread his irony as cruelty. Still, used with an appropriate subject and the right audience, understatement is more powerful than hyperbole.

PUNS

A *pun* is a word employed in two or more senses, or a word used in a context that makes the reader think of a second term resembling it in sound. In the first of the two following examples the pun depends upon different meanings of the same word; in the second, upon one word's sounding like another:

> A cannon-ball took off his legs, so he laid down his arms.
>
> Thomas Hood
>
> During the two previous centuries musical styles went in one era and out of the other. . . .
>
> Frank Muir

While puns resemble one kind of irony in simultaneously using words in different senses, they differ in more important ways. For one thing, a pun is almost exclusively a device of humor. (At least it is so today. In earlier centuries poets and dramatists often employed puns in serious contexts.) Mark Twain, for instance, makes us laugh by punning on the expression "raising chickens":

> Even as a schoolboy poultry-raising was a study with me, and I may say without egotism that as early as the age of seventeen I was acquainted with all the best and speediest methods of raising chickens, from raising them off a roost by burning lucifer matches under their noses, down to lifting them off a fence on a frosty night by insinuating a warm board under their feet.

For another thing, puns, the better ones at any rate, work more like metaphors and similes. They reveal unexpected connections. A good pun not only amuses us, it surprises us by pointing out a significant and hitherto unseen similarity. The humorist S. J. Perelman entitles one collection of his essays *The Road to Miltown, or Under the Spreading Atrophy*. The pun on "atrophy" is effective not only because the word sounds like "a tree" and the phrase echoes a famous line of American poetry ("under the spreading chestnut tree"), but also because in an age given to the wholesale swallowing of tranquilizers like Miltown, atrophy may indeed be spreading.

Because they have become a sign of "low humor," many people think puns are unseemly in modern exposition. That judgment is a bit harsh: a good pun is often worth making. Use them only when you wish a light, informal tone, however. Even then a pun should be clever and revealing. A clumsy or inappropriate pun is worse than none at all.

ZEUGMA

Zeugma (pronounced ZOOG ma) is a special kind of pun involving a verb. It occurs when the same verb is used with two or more objects, either (1) applying to each of them in a different sense, or (2) even when having the same sense, creating an apparently incongruent combination of ideas.

In these two examples, the verb operates in slightly different senses.

The novelist Lawrence Durrell is describing the plight of a maiden being chased by monks whose intentions are not honorable:

> Joanna, pursued by the three monks, ran about the room, leaping over tables and chairs, sometimes throwing a dish or a scriptural maxim at her pursuers.

The zeugma occurs with "throwing," which is used in both a literal sense (with "dish") and a metaphorical one (with "maxim").

In the second example, a definition from Ambrose Bierce's *The Devil's Dictionary,* the two meanings carried by "depressing" are even wider apart:

> **Piano,** n. A parlor utensil for subduing the impenitent visitor. It is operated by depressing the keys of the machine and the spirits of the audience.

An instance of the second case, in which the verb has the same meaning but combines with two objects to create an unusual coupling of ideas, is found in this sentence:

> She left his apartment with tarnished virtue and a new mink.

Zeugma, like puns in general, is a comic figure of speech. At its best zeugma is witty and amusing, and it increases meaning by revealing hidden connections. Not only does Durrell's pairing of dishes and scriptural maxims amuse us: it also leads us to see their equal inefficacy in Joanna's plight.

IMAGERY

An *image* is a word or expression that speaks directly to one or more of the senses, as in this description of the Seine in Paris:

> The river was brown and green—olive-green under the bridges—and a rainbow-coloured scum floated at the sides.
>
> Jean Rhys

Images are classified according to the sense to which they primarily appeal. Visual images, as in the sentence above, are the most common. Next in frequency, probably, are auditory images, which appeal to the ear:

> The [medieval] house lacked air, light, and *comfort moderne*; but people had little taste for privacy. They lived most of their lives on the streets, noisy indeed by day, with pounding hammers, screaming saws, clattering wooden shoes, street cries of vendors of goods and services, and the hand bells of pietists, summoning all to pray for the souls of the dead.
>
> Morris Bishop

Images can appeal to other senses: to smell, taste, touch, even to the sense of movement and balance (these are called kinesthetic images). Here, for instance, is an indictment of the odors of a modern city:

the reek of gasoline exhaust, the sour smell of a subway crowd, the pervasive odor of a garbage dump, the sulphurous fumes of a chemical works, the carbolated rankness of a public lavatory . . . the chlorinated exudation of ordinary drinking water. . . .

Lewis Mumford

Often an image appeals to two or more senses simultaneously, primarily to one, secondarily to the other(s). Thus in Rhys's sentence about the Seine, while the imagery is essentially visual, we can with a little imagination "feel" the water and even "smell" the oily scum.

At their simplest, images re-create sensory experiences. Bishop, for instance, uses auditory images to describe what the ears would have heard in a medieval city. In the following passage the writer re-creates the experience of walking in small streams:

Exploring a streambed can be done on a purely sensual level. How it all feels—moss, wet rock, soft mud under feet; cold fast mountain water or the touch of a sun-warmed gentle brook; water wrapping itself around your ankles or knees, swirling in little eddies, sparkling in small pools, rushing away white and foamy over rapids, or calmly meandering over glistening pebbles.

Ruth Rudner

Rudner's passage also illustrates how images may be mixed to appeal to several senses. Many of her words are tactile: "soft mud under feet," "the touch of a sun-warmed gentle brook," "water wrapping itself around your ankles or knees." Others are visual: "moss, wet rock," "swirling in little eddies," "sparkling," "glistening pebbles." Still others, kinesthetic: "swirling" again, "rushing away," "calmly meandering."

On occasion images can be stretched to mean more than the re-creation of something seen or heard or felt. In the following passage the writer describes a California landscape, the scene of a murderous love affair:

The lemon groves are sunken, down a three- or four-foot retaining wall, so that one looks directly into their foliage, too lush, too unsettlingly glossy, the greenery of nightmare; the fallen eucalyptus bark is too dusty, a place for snakes to breed.

Joan Didion

Literally these images describe the trees and the bark-strewn ground. Yet they suggest unnaturalness and evil too, a morbid aura ominous of death. We cannot say that they "mean" evil or death in any exact sense, as we may say that the vehicle of a metaphor signifies the tenor for which it stands. Nonetheless the images have overtones which cause the passage to vibrate with appropriate feelings, evaluations, judgments.

At times an image acquires a symbolic value, rendering a complex abstract idea in a sharp and striking perception. The novelist and essayist George Orwell uses an image in this way to pass judgment upon a socialist political speaker:

When one watches some tired hack on the platform mechanically repeating the familiar phrases—*bestial atrocities, bloodstained tyranny, free peoples of the world, stand shoulder to shoulder*—one often has the curious feeling that one is not watching a live human being but some kind of dummy: a feeling which suddenly becomes stronger at moments when the light catches the speaker's spectacles and turns them into blank discs which seem to have no eyes behind them.

Images, then, are important to good diction, to the kind of diction that not only informs readers but interests them. When you are writing about something you experienced directly with your eyes or ears or hands or nose, images are essential. Your purpose is to make the reader see and hear and feel. Precision requires that you learn what things are called, that you distinguish not only with your nose but with words between a "sour smell . . . sulphurous fumes . . . carbolated rankness." See how precise the diction is in this sentence describing the foods in a German delicatessen:

> There were rows and rows of hams and sausages of all shapes and colours—white, yellow, red, and black; fat and lean and round and long—rows of canned preserves, cocoa and tea, bright translucent bottles of honey, marmalade and jam; round bottles and slender bottles, filled with liqueurs and punch. . . .
>
> Thomas Mann

A poorer writer might simply have said "there were rows and rows of hams and sausages and canned preserves" and let it go at that.

Good images depend upon the writer's having looked and listened and touched the world and having searched out the words which really name or describe that world. When the words are found, they connect readers to the same reality.

EXERCISE

1. What figures of speech are found in the following passages? Do you think each is effective or not?

A. The tension, on that first fair morning in September when we drove him to school, almost blew the windows out of the sedan.

E. B. White

B. Thanks to technological progress, Big Brother can now be almost as omni-present as God.

Aldous Huxley

C. Another time, we went to Mannheim and attended a shivaree—otherwise an opera—the one called "Lohengrin." The banging and slamming and booming and crashing were something beyond belief, the racking and pitiless pain of it remains stored up in my memory alongside the memory of the time that I had my teeth fixed.

Mark Twain

D. But hunger gradually dried both his throat and his heartstrings. . . .

Lawrence Durrell

E. We lived always in the stretch or sag of nerves, either on the crest or in the trough of waves of feeling.

T. E. Lawrence

F. Both the good and the bad civilization cover us as with a canopy, and protect us from all that is outside. But a good civilization spreads over us freely like a tree, varying and yielding because it is alive. A bad civilization stands up and sticks above us like an umbrella—artificial, mathematical in shape; not merely universal, but uniform.

G. K. Chesterton

G. And I supposed there was some sealed-off hold where the base of our social pyramid rested; where tourists were chained to the kelson under the whips of savage taskmasters, while their flesh was subdued by a diet of weevily biscuit and stale water.

William Golding [Of third-class passengers on an ocean liner]

H. But those who know a really good woman are aware that she is not in a hurry to forgive, and that the humiliation of an enemy is a triumph to her soul.

W. M. Thackeray

I. It was still dark and jumping up in terror he stretched out his hands like the blinded Polypheme seeking for Odysseus.

Lawrence Durrell

J. When the Court Chapel opened there was an oceanic forward crush.

Frederic Morton

K. By our own act we were drained of morality, of volition, of responsibility, like dead leaves in the wind.

T. E. Lawrence

L. Gradually you experience a sensation that is certainly one of the most extraordinary man has ever felt. You are transcending human nature. You feel immeasurably superior to the crawling beings in the miniature world immersed in silence two thousand feet below.

Ad for an airplane

M. His face was webbed; in fact, the wrinkles were so dense that it seemed all expression was caught in a net.

Sharon R. Curtin

N. By the law of Nature, too, all manner of Ideals have their fatal limits and lot; their appointed periods, of youth, of maturity or perfection, of decline, degradation, and final death and disappearance.

Thomas Carlyle

O. She [a third-grade teacher] cooks for the children on the stove that heats the room, and she can cool their passions or warm their soup with equal competence.

E. B. White

P. When the braggart's accomplishments turn out wrong, he is perfectly willing to give someone else the credit.

Student

Q. He is eating an omelette, she has chosen a sole. He watches the bones she removes daintily from her mouth and counts them like so many crimes on her part.

Colette

R. For prose is so humble that it can go anywhere; no place is too low, too sordid, or too mean for it to enter. It is infinitely patient, too, humbly acquisitive. It can lick you up with its long glutinous tongue the most minute fragments of fact and mass them into the most subtle labyrinths, and listen silently at doors behind which only a murmur, only a whisper is to be heard. With all the suppleness of a tool which is in constant use it can follow the windings and record the changes which are typical of the modern mind.

Virginia Woolf

S. The Fourth of July was a death and a resurrection, the very Easter of our national spirit.

Catherine Drinker Bowen

2. Compose two sentences using zeugma, and two others which contain puns.

3. In your next two or three themes include one or two examples each of metaphor, simile, and irony. Identify the passages in the margin.

4. Using the passage by Thomas Mann on page 411 as a model, compose a brief description (about 100 words) of the cheeses in the dairy case of a supermarket or of fruits and vegetables or seafood. Perhaps you would prefer to write about the window display of a hardware store or of a yarn and fabric shop—anyplace will do where a profusion of objects are spread out on display. The essential thing is to be precise: give names, shapes, sizes, colors.

CHAPTER 44

Unusual Words and Collocations

INTRODUCTION

In this chapter we shall consider remarkable nonfigurative diction—that is, diction which does not involve an extension or alteration in the meaning of a word. All the examples use words employed in their literal senses which nonetheless are unusual. In some cases the word is simply very uncommon, one that most of us have to look up. Others employ ordinary words used in extraordinary senses or in unconventional collocations.

(A *collocation* is simply a group of words making a small unit of meaning within the larger framework of a clause or sentence. For example, in the sentence "Ambitious people seek a place in the sun," the phrase "a place in the sun" is a collocation, a conventional and predictable one. In the sentence "Wise people seek a place in the shadows" the phrase "a place in the shadows" is an unusual collocation.)

A writer using such diction—using it effectively, that is—is not just showing off, trying to impress us with what he or she knows. Such diction strikes us not only because it is uncommon but also because it says exactly what the writer wants to say. Like a good simile or metaphor, an unpredictable word or combination of words conveys a fresh and valuable idea or perception or feeling. It stretches our minds to accommodate something new.

Before we look at examples, however, a caution. In exposition the most desirable quality in diction, next to precision, is simplicity. Simplicity does not mean, of course, simple-mindedness. It means that diction ought not to be any more difficult than the subject requires.

Stressing the advantages of uncommon words and collocations may seem to contradict the principle of simplicity. Actually, it is less a contradiction than a qualification. Unusual words, words that send readers to the dictionary, are valuable when they are precise and interesting and surprising—*and infrequent.* An occasional new word or combination pleases and stimulates readers. If every sentence contains one, all including the most dogged will turn wearily away.

Do not be afraid, then, to experiment with odd words and phrases. But keep the experiment under tight control.

UNUSUAL WORDS

In many cases a strikingly uncommon word comes from Greek or Latin:

> The average autochthonous Irishman is close to patriotism because he is close to the earth.
>
> G. K. Chesterton

> The *Morro Castle* drifted broadside on a sand bar, a few yards off the huge lacustrine Convention Hall, at the foot of Sixth Avenue.
>
> William McFee

Autochthonous derives from Greek and means the condition of being a native, one born in a particular region. *Lacustrine,* from Latin and meaning "relating to a lake (or other body of water)," here signifies that the Hall (in Asbury Park, N.J.) stood on pilings in the sea.

In other instances the odd word may be an older, seldom-used English one or a borrowing from a contemporary foreign language or an expression from the technical vocabulary of a profession or business:

> We stood there mumchance and swallowing, wondering what the devil this construction was.
>
> Lawrence Durrell

> For when the Commodore roused his starboard watch at 5:14—having given them an hour and a quarter as a lagniappe—there was a good feeling of having turned a corner unaware. . . .
>
> Christopher Morley

> Very minute are the instructions of the Government for the disposal, wharfage, and demurrage of its dead. . . .
>
> Rudyard Kipling

Mumchance, an old English word seldom heard today, means "silently," and here implies a shocked, stunned silence. *Lagniappe* is borrowed from American Spanish and is common in Louisiana though not elsewhere; it signifies a little gift, something extra thrown in for good will, like the thirteenth roll in a baker's dozen. *Wharfage* and *demurrage,* finally, are terms from commerce and refer to the charges for using a wharf and for delaying a ship or other carrier. Kipling is implying the callousness of the government toward those who died in its service.

All such words should be used with caution. Foreignisms are especially liable to abuse. When you first acquire words like *Weltanschauung* (roughly, "world view," "world philosophy") or *raison d'être* ("reason for being") it is tempting to show them off. Purpose does sometimes legitimize foreignisms. No exact English equivalent may exist, and even if it does, the context may justify the foreign phrase. But be sure the justification is real. Otherwise you will sound pretentious.

UNUSUAL MEANINGS

Now and then unusualness resides not in the rarity of the word but in that of the meaning with which it is used. A writer may restore to a word an older sense, closer to its etymology. Robert Frost, for instance, writing about the United States, speaks of the "land realizing itself westward." Most readers, accustomed to understanding *realize* in the sense of "to understand clearly," must pause a moment to grasp that the poet is employing the term in the older sense of "to make real." Thus the nation made itself "real" as it expanded into the west.

In the following sentence *imagination* does not have its common meaning of "creative faculty" but instead signifies the products of that faculty:

> Universities flourished; scholars wrote their profundities and novelists their imaginations.
>
> Morris Bishop

Occasionally you can make a commonplace word striking by shifting it out of its usual grammatical function. For instance, in this passage the writer uses *indestructible* as a noun in describing the coming of spring:

> Under the spruce boughs which overlay the borders, the first shoots of snowdrops appeared, the indestructible.
>
> E. B. White

NEOLOGISMS

Neologisms constitute a special class of rare words. Literally "new words," they are made up by the speaker or writer. Some are new in the sense of being original combinations of phonemes (that is, sounds). In the following passage James Thurber uses several neologisms to describe the family car being hit by a trolley:

> Tires booped and whoosed, the fenders queeled and graked, the steering wheel rose up like a spectre and disappeared in the direction of Franklin Avenue with a melancholy whistling sound, bolts and gadgets flew like sparks from a Catherine wheel.

Thurber's coinages are *onomatopoeic* (directly imitating sound). In the next example the neologism is formed by adding a suffix which does not conventionally go with the word (and in the process making a pun):

> But once there came to "the grey metropolis" a Finnish lady—a most perfect representative of non-Aryan beauty and anythingarian charm—to whom not only men, but what is more wonderful, most women, fell captive the moment they saw her.
>
> George Saintsbury

Probably most neologisms are novel compound words. An historian, for instance, describes the most remarkable quality of a particular statesman as

> his "you-be-damnedness."
>
> Barbara Tuchman

A traveler in Sicily complained about the crude duckboards placed for tourists around an excavation of beautiful mosaics that

> It was a groan-making thing to do and only an archeologist could have thought of it.
>
> Lawrence Durrell

Such constructions are called *nonce compounds,* in distinction to conventional compounds, which everybody uses, like *teen-ager* and *school-boy.* Nonce compounds are hyphenated by most writers, unlike conventional compounds, some of which are hyphenated while others are written as one word. Occasionally a nounce compound consists of a number of words strung together in a phrase acting as a single grammatical unit (usually a modifier) like the ten-word adjectival in this sentence (it modifies a three-word noun):

> I doubt whether even the breathless, gosh-gee-whiz-can-all-this-be-happening-to-me TV-celebrity-author could cap this shlock classic with another.
>
> Pauline Kael

THE ADVANTAGES OF RARE WORDS

Whether a neologism or a derivative from another language or from an earlier stage of English, a rare word catches our attention. Sometimes this is the only essential function it performs. Chesterton's *autochthonous,* for example, could easily be replaced, with no loss of meaning, by the less rare *indigenous* or by the common *native.* But *autochthonous* has the virtue of grabbing our eye and of making the statement memorable.

Not only are they eye-catching; such words are also interesting. Many people enjoy meeting new words, especially exotic ones. Some authors—Chesterton among them—are well known for their cleverness in deploying odd words:

In the thirteenth century Bishop Henry of Liege had sixty-one children, fourteen of them within twenty-two months, setting perhaps a record for clerical philoprogenitiveness.

Morris Bishop

Every time a new revolutionist gives a show he issues a manifesto explaining his aims and achievements, and in every such manifesto there is the same blowsy rodomontadizing that one finds in the texts of the critics.

H. L. Mencken

On the other hand, the chief advantage in using an unusual word may not be so much that it is stimulating or interesting (though such words always are) as that it is precise and economic, saying what no other single term can express. Learned Greek and Latin words—while they often appear forbidding—are capable of expressing complex ideas very efficiently. *Lacustrine* in McFee's sentence, for instance, takes the place of a long phrase of simpler words. Another example occurs in the following sentence. The writer is discussing the fact that in fifteenth-century England the training of lawyers moved from Oxford to the Inns of Court in London. He describes the change as

a shift which underlies the increasing laicization of literate culture.

Geoffrey Hindley

It would be possible to state the idea conveyed by *laicization* in another way, but it would need a ponderous phrase like "availability to the general public."

Occasionally, too, big words are used for humor. For instance, here an essayist describes greeting an elderly "great lady":

I lifted my hat in the air; and then seizing the cinque-digitated paw of this female, I moved it up and down several times, giving utterance to a set formula of articulated sounds.

Logan Pearsall Smith

By setting *cinque-digitated* alongside *paw* Smith comments with witty irony upon pretension and reality. The novelist Lawrence Durrell also uses big words ironically in the following sentence to mock the kind of writing that leans too heavily upon them, a style he describes as

the complicated tortuosities of platonic phraseologists. . . .

Durrell's parody points out that big words ought not to be a mannerism of style. They are effective in inverse ratio to their frequency. In the examples we have looked at, the rare terms work because they are placed in the context of more everyday diction. Do not trot out an unusual word just to get one up on your readers. Use such diction sparingly and only when it counts, catching the reader's eye for a legitimate reason or expressing an idea precisely and economically.

UNUSUAL COLLOCATIONS

An unusual collacation is an unlikely and remarkable combination of words, each commonplace enough in itself but rarely used with the other(s). This description of a Midwestern industrial plant is an example:

Republic Steel stood abrupt out of the flat prairie.

Howard Fast

We do not think of buildings as "standing abrupt," but for that very reason the diction is memorable, just like the structures it describes rearing dominantly out of the level land.

Here are several other instances:

. . . the crackling sea. . . .

Dylan Thomas

The clammy hauteur of President Hoover. . . .

Arthur M. Schlesinger, Jr.

Under the trees, along the cemented paths go the drifts of girls, sympathetic and charming. . . .

William Golding [Of students at a southern college]

Any grammatical combination may serve as an unusual collocation; a subject and verb, for instance:

But her smile was the *coup de grâce* and her sigh buried him deep.

W. Somerset Maugham

Or verb and complement:

He smiles his disappointments and laughs his angers.

e. e. cummings

UNUSUAL VERBS

Verbs, in fact, are a fertile source of implied meaning when collocated with unlikely subjects or objects:

But the weeks blurred by and he did not leave.

Willard R. Espy

. . . no bird song splintered the sunflecked silence.

Joan Lindsey

Often an unusual verb conceals a comment:

The more we prattle about morality, the more the world shows us how complicated things really are.

Samuel C. Florman

The cops squealed with excitement.

Howard Fast

The stock market crash demolished the crazy structure of Coolidge prosperity and the expatriates began to scurry homeward.

Carey McWilliams

. . . and then the hideous mannequins galumphed with squeaky shoes on stage.

Nancy Mitford

Each of these verbs is chosen because of its adverse connotations. *Prattle* suggests childishness; *squealed,* piglike; *scurry,* rats or insects; *galumph,* comic awkwardness. And each enriches the meaning of its passage, implying considerably more than it literally states.

UNUSUAL ADJECTIVES

Many other unusual collocations involve a modifier (typically an adjective) and its headword, as in Dylan Thomas' "the crackling sea." One variety of such adjectives is called a transferred epithet: a word conventionally applied to one noun or class of nouns used to modify instead something associated with that noun:

. . . a boiling kettle, a steaming pot. . . .

He [the philosopher Herbert Spenser] would sit upstairs in his angry overalls, too angry to come down to luncheon.

Harold Nicolson

It is, of course, water that is boiling and steaming, not the kettle and the pot. Spenser's overalls were not angry, he was—he wore the overalls only when he was angry.

OXYMORON AND RHETORICAL PARADOX

When the oddity of a collocation becomes extreme and apparently contradictory it is called *oxymoron.* A famous example is John Milton's description of hell as "darkness visible." In *oxymoron* the modifier seems somehow to contradict its headword: "darkness," which makes things invisible, cannot itself be "visible." Here are several other examples:

. . . a practical mystic. . . .

Lord Roseberry

. . . delicious diligent indolence. . . .

John Keats

A yawn may be defined as a silent yell.

G. K. Chesterton

A *rhetorical paradox* is an oxymoron writ large. (Oxymoron, in fact, has been defined as a "condensed paradox.") It too expresses a strikingly unusual or apparently contradictory idea. It differs only in being longer and not focusing the contradiction in a modifier and headword:

His soul will never starve for exploits or excitement who is wise enough to be made a fool of.

G. K. Chesterton

Rhetorical paradox and oxymoron are distinct from *logical paradox*. The last is a statement which literally contradicts itself; it asserts that something is simultaneously both true and not true, thus violating what logicians call the law of noncontradiction. A classic case is:

"All Cretans are liars," said a Cretan.

A rhetorical paradox, on the other hand, does not contain a true logical contradiction. It may seem to. Chesterton's statement appears to be self-contradictory. So does this one (the writer is arguing against a proposal for mass schoolteaching by radio):

Instruction will be by experts—the invisible experts, speaking with a voice that is not a voice, delivering to the invisible pupil the canned lesson, courtesy of the advertised product.

E. B. White

But even when a rhetorical paradox does seem to assert two mutually exclusive ideas—as in White's "voice that is not a voice"—the seeming logical contradiction disappears when we realize that a key term is deliberately being used in different senses. Thus in White's sentence "voice" means (1) the literal voice on the radio and (2) a human presence—warm, reassuring, communicative. A "voice" in the first sense is not and cannot be, White argues, a "voice" in the second. There is no real contradiction at all.

In Chesterton's case the paradox depends not upon giving the same word different meanings, but rather upon using two key terms (*wise* and *fool*) in special, thought not unique, senses. By *wise* Chesterton means simple and pure in spirit, unwordly and good; by *fool*, an innocent and trusting victim rather than the self-deluded egotist, the word's more usual meaning.

Another kind of rhetorical paradox is not so much an apparent self-contradiction as an actual contradiction of a commonly accepted belief. The following sentence, which opens a student theme, is an example. It contains no inner contradiction, but it violently disagrees with conventional attitudes toward baseball:

Baseball is an interminable game played by fools who have nothing better to do, for the amusement of idiots who have nothing to do at all.

Paradoxes of this sort may take the form of standing a cliché or popular maxim on its head. Someone remarked, for instance, that the German General Staff "has a genius for snatching defeat from the jaws of victory." Oscar Wilde mocked Victorian morality by reversing the smug judgment that "drink is the curse of the working class"; he put it that

work is the curse of the drinking class.

And an historian remarks upon the gross ineptitude of a British politician:

He had triumphed over all his advantages.

Thomas Pakenham

Oxymoron and rhetorical paradox, then, are especially striking kinds of unusual collocation. When they grow naturally out of the subject and reveal something important about it, they are very effective. Used too often or superficially, without any significant connection to the point, they succeed only in being annoying.

ACCUMULATION, OR PILING UP

Accumulation for our purposes here means stringing together a number of words, all belonging to the same part of speech and grammatically parallel—that is, connected to the same thing. Most commonly the words are a series of verbs serving one subject or of adjectives attached to the same headword:

They glittered and shone and sparkled, they strutted, and puffed, and posed.

Beverley Nichols [About Yugoslavian army officers]

He criticized and threatened and promised. He played the audience like an organ, stroked them and lashed them and flattered and scared and comforted them, and finally he rose on his toes and lifted his fists and denounced that "great betrayer and liar," Franklin Roosevelt.

Wallace Stegner [About the political priest Father Coughlin]

Lolling or larricking that unsoiled, boiling beauty of a common day, great gods with their braces over their vests sang, spat pips, puffed smoke at wasps, gulped and ogled, forgot the rent, embraced, posed for the dickey-bird, were coarse, had rainbow-coloured armpits, winked, belched, blamed the radishes, looked at Ilfracombe, played hymns on paper and comb, peeled bananas, scratched, found seaweed in their panamas, blew up paper bags and banged them, wished for nothing.

Dylan Thomas [About Welshmen on a seaside holiday as observed by a child]

Manipulative, industrious, strangely modest, inexorable, decent, stodgy, staunch, the Habsburgs had come out of Switzerland in 1273.

Frederic Morton [About the reigning house of Austria]

How, people are asking, could four mopheaded, neo-Edwardian attired, Liverpudlian-accented, guitar-playing, drumbeating "little boys" from across the ocean come here and attract the immense amount of attention they did by stomping and hollering out songs in a musical idiom that is distinctly American?

John A. Osmundsen [About the Beatles]

The unusualness and success of such collocations lies not so much in unconventional or paradoxical combinations as in the sheer quantity of words and, of course, in their quality.

MIXED LEVELS OF USAGE

Level of usage refers to the fact that some words have a limited appropriateness. They are suitable, say, for formal but not informal occasions (*pedagogue,* for instance). Contrarily, other words are at home in an informal atmosphere but not a formal one (*prof*). Of course, most of the words we use are acceptable at any time (*teacher*), and level of usage does not really apply to them.

One way of achieving unusualness in diction is to mix words from different usage levels so that learned literary terms rub elbows with colloquialisms and slang:

> Huey [Long] was probably the most indefatigable campaigner and best catch-as-catch-can stumper the demogogically fertile South has yet produced.
>
> Hodding Carter

> American perceptions of empire have decline and fall built in. Decline and fall are both the outcome of and the alternative to empire. Which puts Americans in a fine pickle today.
>
> James Oliver Robertson

Today the distinction between formal and informal styles is not maintained so rigidly as it used to be. Many writers mix literary and colloquial expressions with a freedom that would have been frowned upon a generation or two back. While this freedom is welcome, it poses its own problems. It must be exercised responsibly. The mix must work. It cannot be merely an artificial forcing of an occasional bit of slang to relieve a relatively formal style, or, contrarily, shouldering in big Latinate words now and then to decorate a colloquial style. Always words should be chosen because they say what you want to say.

When the mix does work, a writer achieves not only precision but a pattern of diction interesting in itself. Here, for example, is a critic discussing modern detective fiction:

> The moral fabric of any age, of any society, is a tapestry in which there are strikingly different and even antithetical motifs. Our popular art forms show that the prevailing fashion in heroes runs to the extroverted he-man, the tough guy who saves the world with a terrific sock on the jaw of the transgressors, and the bang, bang of his pistol. But even this generation, so much exposed to philosophies of power, has its hankering for the light that comes from within; and in its folklore there appears, intermittently, a new kind of priest-hero—the psychoanalyst.
>
> Charles J. Rolo

Rolo's language is generally literary and formal: "moral fabric," "antithetical motifs," "transgressors," "philosophies of power," "intermittently," "priest-hero," "psychoanalyst." At the same time he uses colloquialisms: "he-man," "tough guy," "terrific sock on the jaw," "hankering." The diction is thus unpredictable. It surprises and therefore pleases us.

But also the colloquialisms are appropriate to the topic. They suit the kind of detective fiction Rolo is discussing. *Virile male,* for instance, means much the same thing as *he-man,* but it does not convey the strong flavor of the tough private-eye.

Formal and colloquial words may be played off against each other in even more striking fashion. In the following passage the journalist A. J. Liebling describes fans at a prizefight, specifically those rooting for the fighter other than the one he favors:

> Such people may take it upon themselves to disparge the principal you are advising. This disparagment is less generally addressed to the man himself (as "Gavilan, you're a bum!") than to his opponent, whom they have wrongheadedly picked to win.

The contrast is comic between the deliberately inflated diction in which Liebling describes the fans' behavior ("disparage the principal you are advising") and the language they actually use (" 'Gavilan, you're a bum!' ").

EXERCISE

1. What is unusual about the diction of the following sentences? Look for odd words or meanings and unconventional combinations or accumulations.

A. Graceful, athletic and playful, spontaneous and high-spirited, prettily impulsive, generous and affectionate, she possessed a most engaging charm.

D. Harris Willson

B. The slums [of Vienna] sleazing through the town's south and west. . . .

Frederic Morton

C. They advanced, retreated, struck at one another's hands, clutched at one another's heads, spun round alone, caught one another and spun round in pairs until many of them dropped.

Charles Dickens

D. [There were church bells] in the bat-black, snow-white belfries, tugged by bishops and storks. And they rang their tidings over the bandaged town, over the frozen foam of the powder and ice-cream hills, over the crackling sea. It seemed that all the churches boomed for joy under my window; and the weathercocks crew for Christmas, on our fence.

Dylan Thomas

E. The wind shrieks and hisses down the valley sonant and surd, drying up puddles and dismantling the trees.

Annie Dillard

F. Whatever the explanation, it is a bleak era for love, which makes it a time of dull joys, small-bore agonies and thin passions.

Russell Baker

G. It is possible, as I happen to know, to think pacificism a very direct menace to peace.

G. K. Chesterton

H. Here was a suitable enemy—powerful, mysterious, international, aggressive.

Robert Coughlan

I. The satirist's occupational disease is intellectualism, a detachment so poised that it slides into a withdrawn superiority.

Kingsley Amis

J. . . . a bread and butter paradise. . . .

W. M. Thackeray

K. This flashy vehicle was as punctual as death: seeing us waiting at the cold curb, it would sweep to a halt, open its mouth, suck the boy in, and spring away with an angry growl.

E. B. White [Can you guess what is being described?]

L. The best of his work can be found in the worst of his work.

G. K. Chesterton

M. Going to sea became the last resort of the dregs of the waterfront, the vicious, the improvident, the incompetent, and the irresponsible.

William McFee

N. "Fire!" cried Mrs. Prothero, and she beat the dinner-gong. And we ran down the garden, with the snowballs in our arms, towards the house, and smoke, indeed, was pouring out of the dining-room, and the gong was bombilating, and Mrs. Prothero was announcing ruin like a town crier in Pompeii.

Dylan Thomas

O. Conservative, goateed, seventy-year-old Joseph Ransdall, the incumbent, whom Huey dubbed, "Feather Duster," burbled unavailingly.

Hodding Carter

P. It was a big, squarish frame house that had once been white, decorated with cupolas and spires and scrolled balconies in the heavily lightsome style of the seventies, set on what had once been our most select street.

William Faulkner

Q. As a boy he was studious, stubborn, rugged, enduring, persistent, wholly concentrated and with little regard for the comforts or amenities of life.

Roger Burlingame

R. The [Roman] soldier's load weighed more than eighty pounds, which he had to hump for fifteen or twenty miles in all weathers. . . .

Robert Graves

S. I spent some of the quietest Sundays of my life in Uncle Amos's yard, lying under apple trees and not listening to Uncle Amos who was bumbling away at something he did not expect me to listen to at all.

Robert P. Tristram Coffin

T. The performance of the Washington establishment has been pathetic, aimless, ignoble, listless, ignorant, and self-indulgent while the problem has gone from worse to intolerable.

Jack Anderson

2. Modeling your work upon example A in the preceding exercise, complete the following sentences with strings of appropriate adjectives. The blanks are merely suggestive; you are free to use one or two adjectives less or more. Study how Willson groups his modifiers and experiment with similar groupings.

1. _____, _____, _____, _____, _____, he was a dangerous enemy.
2. _____, _____, _____, _____, _____, the party was one of the nicest I ever went to.
3. _____, _____, _____, _____, _____, my uncle was an unusual man.

3. Using item C in the first exercise as a model, describe a crowd at a football game (or any other well-attended event). Begin with "They" and follow the subject with a series of expressive verbs.

4. In your next theme include one or two examples of unusual words and a rhetorical paradox. Indicate the passages in the margin.

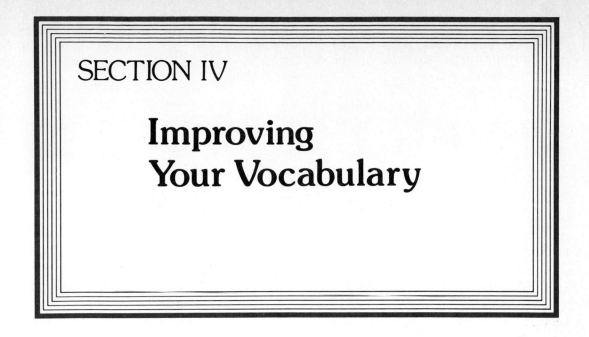

SECTION IV

Improving Your Vocabulary

45

Dictionaries
and Thesauri

INTRODUCTION

Vocabulary is best improved in practice, in the activities of reading and writing. Memorizing lists of words has dubious value. The words are abstracted from any context so that while you may learn a denotation, you acquire little feeling for connotation and level of usage and consequently may misuse the word in subtle ways. Your vocabulary should not be a forced exotic growth but a natural one, enlarging with your knowledge and experience.

Keep a good dictionary handy as you read. When you come upon a strange word, pause and look it up. If you cannot stop or have no dictionary nearby, at least lightly check the margin (assuming the book is your own) or write the word on a piece of paper to remind yourself to consult a dictionary later. Without such a reminder you will probably forget or remember only that there was a word you wished to look up which has now fled your mind. It is best, however, to go to the dictionary before reading on. And as you write, don't be content with "thinking" you know the spelling, grammatical function, or meaning of a word. If you are not sure, look it up. Look up, too, the errors in diction marked on your corrected themes, and should you still not understand why a word is wrong, ask your instructor.

The dictionary is the key to improving your vocabulary. To use it effectively you need to know what a dictionary is and what it contains. In the rest of this chapter we discuss exactly that. Mostly we talk about general dictionaries—those that list all or most of the words in the language. But we shall also glance at thesauri, a special kind of dictionary listing synonyms—that is, words having similar meanings.

GENERAL DICTIONARIES

A general dictionary contains the words currently used by the speakers and writers of a language or which readers are likely to come across in older literature. If it lists all such terms, it is called *unabridged*. If it reduces its contents by omitting many technical or obsolete words, it is an *abridged* edition, sometimes called a desk dictionary.

Two unabridged dictionaries are standard for modern English: *Webster's Third New International Dictionary* (G. & C. Merriam Company) and the *Oxford English Dictionary* (Oxford University Press), familiarly known as the *OED*. We shall return to these massive works a bit later.

Of more immediate concern is the abridged dictionary. Several excellent dictionaries of contemporary American English are suitable for college work. If you do not already own a good desk dictionary, you should purchase one. (Ask your teacher's advice.) A proper dictionary is indispensable, not only for English class, but for all your subjects. Do not succumb to false bargains. The $1.39 special picked up in a drugstore is probably based on a word list long out of date and gives you nothing like the full range of information a good dictionary offers. Even an older copy of a reputable dictionary will be less helpful than the current edition, for it will not include recent terms from science and technology.

Whichever dictionary you own, take an hour or so to read its table of contents and to familiarize yourself with its organization. A typical dictionary contains three parts: the front matter, the actual word list, and the back matter or appendices. Front matter includes all material preliminary to the word list. This varies in different dictionaries, but in all cases it explains how the word list is set up, how to read an entry, what the abbreviations stand for, and so on.

Do not skip the front matter, assuming that because you are familiar with one dictionary you know how all others work. Significant variations do exist. For example, the order in which the several meanings of a word are arranged depends upon decisions made by the lexicographer. *Webster's Seventh New Collegiate Dictionary* (G. & C. Merriam Company) explains that the "order of senses is historical: the one known to have been first used in English is entered first."[1] *The American Heritage Dictionary of the English Language* (Houghton Mifflin Company), on the other hand, does not follow an historical sequence: rather "the first definition is the central meaning about which the other senses may be most logically organized."[2]

In addition to the explanations about the word list, front matter also includes general information about spelling, pronunciation, grammar, and so forth. *The American Heritage Dictionary of the English Language* contains valuable discussions of the origin and history of English, of usage, grammar, spelling, and dialects, and of how computers help to analyze words for the lexicographer. The articles give you an excellent introduction to how contemporary scholars think about language and dictionaries.

1. "Explanatory Notes," 11.5, p. 12a.
2. "Guide to the Dictionary: Order of Definitions," p. xlvii.

Back matter, too, varies from book to book. *Webster's Seventh New Collegiate Dictionary,* for instance, discusses punctuation in its back matter and includes lists of famous persons, of important places, and of colleges in the United States and Canada. *The American Heritage Dictionary of the English Language* does not treat punctuation and inserts the names of people and places into its general word list (colleges it does not identify at all).

Although the front and back matter contains much important information, the chief part of a dictionary is its word list. To use the word list to best advantage, you need to understand how entries are organized and the kind of information they give. To begin with, you should realize that while a dictionary is an authority, it is an authority only in a very special sense. It does not tell you how a word *ought* to be spelled or spoken or used, it tells you how it *is* spelled, spoken, and used. The form and meaning of a word depend upon the speakers and writers of English. Acting as an unconscious collectivity, they—or rather we, all of us—constitute the "authority." Lexicographers collect hundreds, even thousands, of citations for each word they list. From these they determine how a word is actually spoken and spelled, what it is used to mean, and any regional, occupational, or social variations affecting its use. If a lexicographer has personal feelings about the "proper" spelling, pronunciation, or meaning of a word, he or she does not substitute these for the facts revealed by the citations.

An example of this objectivity can be seen in the word *ain't.* Despite the old witticism that "*ain't* ain't in the dictionary," it does appear in good contemporary dictionaries. And it should, for *ain't* is a common word. A dictionary will tell you that it is a contraction of *am not* and that it is regarded as substandard. You may think that *ain't* should never be used at all, or you may feel that it is a useful contraction which should never have been outlawed. But lexicographers, as lexicographers, will not and cannot express either of these opinions; they must simply record the facts about *ain't* as their samples reveal them.

The exact arrangement of information in a typical entry varies from dictionary to dictionary. Faced with conveying much information in little space, lexicographers depend upon numbers, letters, punctuation, and differences in typeface to organize and clarify an entry. Somewhere in the introductory material they carefully explain how they use these typographical devices.

In all dictionaries the entry lists a word according to an explained principle of alphabetization, indicates its spelling (including any variations), stress, syllabication, pronunciation, its grammatical function (noun, verb, adjective, and so on), the various senses it bears, and usually its etymology (its origin and history). Here are two sample entries, each with explanations. The first comes from *Webster's Seventh New Collegiate Dictionary.*

¹hab·it \ˈhab-ət\ *n* [ME, fr. OF, fr. L *habitus* condition, character, fr. *habitus,* pp. of *habēre* to have, hold—more at GIVE] **1** *archaic* : CLOTHING **2 a** : a costume characteristic of a calling, rank, or function **b** : RIDING HABIT **3** : BEARING, CONDUCT **4** : bodily appearance or makeup : PHYSIQUE **5** : the prevailing disposition or character of a person's thoughts and feelings : mental makeup **6** : a usual manner of behavior : CUSTOM **7 a** : a behavior pattern acquired by frequent repetition or physiologic exposure that shows itself in regularity or increased facility of performance **b** : an acquired mode of behavior that has become nearly or completely involuntary **8** : characteristic mode of growth or occurrence **9** *of a crystal* : characteristic assemblage of forms at crystallization leading to a usual appearance **10** : ADDICTION

syn HABIT, HABITUDE, PRACTICE, USAGE, CUSTOM, USE, WONT mean a way of acting that has become fixed through repetition. HABIT implies a doing unconsciously or without premeditation, often compulsively; HABITUDE implies a fixed attitude or usual state of mind; PRACTICE suggests an act or method followed with regularity and usu. through choice; USAGE suggests a customary action so generally followed that it has become a social norm; CUSTOM applies to a practice or usage so steadily associated with an individual or group as to have the force of unwritten law; USE and WONT are rare in speech, and differ in that USE stresses the fact of repeated action, WONT the manner of it.

²habit *vt* : CLOTHE, DRESS

Main entry
  Superscript ¹ indicates that this is the first of two or more homographs (words having the same spelling and sound but used in different senses).
  The dot marks the syllabication. If you must split a word between lines, break it only at a point indicated by a dot.
Pronunciation
  In this dictionary the pronunciation is placed between slash marks and rendered in phonetic symbols (mostly similar in form to letters) whose values are listed at the bottom of each recto (right-hand) page.
  The mark ' indicates stress. It is placed before the accented syllable (that is, the one spoken with greatest force).
Part of speech
  n = noun.
Etymology
  Placed within brackets, the etymology uses capital abbreviations for languages and lower case abbreviations for other words: thus ME = Middle English, OF = Old French, L = Latin, fr. = from and pp. = past participle. Foreign words are italicized and their meanings are given in roman type without quotation marks.
  SMALL CAPS, here and elsewhere throughout the entry, signal that a term should be consulted in its alphabetical place in the word list for further information relevant to *habit.*
Definitions
  In this dictionary definitions are arranged in historical order. Different senses are distinguished by boldface arabic numerals; nuances within the same sense, by boldface lower case letters.
  Archaic is a status label indicating that a word or, as in this case, a particular sense of a word is used very rarely by contemporary speakers and writers.
  Of a crystal is a subject label indicating a special sense of the word in a particular subject or profession, here crystallography.
Synonyms
  A discussion of a group of words similar in sense but subtly different in meaning or usage. After the entry in the main word list of each of the terms in small caps following *habit,* there is a reference to this discussion. Thus at the end of the entry for *custom* you will find "*syn* see HABIT."
Homograph of *habit,* here a transitive verb meaning to clothe, to dress.

The second example comes from *The American Heritage Dictionary of the English Language.*

wake[1] (wāk) *v.* **woke** (wōk) or *rare* **waked** (wākt), **waked** or *chiefly British & regional* **woke** or **woken** (wō′kən), **waking, wakes**—*intr.* **1. a.** To cease to sleep; become awake; awaken. Often used with *up.* **b.** To be brought into a state of awareness or alertness. **2.** *Regional.* To keep watch or guard, especially over a corpse. **3.** To be or remain awake.—*tr.* **1.** To rouse from sleep; awaken. Often used with *up.* **2.** To stir, as from a dormant or inactive condition; rouse: *wake old animosities.* **3.** To make aware of; to alert. Often used with *to: It waked him to the facts.* **4.** *Regional.* **a.** To keep a vigil over. **b.** To hold a wake over.—*n.* **1. a.** A watch; vigil. **b.** A watch over the body of a deceased person before burial, sometimes accompanied by festivity. **2.** *British.* A parish festival held annually, often in honor of the patron saint. **3.** The condition of being awake: *between wake and sleep.* [Middle English *wakien* and *waken,* Old English *wacian,* to be awake and *wacan* (unattested), to rouse. See **weg-**[2] in Appendix.*]

Main entry
The superscript [1] indicates that this is the first of at least two homographs, different words with the same spelling and pronunciation but different senses.

Pronunciation
This is enclosed within parentheses and uses symbols and marks set out in a table at the beginning of the word list.

Part of speech
v. = verb (*intr.* = intransitive and *tr.* = transitive); *n.* = noun.

Inflected forms
For verbs these are the principal parts. (For nouns they would be the singular and plural, for modifiers the comparative and superlative forms.) As listed in this dictionary the principal parts, set in boldface, include the past preterite (*woke*), the past participle (*waked*), the present participle (*waking*), and the third person singular active indicative present (*wakes*). Alternate forms are given for the past and past participle, with the less common following the more common and labeled as *rare* or *chiefly British & regional* (that is, confined to the speakers of a particular geographical area rather than common to all users of English).

Definitions
These are divided into the senses of the verb and of the noun. The former, in turn, are distinguished for both the intransitive and transitive uses of the verb. Within each category the various meanings are ordered, in this dictionary, by beginning with the most common or central. Different senses are marked by arabic numerals in boldface; subdivisions within a particular sense by lower-case letters in boldface. Where useful, brief examples of a sense are given in italics.

Etymology
The etymology, set within brackets, traces the origin of the modern word. Foreign terms are italicized, and their meanings are in roman type without quotation marks. "Unattested" means that no actual record of a form exists,

Usage: The verbs *wake, waken, awake,* and *awaken* are alike in meaning but differentiated in usage. Each has transitive and intransitive senses, but *awake* is used largely intransitively and *waken* transitively. In the passive voice, *awaken* and *waken* are the more frequent: *I was awakened* (or *wakened*) *by his call.* In figurative usage, *awake* and *awaken* are the more prevalent: *He awoke to the danger; his suspicions were awakened. Wake* is frequently used with *up;* the others do not take a preposition. The preferred past participle of *wake* is *waked,* not *woke* or *woken: When I had waked him, I discovered that the danger was past.* The preferred past participle of *awake* is *awaked,* not *awoke: He had awaked several times earlier in the night.*

wake² (wāk) *n.* **1.** The visible track of turbulence left by something moving through the water: *the wake of a ship.* **2.** The track or course left behind anything that has passed: *"Every revolutionary law has naturally left in its wake defection, resentment, and counterrevolutionary sentiment."* (C. Wright Mills). **—in the wake of. 1.** Following directly upon. **2.** In the aftermath of; as a consequence of. [Probably Middle Low German *wake,* from Old Norse *vok,* a hole or crack in ice. See **wegw-** in Appendix.*]

though the form may be inferred from other evidence. *weg-²* refers to a list of Indo-European roots contained in an appendix following the word list. (Indo-European is the name given to the mother language of English and most other western languages, as well as of many in the Near East and India. That language does not exist in any written record. However, linguists can reconstruct many of its words or word elements, collectively called roots, from evidence in languages descended from Indo-European.)
Usage
A discussion of how the word and its various forms are actually used by contemporary speakers. The discussion is illustrated by typical cases, printed in italics.

Main entry of **wake²**
Wake², a homograph of **wake¹**, is a different word with a different meaning.
Quoted citation
Rather than a typical example, this is an actual employment of the word, attributed to a specific writer. It is an example of the kind of citation from which the dictionary maker works. Collecting hundreds or thousands of such specific examples of a word, he or she frames the definition.
Idiom using the word.

UNABRIDGED DICTIONARIES

Occasionally you will come upon a word that is not in your desk dictionary, perhaps because it is too old or too specialized. Turn then to an unabridged dictionary. The standard work for American English is *Webster's Third New International Dictionary* (G. & C. Merriam Company). This is the volume you will find in libraries, usually standing on its own pedestal in a place convenient for public use and open somewhere near the middle. (It should be left like that to protect the binding.)

Webster's Third New International contains more than 450,000 words, including many older expressions and technical terms omitted from desk dictionaries. In addition to the customary explanatory notes, its front matter contains extensive discussions of spelling, punctuation, plural forms, the use of italics, and the handling of compound words. Accompanying the word list are thousands of illustrations (a few in the form of color plates) and numerous tables (the chemical elements, for instance, the Indo-European language family, radio frequencies, time zones, and so on).

Even more massive is the *Oxford English Dictionary,* published by the Oxford University Press in twelve volumes with a supplement, of which two volumes have so far appeared. Several features distinguish the *OED.* It contains more older words than the *New International.* It arranges its definitions in historical sequence and copiously illustrates each sense by dated quotations (a total of about 1,800,000). These begin with the earliest recorded use of a word in a particular sense and include, if possible, at least one instance for every century thereafter until the present (or until the last known use in the case of an obsolete word or meaning). The dated citations make the *OED* indispensable to scholars studying the history of words or ideas.

On the other hand, the *OED* is less useful for American English. For example, someone curious about the meaning of *Chicago pool* or the origin of *OK* will find neither expression in the main word list. Both unabridged dictionaries, then, are necessary to students of English. Probably you will not have frequent occasion to consult either, but you would be wise to learn where each is found in your library.

SPECIAL DICTIONARIES: THESAURI

Special dictionaries are those restricted to some particular aspect of the general language or to the language of a specific group, profession, or region. There are hundreds of such works, too many to list here. In your library you will find a reference section presided over by a librarian trained to handle reference material. If you need, say, a dictionary of Americanisms or a dictionary of medicine, geography, or philosophy, the reference librarian is the person to turn to.

Informative introductions to special dictionaries and reference works in general may be found in *The Basic Guide to Research Sources,* edited by Robert O'Brien and Joanne Soderman (New American Library, 1975) or *Reference Readiness: A Manual for Librarians and Students,* second edition (Linnet Books, 1977). In a later chapter, when we talk about the research paper, we list a sampling of special dictionaries.

For the moment only one kind need concern us. This is the dictionary of synonyms, or thesaurus (the word comes from Greek and means "treasure"). *Synonyms* are words in the same language having similar meanings. *True* or *exact* synonyms have identical senses and are usually alternate names for the same object. In sailboats, for instance, *mizzen* and *jigger* both signify the same sail and are therefore exactly synonymous. Most synonyms, however, are less than exact. For example, *pal* and *friend* overlap to a considerable degree but are not precisely coextensive: any pal is a friend, but any friend is not necessarily a pal. In listing synonyms a thesaurus necessarily obscures this distinction between exact and near synonyms. To discriminate all shades of meaning would result in a vast work of dozens of volumes, too expensive to buy and too cumbersome to use.

Probably the best known thesaurus is *Roget's,* first published in 1852 by Mark Peter Roget, an American physician and professor. Dr. Roget titled his book *A Thesaurus of English Words and Phrases, Classified and Arranged so as to Facilitate the Expression of Ideas and Assist in Literary Composition.* Roget devised a system of grouping words in categories of ideas, each category being given numbers and subdivisions. Thus users searching for words meaning, say, "friendship" could look under the appropriate category. In order to make his book usable from the other direction—that is, from word to category—Roget also included an alphabetical index of words, each keyed to its category by the appropriate number. Early in the twentieth century C.O.S. Mawson simplified Roget's scheme. Neither *Roget* nor *thesaurus* is a copyrighted term, and a number of *Roget's* are available today, some revisions of Roget's original work, others of Mawson's modification, and still others alphabetical listings of words without Roget's "categories."

Besides the various *Roget's* there are other thesauri on the market. The best of these are *Webster's Collegiate Thesaurus* (G. & C. Merriam Company) and *Webster's New World Thesaurus* edited by Charlton Laird (World Publishing Company, 1974).[3]

The limitations of any thesaurus are revealed in the directions given in one edition of a *Roget*: "Turning to No. 866 (the sense required), we read through the varied list of synonyms . . . and *select the most appropriate expression*" (italics added). That matter of selection is crucial, and a thesaurus does not offer much help. For example, among the synonyms listed in one *Roget's* under the category *seclusion/exclusion* are the words *solitude, isolation, loneliness,* and *aloofness.* They are merely listed as alternates with no distinctions made, but, except in a very loose sense, these words are not synonymous and may not be interchanged indiscriminately. *Solitude* means physical apartness, out of the sight and sound of others, a condition not necessarily undesirable; in fact, *solitude* may be used with positive connotations, as in "She enjoys solitude." *Loneliness,* on the other hand, has a more subjective significance, relating to the feeling of being apart; it does not necessarily

---

3. Like *Roget's,* the term *Webster's* is not in copyright and may be used by different, and competing, companies.

imply physical separation—one can feel loneliness in a crowd of Christmas shoppers—and it would never be used in a positive sense. *Isolation* stresses the fact of physical separation, out of connection and communication with others, and is often used when that separation is not voluntary, as solitude may be. *Aloofness,* finally, is self-chosen separation, a deliberate withdrawal from others which may connote a sense of superiority, though it does not have to.

To use these words effectively you need to know considerably more about them than a thesaurus tells you. With many words—those in the example, for instance—a good abridged dictionary provides more help. This is not to say that a thesaurus is a waste of money. Used wisely it can improve your diction. It may remind you of a word you have forgotten, or it may acquaint you with a new word. But before you use that new word, learn more about it. At the very least look it up in a dictionary. Do not simply pluck a fancy word from a list of so-called synonyms and plant it in your prose. It may prove a strange growth.

More useful for college work than a thesaurus is a work published by the G. & C. Merriam Company: *Webster's Dictionary of Synonyms.* It offers a fuller discussion of synonyms than does a thesaurus. For example, where *Webster's Collegiate Thesaurus* uses about one inch of a column for *solitude,* the *Dictionary of Synonyms* spends more than seven inches, carefully distinguishing *solitude* from *isolation, loneliness,* and so on.

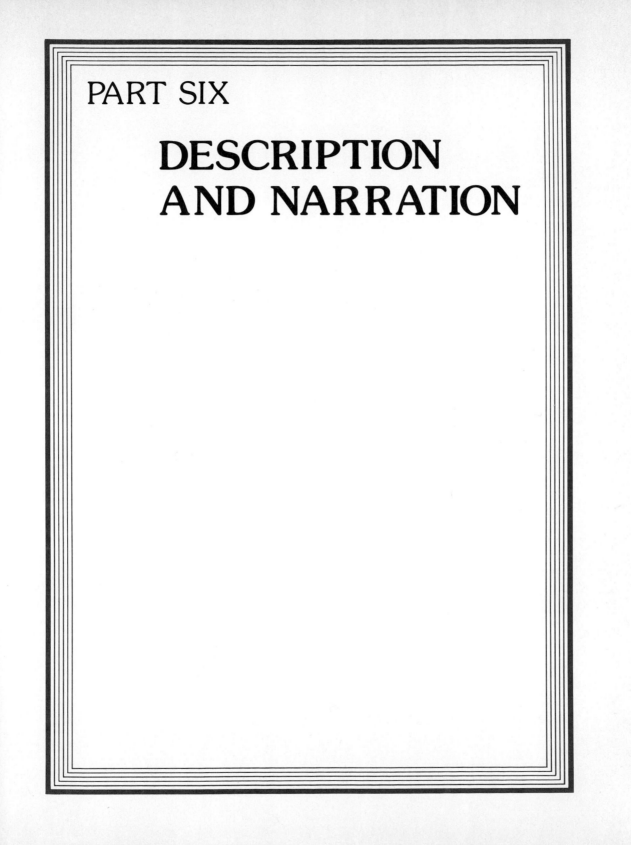

PART SIX

DESCRIPTION AND NARRATION

CHAPTER 46

Description

INTRODUCTION

Description is a cousin to exposition. Both seek to inform the reader. But description embodies sensory experiences—how something looks, sounds, tastes—not, like exposition, what the writer knows or thinks or believes.

While description is not usually the primary purpose of a college paper, it is often important. When you write up an experiment in biology or engineering, say, you will find description necessary. Description may well be needed in an essay for sociology or history. Occasionally English compositions are descriptive. Learning to report what you perceive is, then, an important aspect of writing well.

Most of the time you will report visual perceptions, for vision is the sense writers call upon most. Most of the examples here concern how things look. But description can cover other kinds of perception. It is possible—and sometimes necessary—to describe sounds, smells, tastes, or tactile sensations. In the following passage, for example, the writer uses sounds to describe the beginning of an act of revolutionary violence in China:

> Five shots went off in a nearby street: three together, another, still another. . . . The silence returned, but it no longer seemed to be the same. Suddenly it was filled by the clatter of horses' hoofs, hurried, coming nearer and nearer. And, like the vertical laceration of lightning after a prolonged thunder, while they still saw nothing, a tumult suddenly filled the street, composed of mingled cries, shots, furious whinnyings, the falling of bodies; then, as the subsiding clamor was heavily choking under the indestructible silence, there rose a cry as of a dog howling lugubriously, cut short: a man with his throat slashed.
>
> André Malraux

Whatever sense it deals with, descriptive writing is of two broad kinds: objective and subjective. The difference depends upon whether the emphasis is upon the thing being described or upon the writer's response to it. A writer may set aside those aspects of his perception which are unique to himself and concentrate on describing the object in its own right. This is *objective* description. *Subjective* (also called "impressionistic") description results when a writer deliberately projects his or her feelings into the description, so that readers are made aware of the object not as something in itself, but as something being experienced by a particular observer. In objective description the writer says, in effect, "This is how the thing is." In subjective description he or she says, "This is how it is for me."

Neither kind of description is "truer" than the other. Both are (or can be) true, but they are true in different ways. Objective description is true (or false) in relation to fact; subjective, in relation to feeling or evaluation. One can more easily check on the first kind of truth, of course. We can generally agree whether or not a description fairly represents an object. Subjective description, on the other hand, is "true" not because it reports accurately but because it presents a valuable response. If we do not agree with how a writer responds to an object, we cannot say that the description is false. We can only say that it is not true for us—that is, we do not share it.

Nor are the two kinds of description hard-and-fast categories into one or the other of which any piece of descriptive writing must fall. Objective and subjective description are at opposite ends of a long line. Most cases fall somewhere between, and partake, in varying degrees, of both. Generally, however, one mode or the other will dominate and fix the focus. In scientific and legal writing, for instance, objectivity is desirable. In personal writing, description is likely to be subjective.

In both kinds success hinges upon three things: (1) specific images (that is, sharp details which appeal to one or another of the senses); (2) selection of these details in accordance with a guiding principle; and (3) clear organization of the details.

OBJECTIVE DESCRIPTION

SELECTION OF DETAIL

In objective description the principle of selection exists in the thing itself. The writer must ask: Which details are essential to seeing and understanding? Which are accidental and relatively unimportant? Essential details should make up the bulk of the description. Secondary details may be included selectively if the writer has time and space.

The following description of a fresh-water fish by an eighteenth-century English naturalist is a good example of concentration upon essential details:

> The loach, in its general aspect, has a pellucid appearance: its back is mottled with irregular collections of small black dots, not reaching much below the *linea lateralis,* as are the back and tail fins: a black line runs from each eye down to the nose; its belly is of a silvery white; the upper jaw projects beyond the lower, and is surrounded with six feelers, three on each side; its pectoral fins are large, its ventral much smaller; the fin behind its anus small; its dorsal fin large, containing eight spines; its tail, where it joins to the tail-fin, remarkably broad, without any taperness, so as to be characteristic of this genus; the tail-fin is broad, and square at the end. From the breadth and muscular strength of the tail, it appears to be an active nimble fish.
>
> Gilbert White

White focuses upon those features which enable us to recognize the loach: the size and shape of its tail and fins, the number of feelers on each side of the upper jaw, and so on. Such scientific description is a kind of definition, differentiating a particular thing from others similar to it.

ORGANIZATION OF DETAILS

Objective description, especially the visual kind, often begins with a brief general view, presenting the object in a comprehensive, unde-tailed image. It then analyzes this into its natural parts and treats each in more detail, following an organization inherent in the subject itself. Here, for instance, is a description by a student of a lake in Maine:

> In shape the lake resembles a gently curving S, its long axis lying almost due north-south. The shoreline is ringed with rocks of all sizes, from huge boulders to tiny pebbles—the detritus of the ice age. Beyond the rocks the forest comes almost to the water's edge. Mostly pine and hemlock, it contains a few hardwoods—maple, oak, birch. Here and there an old pine, its roots washed nearly clean of support, leans crazily over the water, seeming about to topple any instant. But it never does; trees fall this way for years.

First we view the lake in its entirety, as a hawk might see it. Then we focus down and move progressively closer to shore. We see the rocks immediately at the water's edge, then the forest, then the various kinds of trees, and finally the old pine leaning over the water. The description, in short, is organized: it moves from general to particular, and it analyzes the visual experience of the lake into three parts—the lake as a whole, the shoreline, and the forest around.

To effect these changes in viewpoint, the writer does not waste time directing us. He does not say, "As we leave the bird's-eye view and come down for a closer look, we observe that the shoreline is ringed with rocks." It is awkward and wordy to turn yourself into a tour-guide when you write description. Better to move about the object implicitly without dragging readers by the hand.

Doing this generally requires a point of view that is impersonal and omniscient: impersonal in the sense that the writer does not refer to himself or herself; omniscient in the sense that nothing is hidden and the writer can range about the object with complete freedom—above, below, around it, inside and out. Readers will follow if a writer has clearly organized what they are supposed to see.

But he or she must organize. Writers of description do not just "see." They analyze what they see and impose a pattern upon it. The twin processes of analyzing and organizing the visual experience stand out clearly in the following sentence by the novelist Joseph Conrad, describing a view of the English coast. In this case the angle of vision does not change as it did in the description of the lake, but the principle of organization is clear:

> Beyond the sea wall there curves for miles in a vast and regular sweep the barren beach of shingle, with the village of Brenzett standing out darkly across the water, a spire in a clump of trees; and still further out the perpendicular column of a lighthouse, looking in the distance no bigger than a lead pencil, marks the vanishing point of the land.

Our view of the scene proceeds from near to distant. Our eye moves outward through a series of receding planes: the sea wall, the beach, the village with its spire and trees across the water, and the lighthouse in the offing.

A more extended example of how objective description organizes what we see occurs in the following passage from an account by Admiral Richard E. Byrd of his sojourn at the South Pole. He is describing the hut in which he spent several months, alone:

> The day's work done, I took the luxury of a meditative inventory; and what I saw was good. The means of a secure and profound existence were all handy, in a world I could span in four strides going one way and in three strides going the other. It was not a bright world. The storm lantern hanging from a nail over my bunk burned dimly; and the gasoline pressure lamp, suspended from the ceiling, seemed to concentrate its brilliance all in one patch, making the shadows seem all the darker. But the dimness was rather to my liking. It gave depth to the room, and, somehow, made my possessions seem bigger.
>
> My bunk, fastened to the north wall, was about three feet off the floor, with the head flush against the eastern wall. At the foot of the bunk, on a small table, was the register, a glass-enclosed mechanism of revolving drum and pens which automatically recorded wind direction and velocity as reported by the wind vane and anemometer cups to which it was electri-

cally connected. The dry cells powering the pens and driving the drum were racked underneath. Across the room, in the southeast corner, was a triangular shelf holding the main combination radio transmitter and receiver, with a key fastened near the edge. The transmitter was a neatly constructed, 50-watt, self-excited oscillator which Dyer had assembled himself, and which was powered by a 350-watt, gasoline-driven generator weighing only 35 pounds. The receiver was a superheterodyne of standard make. Above this shelf was a smaller one holding the emergency radio equipment, consisting of two 10-watt transmitters powered by hand-cranked generators, plus two small battery receivers, each good for about a hundred hours. These were stand-by equipment. And above this shelf was a still smaller shelf holding spare parts for the radio.

The east wall, between the head of the bunk and the radio corner, was all shelves—six, to be exact. The lower ones were stocked with food, tools, books, and other odds and ends. On the top shelves were instruments and chronometers, all placed high and some wrapped in cotton. On the south wall my windproofs, fur mukluks, parka, and pants hung drying from ten-penny nails. Pushed against the middle of the same wall was a food box on which was a portable victrola in a battered green case. The table was also the family board. On the floor in the south-west corner was a box which I called the ice box, since anything put into it would stay frozen. Among other things it then contained were two Virginia hams which my mother had sent me.

The stove was a foot or so out from the west wall, about midway between the door and the triple register. It was an ordinary two-lid, coal-burning caboose stove, except that this one had been converted into an oil-burner by fitting a round burner over the grate and rigging a three-gallon gravity tank to feed it. It burned Stoddard solvent, ranking midway between kerosene and gasoline among the petroleum distillates. A liquid fuel was chosen instead of coal because coal was too bulky to haul. From the stove the stack went straight up to within two feet of the ceiling, where it bent and ran along the wall before passing through a vent above the foot of the bunk. By carrying the pipe across the room in this way, we thought we were providing the equivalent of a radiator; but the scheme was a clumsy makeshift. Two or three pipe sections were lost on the trail, somewhere between Little America and Advance Base; and, the only reserve sections being of a different size, we had used empty five-gallon tins as joints, cut open at the ends to fit. Ingenious as they were, the connections were scarcely air tight. This crude, inoffensive-looking heating plant held for me the power of life and death. Innocent in every rude line, it would nearly kill me a few months hence. And the time would come when I should wonder how I could have been such a fool as not to see what was in plain sight for me to see.

In the first paragraph Byrd shows us a general view of his small, dim world. In the remainder of the passage he analyzes that world in terms of its four walls (an analysis implicit in the subject). We stand in the middle of the room and our eyes move clockwise from north to east to south to west. Assuming Byrd's readers were competent draftsmen and read the paragraph with equal care, we may suppose that they would all make much the same sketch of the shack at Antarctica, just as they would draw similar versions of Conrad's sea view or the student's lake.

DICTION IN OBJECTIVE DESCRIPTION

In objective description words are chosen for the exactness of their denotation, not for the forcefulness of their connotation. There is no emotive language in the passage by Byrd. He is seeing, not reacting or evaluating; and he wants us to see.

Consequently he names things. That is why he uses technical nouns: "anemometer," "oscillator," "chronometer." Where a common noun must be used Byrd specifies it with adjectives: "tenpenny nail"; "battered green case"; "two-lid, coal-burning caboose stove."

Such careful denotative diction renders the scene, enabling readers to visualize it very clearly. Anyone who has the knowledge can see a "tenpenny nail" in imagination more precisely than a "nail"; a "two-lid, coal-burning caboose stove" better than a "stove."

Accuracy like this is especially important in legal and scientific writing. It is vital that a patent attorney, for instance, describe an invention precisely and objectively. And precision is the essence of scientific description. Look again at the passage by Gilbert White back on page 443. White says "six feelers, three on each side," not simply "several feelers." He is careful to differentiate fins by their technical names: "pectoral," "ventral," "dorsal." By using biological terms he achieves precision without sacrificing economy. (He has, of course, the advantage of writing for trained readers.)

Scientific description is not easy to write well. Given enough time to observe and the training to know what to look for, anyone can compose a reasonably accurate description of a fish. But it requires more care to write a description that is accurate and at the same time forceful, interesting prose. It is worth studying White's paragraph to learn how he organizes it and how he gives vitality and movement to his description by using short, direct clauses, with just enough variety in their construction to avoid monotony.

SUBJECTIVE DESCRIPTION

When describing objectively, the observer acts as a camera, recording precisely and impersonally. When writing subjectively, no longer an impartial recorder, he or she becomes part of what is seen or heard. The writer's point of view is usually personal; attitudes and values color the object being described; he or she projects the feelings which the object or scene arouses into the words.

These feelings constitute an impression which is as much a part of the description as the object itself. In fact, it is more: the impression is the principle by which the writer selects and organizes what he or she tells us. Sometimes writers state their impressions directly, as in the following paragraph in which an English writer describes her reaction to the citizens of Moscow:

> I wandered about in the morning and looked at the streets and the people. All my visit I looked and looked at the people. They seem neither happier nor sadder than in the West, and neither more nor less worried than any town dweller. (People in towns are always preoccupied. "Have I missed the bus? Have I forgotten the potatoes? Can I get across the road?") But they appear stupid, what the French call *abruti*. What do they think? Perhaps they don't think very much, and yet they read enormously. I never saw such a country of readers—people sitting on benches, in the metro, etc., all read books (magazines seem not to exist); on the trains they have lending libraries. They are hideously ugly. Except for a few young officers, I never saw a handsome man; there seem to be no beautiful women. They have putty faces, like Malenkov. It is nonsense to speak of Asiatics, Mongol Hordes and so on—the pretty little Tartar guards at Lenin's tomb were the only people I saw with non-European cast of features.
>
> Nancy Mitford

FIXING THE IMPRESSION IN IMAGES

While subjective description often defines the impression directly, it cannot rely upon such abstract statements completely. The writer must anchor his or her feelings in specific images—that is, in concrete details appealing to the senses. An explicit statement of feeling may set them up, but ultimately it is the details, emotionally charged, which give physical reality to the impression.

We can see the connection between a general, direct assertion of feeling and the focusing of feeling in images in the following paragraph about the slums of East London in 1902:

> No more dreary spectacle can be found on this earth than the whole of the "awful East," with its Whitechapel, Hoxton, Spitalfields, Bethnal Green, and Wapping to the East India Docks. The colour of life is grey and drab. Everything is helpless, hopeless, unrelieved, and dirty. Bath tubs are a thing totally unknown, as mythical as the ambrosia of the gods. The people themselves are dirty, while any attempt at cleanliness becomes howling farce, when it is not pitiful and tragic. Strange, vagrant odours come drifting along the greasy wind, and the rain, when it falls, is more like grease than water from heaven. The very cobblestones are scummed with grease.
>
> Jack London

Jack London begins with a broad abstract statement of the impression the slums make upon him: "no more dreary spectacle"; "the color of life is grey and drab"; "everything is helpless, hopeless, unrelieved, and dirty." But in the final sentences he renders the impression in images: "vagrant odours," "greasy wind," "rain . . . like grease," "cobblestones . . . scummed with grease."

As you can see, details are used differently in impressionistic description than in objective. Connotations are more important, and words are charged with emotion. The writer wants to arouse in us responses like his own. To succeed he must do more than tell us what he feels. He must re-create the scene or object, not as it literally is, but in a significantly altered manner. He must select some details, ignore others, exaggerate one image, underplay another, introduce compelling similes and metaphors.

In short, the visual image is refracted through the writer's consciousness and changed. It may emerge idealized, like a landscape by a romantic painter. It may be bent into ugliness, like the distortion in a funhouse mirror. But the result is not how the scene would look to a neutral eye.

Such idealization or distortion is perfectly legitimate. The writer of subjective description signs no contract to deliver literal truth. "Here," he or she says, "is how *I* see it." Yet the description may constitute a more subtle and revealing truth than objective accuracy, as an artist's caricature may reveal a hidden reality about the subject.

To convey subjective truth, then, a writer must embody his or her responses in the details of the scene. Often, in fact, a writer relies exclusively upon such embodiment, making no extended statement of feeling but forcing the scene or object to speak for itself. A simple case is catalogue description, in which the writer lists detail after detail, all contributing to a dominant impression. The following paragraph is a good example (it describes an outdoor market on Decatur Street in New Orleans):

> The booths are Sicilian, hung with red peppers, draped with garlic, piled with fruit, trayed with vegetables, fresh and dried herbs. A huge man, fat as Silenus, daintily binds bunches for soup, while his wife quarters cabbages, ties smaller bundles of thyme, parsley, green onions, small hot peppers and sweet pimientos to season gumbos. Another Italian with white mustache, smiling fiercely from a tanned face, offers jars of green filé powder, unground allspice, pickled onions in vinegar. Carts and trucks flank the sidewalk; one walks through crates of curled parsley, scallions piled with ice, wagonloads of spinach with tender mauve stalks, moist baskets of crisp kale; sacks of white onions in oyster-white fishnet, pink onions in sacks of old rose; piles of eggplant with purple reflections, white garlic and long sea-green leeks with shredded roots, grey-white like witches' hair. Boxes of artichokes fit their leaves into a complicated pattern. Trucks from Happy Jack, Boothville and Buras have unloaded their oranges; a long red truck is selling cabbages, green peppers, squashes long and curled like the trumpets of Jericho. There is more than Jordaens profusion, an abun-

dance more glittering in color than Pourbus. A blue truck stands in sun-
light, Negroes clambering over its sides, seven men in faded jeans, washing-
blue overalls; the last is a mulatto in a sweater of pure sapphire. A mangy
cat steps across a roadway of crushed oranges and powdered oyster-shells.

John Peale Bishop

Not only the individual meanings, but the very profusion of Bishop's
details convey abundance and vitality far more effectively than any
plain statement would. In description it is not possible to overestimate
the importance of specificity. Study how carefully Bishop names colors,
for instance.

While the details in a catalogue description are usually chosen according to an underlying feeling or evaluation, the selection is less rigorous
than in other kinds of subjective description. (Notice that Bishop includes the "mangy cat" and the "crushed oranges.") More often the
writer of subjective description "edits" the scene or object, using fewer
details and only those conducive to the impression he or she wants to
communicate. The novelist Thomas Wolfe, for example, draws a picture
of idealized attractiveness in this description of a modest home:

> On the outskirts of a little town upon a rise of land that swept back
> from the railway there was a tidy little cottage of white boards, trimmed
> vividly with green blinds. To one side of the house there was a garden
> neatly patterned with plots of growing vegetables, and an arbor for the
> grapes which ripened late in August. Before the house there were three
> mighty oaks which sheltered it in their clean and massive shade in summer,
> and to the other side there was a border of gay flowers. The whole place
> had an air of tidiness, thrift, and modest comfort.

The last sentence sums up the scene and constitutes a kind of direct
statement of feeling, as do such modifiers as "neatly," "clean," "gay."
But on the whole it is the images that create the sense of middle-class
fulfillment. Any ugliness is excluded. If the lawn were disfigured here
or there by crabgrass, if a few weeds leered among the flowers, the
facts are discreetly hidden.

Very different are the details—and the impression—in this description of another home by the English writer George Orwell (he is describing the dwellings of miners in the north of England):

> I found great variation in the houses I visited. Some were as decent as
> one could possibly expect in the circumstances, some were so appalling
> that I have no hope of describing them adequately. To begin with, the
> smell, the dominant and essential thing, is indescribable. But the squalor
> and the confusion! A tub full of filthy water here, a basin full of unwashed
> crocks there, more crocks piled in any odd corner, torn newspaper littered
> everywhere, and in the middle always the same dreadful table covered
> with sticky oilcloth and crowded with cooking pots and irons and half-
> darned stockings and pieces of stale bread and bits of cheese wrapped round
> with greasy newspaper! And the congestion in a tiny room where getting

from one side to the other is a complicated voyage between pieces of furniture, with a line of damp washing getting you in the face every time you
move and the children as thick underfoot as toadstools!

Sometimes a writer concentrates upon one or two images which stand
as symbols of the emotion he or she seeks to convey. In the following
passage Alfred Kazin projects into two key images his childhood despair
at stuttering and being forced to attend a special school:

It troubled me that I could speak in the fullness of my own voice only
when I was alone on the streets, walking about. There was something
unnatural about it; unbearably isolated. I was not like the others! At mid-
day, every freshly shocking Monday noon, they sent me away to a speech
clinic in a school in East New York, where I sat in a circle of lispers and
cleft palates and foreign accents holding a mirror before my lips and rolling
difficult sounds over and over. To be sent there in the full light of the
opening week, when everyone else was at school or going about his business,
made me feel as if I had been expelled from the great normal body of
humanity. I would gobble down my lunch on my way to the speech clinic
and rush back to the school in time to make up for the classes I had
lost. One day, one unforgettable dread day, I stopped to catch my breath
on a corner of Sutter Avenue, near the wholesale fruit markets, where
an old drugstore rose up over a great flight of steps. In the window were
dusty urns of colored water floating off iron chains; cardboard placards
advertising hairnets, EX-LAX; a great illustrated medical chart headed THE
HUMAN FACTORY, which showed the exact course a mouthful of food follows
as it falls from chamber to chamber of the body. I hadn't meant to stop
there at all, only to catch my breath; but I so hated the speech clinic
that I thought I would delay my arrival for a few minutes by eating my
lunch on the steps. When I took the sandwich out of my bag, two bitterly
hard pieces of hard salami slipped out of my hand and fell through a
grate onto a hill of dust below the steps. I remember how sickeningly
vivid an odd thread of hair looked on the salami, as if my lunch were
turning stiff with death. The factory whistles called their short, sharp blasts
stark through the middle of noon, beating at me where I sat outside the
city's magnetic circle. I had never known, I knew instantly I would never
in my heart again submit to, such wild passive despair as I felt at that
moment, sitting on the steps before THE HUMAN FACTORY, where little robots
gathered and shoveled the food from chamber to chamber of the body.
They had put me out into the streets, I thought to myself; with their mirrors
and their everlasting pulling at me to imitate their effortless bright speech
and their stupefaction that a boy could stammer and stumble on every
other English word he carried in his head, they had put me out into the
streets, had left me high and dry on the steps of that drugstore staring
at the remains of my lunch turning black and grimy in the dust.

In Kazin's description, selection is extremely important. The passage
focuses down and down to the images of THE HUMAN FACTORY and the
two pieces of salami. Kazin tells us what his feelings were (he is quite

explicit in fact). But he succeeds in communicating the despair of an alienated child because he projects and concentrates this feeling in the salami with its "odd thread of hair . . . turning black and grimy in the dust," and the inhuman little robots endlessly shoveling food in a body that has become a machine. In the world symbolized by these images there is little room for humane values—for love, compassion, tender understanding.

Kazin's paragraph demonstrates the importance in all description of the "crystallizing image," the detail that precipitates the visual experience. The writer of description, we have said, must make readers see (or hear or taste or smell). He or she cannot achieve this simply by listing mechanically every detail that falls within the perceptual field. (Even in catalogue descriptions, like that by John Peale Bishop, we are shown only a portion of what exists to be seen.) These few details, however, must be rendered with such precision that the reader can see them vividly in his mind's eye. Since they are essential details, they will crystallize the scene, making it solid and real. To vary the analogy, we may say that writing description is like developing a photograph. The writer, carefully choosing his or her details and expressing them in compelling images, begins the process; the reader develops the rest in the fluid of his or her own experience. The point to remember about using details is this: select those that are essential to the impression you want to convey, and describe them precisely and concretely so that readers can *see* them.

METAPHOR AND SIMILE IN SUBJECTIVE DESCRIPTION

In addition to selecting and rendering details, the writer of description may also introduce comparisons, often in the form of metaphors and similes. In Bishop's paragraph about Decatur Street, for example, the proprietor is "fat as Silenus" (an ancient god of wine), the leeks "sea-green" with roots "like witches' hair," and the squashes "long and curled like the trumpets of Jericho."

Metaphor is even more central in the following passage about the Great Wall of China. The Wall assumes a monstrous power as it marches over and dominates the land:

> There in the mist, enormous, majestic, silent, and terrible, stood the Great Wall of China. Solitarily, with the indifference of nature herself, it crept up the mountain side and slipped down to the depth of the valley. Menacingly, the grim watch towers, stark and four square, at due intervals stood at their posts. Ruthlessly, for it was built at the cost of a million lives and each one of those great grey stones has been stained with the bloody tears of the captive and the outcast, it forged its dark way through a sea of rugged mountains. Fearlessly, it went on its endless journey, league upon league to the furthermost regions of Asia, in utter solitude, mysterious

like the great empire it guarded. There in the mist, enormous, majestic, silent, and terrible, stood the Great Wall of China.

W. Somerset Maugham

EXAGGERATING DETAILS

Finally, the writer of subjective description may feed his or her reaction back into the scene by distorting or exaggerating details. Mark Twain, who was adept at the art of hyperbole, or exaggeration, tells about a trip in an overland stage in the 1860s. The passengers have spent the night at a way station and Twain describes the facilities for cleaning up the next morning:

> By the door, inside, was fastened a small old-fashioned looking-glass frame, with two little fragments of the original mirror lodged down in one corner of it. This arrangement afforded a pleasant double-barreled portrait of you when you looked into it, with one half of your head set up a couple of inches above the other half. From the glass frame hung the half of a comb by a string—but if I had to describe that patriarch or die, I believe I would order some sample coffins. It had come down from Esau and Samson, and had been accumulating hair ever since—along with certain impurities.

We are not supposed to take this literally, of course. Twain is exercising the satirist's right of legitimate exaggeration, legitimate because it leads us to see a truth about this Western hostel.

PROCESS DESCRIPTION

A *process* is a directed activity in which something undergoes progressive change. The process may be natural, like the growth of a tree; or it may be humanly directed, like an automobile taking shape on an assembly line. But always something is happening—work is being done, a product being formed, an end of some kind being achieved.

To describe a process you must analyze its stages. The analysis will determine how you organize the description. In a simple case, such as baking a cake, the process has obvious, prescribed steps; the writer needs only to observe and record them accurately. On the other hand, complicated and abstract processes—for instance, how a law comes into being as an act of Congress—require more study and thought.

Here is a simple example of a process, a natural one—a small frog being eaten by a giant water bug:

> He didn't jump; I crept closer. At last I knelt on the island's winterkilled grass, lost, dumbstruck, staring at the frog in the creek just four feet away. He was a very small frog with wide, dull eyes. And just as I looked at him, he slowly crumpled and began to sag. The spirit vanished from his eyes as if snuffed. His skin emptied and drooped; his very skull seemed

to collapse and settle like a kicked tent. He was shrinking before my eyes like a deflating football. I watched the taut, glistening skin on his shoulders ruck, and rumple, and fall. Soon, part of his skin, formless as a pricked balloon, lay in floating folds like bright scum on top of the water: it was a monstrous and terrifying thing. I gaped bewildered, appalled. An oval shadow hung in the water behind the drained frog; then the shadow glided away. The frog skin bag began to sink.

Annie Dillard

At the beginning of the description the frog is whole and alive, sitting in the creek; by the end it has been reduced to a bag of skin. This change is the process Dillard describes. It is continuous rather than divided into clearly defined steps. Yet it is analyzed. Verbs, the key words in the analysis, create sharp images of alteration: "crumpled," "collapse," "shrinking," "deflating," "ruck," "rumple," "fall." The similes and metaphors translate an unusual visual experience into more familiar ones: "like a deflating football," "formless as a pricked balloon."

The next example of process description involves an assembly line at a cosmetics plant:

> Cream-jar covers joggle along a moving belt. Six iron arms descend to set paper sealers on sextuplicate rows of cream pots. Each clattering cover is held for a moment in a steel disk as a filled cream jar is raised by a metal wrist and screwed on from underneath.
>
> At the mascara merry-go-round a tiny tube is placed in each steel cup—clink. The cups circle—ca-chong, ca-chong, ca-chong—till they pass under two metal udders. There the cups jerk up—ping—and the tubes are filled with mascara that flows from the vats upstairs in manufacturing. The cups continue their circle till they pass under a capper—plump. The filled, capped tubes circle some more till they reach two vacuum nozzles, then—fwap—sucked up, around and down onto a moving belt.
>
> All along the belt women in blue smocks, sitting on high stools, pick up each mascara tube as it goes past. They insert brushes, tamp on labels, encase the tubes in plastic and then cardboard for the drugstore displays.
>
> At the Brush-On Peel-Off Mask line, a filler picks an empty bottle off the belt with her right hand, presses a pedal with her foot, fills the bottle with a bloop of blue goop, changes hands, and puts the filled bottle back on the line with her left hand, as she picks up another empty bottle with her right hand. The bottles go past at thirty-three a minute.

Barbara Garson

Garson's description provide a fine example of how analysis determines paragraphing. Three products are involved—cream, mascara, and the "Brush-On Peel-Off Mask"—and each is treated in a separate paragraph. For the mascara two are used, marking the two-stage process of the tubes' being first filled and then packaged.

The sentences are also determined by the analysis. Thus the three sentences of the first paragraph distinguish (1) the covers on the conveyor belt, (2) the iron arms placing sealers on the pots, and (3) the fixing of the lids onto the jars. Notice, too, the long sentence in the fourth paragraph; it uses parallel verbs to analyze the filler's movements.

Process description may be either objective or subjective. Both the foregoing examples are relatively objective, though each suggests responses. Even though Dillard's subject is horrifying and she actually expresses her reaction ("it was a monstrous and terrifying thing"), her images are objective. Dillard concentrates on rendering the visual experience in and of itself (which in a case like this perhaps best communicates the horror).

Despite its objective surface, Garson's description also implies a reaction. Her diction—especially the words imitating sounds—suggests the inhuman quality of the assembly line. Her fourth paragraph cleverly hints her feelings about work on the line. The long elaborate first sentence describing the worker's mechanized movements is followed by a brief matter-of-fact announcement that "the bottles go past at thirty-three a minute." The implication makes sensitive readers wince.

EXERCISE

1. Describe a panoramic view in a single sentence, like the one by Joseph Conrad on page 444. You may organize the view from the near to the far, or the other way round, or from side to side. The important thing is to compose the sentence so that the view is *organized,* and to do it implicitly without the obviousness of "Moving our eyes a few feet to the left, we observe. . . ."

2. Describe your classroom or your room at home as Admiral Byrd did his shack at Antarctica. Try to make readers see the room so clearly that they could draw an accurate plan. Despite its apparent simplicity, this is not an easy assignment. Unless you vary the sentence style a little and prune all deadwood, it will prove dull to write and duller to read.

3. Describe a plant, a flower, an insect, a fish with as much objective accuracy as you can. Even though your teacher may not be familiar with them, feel free to employ appropriate scientific terms. If you do not know enough science to handle one of the topics suggested, attempt an objective description of a piece of furniture, a tool, an item of sports equipment, a dress or other article of clothing.

4. Catalogue description like that by John Peale Bishop on page 448 is fun to write if you enter into the spirit of the thing. Go to a department of your supermarket or to a hardware store, a bakery, a toyshop, and describe what you see. Learn the names of things. Do not say simply "cheeses," but "Edam and Camembert and Gorgonzola"; describe colors and shapes and sizes. Behind all of your images there should be a general impression, even if nothing more than a sense of the beauty or variety or profusion of the articles on display.

5. This exercise requires two separate paragraphs, each complete in itself, like those by Thomas Wolfe and George Orwell on pages 449–450, describing similar objects or scenes. The first should idealize the subject you choose; the second should describe an ugly or dilapidated version of something comparable. A number of topics are possible— two neighborhoods in the same city, two towns of the same size but very different atmospheres, two automobiles, playrooms, stores, bars, restaurants. In both paragraphs introduce metaphors or similes that help to convey your impressions.

6. Mark Twain is a discouraging writer to imitate. If you feel daring, try your hand at an exaggerated description of a hotel you stayed in, a tourist attraction you have visited, a fraternity house. You are free to exaggerate and distort, but the hyperbole must reveal an essential truth.

7. Quoted below is part of a paragraph by Virginia Woolf describing a painting by the English artist Walter Sickert. Study it and write a similar description of one of these paintings (or any other your teacher may suggest): *Pope Leo X* by Velazquez, *The Card Players,* by Cézanne, *Gulf Stream* by Winslow Homer, *Guernica* by Picasso, *Gin Lane* by Hogarth. (If your library has copies of Sickert's work, try to find the painting Woolf describes. Comparing her description with the actual painting will help you to see what she has done.) Use any of the techniques of subjective description we have discussed.

> You remember the picture of the old publican [in England, the owner of a bar], with his glass on the table before him and a cigar gone cold at his lips, looking out of his shrewd little pig's eyes at the intolerable wastes of desolation in front of him? A fat woman lounges, her arm on a cheap yellow chest of drawers, behind him. It is all over with them, one feels. The accumulated weariness of innumerable days has discharged its burden on them. They are buried under an avalanche of rubbish. In the street beneath, the trams are squeaking, children are shrieking. Even now somebody is tapping his glass impatiently on the bar counter. She will have to bestir herself; to pull her heavy, indolent body together and go and serve him. The grimness of that situation lies in the fact that there is no crisis; dull minutes are mounting, old matches are accumulating and dirty glasses and dead cigars; still on they must go, up they must get.

CHAPTER 47

Narration

INTRODUCTION

A *narrative* is a meaningful sequence of events told in words. It is sequential in that the events are ordered, not merely random. Sequence always involves an arrangement in time (and usually other arrangements as well). A straightforward movement from the first event to the last constitutes the simplest chronology. However, chronology is sometimes complicated by presenting the events in another order: for example, a story may open with the final episode and then flash back to all that preceded it.

A narrative has meaning in that it conveys an evaluation of some kind. The writer reacts to the story he or she tells, and states or implies that reaction. This is the "meaning," sometimes called the "theme," of a story. Meaning must always be rendered. The writer has to do more than tell us the truth he sees in the story; he must manifest that truth in the characters and the action.

Characters and action are the essential elements of any story. Also important, but not as essential, is the setting, the place where the action occurs. Characters are usually people—sometimes actual people, as in history books or newspaper stories, sometimes imaginary ones, as in novels. Occasionally characters are animals (as in an Aesop fable), and sometimes a dominant feature of the environment functions almost like a character (the sea, an old house).

The action is what the characters say and do and anything that happens to them, even if it arises from a nonhuman source—a storm, for instance, or a fire. Action is often presented in the form of a plot. Action is, so to speak, the raw material; plot, the finished product, the fitting together of the bits and pieces of action into a coherent pattern. Usually, though not invariably, plot takes the form of a cause-and-effect chain: event A produces event B; B leads to C; C to D; and so on until the final episode, X. In a well-constructed plot of this kind we can work back from X to A and see the connections that made the end of the story likely and perhaps inevitable.

Stories can be very long and complicated, with many characters, elaborate plots, and subtle interpenetration of character, action, and setting. In writing that is primarily expository, however, narratives are shorter and simpler. Most often they are factual rather than imaginary, as when an historian describes an event. And often in exposition an illustration may involve a simple narrative. Being able to tell a story, then, while not the primary concern of the expository writer, is a skill which he or she will now and again be called upon to use.

ORGANIZING A NARRATIVE

As with so much in composition, the first step in narration is to analyze the story in your own mind. In the actual telling, the analysis provides the organization. The simplest kind of narrative is the episode, a single event unified by time and place. But even an episode must be organized. The writer must break it down into parts and present these in a meaningful order.

In the following case the episode is the brief landing of a passenger ship at the Mediterranean island of Malta. After describing the setting in the first paragraph, the writer divides his story into two parts: the problems of getting ashore (paragraphs 2 and 3), and the difficulties of returning to the ship (4).

We called at Malta, a curious town where there is nothing but churches, and the only sound of life is the ringing of church bells. The whole place reminded me of the strange towns one often sees in the nightmares of delirium.

As soon as the ship anchored, a regular battle began between the boatmen for possession of the passengers. These unhappy creatures were hustled hither and thither, and finally one, waving his arms like a marionette unhinged, lost his balance and fell back into a boat. It immediately bore him off with a cry of triumph, and the defeated boatman revenged himself by carrying off his luggage in a different direction. All this took place amid a hail of oaths in Maltese, with many suggestive Arab words intermingled.

The young priests in the second class, freshly hatched out of the seminary, turned vividly pink, and the good nuns covered their faces with their veils and fled under the mocking gaze of an old bearded missionary, who wasn't to be upset by such trifles.

I did not go ashore, for getting back to the ship was too much of a problem. Some passengers had to pay a veritable ransom before they could return. Two French sailors, who had got mixed up with churches when looking for a building of quite another character, solved the matter very simply by throwing their grasping boatman into the sea. A few strokes with the oars, and they were alongside, and as a tug was just leaving they tied the little boat to it, to the accompaniment of the indignant shrieks from the owner as he floundered in the water.

Henry de Monfreid

In each of the two main parts of the story de Monfried begins with a generalization and then supports it with a specific instance. The effectiveness of his narrative lies both in the skill with which he analyzes the episode and the precision with which he renders characters and action. The glimpses he gives us are brief, but vivid and filled with meaning: the tumbled passenger "like a marionette unhinged," "the mocking . . . missionary," the shrieking indignation of the greedy boatman thrown into the sea.

Their nightmare quality, which is the dominant note of the setting, unifies these details. But their causal connections are relatively unimportant. For example, the sailors do not toss their boatman into the water because of what other boatmen did earlier to the unfortunate passenger. The two events relate not as cause and effect but more generally in showing the greediness of the Maltese.

In more complicated stories, however, events may well be linked in a plot of cause and effect. A brief example of such a plot appears in this account of a murder in New York occasioned by the Great Depression of the 1930s:

Peter Romano comes from a little town in Sicily. For years he kept a large and prosperous fruit store under the Second Avenue elevated at the corner of Twenty-ninth Street. A few years ago, however, he got something the matter with his chest and wasn't able to work any more. He sold his business and put the money into Wall Street.

When the Wall Street crash came, Peter Romano lost almost everything. And by the time that Mrs. Romano had had a baby five months ago and had afterwards come down with pneumonia, he found he had only a few dollars left.

By June, he owed his landlord two months' rent, $52. The landlord, Antonio Copace, lived only a few blocks away on Lexington Avenue, in a house with a brownstone front and coarse white-lace curtains in the windows. The Romanos lived above the fruit store, on the same floor with a cheap dentist's office, in a little flat to which they had access up a dirty oilcloth-covered staircase and through a door with dirty-margined panes. The Romanos regarded Mr. Copace as a very rich man, but he, too, no doubt, had been having his losses.

At any rate, he was insistent about the rent. Peter Romano had a married daughter, and her husband offered to help him out. He went to Mr. Copace with $26—one month's rent. But the old man refused it with fury and said that unless he got the whole sum right away, he would have the Romanos evicted. On June 11, he came himself to the Romanos and demanded the money again. He threatened to have the marshal in and put them out that very afternoon. Peter Romano tried to argue with him, and Mrs. Romano went out in a final desperate effort to get together $52.

When she came back empty-handed, she found a lot of people outside the house and, upstairs, the police in her flat. Peter had shot Mr. Copace and killed him, and was just being taken off to jail.

Edmund Wilson

The bony structure of Wilson's little story is the time references: "For years he kept. . . . A few years ago. . . . When the Wall Street crash came. . . . By June. . . . On June 11. . . . When Mrs. Romano came back. . . ." This temporal skeleton supports a cause-and-effect plot. The basic elements of such a plot are the exposition, the conflict, the climax, and the denouement.

The term *exposition* has a special meaning with reference to narration. The exposition is that part of the plot which gives us the background information about the characters, telling us what we need to know in order to understand why they act as they do in what is about to unfold. Exposition is usually, but not always, concentrated at or very near the beginning of a story. Wilson's exposition occupies the first three paragraphs, which locate Peter Romano in time and place and tell us necessary facts about his history.

Exposition gives way to *conflict,* the second part of a plot. Conflict involves two or more forces working at cross purposes. (Sometimes this takes place between a character and a physical obstacle such as a mountain or the sea; or it may be internalized, involving diverse psychological aspects of the same person.) In this story the conflict, obviously, occurs between tenant and landlord. The third part of a plot, the *climax,* resolves the conflict: here, the shooting. Finally the plot ends with the *denouement,* the closing events of the narrative: Peter Romano's being carried off to jail.

In the simple and often partial stories you are likely to tell in expository writing, it is not always necessary (or even desirable) that you develop all these elements of a plot in detail. You may need to spend your time upon exposition and conflict—as Wilson does—and treat the climax and denouement very briefly. Or you may wish to slight the exposition and concentrate upon the climax. But in any case you must be clear in your own mind about the structure of your plot and how much of each element your readers need in order to understand your narrative.

In organizing a story, then, you should ask these questions. (1) What is the plot? Specifically this comes down to: What is the climax? What events leading to the climax constitute the conflict? What should be included in the exposition? What events following the climax (the denouement) should be told? (2) What are the salient qualities of the characters and how can these best be revealed in speech and action?

(3) What details of setting will help readers understand the characters and what they do?

MEANING IN NARRATIVE

In asking those questions your guide must be the meaning you want the story to convey. *Meaning* in narrative is a complex and difficult matter, and it may operate on at least three different levels. In some stories the meaning is an abstract "truth"—moral, political, religious—which the characters, the plot, and the setting are carefully designed to carry. Such stories are called *allegories,* and sometimes the literal events do not make a great deal of sense in themselves. The real meaning emerges only when the reader steps up from the narrative surface to the more abstract level of idea.

At the other extreme, the meaning of a story exists on its surface. *Realistic* stories do not require us to read the characters or plot as standing for categories of thought or feeling. De Monfried's account of the landing at Malta is an example. It has a meaning, or rather meanings: Maltese boatmen are greedy; their greed is punished, at least in one instance; young priests are naïve. But these are simply generalizations one draws from what literally happens, not a second, higher level of truth, as in an allegory.

In most stories, however, meaning is neither purely allegorical nor purely realistic. It falls somewhere between those extremes and partakes of both. Such stories are realistic in that characters and events correspond to life as we know or can imagine it, and we can generalize from them to real people. At the same time these stories—like allegories—point to another, more abstract and inclusive level of significance. In Edmund Wilson's story, for instance, we can see both a particular (realistic) and a more abstract (allegorical) meaning. On the literal level the narrative describes the tragedy of two men made desperate by economic frustration, and we may fairly apply it to similar men in similar circumstances. At the same time the story can be seen in Marxist terms as revealing the impersonal forces of the exploiting bourgeoisie and the dispossessed urban proletariat, each the victim of a capitalist economy, and each the victimizer of the other.[1] Stories like this, which operate meaningfully on both a realistic and an allegorical level, may be called—for want of a better term—*symbolic.*

In talking about narrative meaning, then, it is useful to distinguish the three modes we have mentioned: allegorical, realistic, and symbolic. But in practice, a story is rarely one or another. Rather a particular story is, say, realistic tending toward the symbolic or symbolic tending

---

1. The fact that the story may be read as implying this theme does not mean, of course, that the theme is necessarily a true assessment of the way things are.

toward the allegorical. But if you do not insist that any story must be exclusively symbolic *or* realistic *or* allegorical, knowing about these kinds of meaning will help you understand how narrative works.

Whatever its mode, the meaning of a story, if it is to be truly communicated, has to be rendered in the characters and plot and setting. It may, in addition, be announced. That is, the writer may explicitly tell us what meaning he or she sees in the story. Sometimes such a statement of theme occurs at the end of a story (the "moral" at the end of a fable, for instance), sometimes at the beginning, sometimes in between. Thus the following account of the execution in 1618 of Sir Walter Raleigh begins with an announcement of its significance. But the writer does not rest content with telling us the theme. He is careful to select appropriate details of speech and action and to ground his theme in them:

> Immortal in the memory of our race, the scene of Raleigh's death has come to us with its vividness undimmed by the centuries. Everything that had been mean, false, or petty in his life had somehow been sloughed off. The man who went to the block was the heroic Raleigh who all along had existed as Sir Walter's ideal and now was to become a national legend.
>
> He had been lodged in the gatehouse at Westminster. At midnight his wife left him for the last time, and miraculously he lay down and slept for a few hours. Early in the morning the Dean of Westminster gave him his last communion. Afterwards he had his breakfast and enjoyed his last pipe of tobacco. At eight o'clock he started on his short journey to the scaffold erected in Old Palace Yard.
>
> Raleigh, so completely a man of the Renaissance, was inevitably concerned at the time with thoughts of fame beyond death. In his speech from the scaffold he did what he could to protect that fame, assuring his hearers that he was a true Englishman who had never passed under allegiance to the King of France. He was concerned also that men should not believe the old slander that he had puffed tobacco smoke at Essex when the earl had come to die. At the end he concluded:

> And now I entreat that you all will join me in prayer to that Great God of Heaven whom I have so grievously offended, being a man full of all vanity, who has lived a sinful life in such callings as have been most inducing to it; for I have been a soldier, a sailor, and a courtier, which are courses of wickedness and vice; that His Almighty goodness will forgive me; that He will cast away my sins from me, and that He will receive me into everlasting life; so I take my leave of you all, making my peace with God.

> There followed the famous moment in which Raleigh asked to see the axe. The headsman was reluctant to show it. "I prithee, let me see it," said Raleigh, and he asked, "Dost thou think that I am afraid of it?" Running his finger along the edge he mused, "This is sharp medicine, but it is a sound cure for all diseases." There was some fussing about the way he should have his head on the block. Somebody insisted that it should

be towards the east. Changing his position, Raleigh uttered a last superb phrase—"What matter how the head lie, so the heart be right?" He prayed briefly, gave the signal to the headsman, and died.

The headsman needed two strokes to sever the head. After holding it up for the crowd to see, he put it in a red leather bag, covered it with Raleigh's wrought velvet gown, and despatched it in a mourning coach sent by Lady Raleigh. Finally both head and body were buried by her in St. Margaret's Church, Westminster.

C. P. V. Akrigg

Akrigg states his point in the opening paragraph: Raleigh died heroically. In the story itself Raleigh's own words and actions carry that theme. The writer wisely lets them speak for themselves without intruding his own commentary. In effective narrative you must render scenes as you want readers to see them and not labor overlong on telling them why your story is significant. If you create real characters and action, readers will gather the meaning.

It is not even necessary to state the point at the beginning or end of the story (though sometimes, as in the example by Akrigg, it is desirable). Edmund Wilson, for instance, does not tell us what the story of Peter Romano and Mr. Copace means: it is clear enough. Similarly the following brief narrative by Ernest Hemingway, which we saw earlier as an example of understatement, leaves its meaning for readers to infer:

> They shot the six cabinet ministers at half-past six in the morning against the wall of a hospital. There were pools of water in the courtyard. There were wet dead leaves on the paving of the courtyard. It rained hard. All the shutters of the hospital were nailed shut. One of the ministers was sick with typhoid. Two soldiers carried him down stairs and out into the rain. They tried to hold him up against the wall but he sat down in a puddle of water. The other five stood very quietly against the wall. Finally the officer told the soldiers it was no good trying to make him stand up. When they fired the first volley he was sitting down in the water with his head on his knees.

Hemingway's story exemplifies realistic meaning. For while one can read philosophical significance into the horrifying episode, there is no evidence that Hemingway intends us to jump to any philosophy. This, he implies, is simply the way things are; the story is its own meaning.

The narrative also exemplifies "objective" presentation. It concentrates on the surface of events, on what can be seen and heard. Such objectivity is not a refusal to see and convey meaning, as inexperienced readers sometimes suppose. It is rather a special way of communicating meaning.

It can be a very powerful way. Hemingway does not *tell* us that war makes men cruel. He shows us; he forces us to endure the cruelty. The meaning of his brief story is more than an idea we comprehend intellectually. It becomes a part of our experience—not as deep and abiding a part, probably, as if we had actually been there, but nonetheless a reality experienced.

This is what the writer of narrative does at his or her best: re-create events in an intense and significant manner and thus deepen and extend the reader's experience of the world. Of course, in narrative of this rich and powerful kind we are entering the realm of creative literature and leaving behind the simpler world of exposition. Still, all narrative, whether literary or serving the needs of exposition, must have meaning, and that meaning must be rendered in character, action, and setting.

POINT OF VIEW AND TONE IN NARRATIVE

Writers are always in the stories they tell, whether that presence is apparent or hidden. It is apparent in the first-person point of view— that is, a story told by an "I." The "I" may be the central character to whom things are happening. Or "I" may be an observer standing on the edge of the action and watching what happens to others, as de Monfried observes and reports the events at Malta but does not participate in them.

Even though a writer narrates a personal experience, however, the "I" who tells the tale is not truly identical with the author who writes it. The narrative "I" is a persona, more or less distinct from the author. Thus "I" may be made deliberately and comically inept—a trick humorous writers like James Thurber often employ—or "I" may be drawn smarter and braver than the author actually is. And in literary narrative "I" is likely to be even more remote from the writer, often a character in his own right like Huck Finn in Twain's great novel.

The other point of view avoids the "I." This is the third-person story, told in terms of "he," "she," "they." Here the writer appears to disappear, hidden completely behind his characters. We know an author exists because a story implies a story-teller. But that presence must be guessed; one never actually observes it.

Nonetheless the presence is there. Even if not explicitly seen as an "I," the writer exists as a voice, heard in the tone of the story. Through style—the words chosen and the sentence patterns into which they are arranged—the writer of a narrative can imply a wide range of tones: irony, amusement, anger, horror, shock, disgust, delight, objective detachment.

Tone is essential to the meaning of a story. The tone of Hemingway's paragraph, for example, seems objective, detached, reportorial on the surface. He avoids suggesting emotion or judgment—words like "pitiful," "horrible," "cruel," "tragic." Rather his diction denotes the simple physical realities of the scene: "wet dead leaves," "paving," "rain," "shutters," "wall," "puddle," "water," "head," "knees."

The absence of emotive words actually intensifies the horror of the scene. But the objectivity of Hemingway's style is more than rhetorical understatement—though it is that—the trick of increasing emotion by seeming to deny it. The tone also presents a moral stance: a tough-minded discipline in the face of anguish. Men die and men kill one another, and we must feel the horror, feel it deeply; but we must also accept its inevitability and stand up to it and not be overwhelmed by it.

Now all this is implied in Hemingway's style—that is, in the tone of his prose. It is obviously a very important part of what he is saying. Thus style is not merely a way of conveying the meaning of a story; it is a part of meaning, sometimes the vital part.

EXERCISE

Using a first-person point of view, write a simple narrative of four or five paragraphs on one of these topics:

> Freshmen registration
> The arrival (departure) of an airplane or train
> A picnic
> A family departing on their summer holiday

You may cast yourself as a central character or, like de Monfried, as an observer. In either case think about the dominant impression, or theme, you want your story to convey—for example, that registration is a finely tuned madness. You may announce the theme or leave it implicit, but be sure to render it in terms of specific characters saying and doing specific things.

PART SEVEN

PERSUASION

Introduction

THE NATURE OF PERSUASION

Persuasion differs from exposition, description, and narration primarily in the way the writer relates to the reader. In persuasion the writer attempts to affect how the reader thinks or believes or acts. In exposition, on the other hand, and in description and narration in special ways, the writer seeks to inform readers, not to change them.

No fine line can be drawn in this matter of how readers are affected. A poignant story, a frightening description, or a lucid bit of exposition may alter a reader's feelings or beliefs. But that is not usually the purpose of such writing. If a reader is somehow changed, the effect is incidental.

With persuasion, however, the effect upon the reader is the principal purpose. Consequently the writer must pay closer attention to his or her reader than the writer of exposition does. The latter certainly must consider what readers can reasonably be expected to know and whether they can follow complex sentences and long, complicated paragraphs. But the persuasive writer must consider much more—not only the readers' general knowledge, intelligence, and reading skill, but also their values, beliefs, and biases; how they look upon themselves and others; what they fear and what they yearn for.

It does not follow that the writer of persuasion cares nothing for the subject. Of course he or she cares. It does mean that readers must loom larger in the writer's vision. He or she must ask: How do I want to alter my readers' beliefs or actions? What do they believe now? How do they act? What sort of people are they? What kind of persuasion is likely to affect them as I wish to affect them?

KINDS OF PERSUASION

There are two basic modes of persuasion—rational and nonrational—
and several varieties within each. Rational persuasion is called *argument*. The nonrational mode has no generic name; its main types are
satire, eloquence, and pathos.[1] It appeals not to intellect but to feeling,
whereas argument attempts to prove its point reasonably by using logic
or evidence. In composition, incidentally, the meaning of "argument"
should not be confused with its common sense of angry disagreement;
as a form of persuasion, argument is reasoned appeal and has nothing
to do with anger.

The difference between argument and emotive persuasion may be
seen in the following two passages, each concerned with the injustice
of poverty. The first speaks of the English poor in the early 1900s.
The writer, having listed the exact quantities and cost of a week's supply
of food for a city family of husband, wife, and five children, goes on
to comment:

> The diet is, of course, far worse than any day's menu might suggest in
> that the quantities available for each person are so small. Seven people
> took their dinner off $1\frac{1}{4}$ lb. of fish—about $2\frac{1}{2}$ oz. for each person including
> a father who does manual labour and three boys over the age of 7. One
> tin of condensed milk and a pennyworth of fresh suffice the family for a
> week, although the other two children are aged 4 and 2. And though the
> joint [roast] may seem comparatively large, the children are said to have
> only a small piece each on Sunday, since meals for the father to take to
> work with him must be provided out of it for the rest of the week.
>
>   Marghanita Laski

The second example concerns poverty in Appalachia in the 1970s:

> The time will come. Someday us poor is going to overrule. We're gonna
> do it, by the help of God we're gonna do it. I believe it. I honest to God
> do. I believe it. The poor is going to overrule. I've got faith in that.
>
>   Shirley Dalton

While both passages intend to persuade, they are very different. The
first is distanced from its subject both in time (it is about the early
1900s but was written much later) and in the position, intellectual
and social, of the writer. It supports its point, the inadequacy of the
diet of the poor, by facts and figures, making no overt appeal to our
emotions. It is, in short, an argument.

The second is more emotive. The writer has no distance from her
subject: she stands in the middle of it, writing about her own life, here

---

1. The term "persuasion" is sometimes narrowed to mean only the emotive modes. In
that classification "argument" is not a type of "persuasion," but its first cousin. This
usage has the disadvantage of leaving no inclusive term to designate both argument
and the emotive modes. Here we shall use "persuasion" in that inclusive sense, and
regard argument as one of its two chief types.

and now. Her words evoke feelings; they are, like much emotive language, repetitive and strongly rhythmic.[2]

Argumentative and emotional persuasion take very different stances toward the reader. Argument assumes readers to be intelligent and more or less objective, able to follow a logical train of thought, to weigh evidence, and not to be swayed by emotion from accepting the conclusions to which the logic or the evidence points. (That these assumptions are not always justified does not matter.) Argument establishes a kind of equality between writer and reader. By the very act of using logic or evidence the writer says to readers, "We meet, as intelligent people, on a common ground of rationality."

Persuasion, on the other hand, always implies at least the possibility of nonequality and manipulation. This is especially the case when a skillful stylist is calculating the emotive force of his words (a politician more ambitious than principled, for instance, or the writer of a public relations release). Then *the writer* acts rationally (if not very scrupulously) and assumes that his or her readers are irrational or at least vulnerable to emotional appeal.

Three important qualifications must be admitted here. First, to the degree that the writer's feelings are genuine—like Dalton's in the example above—we may take it that no disparity exists between the writer's self-regard and his or her attitude toward readers. Still, writing is a reasoned activity. The very act of writing involves stepping back from emotion, and even genuine feeling, when powerfully expressed in prose, must be focused by a guiding intelligence.

Second, people are not always convinced by argument. Yet it is often necessary to persuade them to do what is right, and the necessity justifies emotive persuasion. Finally, argument itself is not always morally impeccable. It may be misused, as when an unscrupulous writer tries to bamboozle us with specious logic or doctored evidence.

But when we have admitted all this, it remains true that emotive persuasion is of a lower moral order than argument. Even emotional appeals urging us to high and just causes involve manipulation. This does not mean that all such persuasion is to be despised and avoided. At times emotion must be enlisted on the side of the angels. It does mean that decent people, when they write, should use the emotive power of persuasion carefully and responsibly, and that, when they read, they should be on guard against those who use it on them.

Both kinds of persuasion, however, merit our attention. In the next chapter we shall look at argument; in the following one, at emotive persuasion. In practice, of course, the categories are not so watertight. Argument and emotive persuasion often work together, though when they do, one mode or the other dominates.

---

2. Elsewhere in her article Dalton also cites facts about the family diet. Nonetheless she appeals throughout to our feelings.

48

Argument

INTRODUCTION

Argument divides broadly and imperfectly into two types: induction and deduction. *Induction* supports its conclusion upon specific facts. If, for example, you wanted to argue that the man next door had a bad temper, you would cite occasions when he became very angry for inadequate reasons. Such cases would constitute evidence, and your claim would stand or fall according to the strength of the evidence.

The other kind of argument, *deduction,* begins with a broad premise and seeks to work out the conclusion from it. Thus you might argue the case of your neighbor's temper:

All redheads are quick-tempered.
Mr. Johnson is a redhead.
Therefore Mr. Johnson is quick-tempered.

(This form of argument is called a syllogism, and we say more about it later in the chapter.)

In practice we often mix the two kinds of argument. The premises upon which a deductive conclusion rest usually need to be supported by evidence. Is it really true that redheaded people have quick tempers? What is the evidence?

Both types of argument are subject to logical rules. In the case of deduction, they are called rules of inference; in the case of induction, rules of evidence. These are numerous and complicated, and all we can do here is to suggest a few of the rudimentary ones.

In addition to deductive and inductive reasoning, an argument usually includes concessions. *Concession* is admitting premises or evidence contrary to what you hope to establish. Skillful arguers do not ignore the dark side of their claims, hoping that if they say nothing readers will not ask embarrassing questions. Instead they point out the negative evidence and try to neutralize it. Anticipating the objections opponents may be expected to raise is an important aspect of arguing effectively.

The rest of this chapter is devoted to deduction, induction, and concession. In a brief final section we touch upon the practical problems of organizing an argument.

DEDUCTIVE ARGUMENT

THE SYLLOGISM

The classical form of deductive argument is the *syllogism* of Aristotelian logic. (The logic laid out by Aristotle in the fourth century B.C. is only one of several logical systems.) The argument about Mr. Johnson's bad temper being deduced from his red hair takes the form of a syllogism. The syllogism has three parts, or propositions:

	B A
MAJOR PREMISE	All redheads are quick-tempered.
	C B
MINOR PREMISE	Mr. Johnson is a redhead.
	C A
CONCLUSION	Therefore Mr. Johnson is quick-tempered.

Each part of a syllogism is a proposition consisting of a subject and predicate joined by the linking verb (also called the "copula") *is,* or an equivalent verb. The positions of subject and predicate—of which there are a total of six in the syllogism—are filled in by the terms—of which there are three: the *major term* (A, or "quick-tempered"), the *middle term* (B, or "redhead"), and the *minor term* (C, or "Mr. Johnson").

Logicians classify syllogisms into four basic types, or "figures," according to how the middle term functions. For instance, if that term is both the subject of the major premise and the predicate of the minor premise—as in our example—the syllogism belongs to the first figure. The figures are subject to rules governing the manner in which valid conclusions may be inferred. Violating these rules so that a conclusion does not follow constitutes a fallacy.

We have not space here to discuss the various figures and rules and fallacies of syllogistic reasoning. Though we mention several of the common fallacies later, a full treatment of these matters belongs in a textbook of logic. It is important, however, to understand that formal logic, like algebra or geometry, is subject to rule, and that if you use such logic in an argument, you are responsible for knowing the rules.

It is also important to understand that logical validity is not the same as empirical truth. A syllogistic conclusion may be valid in that it follows from the premises according to the rules of inference, and yet it may not be true. Its truth depends upon the truth of the premises. It is valid that Mr. Johnson is quick-tempered. But whether it is true rests upon the truth (1) of the major premise that all redheads are quick-tempered, and (2) of the minor premise that Mr. Johnson has red hair. In this case the minor premise is easy to check. The major premise, however, is impossible to prove: even if all the redheads we know are quick-tempered, we cannot know every redhead in the world and so cannot assert that irritability applies to the entire class. Consequently we cannot be sure that the conclusion about Mr. Johnson is true, even though it is certainly valid.

On the contrary, it is possible for a conclusion to be true despite being logically invalid:

> All redheads are quick-tempered.
> Mr. Johnson is quick-tempered.
> Therefore Mr. Johnson is a redhead.

Now we have looked at Mr. Johnson and he does indeed have red hair: the conclusion is true. But it is not valid. The syllogism is flawed by the mistake logicians call the "undistributed middle" (discussed more fully on pages 476–78).

THE ENTHYMEME

In written argument syllogisms are often condensed into the form called enthymemes. An *enthymeme* is simply a syllogism with one of its parts implied rather than stated:

> Mr. Johnson, having red hair, is quick-tempered.

This enthymeme is a syllogism minus the major premise that all redheads are quick-tempered. It can also happen that the minor premise is left implicit:

> Mr. Johnson, like all redheaded people, is quick-tempered.

Sometimes the conclusion may be left unsaid:

> Mr. Johnson is redheaded, and you know what they say about redheads.

Enthymemes make for convenience in argument, but they multiply the chances for error. It is easy enough to make mistakes even when all the parts of a syllogism are laid out in order. The problem of such reasoning is further complicated by the fact that an argument often consists of a series of syllogisms or enthymemes, so arranged that the conclusion of one becomes a premise in the next.

USING SYLLOGISTIC REASONING

As we have seen, the truth of a syllogistic conclusion hangs upon the truth of the premises. Thus when you use this form of argument, you must begin from premises that cannot easily be disputed. Ideally, they should be axiomatic—that is, self-evident truths. Euclidean geometry, for instance, is based upon axioms (or "postulates," as they are also called), such as "a straight line is the shortest distance between two points" or "the sum of the angles of a triangle must be 180°."

However, in the messy world of social, political, literary, and moral controversy—which is where argument usually finds itself—axioms are hard to come by. Consequently, proof in the rigorous sense of mathematics is rarely possible (positive proof, at least; proving that something is not true is not so difficult). The best a skillful arguer can do is to start from premises which readers accept, even though the propositions are not self-evident.

That is the method, for example, of the Athenian philosopher Socrates (470?–399 B.C.), the teacher of Plato and the central figure in Plato's *Dialogues*. In one of these—*The Republic*—Socrates defines the ideal city-state. He proceeds by asking questions of those who are arguing with him. His questions, of course, are artful premises from which he is able ultimately to deduce the conclusions he wants his fellow disputants to accept.

At one point he hopes to convince them that in the Utopian City dramatic poets would not be allowed to practice their art. To his listeners this seems a perverse ideal, for drama was an important element in the civic life of ancient Athens. Socrates' strategy—a bit oversimplified—is first to get the others to admit that it is shameful to show emotion in public, especially fear, grief, or coarse laughter, a moral proposition which well-bred Athenians would assent to. Next he leads them to agree that dramatic poets simulate grief or ribaldry and thereby induce the audience to give way to such emotions in real life. Thus it follows that the ideal Republic should exclude these poets.

We can lay out the syllogism like this:

MAJOR PREMISE	A condition of the ideal state is that citizens restrain their emotions.
MINOR PREMISE	Dramatic poets induce citizens not to restrain their emotions.
CONCLUSION	Therefore dramatic poets must be excluded from the ideal state.

ATTACKING SYLLOGISTIC REASONING: DISPUTING THE PREMISES

Arguments like that of Socrates can be attacked in two ways: by denying that the premises are true, or by showing that the logic violates rules of inference so that—even if the premises are true—the conclusion need not follow.

In the case of Socrates' banishment of dramatists, the logic is legitimate, so that anyone wishing to argue against his conclusion must attack his premises. These are, of course, vulnerable. Even if one agrees that people should strive to be rational (and one does not have to), it may be argued that rationality is not well served by repressing emotion. It is more reasonable, it might be claimed, to allow an institutionalized, socially acceptable outlet for emotion, like the Athenian theater, to exist. Nor is it self-evident that an audience expresses more emotion in private life after experiencing a play. In fact, the drama may purge them of emotions which, if not harmlessly discharged at the theater, would prove dangerous to the stability of the state.

One way of showing the error of a premise is to push it to extremes, to make it absurd by showing that obviously silly consequences must follow from it. The method is called *reductio ad absurdum.* A humorous example occurs in Lewis Carroll's *Through the Looking Glass.* In the company of the irascible Red Queen, Alice is engaged in a strange game of chess and is a little bored:

> . . . there would be no harm, she thought, in asking if the game was over. "Please, would you tell me——" she began, looking timidly at the Red Queen.
>
> "Speak when you're spoken to!" the Queen sharply interrupted her.
>
> "But if everybody obeyed that rule," said Alice, who was always ready for a little argument, "and if you only spoke when you were spoken to, and the other person always waited for you to begin, you see nobody would ever say anything, so that——"
>
> "Ridiculous!" cried the Queen. "Why, don't you see, child——" here she broke off with a frown, and, after thinking for a minute, suddenly changed the subject of the conversation.

ATTACKING SYLLOGISTIC REASONING: EXPOSING FAULTY INFERENCE

When premises cannot be denied, it may be possible to show that a conclusion is fallacious—that is, does not validly follow from the premises. A number of fallacies exist. Here are a few of the more common.

Contradiction   It is a basic law of logic that if you assert that "A is B" you cannot also assert that "A is not B." Of course, you say, that's obvious. But when, instead of clear symbols like A and B, the terms of an argument are abstractions such as "freedom," "democracy," "economic equality," and so on, it is easier than you may think to violate the law of noncontradiction. The writer Dorothy L. Sayers reveals such a contradiction in the claims made by Marxian socialists:

Equality—even social equality—is conceived [by Socialists] in economic terms; the theory is that economic equality will automatically produce social and political equality.

Now, economic equality cannot, as we have seen, co-exist with economic freedom. It requires that the full control of industry shall be taken over by the state. This means that individual liberty is abolished in the very sphere which is declared by the dogma [of Socialism] to be the sphere of man's and the state's essential nature. The individual must therefore be coerced or cajoled into surrendering not only his economic liberty but all his individual economic values to the state.

Sometimes a contradiction lurks below the surface of an argument:

It is sobering but true that we must stand responsible for the folly and crimes of others, insofar and as long as we remain free to set the standard of national character. Thereupon rests the burden of all our blessings.

While statistics of crime and wantonness are made by a small minority, these few people are nevertheless products of the environment which we, the majority, have created.

Sidney L. De Love

The writer assumes these premises: (1) that some people (criminals) are products of their environment, and (2) that other people ("we, the majority") are responsible for creating that environment. These propositions contradict each other. Both apply to "people," and nowhere does De Love suggest or assume that "criminal people" and "we, the majority people" are essentially different. But to assert that people are products of their environment denies that they have free will and can determine their own actions. To assert that people must accept responsibility for creating their environment and to urge that they create a better one supposes that they *do* have free will and are, within limits, responsible for their actions.

Begging the question　To beg the question (the traditional name of this fallacy is *petitio principii*) is to claim as true a conclusion not proven at all but simply assumed somewhere along the line of your argument. It is an easy mistake because you can assume a conclusion so subtly and so early in an argument that you don't know you have done so.

Exposing a petitio is an effective strategy of refutation. The writer and critic C. S. Lewis shows that the fallacy underlies a claim by the poet T. S. Eliot that only poets can function as true critics of poetry:

Mr. Eliot is ready to accept the verdict of the best contemporary poets on his criticism. But how does *he* recognize them as poets? Clearly, because he is a poet himself; for if he is not, his opinion is worthless [according to his own argument]. At the basis of his whole critical edifice, then, lies the judgment "I am a poet." But this is a critical judgment. It therefore follows that when Mr. Eliot asks himself, "Am I a poet?" he has to *assume* the answer "I am" before he can *find* the answer "I am"; for the answer, being a piece of criticism, is valuable only *if* he is a poet.

The vicious circle   A circular argument is a form of begging the question. In the circular fallacy, premise and conclusion actually depend upon each other. But if the truth of the premise is contingent upon that of the conclusion, and that of the conclusion upon the truth of the premise, the propositions are like two limp pieces of spaghetti trying to prop one another up.

In the following passage the political essayist William F. Buckley exposes a vicious circle in the argument that capital punishment ought to be abolished because it is unusual. (Buckley is overtly summarizing a criticism advanced by Professor van den Haag):

> The argument that the death penalty is "unusual" is circular. . . . What has made capital punishment "unusual" is that the courts and, primarily, governors have intervened in the process so as to collaborate in the frustration of the execution of the law.

Buckley points out that those who claim that the death penalty ought to be abolished because it is unusual are the people who have made it unusual because they think it ought to be abolished. Notice that Buckley's point implies nothing about whether capital punishment should, or should not, be abandoned. It merely says that the claim of unusualness does not make a logically valid argument for abolition.

The undistributed middle   The undistributed middle is a common error in arguments purporting to be logical. It is a particularly insidious fallacy since, on the surface, it seems like an impressive syllogism and may easily fool people untrained in logic. Moreover, it is socially and politically dangerous, as the scholar Jacques Barzun stresses:

> The typical flaw of weak minds—it's almost like a visible crack in a piece of pottery—is the habit of committing the fallacy known as the "undistributed middle." It runs like this: "All Communists are atheists. Mr. Jones is an atheist. Therefore he is a Communist." This is guilt by mere association—of ideas. . . . This fallacy, which is not peculiar to our country, is notoriously a danger to life and limb. It is the root idea of the purge, and it tends to prevail wherever "the people" are in power and feel responsible to cultures, ideas, -ologies. The French Revolution gives massive examples of this abstractionism, which is far more murderous than the simple plan of destroying your actual enemies.

The expression "undistributed middle" requires explanation. As we have seen, a valid syllogism of the first figure has the form:

	B	A
MAJOR PREMISE	All redheads are	quick-tempered.

	C	B
MINOR PREMISE	Mr. Johnson is a	redhead.

CONCLUSION   Therefore Mr. Johnson is quick-tempered.

C   A

(Remember that the conclusion, while it is in this case valid, need not therefore be true.)

In syllogisms of this type the middle term is, by definition, the term expressed in both premises ("redhead"), and it must be distributed. In logic "distributed" means "applied to all members of a class": to say "all redheads are quick-tempered" distributes the term "redheads" with regard to the term "quick-tempered." Since the middle term is properly distributed, the conclusion follows. What applies to redheads in general must apply to Mr. Johnson. If we use circles to illustrate the terms—the traditional way of illustrating syllogisms—we see that circle C, being within circle B, must therefore also be inside circle A:

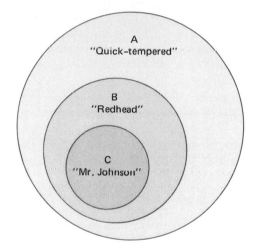

Now consider this "syllogism":

MAJOR PREMISE   All redheads are quick-tempered.

B   A

MINOR PREMISE   Mr. Johnson is quick-tempered.

C   A

CONCLUSION   Therefore Mr. Johnson is a redhead.

C   B

The middle term (now A, or "quick-tempered") is no longer distributed; nowhere is it posited that "all quick-tempered people are redheads." Thus the conclusion is logically fallacious. The failure to distribute the middle term properly can be seen in the circular diagram, where it is obvious that C need not be included within B:

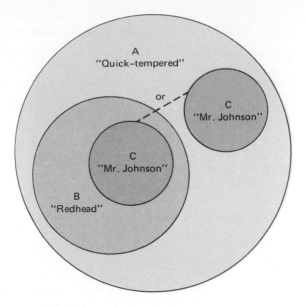

Two alternative circles are necessary for "Mr. Johnson" because while the term must be inside circle A, it *may or may not be* inside circle B. In other words, it is possible for Mr. Johnson to be quick-tempered without being redheaded. Therefore the conclusion "Mr. Johnson is redheaded" is not valid. (It won't do to say, "But he may be." Of course he may, but that is not what the conclusion asserts. The conclusion asserts that he is.)

Equivocation   Equivocation results from ambiguity—that is, using the same term in two different senses. Consider this argument:

MAJOR PREMISE   Democracy is government by the people.
MINOR PREMISE   A single-party Socialist state is a democracy.
CONCLUSION   Therefore in such states people govern themselves.

The equivocation takes place in the term "democracy." In a single-party Socialist state, "democracy" means not so much government by the people as government in the name of and for the people. The point is not that this is an illegitimate sense of "democracy" (whether it is or isn't may be argued endlessly); the point is rather that it is used in a different sense than the term has in the major premise. Consequently what is asserted as true of "democracy" in the major premise does not necessarily apply to "democracy" as it is used differently in the minor premise.

Such ambiguities are sometimes deliberate and cynical tricks. But they are probably more often unintentional self-deceptions. Either way they are fallacious and misleading. Arguments sometimes proceed end-lessly and tediously because the disputants are using a key term to mean different things. Neither person may be deliberately equivocating, but neither sees that an ambiguity exists in their common use of the term.

Such a dispute involves the Cheshire Cat in *Alice in Wonderland*. The Cat has the odd and disconcerting ability to make any part or all of his body disappear and reappear at will. While he and Alice are watching a croquet match on the grounds of the King and Queen of Hearts, the Cat renders all of his body invisible except his head, which sits spectrally among the branches of a tree. The sight annoys the King, who runs to fetch the executioner to behead the Cat. Alice has gone away for a moment:

> When she got back to the Cheshire Cat, she was surprised to find quite a large crowd collected round it: there was a dispute going on between the executioner, the King, and the Queen, who were all talking at once, while all the rest were quite silent, and looked very uncomfortable.
>
> The moment Alice appeared, she was appealed to by all three to settle the question, and they repeated their arguments to her, though, as they all spoke at once, she found it very hard to make out exactly what they said.
>
> The executioner's argument was, that you couldn't cut off a head unless there was a body to cut it off from: that he had never had to do such a thing before, and he wasn't going to begin at *his* time of life.
>
> The King's argument was, that anything that had a head could be be-headed, and that you weren't to talk nonsense.
>
> The Queen's argument was, that if something wasn't done about it in less than no time, she'd have everybody executed, all round. (It was this last remark that had made the whole party look so grave and anxious.)

The equivocation is in the term "behead," which means one thing to the executioner and another to the King. The Queen's argument, by the way, is not a case of equivocation. Rather it is what is sometimes called the "argument of the club"—that is, violence or the threat of violence. (The Cat escapes King, Queen, and executioner, with head and body intact.)

Ignoratio elenchi This formidable Latin phrase translates as "igno-rance of proof." What it means is simple enough: proving the wrong thing. Suppose, for instance, you wish to argue against the death pen-alty. You cite statistical studies which show that capital punishment has no detectable deterrent effect: for example, states without the death penalty have no increase in murder when compared with similar states where the penalty still exists.

The fallacy is that you have not shown proof that the death penalty is wrong. Capital punishment may be justified on other grounds: retributive justice, for example, or cost effectiveness (it is cheaper to execute murderers than to keep them in prison a long time).

Often you can avoid this fallacy by phrasing an argument more carefully. You should not begin by saying that "the death penalty should be abolished because it does not deter murder." You cannot even begin by saying that "the only legitimate reason for retaining the death penalty is the claim that it deters murder, but it does not have a deterrent effect," for that premise still begs the question of whether deterrence *is* the "only legitimate reason."

You must list all the justifications used to support capital punishment and answer each in turn. Then, while you will not have proven absolutely that execution should be abolished, you will have made a respectable case. And no one can say that what you have proven is beside the point.

False dichotomy   A false dichotomy rests on the unreasonable assumption that only two alternatives exist; if one does not hold, the argument runs, the other must. Either-or conditions do exist: if a human being is not alive, then he or she is dead. But more often than not, such alternatives which reduce complex moral, social, or political problems to a simple "either-or" do so fallaciously.

The philosopher Walter Kaufmann points to the frequency of false dichotomy in the arguments urged by apologists for Christianity (an "apologist" in this sense is a justifier or defender):

> Renouncing false beliefs will not usher in the millennium. Few things about the strategy of contemporary apologists are more repellent than their frequent recourse to spurious alternatives. The lesser lights inform us that the alternative to Christianity is materialism, thus showing how little they have read, while the greater lights talk as if the alternative were bound to be a shallow and inane optimism. I don't believe that man will turn this earth into a bed of roses either with the aid of God or without it.

Nonqualification   Nonqualification means that premises are stated too absolutely—that is, without any qualification. You may correct the fallacy (also known by the Latin tag *secundum quid*) by rephrasing the premises. Consider this argument:

Colleges should not be in the business of subsidizing semiprofessional sports.

College basketball is a semiprofessional sport.

Therefore college basketball should be abolished or reformed.

The inference is valid. But the term "college basketball" is too absolute. It should be replaced by a more qualified term, such as "some college basketball." The appropriate conclusion—granted the major premise—is that "some college basketball [that which is semiprofessional] should be abolished or reformed."

Non sequitur Literally *non sequitur* means "it does not follow." We generally use the phrase as a blanket term for any argument in which the conclusion does not follow from the premises and in which there is not even an appearance of logical thought progression. Take the argument "A college education has no value; this country was built by people who never went to college." Even if we grant the truth of the second proposition (and, of course, it is not unreservedly true), it has nothing to do with the conclusion.

Non sequiturs are sometimes deliberately employed for comic effect:

"I'll buy you an ice-cream cone, dear."
"All right, but don't make it strawberry."
"Why don't you like strawberry?"
"Well, that's because Uncle Harry wouldn't go to see the tigers."

There follows a long, intricate chain of cause and effect comically connecting Uncle Harry, tigers, and strawberry ice cream, thus filling in the original non sequitur.

False analogy In logic and in science some types of analogy constitute genuine proof. In general argument, however, the term "analogy" designates a resemblance between two or more things which may help to convince readers that one's ideas are right but which is not proof. We shall call such analogies *rhetorical*. They become fallacies when they are used, not as illustrations or as evidence, but as claims of proof.

Here is a rhetorical analogy used to support the contention that in our time the novel is a dying form of art, worked out and incapable of further development:

It is erroneous to think of the novel—and I refer to the modern novel in particular—as of an endless field capable of rendering ever new forms. Rather it may be compared to a vast but finite quarry. There exist a definite number of possible themes for the novel. The workmen of the primal hour had no trouble finding new blocks—new characters, new themes. But present-day writers face the fact that only narrow and concealed veins are left them.

Jose Ortega y Gasset

Now such comparisons often make an effective strategy in argument. They may persuade us and we say, "Yes, that is how the matter is." But always we should remember that an analogy illustrates an idea, perhaps in very graphic and compelling fashion, but that it is not proof.

It cannot be proof simply because the analogical subject is of a different order than the main subject, and therefore what applies to the first does not necessarily apply to the second. Quarries *do* get worked out and must be abandoned. If the novel were exactly like a quarry, we would be forced to conclude that the same fate would befall it, if not now, then later. The novel, however, is *not* a quarry, and consequently the conclusion does not follow that it must inevitably be "worked out." Neither does it follow, of course, that the art form of the novel may *not* be drying up (to vary the metaphor).

In one sense rhetorical analogies are always false, and when they are offered in lieu of evidence or logic, we should not grant them more worth than they have. Usually, however, we reserve the label "false analogy" for cases where the analogy is glaringly wrong, or is, at least, one you disagree with. Pointing out such misleading analogies in the arguments of others is an effective way to answer them. The philosopher Bertrand Russell takes issue with those who draw a comparison with poison gas to support the claim that the H-bomb will never be used in war:

> Another widespread delusion is that perhaps in a great war H-bombs would not be employed. People point to the fact that gas was not employed in the Second World War. They forget that gas had not proved a decisive weapon even in the First World War and that in the meantime gas-masks had been provided which were a complete protection. Any analogy is therefore entirely misleading.

INDUCTION

Induction begins with particulars and works from them to conclusions. When you deal with a very small class, you may be able to arrive at binding conclusions by means of induction. Suppose a political organization consists of exactly twenty members and that each of these people believes that the best form of government is a king with unlimited power. We may conclude that the organization accepts the principle of absolute monarchy. On the other hand, we could not conclude, on the basis of this evidence, that the group wished to establish a monarchy.

Such closed, finite classes are rare, however, in the kind of controversies over which argument generally arises. We cannot describe with certainty what political principle "Americans" find essential; there are simply too many of us. In actual argument induction rarely leads to proof but rather to conclusions of varying degrees of probability. The probability may approach certainty. It is fair to conclude that "very few Americans believe in absolute monarchy." But we cannot say that "no American believes in absolute monarchy," for even if we asked each one, we could never be sure that some of our first respondents had not changed their minds before the poll was finished.

On the other hand, disproof is often possible by means of induction. People guilty of nonqualification are especially vulnerable to inductive refutation. It would be easy to bring forward evidence refuting the proposition that "most Americans believe in absolute monarchy."

EVIDENCE

In practice an inductive argument usually begins with the conclusion to be supported or denied and then cites reasons why it is or is not true. The reasons constitute evidence, and evidence may take several forms. Often it consists of assertions unsupported by details but which appeal to our common sense, to our awareness of how things are. Again, evidence may consist of one or more illustrations, which stand as typical of a wide number of cases. Best of all is evidence in the form of specific facts, often, though not necessarily, statistical. In a moment we shall look at examples of these kinds of evidence. But first a word about three criteria which all evidence must meet.

1. *Evidence must be accurate and complete.*

This criterion means not only that evidence must be true, it must also be the whole truth. Evidence must not be fabricated or "adjusted," nor may it be selected to include only what fits the argument. If some facts tell against your position, admit them and neutralize them if you can. Hiding contrary evidence by silence is dubious honesty and poor strategy. An opponent will find the dirt under the rug and then your position will be weaker than if you had admitted the embarrassing facts in the first place.

2. *Evidence must be relevant.*

Evidence must truly bear upon the issue. Often relevance is obvious. If unemployment increases and inflation gets worse, this is indisputable evidence that the economy is in trouble. But sometimes relevance must be demonstrated. The evidence will need to be interpreted to show its significance.

It may even be possible to interpret what appears to be negative evidence to give it positive value. For instance, a writer arguing that nineteenth-century Americans commonly believed in a romantic conception of marriage (as distinguished from a practical one such as economic or social advantage) is confronted by the fact that throughout the century the divorce rate constantly rose. At first glance this fact might seem to tell against his claim for the primacy of romantic marriage. However, when interpreted as the writer thinks it should be, this fact turns into evidence supporting his position:

> This increase [in the divorce rate] may seem to belie what has already been maintained—that the nineteenth century was devoted to the ideal of romantic marriage. Actually, more closely considered, the two would seem really to be in harmony. Divorce frequently indicates not a low ideal of marriage but a high one. When people are married by arrangement, they expect rather little of marriage emotionally. They may therefore drift

along indefinitely, and even infidelities, provided they do not become socially notorious, may cause little difficulty. Romantic marriage, however,
is founded upon the expectation of great or even transcendent emotional
return. When this high ideal is unrealized, a clean break, by means of
divorce, often seems the most idealistic solution.

George R. Stewart

Interpreting evidence in a damaging way is an effective strategy for
attacking an argument. Mark Twain once wrote a book popularizing
the claim that Shakespeare's plays were written by Sir Francis Bacon.
(While it has little scholarly support, the notion that William Shakespeare did not compose the plays attributed to him has been held in
various forms by a number of people.) Obviously Twain must prove
that Shakespeare could not have created the plays himself. As evidence
Twain cites the fact that Shakespeare's death did not occasion much
public notice:

> Yet his death was not an event. It made no stir, it attracted no atten-
> tion. . . .
>   His death was not even an event in the little town of Stratford.

Twain then compares this lack of attention with the very different
case of writers in his own time, and specifically with a literary celebrity
like himself. He concludes that Shakespeare was not a celebrated person
and therefore could not have written the plays he is credited with.
   His argument can be reduced to this syllogism:

If a man is a famous writer, his death will be a notable event.

Shakespeare's death was not a notable event.

Therefore Shakespeare was not a famous writer.

But what concerns us is not the deductive logic. It is the evidence
upon which Twain's major premise rests: that Shakespeare's death was
"not an event. . . . it attracted no attention." For one thing the claim
is not quite accurate. But more to the immediate point is a second
consideration: Twain misinterprets the evidence (honestly so) and thus
gives it a false relevance. He assumes several things. First, that Shakespeare's plays had the same kind of national importance they had acquired by Twain's time. Second, that the early seventeenth century
(Shakespeare died in 1616) gave to playwrights the same popular ac-
claim that famous writers of his own day were accorded. Finally, that
similar media—newspapers, magazine, books, and so on—existed for
recording popular events. None of these assumptions is true (the argument in fact is a revealing example of a false analogy). Consequently
the evidence is not relevant: the fact that Shakespeare's death was
not widely commented upon in surviving records of the time in no

way leads us to conclude, even probably, that Shakespeare was not the author of his plays.

3. *Evidence must be sufficient.*

Sufficiency means that the evidence must be weighty enough to support the conclusion. Because the issues with which arguments typically deal are complex, evidence usually exists on both sides. One can, then, bring forward isolated evidence which is both true and relevant, but which, by itself, is not strong enough to support the conclusion.

It is true, for instance, that among college graduates, as a group, there are relatively more political liberals than among nongraduates (setting aside the problem of what "liberal" means). But in isolation this fact is not sufficient to argue that colleges teach liberalism. Even if college did make these people liberal, we need further evidence to show that the effect resulted from deliberate educational policy or practice.

THE ASSERTIVE ARGUMENT

The evidence which must meet the tests of truth, relevance, and sufficiency may be, we said, of three kinds. The first includes assertions that appeal to common sense. In an assertive argument the writer advances general reasons to support his or her position, but not specific facts or examples. Readers accept or reject the general claims to the degree that these accord with their own experience. Here, for instance, is an assertive argument that advertising creates the consumer society:

> In a simpler time, advertising merely called attention to the product and extolled its advantages. Now it manufactures a product of its own: the consumer, perpetually unsatisfied, restless, anxious, and bored. Advertising serves not so much to advertise products as to promote consumption as a way of life. It "educates" the masses into an unappeasable appetite not only for goods but for new experiences and personal fulfillment. It upholds consumption as the answer to the age-old discontents of loneliness, sickness, weariness, lack of sexual satisfaction; at the same time it creates new forms of discontent peculiar to the modern age. It plays seductively on the malaise of industrial civilization. Is your job boring and meaningless? Does it leave you with feelings of futility and fatigue? Is your life empty? Consumption promises to fill the aching void; hence the attempt to surround commodities with an aura of romance; with allusions to exotic places and vivid experiences; and with images of female breasts from which all blessings flow.
>
> Christopher Lasch

Assertions concerning matters of obvious truth may be sufficient evidence. With more disputed and complex issues, however, they have less weight. In such cases an assertive argument should be regarded with a healthy skepticism and not passively accepted (or rejected); suspend your judgment until you have better evidence.

Sweeping generalizations about social and political questions are easy. Sometimes generalizations are all that can be argued because detailed evidence does not exist. Indeed, such assertions may act as useful guides, like scientific hypotheses, to areas where evidence is wanting. At the same time, remember that, however brilliant and compelling it may seem, an assertion on such questions does not make very strong evidence.

THE ILLUSTRATED ARGUMENT

A step above assertive arguments are those supported by illustrations. Examples provide evidence of the truth of a claim, even though one or two are not often enough to make a strong case. In the following paragraph the writer uses an example to back his point that the wisdom of the East is overrated:

> One hears a good deal nowadays about the inherent superiority of "the East" over "the West," because the East has learned the wisdom of contemplation in place of our busy activity. Those of us who go there—if lucky—are handsomely received by a wealthy and cultivated Hindu who, when he is not contemplating, directs several publications that denounce Western materialism as the enemy of civilized life. But as his guest one cannot help noticing that the palace he lives in and the jeweled marvels of his private museum are fairly material goods, balanced by the abject degradation of the naked paupers outside.
>
> Jacques Barzun

The value of an illustration as evidence depends upon how typical it is. Writers will often assert typicality: "A characteristic example . . ." they will say. But the claim must usually be taken on faith by readers, who rarely know enough about the subject to judge whether an example is truly characteristic or an odd case selected because it mortises neatly into the argument. Therefore keep an open mind when a proposition is supported only by examples, unless you have the experience to judge their typicality. As with the assertive argument, the writer's contention may not necessarily be untrustworthy; it may simply need stronger evidence before it can be accepted.

THE FACTUAL ARGUMENT

An indisputable fact makes the best evidence. Facts are especially effective in disproof; generalizations too sweeping in their claims can easily be refuted by citing facts that do not fit them. The Canadian historian Ralph Heintzman calls upon the fact of Russian literature to answer the claim that censorship curtails literary activity. (Heintz-man's argument does not imply that he favors rigorous censorship):

> It is by no means self-evident, as many today assume without adequate reflection, that the existence of censorship inevitably entails a loss of artistic power. As everyone knows, nineteenth-century Russia was subject to a high degree of political censorship (though not nearly as high as in the Soviet Union today), yet it also witnessed one of the greatest flowerings of literary creativity in the history of the world, thus perhaps confirming Northrop Frye's suggestion that literature actually flourishes in difficult circumstances.

Often facts are cast in the form of statistics. Here is a case where scores from the Scholastic Aptitude Test are cited to support the claim that contemporary American education

> has produced new forms of illiteracy. People increasingly find themselves unable to use language with ease and precision, to recall the basic facts of their country's history, to make logical deductions, to understand any but the most rudimentary written texts, or even to grasp their constitutional rights.
>
> Christopher Lasch

In his next paragraph Lasch supports his contention:

> One study after another documents the steady decline of basic intellectual skills. In 1966, high school seniors scored an average of 467 points on the verbal section of the Scholastic Aptitude Test—hardly cause for celebration. Ten years later they scored only 429. Scores on the mathematical part of the test dropped from an average of 495 to 470.

Statistics, however, are not sacrosanct. Like all evidence they are subject to the tests of accuracy, relevance, and sufficiency. In the case of the declining scores in the Scholastic Aptitude Test, the evidence is accurate: during the period cited, scores did go down. On the other hand, the relevance and sufficiency of the evidence are open to dispute.

Consider this passage by another writer, who is arguing against the simplistic claim that the alleged failure of American education has been caused by bad educational theory and practice for which the remedy is "back to basics," the three Rs. (It should be emphasized that a simplistic "back to basics" is not Professor Lasch's solution; his analysis of the ultimate cause of the decline of learning is more complex and subtle.)

> In the August 28, 1977, issue of *The New York Times Magazine,* Frank E. Armbruster, of the Hudson Institute, performed the service of condensing the full Back to Basics argument into a few pages. He stated, first of all, that the ability of American children to read and write had declined over the past ten years; this decline, he said, was the result of the failure of the schools to educate children as well as they used to. For this failure

he blamed school reformers and many of the innovations that had been made in the school system since the nineteen-thirties. . . . Children must, he concluded, learn basic skills and basic information, and the only way to insure that they do is to throw out these innovations and go back to the old-fashioned system of rote work, grading, and drill.

The first problem with the argument is that there is no hard evidence for any of Armbruster's propositions. Several, though not all, indicators show that the ability to read and write has declined over the past ten years, but this recent decline is difficult to assess, given the change in the school population. That this decline is the fault of the schools is simply a guess, since—if the decline exists—there are so many other possible causes, such as the decline of parental discipline or the omnipresence of television. As for the contention that traditional methods are superior to all others, there is some evidence that this is nonsense. In the nineteen-forties, the last time the Back to Basics movement was riding high, a respectable group of educators did a test on college students to determine the effectiveness of progressive education; the test, which continued for eight years, showed that students who graduated from progressive schools had a slightly higher academic achievement in college than students who graduated from traditionally run schools. This one test may not be proof positive—the variables in educational testing are nearly impossible to control—but in all the hundreds of studies done on children over the past few decades it has never been shown that children learn better by drill and rote methods than by any other system. The Back to Basics people may be right about certain of the innovations, but since they condemn opposite theories and forms of education in the same breath—nineteen-fifties-style "progressive education" *and* the New Math and Social Studies—they cannot show how or why, or even whether, they are right. The blanket indictment of all new methods makes for an implausible argument.

Frances Fitzgerald

Thus even though the decline in SAT scores is indisputable, it is not necessarily relevant to the claim that the cause is bad teaching. Nor, even if it is relevant, is it sufficient to prove that claim, not by itself. Its relevance is arguable because of the possibility of counterexplanation. Granted that reading scores have fallen (and that the test accurately measures reading skill), the reason may be the thousands of hours children spend in front of television sets rather than reading. (Nothing develops reading skill like reading). The sufficiency of the statistic is doubtful because it is only one piece of evidence and others suggest that the decline is less great.

Statistics, as you see, are tricky. A sample must be typical and not skewed in such a way as to make the figures unreliable. For example, Ferguson cites a study showing that graduates of progressive schools "had a slightly higher academic achievement in college." But the parents of such students probably included a larger proportion of wealthy college graduates, interested in education and ambitious for their children, than did the parents of the other group. Private progressive schools cost money; public ones are found in well-to-do communities with relatively large populations of successful, professional people. Such people are likely to encourage, or to pressure, their children to do well in school. This may explain why these children do better in college, not simply that they graduated from progressive schools.

SOME FALLACIES OF INDUCTION

Most of the fallacies plaguing deductive argument apply also to induction: false analogy, begging the question, circular argument, false dichotomy, ambiguity, and so on. Such mistakes are called either *material fallacies* (like the first four) because they involve the matter (or factual content) of the argument, or *linguistic fallacies* (like ambiguity) because they involve the meanings of words.[1]

In addition to the material and linguistic fallacies that we have already discussed, others apply primarily to induction.

Argumentum ad ignorantiam (argument to ignorance) This fallacy lies in the claim that either (1) something must be true because it has never been proven false, or (2) it must be false because it has never been proven true. For example: (1) "there must be life after death since no one has ever proven that there isn't"; (2) "there cannot be life after death since no one has ever satisfactorily demonstrated that it exists."

Fallacy of complex (or loaded) questions The classic example is: "Have you stopped beating your wife yet?" The trick consists of demanding either a "yes" or "no" to a question which requires a more complex answer. The question, moreover, is so designed that either "yes" or "no" damages the opponent. Lawyers have been known to exploit this fallacy.

Fallacy of composition This argument assumes that what applies to the parts (either all or some) of a complex entity applies to the whole entity. For example: "Universities must be young in spirit because they contain so many young people."

The converse of the fallacy of composition is the *fallacy of division*: that what applies to a whole applies necessarily to some or all of its parts. "Since a university is dedicated to learning, all its students must be dedicated to learning."

---

1. Two other categories of fallacy are formal and emotional. *Formal fallacies,* which apply to deductive reasoning, are those in which the error occurs in the form of the logic: the undistributed middle is an example. *Emotional fallacies,* some of which we shall consider in the next chapter, appeal not to fact or logic but to the emotions of the audience.

Fallacy of consensus gentium (*"agreement of people"*)   Here one argues the truth of a conclusion because most people believe it, either at the present time or universally. "Throughout history most societies have honored their old and experienced members, so age is deserving of respect."

Fallacy of converse accident   The writer generalizes a conclusion from examples that are not typical. "University professors are always interesting and entertaining. My history professor is a fine speaker and tells very funny stories."

Floating comparison   A floating comparison is the comparative or superlative form of a term which is not connected to anything. "More doctors prescribe aspirin." *More* than what? "More doctors prescribe aspirin [than some other drug]"? If so, what other drug? "More doctors prescribe aspirin [today than they did twenty years ago]"? And so on. This still leaves room for the possibility that most doctors do not prescribe aspirin.

Gambler's fallacy   This fallacy ignores the laws of probability and argues that since something has not happened for a long time, its probability of occurrence thereby increases: "I haven't had a good poker hand all night; I'm due one."

Conversely the same fallacy argues that because a chance occurrence has been repeated over and over the odds against its happening increase. But even if a coin has come up heads ten times running, the probability of its being heads on the eleventh toss remain the same: 1 in 2.

Genetic fallacy   In this argument, also called the *reductive or "nothing-but"* fallacy, something is to be understood solely in terms of its origin or genesis and must be accepted or rejected on that basis. For example, take the fallacious insistence that "the word 'aggravate' should not be used in the meaning of 'annoy' because its original Latin sense was 'to make heavier,' hence 'to amass, pile up.'"

The genetic fallacy ignores or denies the significance of change and tries to control the present in the name of the past.

Hasty generalization   This blanket term covers generalizing from limited or inadequate evidence. More specific types are the fallacies of *composition,* of *division,* and of *converse accident* (discussed above) and the *skewed sample* and *slanting* (discussed below).

Misleading context   Evidence is cited out of its context so that misleading conclusions are drawn. A crude example is excerpting the words "most remarkable" as a favorable comment on a play when what the reviewer actually wrote was: "This play is the most remarkable piece of dramatic turgidity I have ever been forced to endure."

Non causa pro causa (*"there is no cause of the kind that has been given as the cause"*) The argument of a false cause assumes that an event caused an effect simply by association. Many superstitions are instances of non causa: for example, the football fan who wears an old T-shirt to a big game because the last time he wore it the team won.

Post hoc ergo propter hoc (*"after this, therefore the consequence of this"*) This fallacy confuses temporal and causal relationships, arguing that since event B followed event A, A must have been the cause of B. It is much the same fallacy as non causa except that the causal relationship usually has a greater appearance of plausibility. For example, one might argue that because greater sexual freedom often follows a war, war causes the freedom. But perhaps war and the loosening of sexual mores are common consequences of yet another cause, or perhaps the war is a necessary condition of sexual freedom but not a sufficient cause (that is, the change in sexual attitudes may require a war in order to occur, but the war is not the real cause of the change).

Red herring A red herring is a false or irrelevant issue. The fallacy consists of switching the argument from something you can't handle to something you can. Thus a husband accused of inattention to his wife might answer that he was a good provider.

Skewed sample The accuracy of statistical evidence depends upon the polling sample truly representing the larger population about whom the conclusion is made. When a sample is not representative, it is said to be skewed. A famous instance of poor sampling occured in a poll taken in 1936 to predict the Roosevelt-Landon presidential election. A now-defunct magazine, the *Literary Digest,* predicted that the Republican Landon would win by a substantial 57 per cent of the popular vote. The prediction was spectacularly wrong. Landon had only 37.5 per cent and was overwhelmed 523 to 8 in the electoral voting. The *Digest* had selected its respondents from telephone and automobile registrations. In 1936 such lists were disproportionately weighted with well-to-do, conservative voters whose preference for Landon was not shared by the population at large.

Slanting Slanting selects or overemphasizes evidence in favor of a particular conclusion and/or ignores or deemphasizes evidence against it. A newspaper reporter, for instance, who stresses the corruption and laziness of municipal officials but ignores evidence of honesty and hard work is slanting evidence toward the conclusion that politicians are reprehensible.

Special pleading   Special pleading means applying conditions in one case (or rejecting them) but not admitting their application in similar cases. A driver who runs stop-signs on the argument that "the other car wasn't close enough to be endangered" is guilty of special pleading when he curses other drivers who cut him off by failing to stop.

Straw man   A straw man is the misrepresentation of an opponent's argument to make it seem weaker than it is. To argue that "evolution claims we are all descended from monkeys" misrepresents evolutionary theory and makes it easier to argue against.

Tu quoque ("you also")   Tu quoque takes several forms. One is to answer a charge by saying that your opponent is guilty of the same thing: "You say I'm a liar? Well, you're a liar yourself." Such an answer may be true and it may be diverting, but it does not address the issue.

A second form of tu quoque is to accuse an opponent of acting in a manner not consistent with what he or she has said. Thus a thief who praises honesty might be answered by tu quoque. But again the argument is irrelevant. If the devil quotes scripture, it does not invalidate scripture.

Finally, tu quoque may consist of contrasting what someone says now with what he or she said in the past. Politicians are fond of revealing each other's inconsistencies by artfully quoting from speeches made in the past. Such inconsistency, however, has no bearing upon the issue, though it may bear upon the character of the politician.

Unqualified source   This fallacy involves citing evidence from a source not qualified to give it: for example, arguing for or against the policy of high interest rates to cure inflation on the basis of random interviews with downtown shoppers.

REFUTATION AND CONCESSION

REFUTATION

To refute is to disprove, to answer another argument. Often refutation is the basic purpose of arguing. But even when your purpose is to advance and support a positive conclusion, you usually have to anticipate and counter opposing claims.

Some of the examples cited earlier in connection with various fallacies are cases of refutation. On page 475 C. S. Lewis refutes T. S. Eliot's point by showing that his argument hides a petitio principii. Alice refutes the Queen of Hearts by a kind of reductio ad absurdum (page 474).

That is how refutation works. It points out weaknesses in logic or evidence. In the following case the historian Henry Thompson Rowell argues that we do not have sufficient evidence to answer the question: How many people lived in ancient Rome at the beginning of the Christian era? He examines various theoretical estimates that have been made and refutes them one by one. For example:

> . . . the size of the population has been worked out on the basis of the amount of grain imported annually from overseas. Here the amount is calculated on a combination of figures, one of which belongs to a period some sixty years after the Augustan Age. The other indispensable factor in this approach is a knowledge of the average consumption per capita. This we do not have.

CONCESSION

Sometimes an opposing argument cannot be refuted; it is neither fallacious nor based on false evidence. Then it must be conceded. As we said, the wiser course—as well as the more honest—is to admit a weakness, not to step around it and hope readers will miss or politely ignore the hole. You must do more, however, than simply acknowledge the hole. You must fill it in if you can—that is, neutralize the contrary evidence and even perhaps turn it to positive account.

A masterly example of concession is this passage by the English writer Dorothy L. Sayers. She is arguing against the prejudiced view that women are less reliable workers than men because they prefer marriage to careers:

> Now, it is frequently asserted that, with women, the job does not come first. What (people cry) are women doing with this liberty of theirs? What woman really prefers a job to a home and family? Very few, I admit. It is unfortunate that they should so often have to make the choice. A man does not, as a rule, have to choose. He gets both. In fact, if he wants the home and family, he usually has to take the job as well, if he can get it. Nevertheless, there have been women, such as Queen Elizabeth and Florence Nightingale, who had the choice, and chose the job and made a success of it. And there have been and are many men who have sacrificed their careers for women—sometimes, like Antony or Parnell, very disastrously. When it comes to a *choice,* then every man or woman has to choose as an individual human being, and, like a human being, take the consequence.
>
> As human beings! I am always entertained—and also irritated—by the newsmongers who inform us, with a bright air of discovery, that they have questioned a number of female workers and been told by one and all that they are "sick of the office and would love to get out of it." In the name of God, what human being is *not,* from time to time, heartily sick of the office and would *not* love to get out of it? The time of female office-workers

is daily wasted in sympathising with disgruntled male colleagues who yearn to get out of the office. No human being likes work—not day in and day out. Work is notoriously a curse—and if women *liked* everlasting work they would not be human beings at all. *Being* human beings, they like work just as much and just as little as anybody else. They dislike perpetual washing and cooking just as much as perpetual typing and standing behind shop counters. Some of them prefer typing to scrubbing—but that does not mean that they are not, as human beings, entitled to damn and blast the typewriter when they feel that way. The number of men who daily damn and blast typewriters is incalculable; but that does not mean that they would be happier doing a little plain sewing. Nor would the women.

Sayers answers two counterclaims: that women would rather marry than work (in the first paragraph), and (in the second) that newspaper surveys show that women " 'are sick of the office and would love to get out of it.' "

She puts the first of these claims in the form of a rhetorical question: "What woman really prefers a job to a home and family?" and concedes its truth in a forceful reply: "Very few, I admit." But then she argues that the question is misleading. It cannot be answered as simply as its askers suppose. The question ignores the different social and economic situations of men and women. (It depends, in effect, she says, upon a hidden false analogy). The choice women must make, Sayers insists, is different from the choice posed for men. In addition to false analogy, the claim involves at least two other fallacies: it begs the question of whether the difference in choice is necessary or justifiable, and it is a complex (or loaded) question.

The second objection—what the survey shows—reveals statistical slanting. It is always wise to ask about statistical evidence: "What does it *not* show?" In this case the figures do not show how men would respond to the same survey. Men would answer, Sayers argues, much the same as women. (Notice her assertion based upon personal experience.) If we grant the validity of that experience, the claim that women " 'are sick of the office and would love to get out of it,' " even though true, has no relevance to the claim that women are, as workers, essentially less reliable than men.

Thus by meeting the objections that might be urged against her, Sayers disposes of them. Indeed, she turns them around so as to strengthen her own case, and what began as a concession becomes a refutation.

COMPOSING AN ARGUMENT

BEGINNING

As with any piece of writing, you begin an argument by being clear about what you intend to do. Set out what you are arguing for or against, and limit it as narrowly as you can. Don't bite off more than you can chew. Many arguments fail because the writers are overly ambitious. In a four- or five-page paper it is difficult to argue that higher education needs radical revision. That claim can be made (and answered), but it deserves more time and space than you have for a short composition. It is more realistic to confine the topic to a specific issue: that pass-fail, say, is (or is not) better than the five-letter grading used at most colleges, or that the present system of tenure for college teachers ought (or ought not) to be changed.

An important part of framing the issue is to establish whether your argument will be essentially positive or negative. A positive argument urges us to accept an idea or course of action; a negative one, to reject it. Of course, positive and negative arguments imply each other. If you are *for* pass-fail grading, you are probably *against* the five-letter method. Even so, the focus should be clearly one side or the other.

You may wish also to indicate the main lines of the argument: the basic logic in the case of a deductive argument, or, in an inductive one, the kind of evidence you will offer and the order in which it will come. It is especially advantageous in argument that readers have a preview of the plan. But how thoroughly you spell it out depends upon the length and complexity of the paper.

Define your key terms. If you are arguing the question of tenure, make clear what the concept means and all that it entails. Definition is especially important when you must deal with abstract, value-loaded terms such as "democracy," "economic freedom," "equal rights." These expressions have so many shades of meaning that you cannot be sure how your readers will understand them.

Explain the background of your argument. It may not be logically required for what you want to prove, but it will help readers to follow your reasoning. About tenure, for example, you ought to say a word about when, how, and why tenure developed, as well as about the fact that some critics question its value.

This paragraph by the philosopher Brand Blanshard, which begins an argument urging a specific reform of college education, provides a good example of how to do these things:

There is an immense and justified pride in what our colleges have done. At the same time there is a growing uneasiness about their product. The young men and women who carry away our degrees are a very attractive lot—in looks, in bodily fitness, in kindliness, energy, courage, and buoyancy. But what of their intellectual equipment? That too is in some ways admirable; for in spite of President Lowell's remark that the university should be a repository of great learning, since the freshmen always bring a stock with them and the seniors take little away, the fact is that our graduates have every chance to be well informed, and usually are so. Yet the uneasiness persists. When it becomes articulate, it takes the form of wishes that these attractive young products of ours had more intellectual depth and force, more at-homeness in the world of ideas, more of the firm, clear, quiet thoughtfulness that is so potent and so needed a guard against besetting humbug and quackery. The complaint commonly resolves itself into a bill of three particulars. First, granting that our graduates know a great deal, their knowledge lies about in fragments and never gets welded together into the stuff of a tempered and mobile mind. Secondly, our university graduates have been so busy boring holes for themselves, acquiring special knowledge and skills, that in later life they have astonishingly little in common in the way of ideas, standards, or principles. Thirdly, it is alleged that the past two decades have revealed a singular want of clarity about the great ends of living, attachment to which gives significance and direction to a life. Here are three grave charges against American education, and I want to discuss them briefly. My argument will be simple, perhaps too simple. What I shall contend is that there is a great deal of truth in each of them, and that the remedy for each is the same. It is larger infusion of the philosophic habit of mind.

At the end of this we know what Blanshard will argue, the three particular aspects of the problem he will address, and the order in which he will treat them. (The order we can reasonably infer from the arrangement of the "three grave charges" in the opening paragraph.)

ORGANIZING THE MIDDLE

In a deductive argument the structure of the middle follows the order of the premises. First lay out the logic in a preliminary form. Here, for example, is how Robert M. Hutchins—one-time president of the University of Chicago—starts to develop the contention that education must be understood in a liberal sense:

> The obvious failures of the doctrines of adaptation, immediate needs, social reform, and of the doctrine that we need no doctrine at all may suggest to us that we require a better definition of education. Let us concede that every society must have some system that attempts to adapt the young to their social and political environment. If the society is bad, in the sense, for example, in which the Nazi state was bad, the system will aim at the same bad ends. To the extent that it makes men bad in order that they may be tractable subjects of a bad state, the system may help to achieve the social ideals of the society. It may be what the society wants; it may even be what the society needs, if it is to perpetuate its form and accomplish its aims. In pragmatic terms, in terms of success in the society, it may be a "good" system.
>
> But it seems to me clearer to say that, though it may be a system of training, or instruction, or adaptation, or meeting immediate needs, it is not a system of education. It seems clearer to say that the purpose of education is to improve men. Any system that tries to make them bad is not education, but something else. If, for example, democracy is the best form of society, a system that adapts the young to it will be an educational

system. If despotism is a bad form of society, a system that adapts the young to it will not be an educational system, and the better it succeeds in adapting them the less educational it will be.

Hutchins' logic runs like this:

Education must improve people.

Training people to become unquestioning citizens of a totalitarian state does not improve them.

Therefore such training is not education.

The remainder of his argument expands the propositions composing the syllogism. Hutchins explains more exactly what is meant by "improving people" and by the "training" offered in lieu of education in totalitarian societies.

More often than not, the premises of a deductive argument need such expansion and support. This may take the form of explaining why something is true, of the consequences of adopting (or failing to adopt) a particular action, or of illustrations, comparisons, contrasts, and analogies. All these kinds of support constitute reasons why premises should be accepted, and they are a very important element in deductive argumentation.

You may introduce these reasons in an order determined by an inherent pattern—sequence in time, say, or cause and effect. More likely, however, the supporting reasons do not have a convenient, built-in order. Then they should be arranged according to importance. While special considerations may occasionally alter the rule, you will generally find it best to place the most important reason last.

Organize inductive argument in terms of the evidence. Usually you have several items, and the problem becomes how to arrange them. Here again, evidence (assuming no necessary inner pattern) should follow in order of importance, and probably—given a reasonably knowledgeable and attentive audience—a climactic order will prove most effective.

In the course of developing reasons to support premises or of bringing forward evidence, you will meet opposing claims or facts. In ancient rhetoric a separate portion of an argument—the *applicatio*, or refutation—was devoted to the task of confronting objections. In the less formal structure of modern argument, however, we introduce refutations where they are needed rather than gathering them all together.

But while refutations and concessions are necessary, they have to be kept under control. They ought not to become so lengthy and involved or so numerous that they impede or divert the main flow of the argument.

Finally, about the middle of an argument, remember that readers need help to track the turns of thought, more help than with exposition or description. Pay particular attention to connecting words: *thus, as a result, therefore, however, on the other hand, moreover, for instance.* Use connectives very freely in early drafts, even at the risk of sounding repetitious and pedantic. Signaling the logical relationships among your ideas helps you to clarify them. When you revise, however, strike any of these words that seem unnecessary.

CLOSING AN ARGUMENT

Readers expect an argument to end with a final inference or generalization. Even though you told them in the beginning what the conclusion would be, they need to hear it again, now not as a promise, but as a closing statement of what you have shown. Sometimes too, depending upon the subject, the closing may involve a plea that readers act—or refrain from acting—in a certain way.

You usually need not summarize the bare bones of the logic or the major items of evidence in short arguments. As an argument grows more complicated, a summary increases the likelihood of readers' accepting the conclusion. The summary ought, of course, to follow the order in which you presented the logic or evidence in the text.

The following paragraph exemplifies the art of bringing an argument to a close. Henry Thompson Rowell, the historian, argues that there is no method by which we can arrive at a firm figure for the population of ancient Rome:

> The reader must therefore beware of figures reached by any of these methods, used separately or in conjunction with each other. Rome in the Augustan Age was considered a large city by those who knew it. With prosperity, the expansion of the imperial bureaucracy, and the attractions of a metropolis, it probably continued to enlarge its population during the period. But to say that it was inhabited by one million people means no more than that one million seems a comfortable round figure, big enough to be properly imperial and small enough not to be outrageous. If it is right, it is no more than a lucky guess. Smaller or larger numbers may be nearer the truth.

That is how an argument should end: with a clear conclusion, even if it is, as in this case, the paradoxical one that no final assessment is possible.

STYLE AND TONE IN ARGUMENT

Argument has no special style. Like exposition, good argumentation should be clear, forceful, graceful, economic. It does seem, however, that one more easily loses sight of these requirements in argument. Perhaps people become so concerned with ideas that they forget the need to express ideas lucidly and attractively.

Whether or not that is the case, any failure of style is especially damaging in argument. Argument, as we have said, makes more demands upon readers than do other forms of prose, and readers are more quickly turned off by wooden writing. Try then to make sentences forceful. Vary their length. The passage quoted on page 493 by Dorothy L. Sayers shows how effective a varied sentence style is in good argument.

Tone is important. An argument, however reasonable, is never simply a progression of ideas. An argument cannot be abstracted from a particular situation. *Someone* is urging his or her beliefs; *someone else* is being urged. Often the arguer uses a personal point of view and stands forth as a clear personality: that, even more than her syntax, is why Sayers' argument is so impressive. She is there, within her words. So too are Brand Blanshard (page 495) and Jacques Barzun (486).

But even if your point of view must be impersonal—frequently the case in argument—you should still be *in* your prose as a thinking presence. In the next chapter we say a little more about what ancient rhetoricians called the "ethical argument." They did not mean "ethical" in our sense—that is, an appeal to general principles of conduct. Rather the ethical argument is a kind of conviction implicit in the character of the speaker (or writer). The *ethos*—or essential character—of the speaker is more compelling in the nonrational modes of persuasion. But even in argument one feels it. The ethos of those who argue is— or should be—the reasoning mind. The more we sense that mind in the style, the more we shall be impressed by the argument.

EXERCISE

1. Here are two deductive arguments. Develop either one in a paper of five or six pages, arguing either for or against the conclusion. The reasons you urge to support or attack the premises are up to you. The general organization of the argument, however, should follow the form of the syllogism:

   A. The Constitution guarantees citizens the right to bear arms.
      Gun control would deny that right.
      Therefore gun control is unconstitutional.
   B. Anything that promotes cohesiveness in society is good.
      Professional sports promote social cohesiveness.
      Therefore professional sports are socially beneficent.

2. Select one of the following topics and develop an inductive argument of about five or six pages. You may define the issue as narrowly as you like, and you may argue either side. Begin by making your position clear and defining key terms and supply any necessary background information. The body of the paper should consist of the evidence you adduce to support your position, plus any refutations or concessions where these

are needed. Pay particular attention to the arrangement of the evidence, and decide whether you want the most compelling items first or last:

A. Abortion
B. Nuclear energy
C. The Equal Rights Amendment
D. Marijuana
E. Strict separation of men and women in college dormitories
F. Violence in films and on television
G. Censorship of movies, television, and books
H. Electing the president for one six-year term
I. Doing away with state primaries as a method of choosing presidential candidates
J. A government guarantee that every qualified and deserving high-school graduate should have a college education

CHAPTER 49

Persuasion: Nonrational Modes

INTRODUCTION

Nonrational persuasion appeals to emotion rather than to reason. The range of emotional effects at which such persuasion aims is wide: moral indignation, happiness, pity, guilt, anger, laughter, ambition, fear, excitement, self-love, boredom, altruism. By playing upon such feelings the writer intends to change the beliefs and behavior of his or her audience.

Emotive persuasion is all around us. It is the essence of advertising, public relations, political image-making. Even though our schools no longer rigorously train students in rhetoric (the art of persuasion, including argument), the techniques of persuasion work more constantly and more powerfully within our culture than they have ever worked within any other. The force of words has been enormously increased by rapid printing and by radio and television, and to their power has been added that of the visual image.

Unlike argument, emotive persuasion poses a moral problem. It can be used more easily than argument to move us against our will, and even, perhaps, counter to it. Television viewers who respond positively to a picture of a political candidate surrounded by a pretty wife (or attractive husband), two handsome children (one holding a kitten), and a cocker spaniel named Muggsy are reacting to an image. Their emotions are being manipulated, but their intelligence is not involved.

This is not to say that image-persuasion is necessarily nefarious. The image may correspond to reality. The politician may be honorable and capable, the children may be as attractive as they seem, the kitten may belong to the family and not have been borrowed for the occasion.

And realists will argue that if it is necessary to create images to elect good men and women to office, then the end justifies the means. The trouble is that such persuasion places an audience at a disadvantage. Argument provides the ground for its own countering: when reason is appealed to, reason can reply. When emotion is appealed to, it cannot reply; it can only respond or resist.

Perhaps because its appeal is subcortical and difficult to answer, emotive persuasion does have power. That such persuasion entertains in a manner beyond the capacity of argument only increases its power. Effective appeals to the emotions depend upon the skillful—often witty—handling of language, and that very skill is liable to overwhelm our judgment. Subtle irony or unbridled and imaginative insult amuses us, quite apart from whether we think the object deserves our laughter or scorn. When the journalist and social critic H. L. Mencken dismisses grassroots wisdom as "the simian gabble of the crossroads," his diction catches our imagination, even if we have more faith in the common man and woman than he. The phrase may be unfair, but it is witty and memorable. That is its danger.

Another, more subtle danger lurks here. We have remarked on the insistent power of emotional persuasion. But power creates its own counterforce. As people become more and more aware of the manipulations of the image makers, one response is to turn off all messages. Cynicism may protect us against false appeals, but only at the price of ignoring all appeals. To assume that advertisements and press releases and political claims are all lies is no wiser than taking them at face value. Such cynicism devalues all persuasion—legitimate as well as specious.

It is better to be critically intelligent about emotive persuasion. If we understand how it works, we inoculate ourselves against succumbing to it willy-nilly. We need not reject all emotional appeals and the ends at which they aim. Rather we must understand what the end is and how the appeal works so we can accept or reject it on reasonable grounds. We can still laugh at Mencken's wit, but we can also see it as personal opinion, with no extra merit because it is clever.

We need to remember, too, that emotive persuasion may have a worthy purpose. While many questions ought to be settled by reason rather than emotion, on some occasions emotion is the proper and necessary response. There are times to be sad, and times to be happy, and times to be lifted beyond ourselves. Anyone who wishes to write effectively, then, needs to know something about the modes of emotive persuasion.

There are three: satire, eloquence, and pathos. *Satire* evokes moral indignation—sometimes humorously, sometimes angrily—by revealing the difference between how things should be and how they are. *Eloquence* appeals to our ideals, to our sense of virtue and high purpose. *Pathos* arouses pity and compassion.

SATIRE

Originally "satire" referred to a type of Latin poetry which mocked public figures. Then the term expanded to include all literature—prose and poetry—which exposed vice or silliness by mockery. Finally, "satire" came also to mean, more abstractly, the art of such literature— its techniques and methods. We use the word here primarily in this last, abstract, sense.

Satire ridicules folly and viciousness by posing them against wisdom and virtue. The contrast may be explicit, the satirist literally depicting both the evil and the ideal. Or it may be implicit: the writer reveals the folly but expects readers to supply the norm for themselves. Thus a satirist mocking a husband and wife who constantly bicker need not also describe a happy couple: he or she assumes that we know what they are like.

Satire is persuasive in that it aims to convince readers that they should condemn or avoid vice and folly (though occasionally satire expresses such total disgust that it ceases to be moral suasion and becomes a comment on life). To achieve its persuasive purpose, satire relies upon several devices: irony, sarcasm, invective, ironic contrast, and parody.

IRONY

Irony is language used in a manner that contrasts with its conventional sense (see pages 402 ff. for a more full discussion). At its simplest, it is using a word to mean its opposite. When we say of a chicken-hearted person "He's a brave guy," "brave" means "cowardly."

In more subtle form, irony involves not simply a directly opposed sense, but shades and depths of contrary or unexpected meaning. In the novel *Vanity Fair,* for instance, W. M. Thackeray comments of a "good woman" that

> those who know a really good woman are aware that she is not in a hurry to forgive, and that the humiliation of an enemy is a triumph to her soul.

It would miss Thackeray's point to understand him as saying simply that such a woman is "bad" when judged against Christian standards of behavior. She is, of course. But the irony reverberates beyond the particular woman to shake the whole social world. In that world she *is* a "good woman," as society defines "good"; and we are led to see the disparity between what people claim they believe and what they do believe.[1]

Irony finds expression not only in diction but also in sentence style. Thus we might look again at the argument by Dorothy L. Sayers answering the statement that women are less reliable workers than men (see

1. Other ironies work within the words "really know" and "soul." Try to explain them.

page 493). She points out that some women have succeeded remarkably well in jobs considered the prerogative of men. Of the first Queen Elizabeth, a notable instance, Sayers comments:

> It is extraordinarily entertaining to watch the historians of the past, for instance, entangling themselves in what they were pleased to call the "problem" of Queen Elizabeth. They invented the most complicated and astonishing reasons both for her success as a sovereign and for her tortuous matrimonial policy. She was the tool of Burleigh, she was the tool of Leicester, she was the fool of Essex; she was diseased, she was deformed, she was a man in disguise. She was a mystery, and must have some extraordinary solution. Only recently has it occurred to a few enlightened people that the solution might be quite simple after all. She might be one of the rare people who were born into the right job and put that job first.

The syntactic irony is in the brief clauses beginning "She was the tool of Burleigh." Their grammatical simplicity plays ironically upon the "simple-mindedness" of historians refusing to accept a "simple" truth because of their prejudiced view of women.

As a device of persuasion, the subtlety of irony is both its weakness and its strength. Weakness to the degree that readers fail to grasp the writer's true sense, and the more subtle that meaning, the greater the chance of failure. Strength because, given readers alert to it, irony enlists the intelligence and the imagination, drawing the audience in to share the discovery of the folly or wrong. What we discover for ourselves—or think, at least, that we have discovered for ourselves—we are more likely to believe. We are more effectively persuaded by Thackeray's ironic exposure of the gap between the Christian love and charity his society professes and the egocentric, sanctimonious aggression it practices than we would have been by the simple statement that most people really do not forgive their enemies.

SARCASM

Sometimes used for any biting, scathing remark, more narrowly the term *sarcasm* means such a remark expressed as simple irony. Describing a coward as a "brave man" is sarcasm.

A better example occurs in another of Thackeray's sentences. In an essay about King George II of England, Thackeray discusses a scandal concerning the King's wife, Sophia Dorothea. While George was still a prince Sophia took a lover. The affair was discovered. The lover was murdered, and Sophia, separated from George, was imprisoned for thirty-two years. Thackeray is disputing those who claim she was innocent. In a sentence listing other indiscreet ladies, he pours this sarcasm upon Sophia's defenders:

> Yes, Caroline of Brunswick was innocent; and Madame Laffarge never poisoned her husband; and Mary of Scotland never blew up hers; and poor

Sophia Dorothea was never unfaithful; and Eve never took the apple—it was a cowardly fabrication of the serpent's.

INVECTIVE

Invective is abusive language. It persuades by calling names, heaping insults upon what it disputes. Invective differs from irony and sarcasm in that words are used literally; the context demands that we read them in no special sense.

Look at this sentence by the novelist George Orwell. Orwell, a political radical himself, disliked a certain kind of liberal intellectual, those who in our time have been called "the radical chic":

> . . . all that dreary tribe of high-minded women and sandal-wearers and bearded fruit-juice drinkers who come flooding towards the smell of "progress" like blue-bottles to a dead cat.

Orwell's diction is sharpened to strike the raw nerve of prejudice and disgust. Most of his readers will intuitively distrust "sandal-wearers and bearded fruit-juice drinkers," prophets of the counterculture, whose very existence awakens resentment and fear in conventional people. Even more visceral is the feeling aroused by bluebottles. The profoundly disgusting image of bluebottles (bombinating flies that lay their eggs in carrion) buzzing around and settling into a dead cat works upon our horror of death and decay.

All the emotions his imagery arouses will, Orwell hopes, spill onto the liberals he is attacking. There can be no doubt that as writing the passage is skillful and powerful. There can be doubt about its effectiveness as persuasion. Those who share Orwell's disdain will delight in his words. But they do not need to be persuaded. Those he attacks will not cease their buzzing. The uncommitted middle probably will not be convinced by his invective. It is too angry, and public anger makes onlookers turn away in embarrassment.

IRONIC CONTRAST

Ironic contrast consists of showing the ideal and the actual side by side. Here, the writer says, is what we dream or want or expect or deserve; here is what we get. In the novel *Madame Bovary* Gustave Flaubert uses such contrasts frequently to reveal the difference between Emma Bovary's romantic dreams of love and the reality of her life. At one point, for example, she lies in bed, fantasizing about life with her lover, while her husband and baby sleep beside her:

> They would swing in a hammock or drift in a gondola. Life would be large and easy as their silken garments, all warm and starry as the soft nights they would gaze out upon. . . . And yet, in the vast spaces of that imagined

future, no particular phenomenon appeared. The days, all magnificent, were all alike as waves. The vision hovered on the horizon, infinite and harmonious, in a haze of blue, in a wash of sunshine. . . . Then the baby started coughing in its cradle, or Bovary snored more loudly, and Emma didn't get to sleep till morning. . . .

Alan Russell, trans.

A more extended example of ironic contrast occurs in the following passage by the American writer William Saroyan. Saroyan wants to persuade us that movies—back in the 1930s anyhow—are fantasies that have nothing to do with life. Indeed, they turn us away from life until we become lost in false dreams. He describes a sentimental film in which the hero commits a beautiful suicide in a climax of renunciation and reparation. The essay ends with these paragraphs:

Poor Tom. He is sinking to his knees, and somehow, even though it is happening swiftly, it seems that this little action, being the last one of a great man, will go on forever, this sinking to the knees. The room is dim, the music eloquent. There is no blood, no disorder. Tom is sinking to his knees, dying nobly. I myself hear two ladies weeping. They know it's a movie, they know it must be fake, still, they are weeping. Tom is man. He is life. It makes them weep to see life sinking to its knees. The movie will be over in a minute and they will get up and go home, and get down to the regular business of their lives, but now, in the pious darkness of the theatre, they are weeping.

All I know is this: that a suicide is not an orderly occurrence with symphonic music. There was a man once who lived in the house next door to my house when I was a boy of nine or ten. One afternoon he committed suicide, but it took him over an hour to do it. He shot himself through the chest, missed his heart, then shot himself through the stomach. I heard both shots. There was an interval of about forty seconds between the shots. I thought afterwards that during the interval he was probably trying to decide if he ought to go on wanting to be dead or if he ought to try to get well.

Then he started to holler. The whole thing was a mess, materially and spiritually, this man hollering, people running, shouting, wanting to do something and not knowing what to do. He hollered so loud half the town heard him.

This is all I know about regular suicides. . . . The way this man hollered wouldn't please anyone in a movie. It wouldn't make anyone weep with joy.

I think it comes to this: we've got to stop committing suicide in the movies.

PARODY

A *parody* is a mocking imitation of a style of speaking or writing. As a weapon of persuasion, parody shifts the focus from matter to manner; it ridicules not so much what is said as how it is expressed. Of course, the parodist may reply that a coarse or confused or silly way of using words reflects a coarse or confused or silly way of thinking.

In the following example the writer argues that American education is failing partly because of bad educational theory and partly because of public indifference and anti-intellectualism. That last trait he parodies in a diatribe typical—so the writer feels—of too many parents:

> What if your kid *don't* read so good, that don't mean he ain't smart. The Principal up at the high school, he told me that just means the kid ain't got *verbal* intelligence, but he's got *real* intelligence all right. Don't tell *me* that kid ain't got intelligence; jever watch him box my kid? My kid don't outsmart him more 'n half two thirds of the time. Say, 'd I tell you the crack my kid pulled on his old lady last night? Like to batted his ears down, but *jeez* it was smart; afterwards I went out in the kitchen and laughed fit to bust.
>
> John W. Clark

This satire successfully appeals to the prejudices of a particular set of readers, those who dislike men who say "ain't" and "jever" and put down a wife as "the old lady" and laugh at their children's insolence to her. From disapproving of the men, it is a short step to disapproving of their opinions. The parody thus plays upon our prejudices. Of course we should remember that while the connotations of "prejudice" have become almost purely negative, not all prejudices are unreasonable. Most of us are prejudiced against cannibalism and need feel no guilt about the fact. Still, the point here is that Clark's parody aims at arousing latent feelings in his readers concerning a particular type of person.

As with invective, one may wonder whether such parody *converts* others to the writer's beliefs. If it is clever—as Clark's is—we are amused, and if we agree with him we say admiringly, "He's really pinned down that type." But would it change "that type" itself, or would it persuade others who have no strong attitudes on the question?

ELOQUENCE

Eloquence is language used powerfully and fluently to appeal to our nobler emotions—our sense of honor, our love of country, our desire to reach toward virtue. Eloquence is often used elegiacally, to express and evoke feeling on solemn occasions like the deaths of great men and women. Here are two examples: a eulogy by Adlai Stevenson of Eleanor Roosevelt, and Walt Whitman's response to the murder of Abraham Lincoln:

> . . . What we have lost in Eleanor Roosevelt is not her life. She lived that out to the full. What we have lost, what we wish to recall for ourselves, to remember, is what she was herself. And who can name it? But she left "a name to shine on the entablatures of truth, forever."
>
> We pray that she has found peace, and a glimpse of sunset. But today

we weep for ourselves. We are lonelier; someone has gone from one's own life—who was like the certainty of refuge; and someone has gone from the world—who was like a certainty of honor.

The tragic splendor of his death, purging, illuminating all, throws round his form, his head, an aureole that will remain and will grow brighter through time, while history lives, and love of country lasts. By many has this Union been helped; but if one name, one man, must be picked out, he, most of all, is the conservator of it, to the future. He was assassinated— but the Union is not assassinated—*ça ira!* One falls, and another falls. The soldier drops, sinks like a wave—but the ranks of the ocean eternally press on. Death does its work, obliterates a hundred, a thousand—President, general, captain, private—but the Nation is immortal.

Eloquence is not confined to eulogies for the dead. The following is a eulogy for the living. The Reverend Jesse Jackson is urging his hearers to take charge of their own lives, to exert the will to excellence:

My challenge to you today in Cleveland—make a decision. Choose will power over pill power. Make a decision. Choose hope over dope. Make a decision. Choose "I can make it, with or without." Make a decision—say to yourself "Just because it rains I don't have to drown." Make a decision. "I may be in the slum but the slum is not in me." Make a decision—say I can learn—it is possible. Make a decision! Say I ought to learn—it's the moral thing to do. Make a decision. Say I must learn; it is the imperative of now. Make a decision. Repeat this after me: "I AM SOMEBODY! I may be unskilled; but, I AM SOMEBODY." My mind is a pearl—I can learn anything in the world! Down with dope, up with hope. Nobody will save us from us for us but us! I AM SOMEBODY! Right on.

While we tend to think of eloquence as being expressed in literary language, it is really the spirit that counts, not the words. Jackson's diction is commonplace. Only one word might be described as "literary"—"imperative"—and that is not a rare term. His sentence style is short and simple.

Yet his speech is eloquent, and it is rhetorically sophisticated. A rhetorician from ancient Athens or Imperial Rome would recognize techniques he taught to students: *assonance* and *consonance* (the repetition of nearby vowel and consonantal sounds: "choose hope over dope"); *anaphora* (the repetition of initial words: "Choose. . . . Choose. . . . Choose. . . ."); and most of all *epimone* (a refrain: "Make a decision," "I AM SOMEBODY").

His use of such devices does not mean that the Reverend Jackson consciously studied Greek and Latin rhetoric. He may have, of course. But he would not have had to, for his preaching style ultimately goes back to that tradition. In any case some ways of using words suggest themselves to any sensitive and inventive speaker seeking to arouse the emotions of an audience. And they are still effective, as Jackson's speech shows.

Or rather—they are effective for a particular audience. Always the persuasive writer must think in terms of his or her readers, of what kind of appeal will work with them. But before we go into that, let us glance at the last of the three modes of nonargumentative persuasion, pathos.

PATHOS

In ancient rhetoric this term had a broad meaning designating the emotions which a speaker aroused in the audience. Today *pathos* refers more narrowly to pity and compassion. As a term in composition pathos also designates writing which evokes such emotions. (Writing that attempts to do so but ineptly fails is known derisively as "bathos.")

Pathos is often directed toward a particular purpose, for instance to convince us of social injustice. In the following passage the novelist Jack London writes of the plight of the London poor in the early 1900s:

> The unfit and the unneeded! The miserable and despised and forgotten, dying in the social shambles. The progeny of prostitution—of the prostitution of men and women and children, of flesh and blood, and sparkle and spirit; in brief, the prostitution of labour.

Pathos may also express a response not so much to a particular fact as to the general sadness and waste of life. Something of that kind of pathos is felt in the passage below by Walt Whitman. It is called "The Million Dead, Too, Summed Up" and was written at the end of the Civil War and later published as part of Whitman's autobiographical work, *Specimen Days*:

> The dead in this war—there they lie, strewing the fields and woods and valleys and battlefields of the South—Virginia, the Peninsula—Malvern Hill and Fair Oaks—the banks of the Chickahominy—the terraces of Fredericksburg—Antietam bridge—the grisly ravines of Manassas—the bloody promenade of the Wilderness—the varieties of the *strayed* dead (the estimate of the War Department is 25,000 national soldiers killed in battle and never buried at all, 5,000 drowned—15,000 inhumed by strangers, or on the march in haste, in hitherto unfound localities—2,000 graves covered by sand and mud by Mississippi freshets, 3,000 carried away by caving-in of banks, etc.)—Gettysburg, the West, Southwest—Vicksburg—Chattanooga—the trenches of Petersburg—the numberless battles, camps, hospitals everywhere—the crop reaped by the mighty reapers, typhoid, dysentery, inflammations—and blackest and loathesomest of all, the dead and living burial pits, the prison pens of Andersonville, Salisbury, Belle-Isle, etc. (not Dante's pictured hell and all its woes, its degradations, filthy torments, excelled those prisons)—the dead, the dead, the dead—*our* dead— or South or North, ours all (all, all, all, finally dear to me)—or East or West—Atlantic coast or Mississippi valley—somewhere they crawled to die, alone, in bushes, low gullies, or on the sides of hills (there, in secluded

spots, their skeletons, bleached bones, tufts of hair, buttons, fragments of clothing, are occasionally found yet)—our young men once so handsome and so joyous, taken from us—the son from the mother, the husband from the wife, the dear friend from the dear friend—the clusters of camp graves, in Georgia, the Carolinas, and in Tennessee—the single graves left in the woods or by the roadside (hundreds, thousands, obliterated)—the corpses floated down the rivers, and caught and lodged (dozens, scores, floated down the upper Potomac, after the cavalry engagements, the pursuit of Lee, following Gettysburg)—some lie at the bottom of the sea—the general million, and the special cemeteries in almost all the states—the infinite dead (the land entire saturated, perfumed with their impalpable ashes' exhalation in Nature's chemistry distilled, and shall be so forever, in every future grain of wheat and ear of corn, and every flower that grows, and every breath we draw)—not only Northern dead leavening Southern soil—thousands, aye tens of thousands, of Southerners, crumble today in Northern earth.

Whitman is not here a pacifist stressing the horrors of war. He believed, in fact, that the War was necessary to preserve the Union, and that the Union—the United States of America—was necessary to mankind. But he was aware of all that the War had cost, not simply intellectually aware, but knowing in the deepest springs of feeling about "the dead, the dead, the dead."

ETHOS, STYLE, AND THE AUDIENCE

In the last chapter we mentioned the ethical argument. That argument is the character (ethos) of the speaker or writer. In argument the essence of ethos is the reasoning mind. In emotive persuasion it is the writer's feelings.

The man or woman addressing an audience in person can express emotion by tone of voice, gesture, body attitude and movement, laughter, weeping, sad and solemn looks. For that reason emotive persuasion is more effective in speech than in writing. The writer has only words to project ethos. Whitman's repetition of "the dead, the dead, the dead," for instance, is an expression of sorrow, as penetrating and as powerful as the tolling of a bell. Through the words we glimpse the heart of the writer.

It is that glimpse, really, which evokes our emotions. For feeling begets feeling. Words only carry it. The writer's art makes them bear the burden. In persuasion, then, style expresses ethos. At the other end of the equation, style evokes the appropriate feelings in the reader.

Thus in persuasion writers must choose words for their emotional overtones. These may be negative, calling forth disdain and rejection—like "ain't" and "jever" in the parody by Clark (page 507)—or positive, inducing acceptance and reverence—like Whitman's "tragic splendor" and "an aureole that will remain and will grow brighter through time" (page 508).

Sentence style, too, must be charged with emotion. To move deep and tender feelings, sentences themselves must move in a controlled and special way. Their rhythm is slowed and regularized by a clearer marking off of syntactic units, by repetitions, and by metrical runs (the regular grouping of stressed and unstressed syllables typical of poetry).

You have seen these features in several of the previous examples of eloquence or pathos. First, the division of sentences into rhythmic units and the repetitions:

He was assassinated—but the Union is not assassinated. . . .

. . . someone has gone from one's own life—who was like the certainty of refuge; and someone has gone from the world—who was like a certainty of honor.

Second, the metrical runs (ˣ marks an unstressed syllable, ′ a stressed one, and ‵ a syllable of secondary stress), as in these three exemplary passages:

while history lives, and love of country lasts

One falls, and another falls. The soldier drops, sinks like a wave—

but the ranks of the ocean eternally press on.

and someone has gone from the world—who was like a certainty

of honor.

Stylistic features of diction and sentence movement, however, do not exist in and of themselves. Always they are relative to occasion, and a very important aspect of occasion is audience. The writer of persuasion cannot rest content with just expressing his or her own emotion; readers must be made to feel that emotion, and readers of a particular sort, belonging to a specific social class or age group or occupation, having a particular level of education, and particular values and interests. The passage from a speech by the Reverend Jackson (p. 508) is a fine example of a persuasive style adroitly aimed at a specific audience. The short punchy sentences, the repetitions, the colloquial phrase "Right on" (a badge of identity), are all expressions of an ethos which his listeners can admire and identify with.

But words and sentences that turn one group on will turn another off. A gathering of, say, college deans probably would not care much for Jackson's rhetoric (which is not to criticize either college deans or the Reverend Jackson). Modern readers do not rise to Whitman's roll call of Civil War battles. To us Malvern Hill and Fair Oaks and Chickahominy are just names. To Whitman's contemporaries they were sorrow and fear, regret and exultation, danger endured and anguish absorbed.

Always, then, the writer of persuasion must understand the emotion he or she feels, the nature of the reader, and what words and sentence patterns are most likely to evoke the appropriate response. Yet how consciously can this understanding be exercised without sacrificing sincerity? To the degree that a writer is intellectually aware of an emotion and thinks about how to convey it—calculating whether this word or that has the stronger emotional coloring, whether this rhythm or another will be more moving—to that degree we wonder how far we can trust the writer.

The uncertainty is one reason for the moral ambiguity of emotive persuasion. Argument is unambiguous, a game with rules. Readers who know the rules are safe against manipulation and swindle. But in emotive persuasion, clever and cynical writers and speakers can assume an ethos they do not possess and can successfully project into their words emotions they do not feel. They can fool even experienced readers.

On the other hand, sincerity of emotion by itself is not enough. To persuade readers effectively a writer must not only feel, he or she must express feeling in language that will kindle emotion. For experienced and able writers, emotion and stylistic choice probably fuse without much conscious thought or calculation. Insofar as the feelings of such writers are genuine and admirable, we have no reason to doubt or reject their words. The difficulty arises because we can never be sure from the words alone that the emotions are genuine.

EMOTIONAL FALLACIES

In addition to the fallacies discussed in the last chapter in connection with argument, there are others associated with nonrational persuasion. These are called *emotional fallacies* because they appeal to emotion in a way that is irrelevant to the issue. To evoke emotion as Whitman and Jackson do in the examples we have looked it is in no sense fallacious. They intended to arouse feeling in their readers or hearers. But when a writer evokes emotion where more reasoned persuasion is called for, the appeal is a fallacy. Below, in alphabetical order, are some of the more common emotional fallacies.

ARGUMENTUM AD BACULUM ("ARGUMENT TO THE CLUB")

This fallacy plays upon fear. It uses violence or the threat of violence instead of argument. A famous example occurs in the account of the

Peloponnesian War by the Athenian historian Thucydides. The war, fought in the fifth century B.C. between Athens and Sparta, involved many other city states as allies of one or the other of the principals. At one point the Athenians urged the independent island state of Melos to enter the conflict on their side. When the Melians proved reluctant, an Athenian delegation frankly told them:

> . . . the powerful exact what they can, and the weak grant what they must. . . . you are weak and a single turn of the scale might prove your ruin.

ARGUMENTUM AD HOMINEM ("ARGUMENT TO THE MAN")

The fallacy involves abusing the character of an opponent rather than disputing his opinions. Invective—or name-calling—is a variety. An example is this attack upon one-time presidential candidate William Jennings Bryan. Bryan, a religious fundamentalist, served as special prosecutor for the state of Tennessee at the famous "Monkey Trial" in 1925 in which a teacher named Clarence Scopes was accused of violating a state law against teaching Darwin's theory of evolution. Bryan was vigorously attacked in print by the journalist H. L. Mencken in the name of liberalism. Among other things, Mencken wrote about Bryan:

> He was, in fact, a mountebank, a zany without sense or dignity. . . . He seemed only a poor clod like those around him, deluded by a childish theology, full of an almost pathological hatred of all learning, all human dignity, all beauty, all fine and noble things. He was a peasant come home to the barnyard.

ARGUMENTUM AD MISERICORDIAM ("ARGUMENT TO PITY")

The defense counsel who parades his client's weeping wife and blubbering children before the jury, or the accident victim who hobbles into court heavy with bandages and groans loudly enough to be heard, is using the argumentum ad misericordiam.

ARGUMENTUM AD PERSONAM ("ARGUMENT TO THE PERSON")

The appeal is to the particular bias or personal interests of readers. To ignore the claim that a bottle bill will reduce litter, for instance, and answer only that it will increase the cost of drinks, aims at the consumer's desire to save money.

ARGUMENTUM AD POPULUM ("ARGUMENT TO THE PEOPLE")

This fallacy resembles the argumentum ad personam except that the appeal is less to particular interests than to biases and attitudes pervasive throughout a group. To attack an idea as "un-American" or to confirm another as "true Americanism" is an argumentum ad populum, in this case an appeal to patriotic bias. George Orwell's description of a certain type of liberal as "sandal-wearers and bearded fruit-juice drinkers" (see page 505) is another instance of this fallacy. Orwell's words are also an argumentum ad hominem: fallacies often reinforce one another.

ARGUMENTUM AD VERECUNDIAM ("ARGUMENT TO AUTHORITY")

The appeal to authority justifies an idea not on its merits but because a well-known person said it or tradition sanctifies it. Similarly, an appeal may be urged against a proposition because it violates authority or tradition.

For years conservatives justified an isolationist foreign policy by quoting George Washington's advice against foreign entanglements. When Franklin D. Roosevelt ran for a third term in 1940, the decision was widely opposed because it went against the grain of political tradition.

FALLACY OF THE GOLDEN PAST

The appeal is to a time in the past when things were better, to the "good old days." For instance, a critic of contemporary life complains about education that

> people increasingly find themselves unable to use language with ease and precision, to recall the basic facts of their country's history, to make logical deductions. . . .

> Christopher Lasch

The tacit assumption that once upon a time people were able "to use language with ease and precision . . . to make logical deductions" is certainly disputable. These are uncommon skills in any age.

The fallacy of the golden past plays upon the combined moods of nostalgia for the old times and disillusion with the present. Probably the appeal works better with middle-aged and elderly readers than with young ones.

The contrary mistake is the fallacy of the future, approving an idea because it is the "wave of the future."

RIDICULE

A legitimate weapon in satire, ridicule becomes a fallacy when it substitutes for evidence or logic. As Socrates reproves one of his disputants in the dialogue *Georgias*:

> "What's this, Polus? Laughing? Is this a new type of proof, laughing at what your opponent says instead of giving reasons?"

EXERCISE

In our world there are many controversial issues which arouse emotion: nuclear energy, gun control, abortion, income taxes, compulsory military service, welfare, censorship, complete social and economic quality for women, free enterprise, socialism, criminal justice. Select one of these—or another topic if you have one—and compose a brief persuasive essay of four or five pages. You may include argument in the sense of evidence and logical deduction, but you should also attempt to be emotionally persuasive, using some of the devices of satire, eloquence, and pathos discussed in this chapter.

TABLE OF FALLACIES

For convenient reference, the fallacies discussed in the preceding two chapters are here listed in alphabetical order and numbered. If your instructor refers you to, say, "Fallacy 8," you can easily identify it as "begging the question" and turn to pages 475–76, where that fallacy is explained.

	Fallacy	Page
Fal.	1. Argumentum ad baculum ("argument to the club")	512
	2. Argumentum ad hominem ("argument to the man")	513
	3. Argumentum ad ignorantiam ("argument to ignorance")	489
	4. Argumentum ad misericordiam ("argument to pity")	513
	5. Argumentum ad personam ("argument to the person")	513
	6. Argumentum ad populum ("argument to the people")	513
	7. Argumentum ad verecundiam ("argument to authority")	514
	8. Begging the question	475–76
	9. Contradiction	474–75
	10. Fallacy of the complex (loaded) question	489
	11. Fallacy of composition	489
	12. Fallacy of consensus gentium ("agreement of people")	490
	13. Fallacy of converse accident	490
	14. Fallacy of the golden past	514
	15. False analogy	481–82
	16. False dichotomy	480
	17. Floating comparison	490
	18. Gambler's fallacy	490
	19. Genetic fallacy	490
	20. Hasty generalization	490
	21. Ignoratio elenchi ("ignorance of confutation")	479–80
	22. Misleading context	480–91
	23. Non causa pro causa ("there is no cause like the claimed cause")	491
	24. Nonqualification	480–81
	25. Non sequitur ("it does not follow")	481
	26. Petitio principii ("begging of the question")	475
	27. Post hoc ergo propter hoc ("after this, therefore the result of this")	491
	28. Red herring	491
	29. Ridicule	514–15
	30. Skewed sample	491
	31. Slanting	491–92
	32. Special pleading	492
	33. Straw man	492
	34. Tu quoque ("you too")	492
	35. Undistributed middle	476–78
	36. Unqualified source	492
	37. Vicious circle	476

PART EIGHT

THE RESEARCH PAPER AND THE DISCUSSION ANSWER

CHAPTER 50

Gathering, Quoting, and Citing Information

INTRODUCTION

The research paper (or term paper as it is sometimes called) is the most formidable writing challenge you will face as an undergraduate. The challenge does not lie in the writing itself. A term paper is only a long expository essay, essentially similar to the compositions you have been writing, except, perhaps, a bit more formal. Diction, sentence structure, paragraphing, transitions between paragraphs—all the same principles we have studied apply here.

The challenge of the research paper lies in invention—in discovering topics to write about—and in organizing the material you find. Unlike the usual freshman composition, which you can write out of your own experience, the term paper requires considerable research. In some cases—in the sciences, for instance—this necessitates observation and experiment. More likely, however, invention largely arises out of reading.

The first step in constructing a research paper, then, is learning how to find what you want in the library. Next you take and arrange notes from the material you uncover, incorporate those notes into your text, and acknowledge your sources in appropriate footnotes and the bibliography.

We discuss these steps in this chapter. In Chapter 51 we pull everything together by posing a simple research problem and working it through from reference sources to a finished bit of writing.

The writing of research papers really requires a book to itself. Fortunately, several good ones exist, among them:

Barzun, Jacques, and Henry Graff. *The Modern Researcher*. New York: Harcourt Brace Jovanovich, 1957.

Ehrlich, Eugene, and Daniel Murphy. *Writing and Researching Term Papers and Reports: A New Guide for Students*. New York: Bantam, 1964.

James, Elizabeth, and Carol Barkin. *How to Write a Term Paper.* New York: Lothrop, Lee & Shepard, 1980.

Lester, James D. *Writing Research Papers: A Complete Guide,* 3rd ed. Glenview, Illinois: Scott, Foresman, 1980.

Turabian, Kate L. *A Manual for Writers of Term Papers, Theses, and Dissertations,* 4th ed. Chicago: The University of Chicago Press, 1973.

van Leunen, Mary-Claire. *A Handbook for Scholars.* New York: Knopf, 1979.

USING THE LIBRARY

THE CARD CATALOGUE

The key to a library is its card catalogue. This is a bank of drawers located in a central area and containing 3- x 5-inch cards alphabetically arranged listing all the volumes the library holds. In fact it probably contains several cards for each book: a main entry and one or more secondary entries. The main card is filed under the author's last name. In the case of a work composed by many authors, such as an encyclopedia, the main entry is found under the title. Secondary entries are filed according to the first word of the title (excepting the articles *a, an,* and *the*) and also under the general subject or subjects to which the book relates. These subjects are listed near the bottom of the main entry card, and they are, as we shall see, important in research. See page 521 for a sample of a main entry card. Also on page 521 are samples of two secondary entries for the same book. The first is a title card, the second a subject entry.

The cards in a library catalogue may come from three different sources: the library's own cataloguing department; the Library of Congress; and, more recently, a computer cataloguing service to which the library subscribes. The examples given above are of cards supplied by the Library of Congress, our official library, so to speak, where any book submitted for copyright in the United States must be deposited. The cards supplied by these three sources may show minor differences, but all give essentially the same information in the same general format.

While every bit of that information is important to someone, not all of it is vital to you when you research a term paper—the physical size of the book, for instance. You should, however, pay attention to the following items on the card:

The call number   The call number in the upper left hand corner tells you where the book is located in the library. Today most call numbers are based on the system of cataloguing used by the Library of Congress

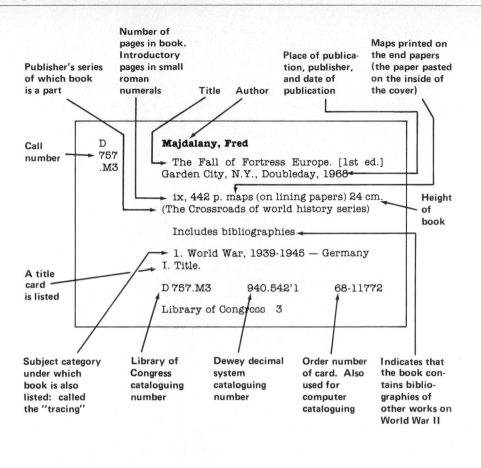

Publisher's series of which book is a part

Number of pages in book. Introductory pages in small roman numerals

Title   Author

Place of publication, publisher, and date of publication

Maps printed on the end papers (the paper pasted on the inside of the cover)

Call number

Majdalany, Fred

The Fall of Fortress Europe. [1st ed.] Garden City, N.Y., Doubleday, 1968

ix, 442 p. maps (on lining papers) 24 cm. (The Crossroads of world history series)

Height of book

Includes bibliographies

1. World War, 1939-1945 — Germany I. Title.

A title card is listed

D 757.M3   940.542'1   68-11772

Library of Congress   3

Subject category under which book is also listed: called the "tracing"

Library of Congress cataloguing number

Dewey decimal system cataloguing number

Order number of card. Also used for computer cataloguing

Indicates that the book contains bibliographies of other works on World War II

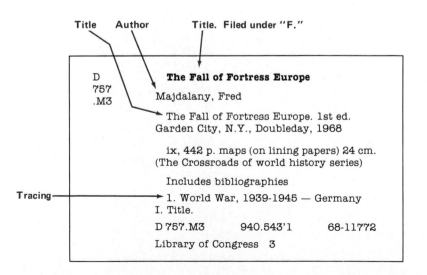

Title   Author   Title. Filed under "F."

D
757
.M3

The Fall of Fortress Europe

Majdalany, Fred

The Fall of Fortress Europe. 1st ed. Garden City, N.Y., Doubleday, 1968

ix, 442 p. maps (on lining papers) 24 cm. (The Crossroads of world history series)

Includes bibliographies

Tracing

1. World War, 1939-1945 — Germany I. Title.

D 757.M3   940.543'1   68-11772

Library of Congress   3

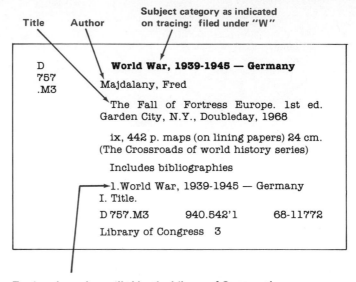

Title   Author   Subject category as indicated on tracing: filed under "W"

D
757
.M3

World War, 1939-1945 — Germany

Majdalany, Fred

The Fall of Fortress Europe. 1st ed. Garden City, N.Y., Doubleday, 1968

ix, 442 p. maps (on lining papers) 24 cm. (The Crossroads of world history series)

Includes bibliographies

1.World War, 1939-1945 — Germany I. Title.

D 757.M3   940.542'1   68-11772

Library of Congress 3

Tracing. In cards supplied by the Library of Congress the tracing is repeated on secondary entries. In cards supplied by a computer service or by the local cataloguer, the tracing may be omitted on these cards, being found only on the main entry.

in Washington. That system divides books into twenty lettered categories according to subject:

A	General works	M	Music
B	Philosophy, religion	N	Fine arts
C	History	P	Language and literature
D	Foreign history	Q	Science
E, F	American history	R	Medicine
G	Geography, anthropology	S	Agriculture
H	Social sciences	T	Technology
J	Political science	U	Military science
K	Law	V	Naval science
L	Education	Z	Library science, bibliography

Some libraries, however, still use the older Dewey decimal system (developed by the philosopher and educator John Dewey). This classifies books under numbered categories:

000	General works	500	Natural science
100	Philosophy	600	Useful arts
200	Religion	700	Fine arts
300	Sociology	800	Literature
400	Philology	900	History and biography

By adding numbers and letters, each system establishes subclasses within the basic categories and succeeds in assigning its own unique number to each book.

In the stacks, books are arranged according to those numbers. It is necessary to have the correct call number when you wish to obtain a book. In many libraries you will not be allowed into the stacks to locate a book for yourself. Instead you will have to fill out a call slip. Be sure to write the call number in full, spacing and aligning it as it is on the card. Include any other information the slip requests: the author's full name (place last name first if required) and the title. Should your library have open stacks, copy the call number, author, and title before you enter them. Don't depend on your memory.

Author and title The card gives the author's full name, even though it may not appear on the title page of the book. If the name on the title page is a pseudonym—"Mark Twain," for example—the card carries the writer's true name—"Samuel Langhorne Clemens." An entry in the card catalogue under the true name refers you to the card listed under the pen name. Many cards also indicate the author's birth date (and death date, if he or she is deceased).

Be sure that in your notes you have the author's full name; the complete title of the book, including any subtitles; and the place and date of publication. You will need this information later for footnotes and bibliography. The name of the publisher should also be noted, although not all instructors will ask you to include it.

Maps, diagrams, bibliographies Notice whether a book contains these. Bibliographies are avenues you can explore to search for other books and articles dealing with your topic.

The tracing The tracing (the list near the bottom of the card of general subjects under which the book is also listed in the catalogue) is especially important. It tells you where you may look in the catalogue to locate other publications bearing on your topic.

REFERENCE WORKS

The card catalogue is only one of the sources of information in your library. Others are to be found in the reference room, presided over by a reference librarian, part of whose job is to advise you where to look for material relevant to your subject. You ought not, however, dump the whole task onto the librarian. He or she will be glad to help you with difficult problems, but you should learn your own way about the reference section and know where basic works are located and how to use them.

Reference works include indexes to periodical literature and newspaper; indexes to government publications; general bibliographies of books; biographical dictionaries; general encyclopedias; almanacs and yearbooks; and encyclopedias, dictionaries, handbooks, and bibliographies restricted to particular fields of knowledge. Numerous reference works are available in each of these categories, far too many to be listed here. You can find more complete information in any of several books dealing with scholarly resources, among them:

Barton, Mary M. *Reference Books: A Brief Guide for Students and Other Users of the Library,* 4th ed. Baltimore: Enoch Pratt Free Library, 1959.

McCormick, Mona. *The New York Times Guide to Reference Materials.* New York: Fawcett, 1979.

Murphy, Robert W. *How and Where to Look It Up.* New York: McGraw-Hill, 1958.

O'Brien, Robert, and Joanne Soderman. *The Basic Guide to Sources.* New York: New American Library, 1975.

Sheehy, Eugene P. *Guide to Reference Books,* 9th ed. Chicago: American Library Association, 1976.

Shores, Louis. *Basic Reference Sources: An Introduction to Materials and Methods.* Chicago: American Library Association, 1954.

Walford, Arthur J. *Guide to Reference Material.* London: Library Association, 1959. Supplement, 1963.

In addition, two of the more general guides to scholarly writing mentioned on page 519 also contain sections listing basic research sources:

Erhlich, Eugene, and Daniel Murphy. *Writing and Researching Term Papers and Reports: A New Guide for Students.* New York: Bantam, 1964.

Lester, James D. *Writing Research Papers: A Complete Guide,* 3rd ed. Glenview, Ill.: Scott, Foresman, 1980.

What follows is a sampling of basic works which will get you started on most problems. Just remember that your library contains an enormous amount of reference material. As you move forward in college, you will become familiar with the works relating to your field.

Indexes to periodicals and newspapers   Periodicals are publications issued at regular intervals. They include both magazines and journals. Magazines commonly contain articles of general interest, not highly technical or scholarly. Journals are usually more technical or scholarly, written for specialists. Both types of periodical are identified by time of publication (year, month, week, or quarter) and by a volume and issue number.

1. *Canadian Periodical Index.* 1928–1947.
   Canadian Index to Periodicals and Documentary Films, 1948–present.
   These list articles published in Canadian magazines and journals.

2. *Catholic Periodical and Literature Index,* 1930–present.
   Indexes articles in over one hundred English-language Catholic periodicals by both author and subject. The general list includes an annotated
   bibliography of books of interest to Catholics, cross-referenced by author, subject, and title.
3. *Essay and General Literature Index,* 1900–present.
   Lists essays—both in collections and published separately—by author,
   title, and subject.
4. *Index to Selected Negro Periodicals,* 1950–present.
   After 1955 this work changed its title to *Index to Selected Periodicals.*
5. *International Index to Periodicals,* 1907–present.
   This index covers more scholarly articles than the *Readers' Guide
   to Periodical Literature* (listed below) and some foreign ones.
6. *The New York Times Index,* 1913–present.
   This work is indispensable to anyone doing research in modern history,
   politics, economics, entertainment, sports, and so on. It indexes stories
   appearing in the late city edition of *The New York Times,* giving
   the date (the date of the story, which is *not* necessarily that of the
   event described) and page and column numbers of the paper. The
   entries also abstract (summarize) the stories, and the *Index* includes
   numerous cross-references.
7. *Nineteenth Century Readers' Guide to Periodical Literature,* 1890–1899.
8. *Poole's Index to Periodical Literature,* 1802–1906.
   Poole's Index is invaluable for tracing magazine articles published
   in the nineteenth century. Entries are listed only by subject (except
   for stories and poems, alphabetized by title). It is therefore difficult
   to locate specific authors.
9. *Readers' Guide to Periodical Literature,* 1900–present.
   The *Readers' Guide* lists articles from about one hundred and thirty-
   five magazines, mostly nontechnical and nonscholarly. Articles are
   indexed by author and subject, and subjects are subdivided.

Indexes to government publications   The four catalogues listed below index publications of the United States government from 1774 to the
present. Because the government is probably the largest publisher in
the world, the mass of material included in these catalogues is enormous.

1. *A Descriptive Catalogue of the Government Publications of the United
   States, 1774–1881.*
2. *Comprehensive Index to the Publications of the United States, 1881–1893.*
3. *Catalog of the Public Documents, 1893–1940.*
4. *Monthly Catalog of United States Government Publications,* 1895–present.

General bibliographies of books   These are lists of books, in some cases
those held by a particular library, in others those published within a
given year or period of years.

1. *Cumulative Book Index,* 1898–present.

   Subtitled *A World List of Books in the English Language,* the *C.B.I.* is a vast bibliography of current books in English (excepting government publications, periodicals, and pamphlets, but including foreign books if these contain English). Entries are by author, title, and subject, arranged in one general list.

2. *General Catalogue of Printed Books.*

   Contains the holdings of the British Museum from 1500 to the present. The British Museum is the equivalent of the Library of Congress, being the copyright library for Great Britain.

3. *Library of Congress Subject Catalog,* 1950–present.

   Indexes by subject the holdings of the Library of Congress.

4. *National Union Catalog,* 1947–present.

   This is the current name of the catalogue of the Library of Congress. It is an author index, and it is useful for finding books your library does not hold.

5. *Paperback Books in Print,* 1955–present.

   Lists all paperbacks available in any given year; contains separate sections for author, title, and subject.

6. *Publisher's Trade List Annual,* 1873–present.

   Alphabetized by publisher, this annual index lists all books in print in the United States. Separate volumes which index works by author, title, and subject are available.

Biographical indexes and dictionaries   These works contain information about notable men and women. Some are confined to people who have died; others to persons still alive.

1. *Biography Index,* 1947–present.

   Indexes biographical material from both books and periodicals on significant people from various periods and places. The *Biography Index* is an important source for any research problem concerning well-known people, whether from our time or from earlier ages of history.

2. *Dictionary of American Biography.*

   Known familiarly as the *DAB,* this multivolume work contains brief biographies of deceased notable Americans. The biographies are the work of scholars and include bibliographies.

   The *DAB* has been abridged in *The Concise Dictionary of American Biography,* which summarizes all the entries in the larger work.

3. *Dictionary of Canadian Biography.*

   Does for Canadians what the *DAB* and the *DNB* (see next item) do for citizens of the United States and Great Britain.

4. *Dictionary of National Biography.*

   The *DNB* consists of brief scholarly biographies of important deceased people who lived in the British Isles and the various colonies. The entries include bibliographies.

5. *Encyclopedia of American Biography,* ed. John A. Garraty.

   Biographies of notable Americans, living and dead. Each entry begins with a summary of the life of the subject. A following essay evaluates the subject's ideas, achievements, and influence.

6. *The New York Times Obituary Index, 1858–1968.*
7. *Notable American Women: A Biographical Dictionary.*

   Modeled on the *DAB*, this work corrects the *DAB's* neglect of women. The biographies begin with 1607 and extend to include women who died in 1950 or before.
8. *Webster's Biographical Dictionary.*

   A one-volume work, Webster's necessarily gives much sketchier information than the *DAB* or the *DNB*. It is, however, international in scope and is a handy desk reference.
9. *Who's Who, 1849–present.*

   Primarily British.
10. *Who's Who in America, 1899–present.*
11. *The International Who's Who.*

    The various *Who's Whos* offer brief biographies of notable living people. The first two have companion volumes for the deceased: *Who Was Who* and *Who Was Who in America*. There are also a number of *Who's Whos* restricted to members of particular professional, ethnic, and religious groups.

General encyclopedias   General encyclopedias consist of alphabetical entries covering numerous aspects of knowledge and endeavor.

1. *Collier's Encyclopedia.*

   The entries in *Collier's* are less scholarly than those in the *Americana, Britannica,* or *New Columbia* encyclopedias (see below). It contains numerous illustrations and is an attractive reference work for students.
2. *Encyclopedia Americana.*

   While it is not restricted to American subjects, the *Americana* is notable for its entires on U.S. history, persons, places, and so on.
3. *New Catholic Encyclopedia.*

   Though it is primarily concerned with subjects concerning Catholicism, the *New Catholic Encyclopedia* also treats matters outside the immediate province of the Church, in art, literature, philosophy, science, and other religions.
4. *New Columbia Encyclopedia.*

   In a single volume, the *Columbia* is shorter than the multivolume encyclopedias, but it is scholarly and reliable. Major articles include bibliographies.
5. *New Encyclopaedia Britannica.*

   The *New Britannica* consists of three parts: a single-volume *Propaedia,* a ten-volume *Micropaedia,* and a nineteen-volume *Macropaedia.* The first is an introduction; it offers an extensive Outline of Knowledge in ten parts, serving as a topical survey of the *Macropaedia.* The *Micropaedia* is essentially a reference work, with articles limited to 750 words or less; cross-references point to relevant information. The *Macropaedia* is educational in intent, consisting of long articles on many fields of knowledge.

Almanacs, yearbooks, and annuals These reference works are issued yearly. Some, like the *Nautical Almanac* or the *Air Almanac,* contain astronomical information relating to celestial navigation. More general almanacs and yearbooks, such as those listed below, supply information about important events of the preceding year.

1. *Americana Annual,* 1923–present.
2. *Britannica Book of the Year,* 1938–present.
   The Americana Annual and the *Britannica Book of the Year* are issued yearly by the *Encyclopedia Americana* and the *New Encyclopaedia Britannica,* respectively. They contain a chronology of the year's events, more detailed accounts of momentous happenings, obituaries, and surveys of countries, states, and cities, as well as information about politics, history, sports, entertainment, and so on.
3. *Canadian Almanac and Directory,* 1847–present.
   Covers a variety of Canadian businesses, governmental agencies, professional groups, and so on. Necessary for anyone seeking information about Canada in any given year.
4. *Catholic Almanac,* 1904–present.
   Lists events and facts relevant to the Catholic Church.
5. *Information Please Almanac,* 1947–present.
   Like *The World Almanac* (see below) the *Information Please Almanac* contains information (often statistical) about politics, government, population, industry and business, sports, entertainment, and many other subjects.
6. *Statistical Abstract of the United States,* 1878–present.
   An annual summary of the statistics collected by various government agencies. The statistics cover economics, politics, and social matters.
7. *Whitaker's Almanack,* 1869–present.
   The best of the British almanacs. Its actual title is simply *Almanack,* but it is commonly called *Whitaker's Almanac* (or just *Whitaker's*) after its London publisher.
8. *Women's Rights Almanac,* 1974, 1975.
   In its own words, this work provides a "handbook for use by all feminists. . . . information and resources for groups and individual women. . . . [and a] record of Women's activities."
9. *The World Almanac and Book of Facts,* 1868–present.
   Covers the same areas as the *Information Please Almanac* (see above), but is somewhat more thorough. It is well indexed and includes a chronology of the significant events of the year.

Special encyclopedias, dictionaries, handbooks, and bibliographies These works are restricted to a particular field (chemistry, for example, or the theater). They vary in content but generally define technical terms in the discipline, discuss key concepts or problems, identify individuals notable in the field, and so on.

Bibliographies list books and articles. Those included here confine themselves to publications relating to a particular area of knowledge.

Such bibliographies are very important in research. Bibliographies of significant books and articles also exist for each subject you study in college. Some of these are issued annually, either as separate publications or in learned journals. Some are collections covering a fixed number of years. When you begin to concentrate in one area of study, you should learn the bibliographical resources available for that discipline.

In addition to separate bibliographies and those published in journals, encyclopedias and handbooks and dictionaries of specific subjects often contain bibliographies after their articles; and many scholarly books and articles include a bibliography or, as it is called in scientific writing, a list of references.

Some bibliographies are annotated—that is, the entries are followed by a brief summary or evaluation of the work.

Special reference works are numerous. What follows is only a sampling from each category. It is more important at the moment that you understand such references exist than that you try to remember particular titles. For a fuller treatment see the reference guides mentioned on page 524. The list in *Writing Research Papers: A Complete Guide* by James D. Lester is especially complete. What follows here will at least get you started.

ART

1. *Encyclopedia of World Art.*
2. *Larousse Encyclopedia of Art.* Ed. Rene Huyhe.

Bibliographies
1. *Art Index,* 1929–present.
2. *Guide to Art Reference Books.* Mary W. Chamberlin.

BUSINESS AND ECONOMICS

1. *American Business Dictionary.* Harold Lazarus.
2. *Dictionary of Modern Economics.* Byrne J. Horton, et al.
3. *Encyclopedia of Banking and Finance,* 7th ed. Glenn C. Munn. Rev. and enlarged by F. L. Garcia.

Bibliographies
1. *Business Books in Print,* 1973–present.
2. *Business Periodicals Index,* 1958–present.
3. *International Bibliography of Economics,* 1952–present.

EDUCATION

1. *American Universities and Colleges,* 11th ed. Ed. W. Todd Furniss.
2. *Dictionary of Education.* Carter Victor Good.
3. *The Encyclopedia of Education.* Ed. Lee C. Deighton.
4. *Encyclopedia of Educational Research.* Ed. Chester W. Harris.

5. *The Handbook of Private Schools: An Annual Descriptive Survey of Independent Education,* 1914–present.
6. *How to Locate Educational Information and Data.* Carter Alexander and A. J. Burke.
7. *Patterson's American Education.*
8. *Who's Who in American Education,* 1928–present.

Bibliographies
1. *Education Index,* 1929–present.
2. *The New York University List of Books in Education.* Comp. and ed. Barbara S. Marks.

HISTORY

General
1. *The Cambridge Ancient History.*
   The Cambridge Medieval History.
   The Cambridge Modern History.
2. *The Cambridge Economic History of Europe.*
3. *An Encyclopedia of World History,* 4th ed. William L. Langer.

Bibliographies
1. *Guide to Historical Literature.*
2. *International Bibliography of Historical Sciences,* 1926–present.

The United States
1. *Album of American History.* James Truslow Adams, et al.
2. *The Encyclopedia of American History,* enlarged and updated. Ed. Richard B. Morris and Henry Steele Commager.
3. *Harvard Guide to American History.* Ed. Oscar Handlin et al.
4. *The Negro Almanac,* 2nd ed. Ed. Henry A. Ploski and Ernest Kaiser.
5. *The Oxford Companion to American History.*
6. *Reference Encyclopedia of the American Indian.* Ed. Bernard Klein and Daniel Icolari.

Bibliographies
1. *Bibliographies in American History,* rev. ed. Henry P. Beers.
2. *The Negro in America: A Bibliography.* Ed. Elizabeth M. Miller and Mary Fisher.
3. *Writings on American History,* 1902–present.

Great Britain
1. *Dictionary of English History.* Sidney Low and F. S. Pulling.
2. *Handbook of Dates for Students of Modern History.* Christopher R. Cheney.
3. *Oxford History of England.*

Bibliographies
1. *Bibliography of British History.* Ed. Stanley Pargolis and D. J. Medley.
2. *Bibliography of British History: Tudor Period, 1485–1603.*
   _____: *Stuart Period, 1603–1714.*
   _____: *The Eighteenth Century, 1714–1789.*
3. *Writings on British History.* A. T. Milne.

Other areas

1. *Asia: A Selected and Annotated Guide to Reference Works.* Comp. G. Raymond Nunn.
2. *Concise Dictionary of Ancient History.* Ed. Percival C. Woodcock.
3. *Concise Encyclopedia of Arabic Civilization: The Arab East and the Arabic West.* Stephan Ronart and Nancy Ronart.
4. *An Encyclopedia of Latin American History.*
5. *Handbook of Latin American Studies.*
6. *The Oxford Classical Dictionary,* 2nd ed. Ed. N. G. L. Hammond and H. H. Scullard.
7. *The Oxford Companion to Canadian History and Literature.* Ed. Norah Story.

LITERATURE

General

1. *Columbia Dictionary of Modern European Literature.* Ed. Horatio Smith.
2. *Dictionary of World Literature.* Ed. Joseph T. Shipley.
3. *Encyclopedia of World Literature in the 20th Century.* Ed. Wolfgang Bernard Fleischman.
4. *Motif-Index of Folk Literature: A Classification of Narrative Elements in Folktales, Ballads, Myths, Fables, Medieval Romances. . . .* Stith Thompson.
5. *The Penguin Companion to Classical, Oriental and African Literature.* Ed. D. M. Lang and D. R. Dudley.
6. *The Penguin Companion to European Literature.* Ed. David Daiches.
7. *The Reader's Companion to World Literature,* 2nd ed. Ed. Lillian Herlands Hornstein, G. D. Percy, and Calvin S. Brown.
8. *Short Story Index,* 1953–present. Dorothy E. Cook and Isabel S. Monro.

Bibliography

1. *Bibliography of Comparative Literature,* 3rd ed. Fernand Baldensperger and Werner P. Friederich.

The United States

1. *Cambridge History of American Literature.*
2. *Concise Dictionary of American Literature.* Ed. Robert F. Richards.
3. *Guide to American Literature and Its Backgrounds Since 1890,* 3rd rev. ed. Howard Mumford Jones.
4. *Literary History of the United States,* 3rd ed. Robert E. Spiller et al.
5. *The Oxford Companion to American Literature.* Ed. James D. Hart.
6. *The Penguin Companion to American Literature.* Ed. Malcolm Bradbury, Eric Mottram, and Jean Franco.

Bibliographies

1. *Annual Bibliography of English Language and Literature,* 1920–present.
2. *Articles on American Literature, 1900–1950.* Lewis G. Leary.
3. *Bibliographical Guide to the Study of the Literature of the U.S.A.,* 2nd ed. Clarence Gohdes.
4. *Bibliography of American Literature,* 1955–present. Jacob Blanck.

   5. *The MLA International Bibliography,* 1956–present.
   6. *A Reference Guide to English Studies.* Donald F. Bond.

Great Britain
   1. *Cambridge History of English Literature.*
   2. *A Literary History of England.* Ed. Albert C. Baugh.
   3. *The Oxford Companion to English Literature.* Ed. Sir Paul Harvey.
   4. *The Oxford History of English Literature.* Ed. Frank P. Wilson and Bonamy Dobree.
   5. *The Penguin Companion to English Literature.* Ed. David Daiches.

Bibliographies
   1. *Annual Bibliography of English Language and Literature,* 1920–present.
   2. *Cambridge Bibliography of English Literature.* Ed. F. W. Bateson.
   3. *The Concise Cambridge Bibliography of English Literature.* Ed. George Sampson.
   4. *The MLA International Bibliography,* 1956–present.
   5. *A Reference Guide to English Studies.* Donald F. Bond.
   6. *The Year's Work in English Studies,* 1919–present. English Association.

Poetry
   1. *Explicator,* 1942–present.
   2. *Princeton Encyclopedia of Poetry and Poetics,* 4th ed. Ed. Alex Premminger, Frank J. Warnke, and O. B. Harrison.

Bibliographies
   1. *Granger's Index to Poetry,* 5th ed. Ed. William F. Bernhardt.
   2. *Poetry Explication: A Checklist of Interpretations Since 1925 of British and American Poems Past and Present.* George Arms and Joseph M. Kuntz.

Drama and film
   1. *The Filmgoer's Companion,* 3rd ed. Ed. Leslie Halliwell.
   2. *McGraw-Hill Encyclopedia of World Drama.*
   3. *The New York Times Film Reviews,* 1913–present.
   4. *The Oxford Companion to the Theatre,* 3rd ed. Ed. Phyllis Hartnoll.
   5. *The Penguin Dictionary of the Theatre.*

Bibliographies
   1. *A Bibliography of Theatre Arts Publications in English, 1963.* Ed. Bernard F. Dukore.
   2. *The Dramatic Index for 1909–1949: Covering Articles and Illustrations Concerning the Stage and Its Players in the Periodicals of America and England and Including Dramatic Books of the Year.*
      Reference to this 42-volume work is made easier by a 2-volume guide: *Cumulated Dramatic Index,* 1909–1949.
   3. *Index to Plays,* 1800–1926. Comp. Ina T. E. Firkens. Supplement, 1935.
   4. *Play Index, 1953–1960: An Index to 4592 Plays in 1735 Volumes.* Ed. Estelle A. Fidell and D. M. Peake.
   5. *Reference Books in the Mass Media: An Annotated Selected Booklist Covering Book Publishing, Broadcasting, Films, Newspapers, and Advertising.* Eleanor Blum.

MATHEMATICS

1. *Mathematics Dictionary: Multilingual Edition.* Ed. Robert C. James, Edwin F. Beckenbach, et al.
2. *The Universal Encyclopedia of Mathematics.*

Bibliography
1. *Guide to the Literature of Mathematics and Physics Including Related Works on Engineering Science,* 2nd ed. N. G. Parke.

MUSIC

1. *Baker's Biographical Dictionary of Musicians,* 4th ed. Theodore Baker.
2. *A Dictionary of Musical Terms in Four Languages.* William James Smith.
3. *Encyclopedia of Popular Music.* Irwin Stambler.
4. *Folksingers and Folksongs in America,* rev. ed. Ray M. Lawless.
5. *Grove's Dictionary of Music and Musicians,* 3rd ed. Sir George Grove. Ed. H. C. Colles.
6. *Harvard Dictionary of Music,* 2nd ed. Ed. Willi Apel.
7. *International Cyclopedia of Music and Musicians,* 9th ed. Oscar Thompson.
8. *Musician's Guide.*
9. *The New College Encyclopedia of Music.* Jack A. Westrup and F. L. Harrison.
10. *The New Encyclopedia of Opera.* Davis Ewen.
11. *New Oxford History of Music.*
12. *The Oxford Companion to Music,* 9th ed. Percy A. Scholes.

Bibliographies
1. *A Check-List of Publications of Music.* Anna Harriet Heyer.
2. *Music Index,* 1949–present.

MYTHOLOGY

1. *Funk & Wagnalls Standard Dictionary of Folklore, Mythology and Legend.* Ed. Maria Leach.
2. *Larousse World Mythology.* Ed. Piere Grimal.
3. *Motif-Index of Folk Literature: A Classification of Narrative Elements in Folktales, Ballads, Myths, Fables, Medieval Romances. . . .* Stith Thompson.

Bibliographies
1. *Index to Fairy Tales, Myths, and Legends,* 2nd ed. Mary Huse Eastman.
2. *Religions, Mythologies, Folklores: An Annotated Bibliography.* Katherine S. Diehl.

PHILOSOPHY

1. *The Concise Encyclopedia of Western Philosophy and Philosophers.* Ed. J. O. Urmson.

2. *The Dictionary of Philosophy.* Ed. Dagobert D. Runes.
3. *Dictionary of Philosophy and Psychology.* James Mark Baldwin.
4. *The Encyclopedia of Philosophy.* Ed. Paul Edwards.
5. *A History of Western Philosophy.* Bertrand Russell.

Bibliographies
1. *Bibliography of Philosophy,* 1933–1936.
2. *Bibliography of Philosophy, Psychology, and Cognate Subjects.* Benjamin Rand.
    Goes only to 1900, but still useful.
3. *The Philosopher's Index: An International Index to Philosophical Periodicals,* 1967–present.
4. *Philosophic Abstracts.*

POLITICAL SCIENCE

1. *The Almanac of American Politics.* Michael Barone, Grant Ojifusa, and Douglas Matthews.
2. *The American Political Dictionary.* Jack C. Plano and Milton Greenberg.
3. *A Dictionary of Politics.* Ed. Walter Laqueur.
4. *Encyclopedia of Modern World Politics.* Walter Theimer.
5. *New Dictionary of American Politics.* Ed. Edward C. Smith and Arnold J. Zurcher.
6. *Political Handbook and Atlas of the World.*
7. *The Statesman's Year-Book.*

Bibliographies
1. *The Literature of Political Science: A Guide for Students, Librarians, and Teachers.*
2. *Materials for the Study of Federal Government.* Dorothy C. Tompkins.
3. *Student's Guide to Materials in Political Science.* Laverne Burchfield.

PSYCHOLOGY

1. *A Comprehensive Dictionary Of Psychological and Psychoanalytical Terms.* Horace B. English and Ava C. English.
2. *A Dictionary of Psychology.* James Drever.
3. *Encyclopedia of Psychology.* Ed. Philip L. Harriman.

Bibliographies
1. *Annual Review of Psychology,* 1950–present.
2. *Handbook of Psychological Literature.* C. M. Louttit.
3. *The Harvard List of Books in Psychology,* 1949. Supplement, 1958.
4. *Psychological Abstracts,* 1927–present.
5. *Psychological Index,* 1894–1935.

RELIGION

1. *The Catholic Encyclopaedic Dictionary.* Ed. Donald Attwater.
2. *The Concise Encyclopedia of Living Faiths.* Ed. Robert C. Zaehner.

3. *Dictionary of the Bible.* Ed. James Hastings. Rev. Frederick C. Grant.
4. *A Dictionary of Comparative Religions.* S. G. F. Brandon.
5. *An Encyclopedia of Religion.* Ed. Vergilius T. A. Ferm.
6. *Encyclopaedia of Religion and Ethics.* Ed. James Hastings.
7. *The Interpreter's Bible.* George A. Buttrick, et al.
8. *The Interpreter's Dictionary of the Bible.* George A. Buttrick et al.
9. *New Catholic Encyclopedia.*
10. *New Schaff-Herzog Encyclopedia of Religious Knowledge.* Philip Schaff. Ed. Samuel Jackson et al.
11. *The Oxford Dictionary of the Christian Church.* Ed. F. L. Cross.
12. *Shorter Encyclopedia of Islam.* Ed. H. A. R. Gibb and J. H. Kramers.
13. *Twentieth Century Encyclopedia of Religious Knowledge.*
      A Supplement to the *New Schaff-Herzog Encyclopedia,* listed above.
14. *The Universal Jewish Encyclopedia.*

Bibliographies

1. *A Basic Bibliography for Ministers,* 2nd ed.
2. *Bibliographical Guide to the History of Christianity.* Ed. Shirley J. Case.
3. *A Bibliography of Bibliographies in Religion.* John Graves Barrow.
4. *The Catholic Periodical Index: A Guide to Catholic Magazines.*
5. *A Critical Bibliography of Religion in America.* Nelson R. Burr.
6. *History of Christianity.* William Henry Allison.
7. *Index to Religious Periodical Literature.* Stillson Judall.
8. *International Bibliography of the History of Religions,* 1952–present.
9. *Religions, Mythologies, Folklores: An Annotated Bibliography,* 2nd ed. Katherine Smith Diehl.

SCIENCE

General

1. *Chambers's Technical Dictionary,* 3rd ed. Ed. C. F. Tweney and L. E. C. Hughes.
2. *A Dictionary of Named Effects and Laws in Chemistry, Physics and Mathematics.* D. W. G. Ballentyne and L. E. Q. Walker.
3. *The Harper Encyclopedia of Science,* rev. ed. Ed. James R. Newman.
4. *McGraw-Hill Encyclopedia of Science and Technology.*
5. *Van Nostrand's Scientific Encyclopedia,* 4th ed.

Bibliographies

1. *Air Force Scientific Research Bibliography, 1950–1956.* Ed. G. Vernon Hooker et al.
2. *Applied Science and Technology Index,* 1913–present.
3. *Science Abstracts,* 1898–present.
4. *Science Reference Sources,* 4th ed. Frances B. Jenkins.
5. *Scientific, Medical, and Technical Books Published in the United States of America,* 2nd ed. Ed. R. R. Hawkins.

Astronomy and space science

1. *Encyclopedia of Astronomy.* Gilbert E. Satterwaite.
2. *The New Space Encyclopedia: A Guide to Astronomy and Space Exploration,* 3rd ed. Ed. M. T. Bizony.

Bibliographies
1. *Air Force Scientific Research Bibliography, 1950–1956.*
2. *A Guide to Information Sources in Space Science and Technology.* Bernard M. Fry and Foster E. Mohrhardt.

Biology
1. *Dictionary of Biological Terms,* 8th ed. Isabella F. Henderson.
2. *The Encyclopedia of Biochemistry.* Ed. Roger J. Williams and Edwin M. Landsford.
3. *The Encyclopedia of the Biological Sciences.* Ed. Peter Gray.
4. *Encyclopedia of the Life Sciences.* Ed. Jean Rostand and Albert Delaunay.

Bibliographies
1. *The American Yearbook: A Record of Events and Progress,* 1911–present.
2. *Biological Abstracts,* 1916–present.
3. *Guide to the Literature of the Zoological Sciences,* 5th ed. R. C. Smith.

Chemistry and physics
1. *American Institute of Physics Handbook.* Ed. Dwight E. Gray et al.
2. *The Condensed Chemical Dictionary.* Arthur Rose and Elizabeth Rose.
3. *A Dictionary of Applied Physics.* Sir Richard T. Glazebrook.
4. *The Encyclopedia of Chemistry.* George L. Clark et al.
5. *Handbook of Chemistry,* 1934–present. Norbert A. Lang.
6. *Handbook of Physics.* Edward O. Condon and Hugh Odishaw.
7. *Thorpe's Dictionary of Applied Chemistry,* 4th ed. Jocelyn F. Thorpe and M. A. Whitely.

Bibliographies
1. *Chemical Abstracts,* 1907–present.
2. *A Guide to the Literature of Chemistry,* 2nd ed. Evan J. Crane.
3. *Guide to the Literature of Mathematics and Physics,* 2nd rev. ed. Nathan G. Parke III.
4. *A Library Guide for the Chemist.* Byron A. Soule.

Geology and climatology
1. *A Dictionary of Geology,* 3rd ed. John Challinor.
2. *Standard Encyclopedia of the World's Mountains.*
3. *Standard Encyclopedia of the World's Oceans and Islands.*
4. *Standard Encyclopedia of the World's Rivers and Lakes.*
5. *The Water Encyclopedia.* David K. Todd.
6. *World Survey of Climatology.* Ed. H. E. Landsberg.

Bibliography
1. *Guide to Geologic Literature.* Richard M. Pearl.

SOCIAL SCIENCE

1. *The Concise Encyclopedia of Archaeology.* Ed. Leonard Cottrell.
2. *A Dictionary of Anthropology.* Charles Winick.
3. *Dictionary of Social Science.* Ed. John T. Zadrozny.

4. *A Dictionary of the Social Sciences.* Ed. Julius Gould and William L. Kolb.
5. *Encyclopedia of Social Work.* Ed. Harry L. Lurie.
6. *Encyclopaedia of the Social Sciences.* Edwin R. A. Seligman and Alvin Johnson.
7. *A Guide to the Study of the United States of America.* Ed. Roy P. Basler et al.
8. *International Encyclopedia of the Social Sciences.* David L. Sills.
9. *A Reader's Guide to the Social Sciences.* Ed. Berthold F. Hoselitz.

Bibliographies
1. *Checklist of Books and Pamphlets in the Social Sciences.* New York State Library.
2. *International Bibliography of Social and Cultural Anthropology.*
3. *International Bibliography of Sociology.*
4. *The Literature of the Social Sciences.* Peter R. Lewis.
5. *London Bibliography of the Social Sciences,* 1931–present.
6. *Research Materials in the Social Sciences.* Ed. Jack A. Clarke.
7. *Research Resources: Annotated Guide to the Social Sciences.* Comp. John B. Mason.
8. *Sources of Information in the Social Sciences: An Annotated Bibliography.* Carl M. White et al.

SPORTS

1. *The Baseball Encyclopedia: The Complete and Official Record of Major League Baseball.*
2. *The Encyclopedia of Sports,* 4th rev. ed. Ed. Frank G. Menke. Rev. Roger Treat.
3. *Everything You Want to Know about Hockey,* rev. ed. Brian McFarlane.
4. *The Modern Encyclopedia of Basketball,* rev. ed. Ed. Zander Hollander.
5. *The New York Times Guide to Spectator Sports.* Leonard Koppett.
6. *The Official Encyclopedia of Football,* 11th rev. ed. Ed. Roger Treat. Rev. Suzanne Treat.
7. *Sports Rules Encyclopedia.* Comp. and ed. Jess R. White.
8. *Who Was Who in American Sports.* Ralph Hickok.

TAKING NOTES

Once you have located a useful book or article, the next step is to extract the information you want. This means reading and taking notes.

As to reading, you should, ideally, read all of any book or article you cite. It is sound scholarly advice "never to quote a book you haven't read completely." But counsels of perfection cannot always be obeyed. Most of us do not have enough time or devotion. But if you cannot read the entire work, do more than merely skim key passages you located through the index. Read enough of the context to be sure you understand what the writer intends. Explore the index for topics related to yours. (If there is no index, you will have to search through the book by reading rapidly until you come across pertinent sections.) Familiarize yourself with the book's general contents, and learn what you can about the background of the writer so you are aware of any biases that may influence his or her views. Check for bibliographies, and remember that sometimes they appear at the end of each chapter rather than at the end of the volume.

Notes are best taken using an orderly, rational method. First supply yourself with an adequate supply of file cards. Don't scribble notes on envelopes, laundry lists, or receipts. Some people like 3- x 5-inch cards, which are easy to handle. Others prefer the larger 5 x 8, capable of holding more but bulkier to carry and store. Whichever you choose, stick to them and don't mix cards of different sizes.

Leave enough space at the top of the card to add the headings that will help you later to arrange your cards. Use a pen or reasonably hard pencil that will not smudge with the repeated shuffling of the cards.

Copy the passage exactly. You can paraphrase and edit later. Put quotation marks around the passage to distinguish it from any comments or paraphrases you may add. If the passage contains anything in quotation marks, change these to single form (assuming that you have used double marks around the passage itself). Don't depend upon memory to clarify references that the context of the book makes clear. Include the explanation on the card, enclosing it in brackets if you place it inside the quotation. Use ellipses to mark the omission of nonessential matter (and be sure it *is* nonessential).

Follow the principle of one note per card and, if possible, one card per note. If you crowd several notes from the same source onto a single card, you will find the economy troublesome later when you begin to arrange your notes according to the organization of your paper. When a note is long enough to require two or more cards, number the cards and identify them as belonging together. If a quotation occupies two or more pages in your source, indicate in your note where new pages begin.

Include on each card the information you will need for footnotes and bibliography: the author's name as it appears on the title page; the full title; the title of the periodical (in the case of an article) and its date, volume, and issue number, as well as the pages on which the article begins and ends; and the page(s) on which you found the passage. For a book, you will also need the place and year of publication and the publisher. Copy the publisher's name exactly as it appears on the title page, although in an actual note you may be able to use a short form: "Harper & Row," for instance, instead of "Harper & Row, Publishers."

Here are two sample notes taken for a paper on the opposition (or lack of it) by the German churches to Hitler during the 1930s:

"In fact, both churches [the Evangelical Lutheran and the Catholic], gave massive support to the regime. The Catholic bishops welcomed 'the new, strong stress on authority in the German state'. . . . twelve Evangelical leaders . . . pledged 'the leaders of the German Evangelical Church unanimously affirm their unconditional loyalty to the Third Reich and its leader.' "

Paul Johnson, A History of Christianity (New York: Atheneum, 1976), p.488.

"Only the free sects stuck to their principles enough to merit outright persecution. The bravest were the Jehovah's Witnesses, who proclaimed their outright doctrinal opposition from the beginning and suffered accordingly. They refused any cooperation with the Nazi State which they denounced as totally evil. The Nazis believed they were part of the international Jewish-Marxist conspiracy. Many were sentenced to death for refusing military service and inciting others to do likewise; or they ended in Dachau or lunatic asylums. A third were actually killed; 97 per cent suffered persecution in one form or another."

Paul Johnson, A History of Christianity (New York: Atheneum, 1976), p. 489.

ARRANGING YOUR NOTES

You will probably wind up with several dozen note cards and maybe even a hundred or more. Spend time studying and organizing your notes. You will, in effect, be organizing your paper.

Of course, the task of organization has begun before this. As you explore various books and articles, focal points will have become apparent, suggesting the major lines of your paper. And as you study and shuffle notes, these divisions will become clearer and perhaps others will emerge.

Write a summarizing heading at the top of each card, placing the note in the appropriate category and indicating the specific point it makes. Try to begin each heading with a key conceptual term; this enables you to alphabetize cards according to important ideas, people, places, events, and so on. Often you will use the same heading for several cards. In that case, assign each card in the group its own number.

The two sample notes about the churches in Hitler's Germany might be headed like this:

> Subservience of state churches to Hitler (1)

> Opposition of Jehovah's Witnesses to Hitler

When a note contains several key ideas, it is helpful to add other cards to the file, cross-referring to that note. The first of our examples talks about the Evangelical Lutheran Church and the Catholic Church, and it should be cross-filed on two other cards like this:

> Evangelical Church — subservient to Hitler
>
>   See card: "Subservience of state churches to Hitler (1)"

> Catholic Church — subservient to Hitler
>
>   See card: "Subservience of state churches to Hitler (1)"

The second of our sample notes might be referred to on a separate card headed "Jehovah's Witnesses—opposition to Hitler."

You will facilitate your task of arranging notes if you invest in a cheap plastic file box and file guides. File guides are heavy cards—commonly in a distinctive color—the same size as your note cards but with a projecting tab across the top. In some the tab runs all the way across, in others a third of the way, and in still others a fifth. The latter two are called third-cut and fifth-cut, and their tabs are staggered for visibility.

You can write your organizational categories on the tabs, using wide ones for the most basic and third- or fifth-cut for subdivisions within these. Such an organization makes a file very easy to use. You can arrange new notes in the appropriate place, add categories as you think of them, and quickly remove all the notes relating to a particular topic.

INCORPORATING NOTES INTO YOUR PAPER

Working up notes into your paper presents special problems. First of all, you should employ direct quotation with restraint. A research paper ought to be more than a patchwork quilt of quotes. You must absorb the ideas and information you have gathered, thoroughly making them your own knowledge, then expressing them in your own words.

When you deal with specific details, however, and especially controversial ones, you will need to use notes more substantively. You may do this either by quotation or by paraphrase.

QUOTATION

Sometimes quotations are introduced after a colon, and sometimes they are integrated into the grammar of the sentence without any punctuation. The first method is more formal. Suppose you wish to quote Professor Johnson's point about the Jehovah's Witnesses in Germany:

> Among the Protestants the Jehovah's Witnesses were especially notable for their resistance to Hitler. Professor Paul Johnson points out: "They refused any cooperation with the Nazi State which they denounced as totally evil. . . . A third were actually killed; ninety-seven per cent suffered persecution in one form or another."[1]

Or the quotation might be integrated into your sentence, in which case no colon is required:

> Among the Protestants the Jehovah's Witnesses were especially notable for their resistance to Hitler. According to Professor Paul Johnson, "they refused any cooperation with the Nazi State which they denounced

> as totally evil. . . . A third were actually killed;
> ninety-seven per cent suffered persecution in one form
> or another."[1]

This method has the advantage of avoiding the awkward break made by the colon and of weaving the quoted material more smoothly into your text.

Notice that in both cases the borrowed words are enclosed in quotation marks; omitted matter is signaled by an ellipsis, using four dots rather than three since the omission includes an entire sentence; and each quotation is followed by a superscript, the small number placed outside and above the closing quote. In an actual paper the superscript would indicate a footnote identifying the source. (More about footnotes in a few pages.)

There is, however, one slight difference between the two quotations. The first preserves the capital "T" of the original; the second uses lower case. It is conventional to retain an initial capital in quotations introduced by a colon or comma, but to alter an original capital to a small letter if the quote is integrated into the syntax of the containing sentence (unless, of course, the capitalized word is a proper name or adjective).

A relatively long quotation is usually indented (in printing it is set in smaller type). Indented quotations do *not* take quote marks, the indentation itself being a sufficient signal. A rough rule is to indent quotations of fifty words or longer (or four or more lines). Lines of poetry, however, are usually set off and indented, even though they are less than fifty words or four lines.[1]

In typescript indent the left-hand margin three spaces to the right of the indentation for a new paragraph. You may wish also to indent the right-hand edge, though it is more convenient simply to run the quote to the regular margin on that side.

If you were to quote most of the note from Professor Johnson's book, you would indent it like this:

> Among the Protestants the Jehovah's Witnesses were
> especially notable for their resistance to Hitler.
> Professor Paul Johnson writes that they
>
> > proclaimed their outright doctrinal opposition from
> > the beginning and suffered accordingly. They refused
> > any cooperation with the Nazi State which they
> > denounced as totally evil. . . . Many were sentenced
> > to death for refusing military service and inciting
> > others to do likewise; or they ended in Dachau [a

---

1. If two or more lines of poetry are worked into the text, it is customary to use quote marks, to separate the lines by a slash, and to preserve the capitalization of the first word in each line:

Andrew Marvell spoke for all of us when he complained "But at my back I alwaies hear/Times winged Charriot hurrying near."

```
Nazi concentration camp] or lunatic asylums. A
third were actually killed; ninety-seven per cent
suffered persecution in one form or another.[1]
```

Like those placed in the text, indented quotations may be introduced formally after a colon or integrated into the grammar of the containing sentence, as with the example above. And of course the quotation must be accompanied by a superscript indicating a footnote.

PARAPHRASE

It is not always necessary or desirable to quote a source. Often facts and ideas can be paraphrased—that is, expressed in your own words. But do not change the substance of the original. Johnson's note could be paraphrased:

```
Among the Protestants the Jehovah's Witnesses were
especially notable for their resistance to Hitler.
They refused to cooperate with the Nazis and suffered
as a result, many being killed or imprisoned.[1]
```

Notice that in this case a superscript follows the paraphrase. It is not easy to draw a line between paraphrases which need footnotes and those which do not. If a fact is generally known or an idea commonly held, it is not usually necessary to acknowledge a specific source. If, on the other hand, the information is specific or if it is controversial or an individual opinion, you must footnote it. Although the words have not been literally borrowed, the fact or idea has.

CONCLUSION

When introducing quotations, avoid awkward repetitions; do not say over and over "As Professor X writes. . . ." Whenever possible, integrate the quotation into your own sentence and vary the manner of the integration. Paraphrase when the exact words are not essential.

Above all, maintain a reasonable balance between quotation (and paraphrase) and your own words. A research paper requires quotation and footnotes. You must show that you have read thoroughly in the subject, that you have taken notes, and that you know how to incorporate this material into your paper and to acknowledge sources.

At the same time, you must also demonstrate that you have done more than stitch together what others have discovered and thought. Research papers are good only to the degree that writers make what they have read part of their own knowledge and express it in their own way, showing that they have really understood what they set out to study.

FOOTNOTES

INTRODUCTION

In.

A footnote is an explanation or reference placed, traditionally, at the bottom of the page, though today notes are often gathered together on a separate page at the end of the paper, chapter, or book, a practice more convenient for writer, typist, and printer.

Footnotes are of two kinds: content notes and citations. The first explain or amplify points in the text; the second identify sources. Explanatory notes are occasionally necessary. But they should be used sparingly and not allowed to clutter a paper. If a point is important enough to warrant explanation or further development, that discussion should usually be worked into the text. If the matter is not that significant, noting it is mere pedantry and distracting to the reader.

Citations—footnotes identifying sources—are another story. It is a principle of scholarship to acknowledge where you got facts and ideas, as a matter of simple honesty, as evidence that you are not talking through your hat, and as a convenience to readers wishing to consult sources for themselves.

Footnotes citing sources have a standardized form. When you write a term paper, you have a responsibility to learn the conventions of footnoting as they apply in your subject and to follow them exactly. These conventions become quite complicated because of the variety of books and articles you may have to quote. Here we can treat only the rudiments of footnote form. You will find more thorough discussions in any of the books on writing term papers listed on pages 519–20. In addition, the following two works give authoritative advice on composing footnotes and bibliographies:

A Manual of Style, 12th edition, revised. Chicago: The University of Chicago Press, 1969.
The MLA Style Sheet. New York: Modern Language Association, 1970.

PLACEMENT OF FOOTNOTES

A footnote consists of two parts: the superscript and the note proper. The superscript—or "superior figure"—appears in the text. In scholarly work it is an arabic numeral (with no following period). Do not use asterisks, daggers, and so on. The superscript appears above and immediately outside of the closing punctuation (the final quote or the period), from which it is not separated by a space.[2]

It is simplest to number notes consecutively from first to last throughout the entire paper (or chapter in the case of a longer work). Other

---

2. This is an example.

methods include: beginning a new set of numbers with each page, or numbering in series of 1 through 9 and then starting over. These systems avoid the awkwardness of numerous two- and three-digit superscripts. But they cause other complications.

In typescript it is easiest to relegate the notes themselves to a separate sheet, placed at the end of the paper, each note being identified by the number used in the superscript. Readers find it more convenient to have notes at the foot of the page, but to place them there demands considerable experience and forethought on the part of the typist. If you do put them there, it is good practice to separate the note from the text by a 12- or 15-space bar:

> 1. George Orwell. The Road to Wigan Pier (New York: Harcourt Brace Jovanovich, 1958).

Notice that the first line of a footnote is indented by several spaces. It is considered poor practice to continue a note from the bottom of one page to the bottom of the next. Hence if you are placing notes at the foot of the page, you must think ahead and end the text with sufficient paper to accommodate notes.

Occasionally notes can be written directly in the text immediately below the reference, separated from the text by lines drawn across the page both above and below the note.

But unless your instructor asks that notes be arranged in one of these other ways, gather them together on their own page at the conclusion of the paper.

FOOTNOTE FORM

Here are two sample footnotes, the first citing a book, the second an article:

> 1. Jonathan Raban. Old Glory: An American Voyage (New York: Simon & Schuster, 1981), p. 135.
> 2. Karl W. Dykema, "Where Our Grammar Came From," College English 22 (1961), 455–465.

You would use complete notes such as these the first time you refer to a publication. Since there are slight differences in form between citations for books and those for articles, let us consider them separately.

FOOTNOTE FORM: BOOKS

1. *The note number.* This must correspond to the superscript. It precedes the note proper and is followed by a period. A single space is left between the period and the first word of the note. Notice that the first line is indented five to eight spaces. (This style refers primarily

to typescript. Material actually in print has many different acceptable forms.)

2. *The author's name.* The name should be in the form in which it appears on the title page, Christian name(s) or initial(s) first. If the first names consist only of initials, you will help readers by adding, in brackets, the letters to make the full name (assuming that you know them—they should be in the main entry in the card catalogue). The name C. O. Sylvester Mawson would appear:

```
C[hristopher] O[rlando] Sylvester Morton.
```

The information may save readers the tedious task of searching the card catalogue, especially with common names such as Smith or Brown.

However, in the case of well-known authors who habitually use initials, it is not necessary to indicate the full name: "G. K. Chesterton" or "W. H. Auden" is sufficient. Observe, incidentally, that a space is left between initials.

When a book has several authors, include them all as their names appear on the title page, separating the names with commas as in any series. Should there be numerous authors, it is simplest to mention only the first name followed by the abbreviation et al., Latin for "and others."

If the author is the editor of the book and is so identified on the title page, follow his or her name with "ed.":

```
1. W. H. Auden and Norman Holmes Pearson, eds.,
Poets of the English Language, 5 vols. (New York:
Viking, 1950).
```

3. *The full title, underlined.* This means the title as it appears on the title page, which may be more than you find on the spine or cover. If there is a subtitle, include it, preceded by a colon:

```
1. Edmund Wilson, The American Earthquake: A
Documentary of the Jazz Age, the Great Depression,
and the New Deal (Garden City, N.Y.: Doubleday, 1958).
```

Separate the title from the author's name by a comma, and capitalize the title as shown in the book. Should that not be indicated (all the type might be in capital letters, for instance), capitalize the first and last words and all those in between except articles, prepositions, and coordinating conjunctions. (An article after a colon introducing a subtitle *is* capitalized, as in the example immediately above.)

In the case of revised or numbered editions, the information is placed after the title, distinguished from it by a comma but not underlined. For "edition" use "ed."; for "revised," "rev."; and for an ordinal number, the arabic numeral followed by the ordinal suffix—"2nd," "3rd," "10th":

```
1. Sylvia Ziskind and Agnes Ann Hede, Reference
Readiness, 2nd ed. (Hamden, Ct.: Linnet, 1977).
```

If the work cited is in several volumes and the reference is general (that is, not to a specific page), the number of volumes should follow the title (or the edition, assuming one is named):

> 1. Henry Sweet, A New English Grammar, Logical and Historical, 2 vols. (Oxford: Clarendon Press, 1891).

Sometimes you may wish to refer to the preface or introduction of a book rather than to the main text. Place the title of that section, as given in the publication, before the main title. It appears in quotes, the main title is underlined, and the two are separated by a comma, set inside the closing quote:

> 1. [George] Bernard Shaw, "Preface," Back to Methuselah (New York: Oxford, 1947), p. xiii.

"George" is in brackets because the playwright's name appears on the work simply as "Bernard Shaw." The page reference is given in lower case Roman numerals because the Preface is so numbered.

Sometimes, too, you may refer to an introduction (or essay) by one author contained in a book by another. The footnote is handled like this:

> 1. Alexander Woolcott, "Forward," Charles Dickens, The Last of the Great Men, by G. K. Chesterton (New York: The Press of the Readers Club, 1942), p. xi.
> 2. Angela Carter, "The Language of Sisterhood," The State of the Language, ed. Leonard Michaels and Christopher Ricks (Berkeley: University of California Press, 1980), 226–234.

In the second example the citation is to the essay in its entirety; both the first and the last pages are cited to locate the piece in the book. Notice that no "pp." is required. Had the reference been to a specific passage in Carter's essay, the note would have ended:

> . . . (Berkeley: University of California Press, 1980), p. 229.

4. *Name of editor, translator, and so on.* If any person other than the author has contributed to the book as editor or translator, his or her name should follow the title, after an appropriate abbreviation ("ed.," "trans.," and so on), the whole preceded by a comma:

> 1. C[hristopher] O[rlando] Sylvester Mawson, Dictionary of Foreign Terms, rev. Charles Berlitz (New York: Thomas Y. Crowell, 1975).
> 2. Émile Zola, Thérèse Raquin, trans. L. W. Tancock (Harmondsworth, Middlesex: Penguin, 1962).

However, it is not necessary to include the name of anyone who has written a special introduction to the book, unless you are citing that introduction. In that case the name of the writer of the introduction comes first.

If the author is identified as "editor," then his or her name begins the note, followed (after a comma) by "ed.," as explained on page 546.

5. *The place and year of publication and the publisher.* These are in parentheses and separated from each other by a comma. In the case of major publishing centers you need not specify country or state: simply New York, London, Paris, Boston, Chicago, and so on. With lesser known cities or cities that might be confused with other places of the same name, use a state or regional designation, as in the Zola example or for a book published in, say, London, Ontario. Some publishers have editorial offices in several cities and list them all on the title page. Do not repeat each name; the first city is sufficient. Put the publisher between the place and year of publication. Separate it from the former by a colon, and the latter by a comma:

> 1. Oscar Handlin, <u>Truth in History</u> (Cambridge, Mass.: Harvard, 1979).

If the place and year of publication end the note, as they would in a reference to a work generally, place a period after the closing quote.

7. *Page reference.* Mostly you will refer to a passage on a specific page, and your note will end with the page number(s). Place a comma after the parenthesis closing the place and date, and follow it with "p." for "page" (or "pp." for "pages") and the appropriate number or numbers. If the passage extends over two or more pages, write both numbers completely:

> 1. Frances Fitzgerald, <u>America Revised: History Schoolbooks in the Twentieth Century</u> (Boston: Little, Brown, 1979), pp. 198–199.

If the reference is to a page marked by a roman numeral—as in an introduction or preface—use the roman form; it will generally be in lower case.

When citing a passage in a work consisting of several volumes, identify the volume number with a capital roman numeral, before the page reference, separated from it by a comma. It is not necessary to use an abbreviation for "volume" or even, in this particular case, for "page":

> 1. Edgar Johnson, <u>Charles Dickens: His Tragedy and Triumph</u> (New York: Simon & Schuster, 1952), I, 67.

Should the reference be to a footnote, that fact must be indicated after the page number and separated by a comma. Use "n." to abbreviate "note":

> 1. Frank J. Sulloway, <u>Freud, Biologist of the Mind: Beyond the Psychoanalytic Legend</u> (New York: Basic Books, 1979), p. 112, n. 6.

After the final number close off the note with a period.

FOOTNOTE FORM: ARTICLES

1. *Number and author's name.* These are handled as in footnotes for books.

2. *Title.* With an article, two titles are needed: that of the article itself and that of the periodical in which it appears. They appear in that order, the first enclosed in quotes, the second underlined, and they are separated by a comma placed inside the closing quote:

```
        1. Rudolph Von Abele, " 'Ulysses': The Myth of
Myth," PMLA. . . .
```

PMLA is not an abbreviation; it is the official title of the publication of the Modern Language Association. However, it is permissible to abbreviate titles according to accepted practice (but only if such practice exists). For example, the *Journal of Modern History* is commonly abbreviated *J. Mod. Hist.* Notice that a period follows each shortened item, that the abbreviated title is underlined, and that words are capitalized just as in the full title.

3. *Volume number of the periodical.* The volume number follows the title, separated from it by a comma. Modern practice is to write volume numbers in arabic, even though the journal itself may use roman; "vol." or "v." is not needed:

```
        1. Rudolph Von Abele, " 'Ulysses': The Myth of
Myth," PMLA, 69. . . .
```

The issue number of the journal is not generally necessary. It becomes so if the pagination begins anew for each issue (many journals continue their page numbering through an entire year) *and* if the month or quarter of the issue is not mentioned.

4. *The month (or quarter—"Spring," "Winter," and so on) and year of the journal.* These are enclosed in parentheses and *not* separated by a comma. Nor does a comma come between the opening parenthesis and the volume number:

```
        1. Rudolph Von Abele, " 'Ulysses': The Myth of
Myth," PMLA, 69 (June 1954). . . .
```

In the case of weekly or monthly periodicals which begin pagination anew with each issue, the date should be written in full as it appears on the cover (including day of month if that is given). With such dated weekly or monthly magazines, it is *not* necessary to cite volume or issue numbers:

```
        1. John Lardner, "A Reporter at Large: Anzio,
February 10th," The New Yorker, February 26, 1944, pp.
48-58.
```

Notice that in such a footnote the date is not enclosed in parentheses.

5. *Page number.* A footnote for an article always requires some page reference. It follows the date and is separated from it by a comma.

When you are citing an article in its entirety, supply both first and final pages, writing the numbers in full. When you are citing a specific passage, simply identify the page(s) on which it occurs. In either case, no abbreviation for page is required (unless there is no volume number for the publication, an unlikely possibility):

> 1. Rudolph Von Abele, "'Ulysses': The Myth of Myth," <u>PMLA</u>, 69 (June 1954), 358–364.

Or, if citing a particular passage:

> 1. Rudolph Von Abele, "'Ulysses': The Myth of Myth," <u>PMLA</u>, 69 (June 1954), 360.

6. *Close the footnote with a period.*

FOOTNOTE FORM: SHORTER VERSIONS

Under certain conditions it is not necessary to give all the information specified above, not, at least, in the footnote.

1. *Information found in the text which need not be repeated.* The notes we have seen so far have been initial citations—that is, first references to a book or an article—in which all the required data goes in the note. Whenever the text supplies any of that information, you need not state it again in the footnote. Suppose, for instance, that the first time you quote a book you mention the author's full name in your introduction; your note would then omit it:

Professor Paul Johnson points out that the Jehovah's Witnesses were almost alone among professed Christians in resisting Hitler.[1]

---

1. <u>A History of Christianity</u> (New York: Atheneum, 1976), p. 489.

If your text had read "Professor Paul Johnson points out in *A History of Christianity* that the Jehovah's Witnesses . . .[1]", your note would be simply:

> 1. (New York: Atheneum, 1976), p. 489.

However, it is awkward to refer to full titles very often. They are best put in footnotes.

2. *Subsequent reference.* Often you will cite the same book or article several times. It is not necessary to repeat all the information given in the initial footnote. Place and date of publication should be omitted. The author's name and the title may or may not be required.

Assuming that the author's name is not mentioned in the passage demanding the footnote, that you have cited only one work by that author, that the immediately preceding footnote is not to that work, and that you cite no other author with the same surname—assuming all this, your note should consist only of the writer's last name plus the page reference:

```
It has been pointed out that the Jehovah's Witnesses
were almost alone among professed Christians in
resisting Hitler.¹
```

> 1. Johnson, p. 489.

If you have cited two or more publications by the same author, a subsequent note must then include the title:

> 1. Johnson, A History of Christianity, p. 489.

A subtitle need not be included in such a case, and it is permissible to use a short form of the title (but not an abbreviation) if it is clear.

Should there be two or more authors with the same surname among your sources, you will have to give the Christian name or initial in subsequent references to avoid confusion.

The Latin abbreviation op. cit. (*opere citato,* "in the work cited") is sometimes used in place of the title in subsequent notes, though obviously it can be employed only when you are citing a single publication by that author. It need not be underlined:

> 1. Johnson, op. cit., p. 489.

Most contemporary scholars, however, simply repeat the title.

Another Latin abbreviation is more common: Ibid. (*ibidem,* "in the same place"). Ibid., which is no longer underlined, is used for a subsequent reference *only* when the work was cited in the immediately preceding footnote. If the reference is to the same page, a simple Ibid. (capitalized) is enough. If it is to a passage on another page, the abbreviation is followed by the page number:

> 1. Johnson, A History of Christianity, p. 489.
> 2. Ibid.
> 3. Ibid., p. 493.

The abbreviation loc. cit. (*loco citato,* "in the place cited") has the same meaning and function as Ibid. and is rarely employed today.

FOOTNOTE FORM: SCIENTIFIC WRITING

We have been discussing the kind of footnoting traditionally employed in literary and historical scholarship. The sciences, however, and philosophical writing often use a different system called "name-and-year citation." At the end of the paper a list of references, arranged alphabetically by the authors' last names, includes the titles of the books and articles and the years of their publication. If several works by the same writer are cited, they appear chronologically. The entries in the list may or may not be numbered.

A source is identified directly in the text by a parenthetical enclosure placed immediately after the information or quotation. If the list of references is not numbered, the parentheses contain the author's last name, the year of publication, and the page number (in those cases where a specific passage is cited). Should the name of the writer be mentioned in the text, only the date appears in the parentheses. (When the name is mentioned, the parenthetical citation occurs immediately after.) Here are several examples from *Sociobiology: The Abridged Edition* by Edward O. Wilson:[3]

> Several theoretical circumstances can be conceived under which competition is perpetually sidestepped (Hutchinson, 1948, 1961).
>
> Although the effect is generally thought of as a vertebrate trait, it also occurs in such insects as bumblebees and crickets (Alexander, 1961; Free, 1961a).
>
> Linsenmair and Linsenmair (1971) discovered that each burrow is occupied by an adult pair that remain mated for life.

In each case the writer's name—whether in the text or in the parentheses—refers us to a publication in the list of references, and that publication is further specified by the year. Thus for the first citation we would find in the references the entry:

    Hutchinson, G. E. 1948. Circular causal systems in
    ecology. Annals of the New York Academy of Sciences,
    50 (4): 221–246.

In that same example the fact that two dates are given indicates that two works by Hutchinson are being cited.

The second example cites a pair of references, separated by a semicolon. In the second of these the "a" following the year signifies that the writer published at least two titles in that year, of which this is the first.

In the third example only the year is given because the writers' names are mentioned in the text.

Should a citation to a particular page be required, it would be handled like this:

    (Hutchinson, 1948, p. 225).

Sometimes the entries in the list of references are numbered, and in that case the year of publication is omitted. The citation consists

---

3. (Cambridge, Mass.: Harvard, 1980). The three examples are from pages 120, 123, and 165, respectively.

simply of the writer's name (when it is not specified in the text), the number of the work in the list, and (if necessary) the page:

(Hutchinson, 15)

or:

(Hutchinson, 15, p. 225).

All in all, this method of citing references has much to recommend it—clarity, simplicity, inexpensiveness. However, it is *not* used in most academic writing outside the sciences.

THE BIBLIOGRAPHY

INTRODUCTION

Bibliog.

In scholarship a *bibliography* is a list of publications cited or consulted. It may be restricted to works actually referred to in the text or footnotes, or it may include all the material the writer consulted, even though some items may not have been cited. When you compose a research paper, be sure you understand what the instructor expects your bibliography to include.

Bibliographies may be annotated—that is, each entry accompanied by a brief summary or evaluation. In term papers, however, annotations are not generally required.

Usually bibliographies are placed on a separate page or pages at the end of the paper, though in long works brief bibliographies may follow each section or chapter.

Wherever it is placed, a bibliography is alphabetized according to the authors' last names. Publications without specific authors are arranged according to the first word of the title (excepting articles). In the case of two or more publications by the same writer, order is determined by title; and in the case of two or more authors with identical surnames, by their Christian names.

BIBLIOGRAPHIC FORM: BOOKS

A bibliographic entry includes the same information as a footnote: the author's name, the full title, the place and date of publication, and publisher (though in the case of books there is no need for a page reference). However, this information is arranged and punctuated somewhat differently.

1. *The writer's name.* The surname comes first (different than in a footnote), then a comma and the first name(s) and/or initial(s). The name is followed by a period, not with a comma as in a footnote. The full name should be used as it appears on the title page:

Dillard, Annie. Pilgrim at Tinker Creek. New York:
    Harper & Row, 1974.

In the case of writers who use only initials, it is a courtesy to readers
to supply the complete name, putting the missing letters in brackets
(though this need not be done for celebrated authors conventionally
known by their initials—T. S. Eliot, for example):

Naipaul, V[idiadhar] S[urajprasad]. The Loss of El
    Dorado. London: Readers Union, 1970.

This information, incidentally, should be available in the card catalogue,
and, in recently published books, it is often given on the reverse of
the title page in the Library of Congress cataloguing data.

When a book has two or more authors, they should all be listed (unless
they are very numerous, when a simple "et al." may follow the first
name). However, only the first name should be reversed, the others
being in normal order:

Ehrlich, Eugene, and Daniel Murphy. Writing and
    Researching Term Papers and Reports: A New Guide
    for Students. New York: Bantam, 1964.

When the author is an editor or translator, an appropriate abbreviation should follow the name, separated from it by a comma:

Michaels, Leonard, and Christopher Ricks, eds. The
    State of the Language. Berkeley: University of
    California Press, 1980.

If you list two or more works by the same writer, it is not necessary
to repeat his or her name. For subsequent entries, use a continuous
underscore of about twelve spaces in place of the name and follow it
with a period. The works should be alphabetized according to the first
words of their titles (excluding articles):

Berlin, Isaiah. Personal Impressions, ed. Henry Ward.
    New York: Viking, 1980.
——————————. Russian Thinkers, ed. Henry Hardy
    and Aileen Kelly. New York: Viking, 1978.

2. *The full title.* Give the title as it appears on the title page. Subtitles
should be included, preceded by a colon. The title is underlined and
closed by a period:

Garson, Barbara. All the Livelong Day: The Meaning and
    Demeaning of Work. New York: Penguin, 1975.

If a single entry involves two titles by the same author—say an introduction, essay, or story within a larger work—both must be given: first
the title of the shorter piece, set in quotes; then that of the containing
work, underlined. The two are separated by a period (sometimes a
comma is used instead):

> O'Connor, Flannery. "A Temple of the Holy Ghost." <u>A
> Good Man Is Hard to Find: Ten Memorable Short
> Stories</u>. New York: Doubleday, 1970.
> Sayers, Dorothy L. "'. . . And telling you a story': a
> Note on The Divine Comedy." <u>Essays Presented to
> Charles Williams</u>. London: Oxford, 1947.

When the shorter title is contained within a book written or edited by someone else, the name of the second author follows the book, preceded by any necessary abbreviation ("ed.," "comp.[iler]," and so on), the whole set off from the title by a comma:

> Burgess, Anthony. "Dubbing." <u>The State of the
> Language</u>, ed. Leonard Michaels and Christopher
> Ricks. Berkeley: University of California Press,
> 1980.

The name(s) of the author(s) of the containing work is not reversed.[4]

3. *The edition.* When a work is an edition other than the first, the fact should be indicated. The edition number follows the title, distinguished from it by a period. It is not underlined:

> <u>A Manual of Style</u>. 12th ed. Chicago: The University of
> Chicago Press, 1969.

No author is given in that example because none appears on the title page. In an actual bibliography the work would be alphabetized by "Manual."

4. *Volume number.* For any work published in two or more volumes, the number of volumes should be stated after the title, from which it is set off by a period. Arabic numerals are used for the number and "vols." for "volumes," and the abbreviation is followed by a period:

> Krapp, George Philip. <u>The English Language in America</u>.
> 2 vols. New York: Ungar, 1925.

5. *Series to which the work belongs.* Occasionally scholarly works are published as part of a series. The name of the series should follow the title, after a period. Then the volume number of the work in that series (if it has a number), in either arabic or roman, according to the original:

> Schlesinger, Alfred Cary. <u>Boundaries of Dionysus:
> Athenian Foundations for the Theory of Tragedy</u>.
> Martin Classical Lectures, XVII, Cambridge, Mass.:
> Harvard, 1963.

---

4. When the primary listing in the bibliography is to the larger work, then it must be alphabetized under the name of its writer or editor:

> Michaels, Leonard, and Christopher Ricks. <u>The State of the Language</u>.
> Berkeley: University of California Press, 1980.

6. *Place and year of publication and publisher.* These are not enclosed in parentheses (another difference from footnote form) but are separated by a comma. When a publisher lists several cities, mention only the first. Should there be no place of publication given, write "n.p."; no date, "n.d." Set the publisher between place and date, preceded by a colon and followed by a comma:

> Stansky, Peter, and William Abrahams. Orwell: The
>   Transformation. New York: Knopf, 1980.

7. *Close the entry with a period.*

BIBLIOGRAPHIC FORM: ARTICLES

For an article the bibliographic entry requires the titles of both the article and the journal. The first is quoted, the second underlined, and the two are separated by a period.

Following the title of the periodical, its volume number (if any) is given, preceded by a comma. The number is usually in arabic in contemporary practice, even though the journal itself may use roman.

Next comes the date. In the case of periodicals that number pages consecutively for the year, the year alone is sufficient; it is enclosed in parentheses and not otherwise punctuated. In the case of weekly or monthly publications which number pages anew with each issue, it is necessary to specify the month or week; these are distinguished from the volume or title by a comma and *not* set in parentheses.

After the date and following a comma, the first and final pages of the article are stated. The abbreviation "p." or "pp." is not required. The entry is closed by a period:

> Ribner, Irving. "The Tudor History Play." PMLA, 69
>   (1954), 591–609.
> Wilford, John Noble. "Space and the American Vision."
>   The New York Times Magazine, April 5, 1981, 53–66,
>   118–120.

Two sets of pages references are given for the second of those examples because the article is continued on subsequent pages of the magazine.

BIBLIOGRAPHIC FORM: SCIENTIFIC WRITING

Just as the method of citing references differs in scientific writing, so does the form of the bibliography. In fact the term "bibliography" is rarely used in contemporary science papers, but rather "List of References" or "Literature Cited." Such lists are handled somewhat differently from one science to another, and even within the same discipline.

You should consult your teacher when writing a paper in a science.[5]

In general the list will be alphabetized by the writers' last names, these being placed first in the entry and followed by the first name(s) or initial(s). A period closes the name. The year of publication follows. If the same writer has published two or more works in the same year, these are distinguished by lower-case letters immediately after the years: "1970a, 1970b," and so on.

The title follows the year, separated from it by a period. In the case of an article, its title precedes that of the journal, and the two are set apart by a period. In some sciences it is the practice to capitalize only the first word of a title and not to enclose the titles of articles in quotes or to underline those of books. However, in these matters of format, usage varies; you should consult your teacher or a stylesheet for your science.

After the title of a journal the volume number is given in arabic (without any abbreviation for "volume"). The issue number may follow this in parentheses. Then, after a colon or comma, the first and last pages of the article (without an abbreviation for "page").

After the title of a book, the publisher is usually identified, separated from the title by a period. Then, following a comma, the place of publication. The entry is closed by a period.

Here are the two examples from biology, but, once again, remember that style for mathematics, psychology, physics, chemistry, geology, and so on will vary:

> Hutchinson, G. E. 1948. Circular causal systems in ecology. Annals of the New York Academy of Sciences, 50 (4): 221–246.
> Wilson, Edward O. 1980. Sociobiology: The abridged edition. The Belknap Press, Cambridge, Mass.

---

5. Clear summaries are to be found in James D. Lester, *Writing Research Papers: A Complete Guide*, 3rd ed. (Glenview, Illinois: Scott, Foresman, 1980); and in Kate L. Turabian, *A Manual for Writers of Term Papers, Theses, and Dissertations*, 4th ed. (Chicago: The University of Chicago Press, 1973).

CHAPTER 51

A Sample
Research Project

CHOOSING A TOPIC

To write a successful research paper, you must choose the topic wisely. Don't be too ambitious. An undergraduate paper cannot encompass vast issues like the New Deal or the Vietnam War or feminism. These are important subjects, but you cannot handle them in fifteen or twenty pages—not, at least, without the years of study and thought such synopsizing demands.

You had better settle upon something smaller, more manageable. First, of course, it should be important and interesting, worth your time and trouble and the reader's. Second, it should be clearly defined, with a beginning and end and focal points upon which to center a discussion. If you are interested in the New Deal, consider, say, the attempt by President Roosevelt in 1936 to enlarge the Supreme Court in order to accommodate his liberal policies. If you want to write about Vietnam, you might concentrate upon the domestic opposition to the war or, even more specifically, upon the riots at the Democratic National Convention in Chicago in 1968. If you are concerned with feminism, you could discuss arguments for and against the ERA. Such topics involve larger issues, yet are relatively small and self-contained, allowing you to focus on specific problems and people and events.

Third, a suitable topic should have been the occasion for discussion so that documents about it exist. Discussion usually results from controversy, and while controversy is not essential to a term-paper topic, it is desirable. Controversy implies a focal point, a "problem" around which the paper can be organized. Was President Roosevelt morally justified in attempting to pack the Court? Was it politically foolish and unrealistic, or a necessary and calculated risk? Were the rioters in Chicago acting on genuine moral principles to protest an unjustifiable war; were they using resistance to the war as a cloak for subverting civil authority; or were they hooligans throwing rocks under the pretext of moral action? Did all of these motives come together in the riots? Is the ERA desirable? Is it necessary to secure equality of opportunity for women? Questions like these give focus to a term paper, and focus is absolutely necessary if your paper is to be written well.

To take a particular example, suppose you are studying the Second World War in a modern history course. You have become interested in the military operations and decide to write about one of these. You ought to avoid such subjects as the Battle of the Atlantic, the Normandy invasion, or Hitler's invasion of Russia, military events too large and complex for a term paper to contain.

What you want is a smaller battle: important, clearly defined, and, if possible, controversial. There are a number of possibilities: the Japanese conquest of Singapore; certain landings on Pacific islands, such as Tarawa or Saipan, by U.S. forces; the naval actions of the Coral sea or Midway; the loss of a German army to the Russians at Stalingrad; the Dieppe raid; the Anzio landing.

You settle upon the last—the landing of the VI Corps (consisting of one British and one American division) of the U.S. Fifth Army at Anzio in January 1944. Anzio is on the Tyrrehenian (southwest) coast of Italy some thirty-five miles south of Rome and about sixty miles above the German defenses which had stalled British and American armies driving up the Italian peninsula. These defenses—the Gustav Line—were created by Field Marshal Albert Kesselring, one of the ablest generals of the war, and were anchored on the famous medieval abbey of Monte Cassino, located on a hilltop overlooking the town of Cassino and the Liri Valley.

The purpose of the Anzio landing was to force the Germans to withdraw from the Gustav defenses by threatening their lines of supply and their rear. The operation did not accomplish this purpose, at least not immediately. Why it did not occasioned considerable controversy.

Anzio, then, is a good topic. It is relatively small and well defined—the battle began on January 22, 1944, and ended four months later on May 25 when the VI Corps finally broke out of the beachhead. And it poses clear questions of what went wrong and why—questions which, while they may not be susceptible of final answers, still provide a clear structure for a term paper.

How would you go about writing that paper?

LOOKING FOR SOURCES

THE CARD CATALOGUE

You begin with the card catalogue of the library, specifically with the heading "Anzio." Don't give up if all you find is a tourist guide dated 1910. Look further along; what you want may be indexed under "Anzio, battle of, 1944" or "Anzio beachhead, 1944." You should find two or three titles at least, perhaps:

> *Anzio Beachhead (22 January—25 May 1944).* American Forces in Action Series. Washington, D.C.: Historical Division, Department of the Army, 1947.
>
> Blumenson, Martin. *Anzio: The Gamble that Failed.* Philadelphia: Lippincott, 1963.
>
> Vaughan-Thomas, Wynford. *Anzio.* New York: Holt, 1961.

While "Anzio" is the first subject heading you should consult, it is not the only one. Look under the larger heading "World War II." There you will locate general histories containing brief accounts and evaluations of Anzio, such as:

> Churchill, Winston S. *The Second World War,* 6 vols. Boston: Houghton Mifflin, 1951.
>
> Dupuy, R. Ernest. *World War II: A Compact History.* New York: Hawthorn, 1969.
>
> Liddle Hart, B. H. *History of the Second World War.* New York: Putnam, 1971.

Look also under "U.S. Army." You should find an official history in many volumes entitled *United States Army in World War II.* Several of the volumes are devoted to the fighting in the Mediterranean, and one of these includes an account of the Anzio operation:

> Blumenson, Martin. *Salerno to Cassino. United States Army in World War II: Mediterranean Theater of Operations.* Washington, D.C.: Office of the Chief of Military History, United States Army, 1969.

These books will get you started. As you read you will gather other subject headings with which you can return to the card catalogue for more information. Check, for example, the names of the generals involved: John Porter Lucas, the initial commander of the VI Corps; Lucian K. Truscott, who relieved him; Mark Clark, commander of the Fifth Army and immediate superior to Lucas and Truscott; Sir Harold Alexander, the British general superior to Clark; Dwight D. Eisenhower, supreme commander in the Mediterranean when Anzio was first planned; Winston S. Churchill (as subject, rather than author); Albert Kesselring, the German commander in Italy; Siegfried Westphal, Kesselring's Chief of Staff; Ernst von Mackensen, the German general who directed the counteroffensive which came close to driving VI Corps back into the sea. Most of these men wrote about Anzio in memoirs composed after the war, and some of them, such as Eisenhower and Churchill, were subjects of books.

Use imagination in exploring the catalogue. Don't stop with general histories of the war. If your library categorizes the campaigns further, you may find something under "World War II—Italian Campaign." Remember, too, that Anzio was planned to relieve Cassino, and books about Cassino may mention the landing. In fact there is a lucid discussion of Anzio in just such a book:

Majdalany, Fred. *Cassino: Portrait of a Battle.* London: Longmans, 1957. (Published in the United States as *The Battle of Cassino.*)

You would never find that account, however, unless you thought to look under "Cassino" and leafed through Majdalany's book; it would not be cross-indexed under "Anzio" in the card catalogue.

Most of the books that you can find will contain bibliographies pointing you toward additional sources. You cannot expect your library to hold all these sources, but often you can obtain books and articles from other libraries by special loan.

NEWSPAPER AND PERIODICAL INDEXES

Not everything written about Anzio appears in book form. Much has been published in newspapers, magazines, and scholarly journals. These sources are available to you through special indexes.

1. The *New York Times Index*

For any event of national or international significance since 1851, the *New York Times Index* is an indispensable bibliographic tool. An annual subject index, it contains numerous cross-references, and these make using it a bit confusing until you understand how they work. As with any index, you should spend a few minutes reading the front matter which explains how entries are set up. For any subject the *Times Index* lists, you will be able to locate the issue(s) of *The New York Times* in which the story appears, the page number, and the column. But you may not find the information under the first entry you consult.

For our project you would begin by looking under "Anzio" in the *Index* for 1944. (A careful scholar would check each subsequent year to the present to catch any postwar discussions of the operation.) You would read:

ANZIO action. See World War II—Mediterranean Front, Ja 23 ff in ¶ 1, F 3, 6 ff in ¶ 19, ¶ 35, ¶ 58, My 1–4, 6–14, 16, 19, 21–26, 30 in ¶ 75, Je 3, 8, 21 in ¶ 91, Jl 16 in ¶ 109, Ag 3, 11, 22 in ¶ 124, N 3 in ¶ 165, D 10, 16 in ¶ 177.

The ¶ numbers refer to sections in the entry "World War II—Mediterranean Front"; the dates, to issues of *The New York Times* carrying stories about Anzio. For example, "Ja 23 ff in ¶ 1" means that in section

1 of "World War II—Mediterranean Front" you should look for the dates January 23 and the next several days. (The abbreviation "f" means "following," that is, the immediately next day or, in the case of a book, page; "ff" is the plural and means the next two or more days or pages.) "F 3, 6 ff in ¶ 19, ¶ 35, ¶ 58" translates "look for the dates February 3, 6, 7, 8 (and perhaps 9, 10, and further) in sections 19, 35, and 58."

Turning next in the *Times Index* for 1944 to the entry "World War II—Mediterranean Front," you find many pages of numbered sections. Each section cited in the "Anzio" entry must be followed up, under the given dates, to find everything published in the year 1944 relevant to Anzio.

Page 563 is a reprint of one section. Under "Anzio" the next-to-last reference was "N 3 in ¶ 165"; ¶ 165 is marked with a check.

Work down the dates (the first is N[ovember] 1) until you come to "N 3, 11:2–4." That entry means *"The New York Times,* November 3, 1944, page 11, columns two through four." There you will find the full story synopsized immediately *before* the date N 3. It is easy to become confused about the relative positions of date and summary and suppose that the synopsis follows the date rather than preceding it, a mistake which leads to fruitless and frustrating searches.

This particular synopsis includes matters other than Anzio; what is pertinent is this:

> Brit Gen Alexander summarizes Italian campaign; acknowledges stalemate; discusses cost to both sides; cites Salerno, Anzio, Cassino and other operations; lauds Kesselring skill. . . .

Next you should consult the November 3 edition of *The New York Times.* Your library probably has microfilms of the paper, and it will be a simple matter to fit the proper reel onto the machine, focus, and wind till you come to page 11 of the November 3 issue. In columns 2 through 4 is a report of a press conference held by General Alexander. Among his comments is this assessment of Anzio:

> . . . he [General Alexander] declared that the Anzio operation while not 100 per cent successful had played a "vital if not decisive" role in subsequent progress.

The statement bears on the question of whether Anzio is to be judged a failure or a success, and therefore should be transcribed verbatim as a note, with the source information on the card that it was expressed by General Harold Alexander at a news conference and reported in *The New York Times,* Nov. 3, 1944, p. 11, cols. 2–4.

Of course you shouldn't read every issue referred to in the "Anzio" entry, not unless you have considerable time and the true scholar's compulsion to leave no stone unturned. But you ought to look at all the synopses and judge from these which are worth following up.

WORLD War II—Mediterranean Front—Cont

1164—GREECE: Brit troops and Greek patriots drive
Gers from Kozane, 240 mi north of Athens; advance to
Arnissa, 7 mi from Yugoslav border; fleeing Gers reptd
heading for Albania; Ger Salonika garrison strength
noted; US paratroop operations during past 6 mos revealed; Brit paratroopers watching Athens liberation event
illus; Greek Natl Dem Army (EDES) battles Gers, Yanina
area; Ital invasion anniv events, Athens, cited; Adm
Cunningham and Gen Wilson leave Athens, N 1,7:1,2;
Brit patrols reach Salonika; Berlin repts Gers evacuate
port, N 2,4:4; Brit Admiralty repts on Aegean waters
activity since Sept 15; force under Brit Vice Adm Raw-
lings includes Brit, French, Greek and Polish vessels,
N 2,11:6; Brit liberate Salonika; narrow Ger Vardar Valley escape route; patrols drive north of Salonika to within 50 mi of Bulgarian border; Gers rept reinforced Tilos
(Piscopi) Is, Dodecanese group, garrison repulses Brit and
US landing units; rept fighting continues, Melos Is, N 3,
7:2; Brit capture Florina, 5 mi from Yugoslav frontier;
derail Ger supply train; Salonika war damage described;
pub welcome cited, N 4,4:1; Greece liberation completed;
additional Brit troops land, Salonika area, N 5,1:7; Salonika port repair begun, N 5,19:4; other war damage
cited, N 6,9:3; Aegean cleared of Ger shipping; mopping-
up completed between Sept 9 and Oct 27 by Brit-com-
manded force including 1 Polish and 2 Greek destroyers,
N 8,21:6; Brit Commandos landing, Salonika, illus; Brit
Navy and Engineers Piraeus port repair role cited, Eden s,
Commons, N 9,6:2,5; Brit-Polish destroyer flotilla sinks
U-boat, Aegean, N 11,5:3; Aegean route to Salonika cleared
of mines 10 days after city liberation; Berlin repts Aegean
island garrisons successful evacuation, N 12,19:3; RAF
activity against Aegean shipping revealed, N 13,5:7; Brit
naval craft sink 1 Ger lighter; damage another, off Alinnia
Is, N 15,4:6; Brit landing reptd, Melos Is, N 16,4:1; Commandos withdraw from Melos after taking prisoners and
inflicting casualties, N 17,5:6; 1st Ital soldiers and officers group reptd repatriated from Greece, N 26,14:2

1165—ITAL MAINLAND ACTION: US 92d Div Negro
troops role discussed; Brit 8th Army Indian troops consol
Ronco River bridgehead, Meldola area, 7 mi south of
Forli; Polish units Forli drive meets strong Ger resistance, Caminato; US 5th Army south and southeast of
Bologna repts patrol activity; US troops take 100 prisoners, Castellaccio area; Brazilian troops on Ligurian coast
take unidentified mt and Calomini; Brit Adriatic area
troops advance slightly; Allied planes strike Ger communication lines; Canadian patrols reach point 2½ mi
from Ravenna; H W Baldwin discusses stalemate; cites
terrain, weather, Ger resistance and skill, insufficient Allied ground strength and heterogeneous troops as major
causes of Allied delay; lists Allied divs, N 1,10:2,3; Brazilian troops role cited; morale discussed; 5th Army US
troops repulse several Ger counterattacks south of Bologna; Ger Castel San Pietro assault cited; Brit patrols
push toward Ravenna; drive Gers from 2 positions north
of Bevano River; other 8th Army forces improve Ronco
River bridgehead; capture Meldola, Brazilian forces take
La Rocchetti, Lama di Sotto, Pradoscello and Monte San
Quireco; rain hampers aerial and ground activity, N 2,
6:1,3; 8th Army units take half of Forli airport; 3 columns
drive on Forli; Gers counterattack below Bologna; US
troops clear Casetta; repulse Ger attacks, Castellaccio;
Brazilians withdraw slightly, Catagnana area; Ger Marshal Kesselring reptd wounded, Bologna area; Brit Gen
Alexander summarizes campaign; acknowledges stalemate; discusses cost to both sides; cites Salerno, Anzio,
Cassino and other operations; lauds Kesselring skill; cites
polyglot army manpower needs, N 3,11:2-4; Brazilian pilots
role cited, N 4,3:1; Brit battle for Forli airfield; Rome
radio repts airdrome capture; rains halt 5th Army action,
Bologna area; Canadian troops reptd stalled 5 mi from
Ravenna; ltd air activity reptd; Brazilian pilots see 1st
action; Swiss rept Gers build northeast Italy defense line
paralleling Austrian frontier, N 4,4:8; Lt Gen Leese leaves
post as Brit 8th Army comdr, N 4,6:6; US Thunderbolts
take heavy toll of Po Valley rolling stock, N 5,6:1; Kesselring reptd badly hurt, N 5,14:4; 8th Army troops consol
Forli positions; continue battle for airport; US, S African
and Brit troops repulse reinforced Gers counterattacking
below Bologna; incessant rain cited, N 5,16:3; US bombers
blast Brenner Pass ry and 2 rys connecting Italy with
Yugoslavia, N 6,3:1; newly-equipped Ital Army to join
5th and 8th Armies; clear weather enables engineers to
improve supply rds below Bologna; permits Allied plane
action; Negro units attack Ger hill position, Bologna area;
Brazilians repulse Ger counterattack, Calomini, N 6,6:4;
US 15th Air Force bombers attack Bolzano, Brenner Pass
ry, N 7,5:1; Ger and Allied patrols reptd active below
Bologna; Ger planes strafe Allied positions; Gers hold
tenaciously to Ronco and Forli airfield; Allied warships
bombard French-Ital border targets, N 7,15:6; S African
servicewomen role cited, N 7,20:3; Liberators attack Brenner Pass targets, N 8,19:6; 8th Army troops aided by
medium bombers move toward Monte Maggiore and San Martino; Gers keep up continuous fire against 5th Army be-

low Bologna, N 8,22:5; Poles progress between Rabbi and
Montone Rivers; capture Monte Casaluda and other hills;
Gers cling to Forli airfield despite Allied bombings; shell
Brit approaching Forli; other 8th Army troops clear San
Rufillo; approach Dovadola; 5th Army patrol activity in-
creases; Negro troops role cited, N 9,6:6; Churchill on
Alexander campaign into Po Valley, London Lord Mayor
luncheon s; cites Polish, Canadian and US troops role,
N 10,4:5; 8th Army troops approach Bussecchio, 1 mi
southeast of Forli; Gers rept no break-through made;
fighting for airfield continues; other Brit troops lead into
Forli from south; take Grisignano and San Martino;
Polish troops take Dovadola; Gers reptd withdrawing
from Forli; 5th Army front repts patrol and artillery
activity; Gers shell Monte Grande and Monte Cerere
areas; Negro units fight for western coastal sector hills,
Massa area; take Fabbiano and Basati, N 10,5:6; US destroyers bombard Ital-French frontier targets, N 10,7:4;
Brit occupy Forli; Gers destroy Montone River main
bridge; dig in on west bank; other Brit troops crossing
Rabbi River meet heavy Montone River shellfire; Polish
troops take Monte Bora; strong Ger Castrocaro position
noted; Ital Govt asks combattants to spare Ravenna;
Negro units take Azzanno, Terrinca and Lavigliani; Ger
Adige River defenses reptd near completion; map, N 11,1:7;
8th Army tanks entering Forli illus; Ger knowledge of
Allied intention to bomb Kesselring hdqrs, Frascati, Sept
7, '43, noted, N 11,5:2,6; Brit troops halted at Ravaldino
Canal outside Forli; snow hampers communications; Ger
evacuate Castrocaro; hold hills to west; Adriatic coastal
positions remain unchanged; Gers hold Coccolia and Gambellara, Ronco River area; 5th Army patrol activity continues, Bologna sector; Indian units occupy Monte Bu-
drialto, Faenza area, N 12,21:1; 15th Air Force raids
bridges in north, N 12,36:1; 8th Army establishes bridgehead across Nuovo Canal, Forli area; other units move into
Gambellara; US troops make slight gains, Bologna area;
Forli patriots role cited; Gers muster 11 planes in strongest attack in wks, Loiano; US soldiers attitude toward
Itals and their language discussed, N 13,4:1; ed, N 17,18:4;
Gers disguised as women fire on 8th Army scout car,
Ronco River area, N 13,4:3; Lt Gen Clark awards 5th Army
plaque to K Cornell for contributions to troops morale, N
13,5:4; 400 Czech puppet troops quit fight against parti-
sans, northern area, N 14,5:6; 8th Army units cross Ghiaia
Canal, 4 mi south of Ravenna after Ger withdrawal;
rept slight progress north of Forli; Gers shell Ronco
River east bank; 5th Army repts ltd patrol activity;
Alexander orders Ital patriots to halt organized operations, N 14,10:2; MAAF raids Casarsa rr bridge, N 14,
13:2; Brit win coastal highway bridge, Ghiaia Canal; capture San Tome, Forli area; Desert Air Force aid noted;
Brit occupy San Varano; cross Montone River at 2 points;
Gers seen making strong Montone River stand; Allies take
hills south of Rimini-Bologna highway; Poles take Monte
Casole in 2-mi advance; US fighter planes strike Ger oil
transfer points, Mantua area; snow and mud illus, N 15,
14:1,2; Brit advance from Montone River bridgeheads;
meet stiffened resistance beyond San Tome; seize foot-
hold on Highway 9 leading to Bologna; capture Poggiolo
and Villagrappa; battle Gers northwest of San Varano,
Forli southern outskirts; Polish units occupy Monte Cereto;
little activity reptd south of Ravenna; patrols probe
Ger positions, Ghiaia Canal area; Indian troops take
Monte San Bartolo, 5th Army front, N 16,6:1; US Capt
Voll downs 4 Ger planes, northern area, N 17,2:4; terror
grips Bologna, N 17,3:6; Liberators attack Piacenza, Bres-
cia and Bologna areas, N 18,3:5; Brit and Indian troops
occupy deserted Modigliana, Faenza area; gain on Route
9 from Forli; rain, cold and hail halt other forces; Brit
cross Bolzino River; advance to Castiglione outskirts; Gers
flood large Adriatic coastal area below Ravenna;
Alexander order barring mil activity angers Ital partisans; press comment, N 18,8:5,6; Liberators bomb Udine,
Aviano, Villafranca, Verona and Vicenza targets, N 19,
4:2; patriots blow up Ger ammunitions dump, Lanterna
Tunnel, Genoa area, N 19,11:4; Polish troops seize Monte
Fortina; other forces meet severe resistance below Faenza;
Brit tanks and infantry strike toward Route 9 from Villagrappa; Ger planes drop incendiaries on Forli; Brit clear
Gers from Molinaccio outpost, Adriatic area; weather halts
other activity, N 19,16:1; 5th Army Brit Grenadier Guards
crossing bridge illus, N 19,17:2; partisans plight as result
of Alexander order discussed; partisans destroy Ger
motorized column, Casola area; Brazilian anti-tank unit
crossing Serchio River bridge illus; 8th Army meets new
delaying line, Adriatic sector; Poles battle fiercely, Monte
Fortino area; Brit capture Golfara; US patrols penetrate
Ger Bologna positions; troops engage in bitter fighting;
Brazilians shifted to area 20 mi southwest of Bologna,
N 20,8:2,3,6; Gers recapture Monte Fortino; operations
elsewhere ltd to patrols, N 21,5:7; Ger Lt Gen Westphal
role cited, N 21,6:4; lr citing Florentines vain effort to
halt Ger destruction, N 21,24:7; patriots harass Gers,
French-Ital frontier area, N 22,6:3; weather limits armies
to patrol activity; US bombers strike Po River bridge
over 12-hr period, Ostiglia area, Mussolini Fascist forces
to be cut, N 22,8:5,6; US bombers hit rr objectives in

2. The *Readers' Guide to Periodical Literature*

The most useful source for locating articles appearing in magazines is the *Readers' Guide to Periodical Literature,* an annual index of about 160 magazines of a general rather than scholarly or technical nature. Articles are alphabetized in one general list by both subject and author (occasionally also by title), and the *Guide* is well cross-indexed.

If you knew the name of a writer of a magazine piece on Anzio but did not know where it appeared, you could look under his or her name. Or if you knew a title, you could determine both the author and the magazine. In a project like the one we are imagining, however, you would turn to the heading "Anzio" in the volume for 1944. (Prior to 1966 the *Readers' Guide* was cumulated in two-year segments; thus the appropriate volume would be 1943–1945.) As with searching the *Times Index,* careful scholarship would require that you look through each year of the *Guide* from 1944 to the present.

In the volume for 1943–1945 you would see under "Anzio" this entry:

ANZIO beachhead. See World War II—Campaigns and battle—Italian front.

Page 564 is a reprint of one page from the *Readers' Guide.*

Notice that the articles are arranged alphabetically by title. The checks have been added to mark articles about Anzio of primary interest (two checks) and of secondary (one). (Needless to say you ought not to mark the *Reader's Guide* in your library.) Not all the pieces single-checked would mention Anzio, but their titles are suggestive and worth investigating.

Here is one of the entries, labeled to identify its parts:

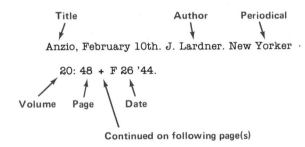

Your library will likely have a separate catalogue for periodicals (it may be called the serials list). Look there under the Ns to locate *The New Yorker.* In the issue for February 26, 1944, you will find the article by John Lardner, "A Reporter at Large: Anzio, February 10th." Your library may have microfilms of the magazine or you may have to find the actual magazine in volume 20 in the stacks.

When you have finished the article, you may well decide that it con-

WORLD WAR II--CAMPAIGNS AND BATTLES see also specific campaigns and battles.

Armor versus paratroopers. Lt Col H. Harmeling, Jr. Combat Forces 4:21-24 F '54

Assignment to Catastrophe. vol. 1:Prelude to Dunkirk, by Maj Gen Sir Edward L.Spears. Review. Combat Forces 5:53 D '54

The axis offensive in central Tunisia - February 1943. Brig Gen Paul M.Robinett. il maps Armor 63:7-17 My-Je '54

The Battle of the Scheldt, 1944. map Can Army J 8:3-13 O '54

Battlefield intelligence. H.W.Baldwin. Combat Forces 3:30-41 F '53

Brief outline of the Burma campaign of 1941-45. Brig M.R.Roberts. map Army Q 68:180-197 Jl '54

The Campaign on New Britain, by Lt Col F.O.Hough and Maj J.A.Crown. Review. Marine Corps Gaz 37:58 Je '53

The capitulation of Hamburg, 3rd May 1945. Dr. J.K.Dunlop. RUSIJ 99:80-89 F '54

The Chindit operations of 1944. Lt Col P.W.Mead. map RUSIJ 100:250-262 My '55

Crete (New Zealand in the Second World War 1939-1945) by D.M.Davin. Review. Army Q 68-8 Ap '54

D plus ten years. Maj Gen H.W.Blakeley. illus Combat Forces 4:23-27 Jl '54

The Dunkirk "Halt Order" - A further reassessment. Capt B.H.Liddell Hart. Army Q 69:207-213 J '55

East Africa Force operations, January to July 1941. Lt Col H.W.C.Stethem. il Can Army J 9:40-56 O '55

The Fall of the Philippines, by L. Morton. Review. Marine Corps Gaz 38:60 F '54; il Armor 63:57-61 Jl-Ag '54

Five Ventures, by Christopher Buckley. Review. Can Army J 9:87-90 Jl '55

The gateway to Leningrad. E.Raus. map Military R 34:99-109 S '54

The German Balkan campaign of 1941 (a digest). Gen Kurt von Tippelskirch. Military R 35:85-99 N '55

The German defeat in Russia. Sqdn Ldr P.C.Lambert. RUSIJ 98:95-101 F '53

Guard officers to attend D-Day observance. Nat Guard 8:16 Mr '54

The Hammelburg mission. Capt Martin Blumenson. map Military R 35:26-31 My '55

History of the Second World War--The Mediterranean and Middle East, v.1, by Maj Gen I.S.O. Playfair and others. Review. Army Q 68:135 Jl '54; Combat Forces 5:59-60 O '54

History of the Second World War: The War in France and Flanders, by Maj L.F.Ellis. Review. Combat Forces 5:53 D '54; Army Q 68:7-8 Ap '54

History of United States Naval operations in World War II; vol. 9: Sicily-Solerno-Anzio, by Samuel E. Marison. Review. Combat Forces 5:54-55 D '54

Invasion (operations of airborne units in invasion of France, 5-7 June 1944). map diag Inf Sch Quart 44:(7)-18 Ja '55

Invasion! The D-Day Story in Pictures, by The Daily Express, London. Review. Roundel 7:25 Ap '55

Iwo Jima:Amphibious Epic, by Lt Col W.S.Bartley. Review. Marine Corps Gaz 38:62 Ag '54

The Japanese seizure of Guam. T.Wilds. map Marine Corps Gaz 39:20-23 Jl '55

Jibouti and Madagascar in the 1939-45 war. Brig W.A.Ebsworth. RUSIJ 98:564-568 N '53

The last mounted charge. Lt Col J.D.Lunt. Army Q 67:245-248 Ja '54

Left flank at Iwo. Maj F.E.Haynes. il Marine Corps Gaz 37:48-53 Mr '53

Leyte: the Return to the Philippines, by M.H.Cannon. Review. Armor 64:57-61 Mr-Ap '55

Mass employment of armor. Col Wm.Darien Duncan. il tab Armor 63:10-20 Mr-Ap '54

Mobility unused. Maj Gen R.W.Grow. il map Military R 32:18-24 F '53

New Guinea and the Marianas, volume VIII of History of United States Naval Operations in World War II, by S.E.Morison. Review. Marine Corps Gaz 37:59 S '53

98th chemical mortar battalion. Alexander Batlin. il Armed Forces Chem J 7:50-52 O '53

Normandy, 1944. Rear Adm K.Assmann. map Military R 34:86-93 F '55

North-West Europe, 1944-1945, by John North. Review. Can Army J 8:92-94 Ja '54

Official History of New Zealand in the Second World War, 1939-1945. Review. Can Army J 8:89-92 Ja '54

Okinawa. Lt Col H.E.Fooks. RUSIJ 98:389-400 Ag '53

Operation Schmidt. Lt Col B.G.Taylor, Jr. map Military R 34:30-39 Ag '54

The Ramagen bridgehead. il Armor 64:39-42 Mr-Ap '55

The Recapture of Guam, by Maj O.R.Lodge. Review. Marine Corps Gaz 38:63 Je '54

Reflections on Anzio. Lt P.Cookley. Military R 33: 96-100 O '53 ✓

The Rommel Papers, edited by B.H.Liddell Hart. Review. AF Times 13:10 My 30 '53; Marine Corps Gaz 37:59 S '53

Service corps operations in the Aleutians. Maj R.D. Shaneman. Can Army J 8:129-131 Jl '54

Stilwell's Mission to China, by C.F.Romanus. Review. Combat Forces 4:53 Ag '53

Strategic withdrawals. G.Blumentritt. Military R 33: 24-34 S '53

The Supreme Command. United States Army in World War II; European Theater of Operations, by Forrest C.Pogue. Review. Signal 9:70 N-D '54

Surprise package. Lt Col J.W.Medusky. il Inf Sch Q 45:60-69 Ap '55 tains nothing you can use. Lardner concentrates upon the GI's view of the beachhead and stresses the crowded, tenement-like conditions and the soldiers' ability to adapt to such an existence. He says nothing significant about the strategic purpose of the landing or why it did not immediately achieve that purpose. (In scholarship you often spend time investigating sources that ultimately don't help you.)

On the other hand, the piece by Lardner in *Newsweek,* May 22, 1944, would prove more useful, as would two articles on the basic strategy of the Italian campaign, one by Admiral W. V. Pratt ("What Broke Down in Italy," *Newsweek,* April 10, 1944), the other by C. G. Paulding ("We Can Still Choose: Is the Italian Front Essential to Victory," *Commonweal,* February 18–25, 1944).

3. The *Air University Periodical Index*

The *Readers' Guide* is indispensable for researching general periodical literature. But you need more specialized indexes for scholarly and technical articles. For every area of scholarship, annual indexes and bibliographies list publications in that field. If you are interested, for example, in what has appeared on the artist Picasso, you would consult the *Art Index.* If you are writing something about schools, you would look in the *Education Index*; if about a subject in American or British literature, in the *MLA International Bibliography.* A reference librarian will help you locate such bibliographic resources for the discipline in which you are working.

For military subjects the appropriate source is the *Air University Periodical Index,* a biannual index of articles appearing in a variety of military and naval journals, such as the *Infantry Journal* or the *Military Review.* The *Air Index* is not as well cross-referenced as the *Times Index* or the *Readers' Guide.* For example, in the volume for 1953–1955 you will find no listing under "Anzio." But always double-check. In the blanket entry "World War II—Campaigns and Battles," you will strike pay dirt (see opposite page).

The entry is read like this:

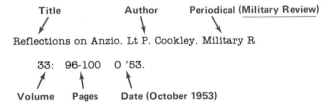

Check the serials list in the library to locate the *Military Review.* When you find Cookley's article in the October 1953 issue you will notice that it is not an original publication but a digest of an article that first appeared in an Irish journal.

The essay contains considerable information relevant to the strategic purpose of Anzio and whether or not it can be judged a success. Copy significant statements on note cards, using the writer's exact words and indicating any omissions or editorial additions. Follow the principle of one note per card and—if possible—one card per note. It is tedious but wise to identify the source fully on each card because ultimately the cards will be separated as you sort them in your file.

A sample note from Cookley's article might look like this:

"Anzio . . . did comply with the main aim in Italy, namely to tie down more German divisions. It succeeded in this purpose to a greater extent than its planners visualized."

P. Cookley, "Reflections on Anzio," Military Review, XXXIII (October 1953), 96.

In identifying the source, use the correct form either for a bibliographic entry or, as here, for a footnote. Notice that space has been left at the top of the card. This is for the heading under which you will file the card later.

When you have finished with Cookley's article and taken all the notes you want, make out a source and summary note. At the top of a fresh card write "SOURCE AND SUMMARY" to identify the card for the file. Then put down the author, title, and so on as you will use it in the bibliography of your paper. (Copy the information about the Irish journal, even though you will probably not have to include this in an actual bibliography.) Compose, *in your own words,* a summary and commentary upon the article. Doing so will clarify the writer's ideas in your own mind.

Just as you need imagination in exploring the card catalogue, so you need it in searching through such sources as the *Time Index,* the *Readers' Guide,* and the *Air Index.* Again, don't quit after you've looked under "Anzio"; not everything published about the operation has that magic word in its title. Check for entries under "Clark," "Lucas," and "Truscott" (the American generals) or "Kesselring" (the German). The *Air Index* has a subcategory under "World War II" called "Personal Narratives." Search for titles containing the names of units that fought

SOURCE AND SUMMARY

   Cookley, P. "Reflections on Anzio," Military Review,
XXXIII (October 1953), 96-103. Digested by the
Military Review from an article in "An Cosantoir"
(Ireland) June 1953.

   Cookley judges Anzio against the broad strategy of
the Italian campaign and stresses its value in forcing
Germany to keep divisions in Italy rather than
transfer them to France.
   He attacks the criticism that General Lucas was
derelict for not advancing immediately to the Alban

                                                      (over)

Hills, arguing with military logic that the minimum
number of troops necessary for such an advance were
not available.
   However, he does criticize Lucas for not moving
quickly and aggressively to take the towns of
Campoleone and Cisterna, important to the German
lines of communication and supply.
   Might Cookley's earlier argument be used against
him here? Perhaps even this more limited goal was
beyond the strength of the initial force?

at Anzio. For instance, if you saw the title "The Third Division in
Italy," you might well read the article because that division was one
of the two which made the initial landing.

GOVERNMENT DOCUMENTS

The U.S. Government is probably the largest publisher in the world.
The chances are that for any subject you might write on, government
publications exist. They are not always easy to find, however. Most
references libraries, which is what a university library primarily is,
have a documents department whose staff can guide you through the
maze of government material (including that issued by individual states,
the United Nations, and various foreign countries).

Two titles published by the federal government about Anzio have
already been mentioned: *Salerno to Cassino,* a volume in the official
history, *U.S. Army in World War II,* and *Anzio Beachhead (22 January–*

25 May 1944), a volume in the American Forces in Action Series. There are many other publications in military history. Here, for example, are two, among many, that would prove useful:

Arnold, Louise. *Special Bibliography 16, Vol. IV: The Era of World War II, Mediterranean Theater of Operations.* Carlisle Barracks, Pennsylvania: U.S. Army Military History Institute, n.d.

Jessup, John E., Jr., and Robert W. Coakley. *A Guide to the Study and Use of Military History.* Washington, D.C.: Center of Military History, United States Army, 1979.

The first contains a bibliography of sixteen titles about Anzio (pp. 181–182), and the second, while it says nothing directly about the battle, contains a list of military bibliographies and guides which would point you toward sources of information (pp. 443–444).

BOOK REVIEWS

Book reviews offer valuable information; you should check the reviews of any books which are essential sources. This is not as difficult as it may sound. Your library will contain the *Book Review Digest* (1905–present), a monthly publication cumulated in an annual volume. It synopsizes reviews of books published in the United States so long as the reviews have appeared in a required number of specified periodicals. Any important book published in this country during the twentieth century is probably listed in the *Book Review Digest* (*BRD*). Fourteen reviews of *The Battle of Cassino* by Fred Majdalany are summarized in the volume for March 1957–February 1958.

In addition, scholarly indexes often list reviews of books published in their field. The *Air University Periodical Index* for 1958 cites ten reviews of the Majdalany book in various military journals.

At the very least, book reviews, especially those by professionals, reveal a good deal about a book's reliability. The favorable reception of Majdalany's study, for instance, implies the soundness of his information and arguments. At their most useful, reviews may contain facts or ideas important to your topic.

ORGANIZING YOUR NOTES

When you began your research, you had some ideas about the Anzio operation, some knowledge of why it was mounted and what happened. As you read, your knowledge grows and key questions become more sharply defined. It is important that you settle on one or two or three of these and that you concentrate your note-taking and discussion upon them. This will save you from a common fault of research papers—diffuseness and lack of focus.

In the case of Anzio, the key questions include these:

1. Who planned the operation?
2. Why?
3. What went wrong and why?
4. Is Anzio to be judged a success or a failure?

None of these is as simple as it sounds; each entails more precise and subtle questions:

1. Who planned Anzio? analyzes into: What part did Churchill play? Eisenhower? Alexander? Clark?
2. Why? To what extent were the motives of these men political? To what extent military?
3. What went wrong and why? Did Anzio fail to achieve its immediate objective because the basic strategy was ill-conceived? Because the operation, while strategically sound, was poorly planned? Because, though feasible and well-planned, the assault was improperly executed? Or because the German response was simply more rapid and effective than the Allies anticipated? If the last is the case, could the planners have been reasonably expected to gauge the German reaction more accurately?

   In personal terms the questions become: To what degree was the responsibility Churchill's and Alexander's? To what degree Clark's? To what degree Lucas'? Was the problem partly in the rivalries and distrust between American and British generals?
4. Is Anzio to be judged a success or a failure? involves the deeper question: How are success and failure in war to be measured? Can an operation that fails in its immediate aim still be considered successful because of the benefits that ultimately follow from it?

Analyzing notes, as you see, is largely a matter of asking the right questions and of drawing useful distinctions. The questions and distinctions provide you with a scheme for organizing your notes and ultimately your paper.

But before you can begin to write, you must get your notes into usable form. Buy a file box and a set of file guides as suggested in the last chapter (page 541). Begin by sorting notes into groups in terms of the questions you have asked. Give each group an appropriate heading; for example:

Alexander	Kesselring
American-British relation	Lucas
Beachhead conditions	Planning of Anzio
Churchill	Source and summary cards
Eisenhower	Strategic purpose of Anzio
Failure of Anzio	Success of Anzio
German response	Units at Anzio
Italian campaign	

These headings should be written on the tabs of the file guides (use either full- or third-cut) and also at the top of each card as you place it into its category. In cases where the same note might fall into two categories, file it where you think it primarily belongs and place a second card in the other category, cross-referring to the first.

Where you have a number of notes, you will probably want to subdivide the category. For example, numerous notes on General Lucas might be further analyzed:

Criticism of Lucas
Defense of Lucas
Dismissal of Lucas
Pessimism of Lucas

Arrange these subheadings under "Lucas," using file cards of shorter cut and, if possible, of different color. You should also add the subheadings to the tops of the cards.

Now you have a working file. You can locate notes as you want them, and when you've finished with a card, you can put it back where it belongs and where you'll be able to find it quickly when you need it again.

WRITING THE PAPER

STUDYING YOUR NOTES

Before beginning to write, you should read your notes carefully and think about what they say. As you study them, keep in mind the distinction between primary and secondary sources. A *primary source* is a document (or recorded statement) originating from an eyewitness, actually present at the event he or she describes. A *secondary source* derives from someone not present at the event but writing or speaking about it from some distance in time or place.

Primary sources are essential to historians. A professional military historian writing about Anzio would consult the Modern Military Records Division of the National Archives in Washington, where the enormous mass of written material connected with the fighting in World War II—planning papers, field orders, after-action reports, and so on— is stored. Even some captured German records are available on microfilm.[1]

Obviously no undergraduate writing a term paper can be expected to sift through all those documents. Most of your sources will be secondary: Majdalany's *The Battle of Cassino* or Liddell Hart's *History of the Second World War*. Some sources will fall in between—Churchill's

---

1. If you are interested, such primary sources are discussed in the back matter of the volumes making up the official history, *U.S. Army in World War II.*

The Second World War or Eisenhower's *Crusade in Europe*. These men played double roles: as actors in the war and as historians writing about it afterwards. To the extent that they discuss events in which they actively participated, they may be regarded as primary sources, though at a remove. For instance, what the official records show that Churchill said at the Tunis Conference on Christmas Day, 1943, when Anzio was decided upon (if the records exist), is a more immediate primary source than what Churchill the historian reports himself as saying in *The Second World War.*

Whether primary or secondary, sources ought not to be accepted uncritically. Do not assume that just because the books and articles from which you have drawn material were written by eyewitnesses or by professional historians, their information and judgments are unquestionable. Eyewitnesses do not see or hear everything; sometimes they misinterpret what they do observe; and sometimes they misrepresent it, deliberately or accidentally.

Generals and politicians writing memoirs years after a campaign want to shine the best possible light upon their actions. For example, the same incident has been remembered quite differently by two men involved, Sir Alan Brooke, Chief of Staff of the British Army, and Dwight D. Eisenhower, Supreme Commander of the Allied Army. During the final years of the war, the two had disagreed about the strategy for defeating Germany. Brooke—and the British generally—favored a "spearhead," a concentration of force at a narrow point to break through the German line. Eisenhower—and the Americans—preferred a broader, more diffuse effort, maintaining pressure along as wide a front as possible. The American strategy prevailed, but not without frequent grumbling from the British.

When a British army group under General Montgomery forced the Rhine two weeks after the U.S. Army had crossed at Remagen, the following incident occurred:

> Churchill and Brooke flew out from Britain to watch Montgomery's assault crossing, and crossed the Rhine themselves later in the day. Eisenhower in his memoirs quotes Brooke as turning to him that day and saying: "Thank God, Ike, you stuck by your plan. You were completely right and I am sorry if my fear of dispersed effort added to your burdens. The German is now licked. Thank God you stuck by your guns." But Brooke, in his memoirs, comments:
>
>> To the best of my memory, I congratulated him heartily on his success and said that, as matters had turned out, his policy was now the correct one; that with the Germans in defeated condition no dangers now existed in a dispersal of effort. I am quite certain I never said to him "You were completely right," as I am still convinced that he was "completely wrong."[2]

2. Don Cook, "General of the Army Dwight D. Eisenhower," *The War Lords: Military Commanders of the Twentieth Century* (Boston: Little, Brown, 1976), pp. 536–537.

Probably neither man lied. Caught up in the emotions of the moment, Brooke may have said more than he remembered; Eisenhower may have remembered meanings Brooke did not express. In any event these two primary sources are in flat disagreement about a matter of fact. Obviously one is suspect, perhaps both.

On the other hand, it does not follow that secondary sources written by historians are always more trustworthy. Good historians obtain and use evidence scrupulously. But even they cannot completely escape the bias of their beliefs and values. British commentators on Anzio, for instance, are more prone than American ones to feel that an immediate breakout at Anzio would have forced a German withdrawal and to blame the failure of such a breakout upon General Lucas or General Clark—both Americans. That position is laid out by Winston Churchill in this assessment of Anzio:

> But now came disaster, and the ruin in its prime purpose of the enterprise. General Lucas confined himself to occupying his beachhead and having equipment and vehicles brought ashore. General Penney commanding the British 1st Division was anxious to push inland. His reserve brigade was however held back with the Corps.[3]

The other view, shared by most professional soldiers, American and British, is that an immediate breakout, while possible, would have been militarily risky, probably ending in disaster. For example, the American general who succeeded to the command of the VI Corps after Lucas was relieved argues:

> I suppose that arm chair strategists will always labor under the delusion that there was a "fleeting opportunity" at Anzio during which some Napoleonic figure would have charged over the Colli Laziali [the Alban Hills], played havoc with the German lines of communication and galloped on into Rome. Any such concept betrays lack of comprehension of the military problems involved. It was necessary to occupy the Corps beachhead line to prevent the enemy from interfering with the beaches.[4]

Neither Churchill nor Truscott, of course, is the ideal "objective" historian. Both were important participants in the event—Churchill as the prime mover of Anzio, anxious for a solution to a military stalemate that was embarrassing him politically; Truscott as the inheritor of a major role which had belonged to a friend and from which that friend had been, Truscott felt, unfairly dismissed. Yet writing in an historian's role, they do contradict each other about what is—finally—the real question concerning Anzio: Should the VI Corps have moved

3. *The Second World War.* Vol. 5: *Closing the Ring* (Boston: Houghton Mifflin, 1951), p. 481.
4. L[ucian] K. Truscott, Jr. *Command Missions: A Personal Story* (New York: Dutton, 1954), p. 311.

inland quickly and could it have forced a German withdrawal from Cassino if it had done so?

Even more disinterested and professional historians get caught in the same contradictions. The distinguished English military historian B. H. Liddell Hart, for instance, writes this about Anzio:

> Churchill's hope of a speedy thrust from Anzio to the Alban Hills was nullified by Lucas's obstinate determination, backed by Mark Clark, to concentrate on consolidating the beachhead before thrusting inland. But in view of the German's swift reaction and superior skill, along with the clumsiness of most of the Allied commanders and troops, Lucas's super-caution may have been a blessing in disguise. An inland thrust in such circumstances could have been an easy target for flank attacks, and have led to disaster.[5]

It is a curious statement and perhaps an example of British bias. Words like "obstinate determination" and "super-caution" imply an indictment of Lucas for failing immediately to exploit the weakness of the initial German reaction to the landing. At the same time Liddell Hart admits that a move inland—necessarily undermanned—would have been extremely risky. Now either events justified Lucas's caution or they didn't; one cannot have it both ways. Moreover, the phrase "blessing in disguise" denies to Lucas even the credit for the build-up which defeated the German offensive against the beachhead. And if Liddell Hart's claim is true that the Allied troops and commanders were militarily less capable than the Germans, is it not possible that Lucas, too, understood the limitations of his force and realized the dangers of an offensive thrust under such conditions? In fact, evidence from Lucas' diary shows that he was well aware of the weakness and lack of training of his force.

This is not the place to argue the question. The example, however, does suggest how you must work with your notes. Don't just shuffle them about and copy them neatly. Think about them. Look for disagreements. These provide focal points for discussion. When you find a contradiction, formulate it as clearly as you can. Then seek evidence supporting both sides and consider what resolution is possible.

Remember that evidence must be pertinent. Historians who have read the diary kept by General Lucas, for example, point out that he was depressed and pessimistic about the Anzio operation. But this fact does not bear upon the question of whether a breakout was militarily feasible. It does, on the other hand, bear upon whether Lucas' superiors were justified in relieving him.

Remember, too, that evidence must be more than reassertion. An historian who simply echoes Churchill's charge against Lucas, as does Liddell Hart, does not confirm it. The statement by the German general

5. *History of the Second World War* (New York: Putnam, 1971), p. 529.

Siegfried Westphal, however, that "the road to Rome was open, and an audacious flying column could have penetrated to the city"[6] provides independent evidence that a thrust might have succeeded. Westphal's remark, however, cannot be construed as an indictment of Lucas unless first it can be shown that Lucas knew, or should have known, how thinly spread the German forces were.

DRAFTING THE PAPER

As you organize your notes, you are approaching the writing process. You should by now have arrived at a broad working outline of the major sections of the paper, perhaps something like this:

I. Introduction: background to Anzio
II. Origin of the operation
III. The plan
IV. The unsoundness of the plan
V. The question of Lucas' responsibility for the failure to break out
VI. Conclusion: Anzio—success or failure?

You have the paper blocked out and can begin to draft. Let's take several sections—say the Introduction and the middle portions III and IV—and see what the finished product might look like.

A sample Introduction   Like any beginning, the opening paragraphs of a term paper must make clear what you plan to do. This means identifying your subject and indicating, in a general and rather subtle way, how the paper will be organized. In addition, the greater length and complexity of a term paper, when compared to the usual English theme, may require that you spend more time explaining what readers need to know about the subject in order to understand specific ideas and arguments as you develop them. In the case of Anzio you will have to identify the time and place of the battle and summarize the action briefly.

Begin by making a slightly more detailed outline of the Introduction:

I. Introduction
    A. The landing and subsequent action
    B. What the planners expected
    C. The questions posed by the failure of the force to attain the objectives set for it

The Introduction developed from that outline might read like this:

---

6. Quoted in Martin Blumenson, *Anzio: The Gamble that Failed* (Philadelphia: Lippincott, 1963), p. 142.

THE ANZIO OPERATION

The Introduction has three functions: (1) to describe the action at Anzio briefly, (2) to focus the questions with which the paper will deal, and (3) to indicate the broad plan of the paper.

The first of these tasks is accomplished in the opening six paragraphs. The first three of these tell just enough about Anzio so that readers will understand, in a general way, what happened without being burdened with details which, at this point, would be irrelevant and confusing.

One of the most controversial actions of World War II took place during the first four or five months of 1944 at an Italian coastal town thirty-five miles south of Rome—Anzio. There, on January 22, the VI Corps of the U.S. Fifth Army landed in an attempt to get behind the German defenses anchored at Cassino, some sixty miles to the south, which had stalled the drive north of the Allied 15th Army Group commanded by the British general Sir Harold Alexander. The hope was that the landing would force the Germans to withdraw from Cassino by threatening their rear.

The VI Corps was commanded by the American general John Porter Lucas, and consisted of one U.S. division (the 3rd) and one British (the 1st) plus several battalions of special troops. In all, the initial force totaled about 36,000 men.

Surprise was complete. The Germans offered no significant resistance and no serious opposition for the next two days, during which the VI Corps fortified the beachhead and landed more troops and many supplies. After that, the commander of the German 10th Army, Field Marshal Albert Kesselring, established a strong defensive force to contain the beachhead. The VI Corps, despite bitter fighting with heavy casualties, was unable to break out. Toward the end of February the Germans mounted a strong offensive to drive the Allied force back into the sea. It failed and the beachhead was secure. However, the stalemate continued until the end of May, when the main units of the 15th Army Group broke the defenses at Cassino and forced a German withdrawal, enabling the VI Corps finally to move inland.

Here, in paragraphs 4, 5, and 6, the emphasis shifts from what did happen to what the planners wanted to happen;

The planners of Anzio had expected the landing force to move inland as quickly as possible. The objectives they had set were to

it also sketches in part of the
background of Anzio.

capture the towns of Cisterna and Campoleone;
to occupy the Colli Laziali (the Alban
Hills), a high ground about twenty miles·
inland and fifteen miles south of Rome; and,
if possible, to move into the city itself.

The Colli Laziali was of great strategic
importance. It controlled the southern
approaches to Rome and also the two major
north-south highways, upon which the German
army at Cassino depended for reinforcements
and supplies. Cisterna and Campoleone were
also important, serving as German
communications centers and supply depots.

The VI Corps did not attain these
objectives. On the contrary, the Germans
succeeded in bottling up the force and came
near to forcing it to evacuate. Even though
the Corps managed to hang on to the
beachhead, the Germans were able to direct
constant artillery fire upon the troops by
virtue of holding the high ground immediately
inland from the landing area.

The final two paragraphs clarify the
specific topics the paper will discuss.
They also indicate the general plan, both
by the order of questions in the seventh
paragraph and by the organizational
statement in the eighth.

The failure of the Anzio operation to
achieve the immediate objectives its planners
had set for it poses the question Why? Was
the plan itself strategically unsound? Was it
inadequately mounted? Was General Lucas too
cautious, wasting time building up the
beachhead when he should have exploited the
initial German weakness by taking the Alban
Hills? Was Anzio, finally, a complete
failure, or did the mere existence of the
beachhead constitute a strategic victory of
sorts, posing a threat to the German rear
which finally contributed to the retreat of
the 10th Army from Cassino?

Notice how this paragraph sets up the
immediately following section and
indicates how it will be organized.

These are the questions this paper will
attempt to answer. First, however, it is
necessary to understand how the idea of a
landing at Anzio originated, who were the
prime movers of that idea, and how it fitted
with the strategy of the Italian campaign
and, beyond that, with the more basic
strategy for defeating Germany.

A sample draft of sections III and IV An introduction does not generally need the support of evidence and quotations with accompanying footnotes. The main portion of a term paper, however, does require documentation. Consequently you should sort out your notes before attempting to draft your middle sections. In the paper on Anzio, let us use sections III and IV: "The plan" and "The unsoundness of the plan." But first we should look at the notes upon which they are based. The notes are reproduced as they would appear on 3 x 5 cards (except that the cards have been numbered for easy reference).

STRATEGIC PURPOSE (1)   1

"Alexander, prodded by Churchill, planned a frontal attack against the Gustav Line, to be assisted by an amphibious diversion at Anzio, sixty miles north. Both men liked the idea so well that the initial roles of the two assaults were shifted; Anzio became the main effort — its objective to cut enemy communications with the north — the Gustav Line assault, the diversion."

R. Ernest Dupuy, World War II: A Compact History (New York: Hawthorne Books, Inc., 1969), pp. 167-168.

STRATEGIC PURPOSE (2)   2

"On the 24th [Dec., 1943], the Chiefs of Staff sent me a detailed statement of their ideas, and a draft which they proposed to send to their Washington colleagues. They favored the plan [for Anzio], but feared we should never win American consent.
"Their conclusions were:
' . . . for the launching of a two-divisional amphibious assault designed to enable Rome to be captured and the armies to advance to the Pisa-Rimini line, and that instructions to that effect should be issued forthwith.' "

(over)

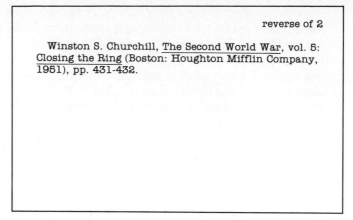

reverse of 2

Winston S. Churchill, <u>The Second World War</u>, vol. 5: <u>Closing the Ring</u> (Boston: Houghton Mifflin Company, 1951), pp. 431-432.

STRATEGIC PURPOSE (3)   3

"Anzio had originally been designed to rescue the Cassino front from deadlock."

Fred Majdalany, <u>Cassino: Portrait of a Battle</u> (London: Longmans, Green & Co., 1957), p. 92.

STRATEGIC PURPOSE — German view   4

"And said Hitler: 'If we succeed in dealing with this business down there [Anzio], there will be no further landing anywhere.'"

Quoted in Anthony Cave Brown, <u>Bodyguard of Lies</u> (New York: Harper & Row, 1975), p. 420.

Source Walter Warlimont, <u>Inside Hitler's Headquarters</u>, 1939-45, trans. R. H. Barry (New York: Praeger, 1966), p. 410.

THE PLAN (1) 5

"The Anzio plan was simple. An Anglo-American force of two divisions would land and strike toward the Alban Hills — seventeen miles inland and fifteen miles from Rome. This hill mass dominates both the main roads — Highways Six and Seven — which were supplying the main German front. It was hoped that the landing in his rear would cause Kesselring to withdraw forces from the Cassino front to cope with the beachhead, and so make it easy for the main force of Fifth Army to break through at Cassino and join up with the landing force. The full plan then provided for

(over)

---

reverse of 5

the Fifth Army to launch a series of attacks on the main front at Cassino immediately prior to the Anzio landing."

Fred Majdalany, Cassino: Portrait of a Battle (London: Longmans, Green & Co., 1957), p. 40.

THE PLAN (2) 6

"This [the crossing of the Rapido River before Cassino] was the main thrust designed to by-pass Cassino and break into the Liri Valley in conjunction with the Anzio landing."

Fred Majdalany, Cassino: Portrait of a Battle (London: Longmans, Green & Co., 1957), p. 61.

THE PLAN — Clark's change in (1)   7

"In his directive to Clark, Alexander was crystal clear in what he required of the 5th Army. It was to attack the Gustav by crossing the Rapido and Garigliano rivers, and draw upon itself the German reserves positioned on the Tyrrhenian coast around Rome. When it was established that these reserves had been committed, Clark was to land the U.S. 6th Corps, an Anglo-American force, at Anzio and Nettuno. The Corps was to strike out and take the Alban Hills, 5th Army was to break through the Gustav, join up with the Corps, and in capturing Rome the Allies were to cut off and destroy the

(over)

---

reverse of 7

German 10th Army. The entire operation was, grandiosely, called the 'Battle for Rome.' But Clark was skeptical of Alexander's orders; he remembered Salerno. Accordingly, he instructed the commander of the 6th Corps to 'advance on' — not take — the Alban Hills; and before doing that, he was to 'secure a beachhead,' and 'advance as soon as he felt himself able to do so.' There was no order to exploit the landing rapidly."

Anthony Cave Brown, Bodyguard of Lies (New York: Harper & Row, 1975), pp. 417-418.

THE PLAN — Clark's change in (2)   8

"General Clark translated General Alexander's desires as follows: 'Mission. Fifth Army will launch attack in the Anzio area. . . . a) To seize and secure a beachhead in the vicinity of Anzio. b) Advance on Colli Laziali [Alban Hills].' What seemed on the surface to be perfectly clear — a mission to be executed in two logically connected parts — was in reality deliberately vague on the second portion. The VI Corps was to establish a beachhead, but was it then to advance toward the Alban Hills or to the Alban Hills?"

Martin Blumenson, "General Lucas at Anzio," Command Decisions, ed. Kent Roberts Greenfield

(over)

reverse of 8

(Washington, D.C.: Office of the Chief of Military
History, Department of the Army, (1960), p. 330.

Blumenson's source for Clark's order is Fifth Army
FO 5, 12 Jan 44. See Modern Military Records Division,
The National Archives, Washington, D.C.

THE PLAN — Clark's change in (3)   9

"In a conference on 12 January 1944 the G-3 of 5th
Army [Training and Operations], Brig. Gen. Donald W.
Brann discussed with Lucas "the vague wording of
the projected advance 'on' the Alban Hills. Brann
made it clear that Lucas' primary mission was to seize
and secure a beachhead. This was all the Fifth Army
expected. Brann explained that much thought had
gone into the wording of the order so as not to force
Lucas to push on to the Alban hill mass at the risk of
sacrificing his corps. Should conditions warrant a
move to the heights, however, Lucas was free to take
advantage of them. Such a possibility appeared slim to

(over)

---

reverse of 9

the Fifth Army staff, which questioned Lucas' ability
to reach the hill mass and at the same time hold the
beachhead to protect the port and landing beaches."

Martin Blumenson, "General Lucas at Anzio,"
Command Decisions, ed. Kent Roberts Greenfield
(Washington, D.C.: Office of the Chief of Military
History, Department of the Army, 1960), pp. 335-336.

THE PLAN — optimism about (1)   10

"The jubilation that the decision for Anzio had brought to the Fifth Army on 8 January was part of a general surge of optimism that spread throughout the higher levels of the theater command. The deadlock in southern Italy seemed about to be split wide open."

Martin Blumenson, Salerno to Cassino. United States Army in World War II: The Mediterranean Theater (Washington, D.C.: Office of the Chief of Military History, United States Army, 1969), p. 322.

THE PLAN — optimism about (2)   11

Alexander thought that Anzio "would so disrupt the German defenses that the Fifth Army could move quickly into Rome."

Martin Blumenson, Salerno to Cassino. United States Army in World War II: The Mediterranean Theater (Washington, D.C.: Office of the Chief of Military History, United States Army, 1969), p. 293.

THE PLAN — POOR — no provision for exploiting   12
                        early success (1)

"It [the Anzio operation] erred on the side of caution in giving initial priority to a supply build-up against counter-attack instead of allowing for an early landing of elements (e.g. light armoured forces) that could have exploited the immediate landing more rapidly."

Fred Majdalany, Cassino: Portrait of a Battle (London: Longmans, Green & Co., 1957), p. 78.

THE PLAN — POOR — no provision for exploiting   13
early success (2)

"There was no plan for the rapid exploitation of
initial success."

G. R. D. Fitzpatrick, "Anzio and Its Lessons,"
Military Review, XXXI (April, 1951), 100.

THE PLAN — POOR — no provision for exploiting   14
early success (3)

"One may wonder why higher headquarters did not
insist on exploitation planning."

G. R. D. Fitzpatrick, "Anzio and Its Lessons,"
Military Review, XXXI (April, 1951), 100.

THE PLAN — POOR — underestimated Germans (1) 15

"It [the Anzio operation] underestimated the
strength of the defenses at Cassino in its assumption
that relatively small forces could break through
there."

Fred Majdalany, Cassino: Portrait of a Battle
(London: Longmans, Green & Co., 1957), p. 78.

THE PLAN — POOR — underestimated Germans (2) 16

"It [the Anzio operation] over-estimated the
consternation that a landing in their rear would cause
the Germans — overlooking the fact that another
army of eight divisions was uncommitted between
Rome and the north. Mark Clark, for instance, told
Truscott that if the force merely held a beachhead at
Anzio, he believed that it would cause the Germans so
much concern that they would withdraw from the
southern front."

Fred Majdalany, Cassino: Portrait of a Battle
(London: Longmans, Green & Co., 1957), p. 78.

THE PLAN — POOR — underestimated Germans (3) 17

" . . . all hands seemed to overlook the fact in Italy
the Allies were opposed by Kesselring, one of the
cleverest defensive commanders in the entire German
Army."

R. Ernest Dupuy, World War II: A Compact History
(New York: Hawthorne Books, 1969), pp. 168-169.

THE PLAN — POOR — VI Corps too weak (1)   18

"The plan was a good one, but it was conceived on
too small a scale. A force of three divisions was not
strong enough to open the road to Rome . . . ."

Nigel Nicolson, "Field-Marshal The Earl
Alexander," The War Lords: Military Commanders of
the Twentieth Century, ed. Field Marshal Sir Michael
Carver (Boston: Little, Brown, 1976), p. 340.

THE PLAN — POOR — VI Corps too weak (2)   19

" . . . the Anzio operation could not be mounted in strength sufficient for its importance; the most critical Allied shortage was landing craft, and all that could be produced in America and England were needed for the build-up of Operation Overlord [the cross-Channel invasion]. . . ."

R. Ernest Dupuy, World War II: A Compact History (New York: Hawthorne Books, 1969) p. 168.

THE PLAN — POOR — VI Corps too weak (3)   20

"Colli Laziali [the Alban Hills] was some 20 miles inland from Anzio. To hold this hill feature against any determined attack would require at least three divisions. To secure the base and lines of com-munication would call for another two divisions. This did not provide for any reserve which is a vital necessity. The VI Corps commander at this stage had at his disposal less than a third of the forces for such a venture."

P. Cookley, "Reflections on Anzio," Military Review XXXIII (October, 1953), 98.

THE PLAN — POOR — VI Corps too weak (4)   21

"As Field Marshal Kesselring said, 'The landing force was initially weak, only a division or so of infantry and without armor. It was a half measure as an offensive that was your basic error.' "

Quoted in P. Cookley, "Reflections on Anzio," Military Review, XXXIII (October, 1953), 97.

No source given for Kesselring's remark.

THE PLAN — POOR — strategically unsound (1)   22

"But Alexander and Churchill in planning Anzio overlooked several vital factors. In the first place, the two operations [Anzio and the forcing of the Rapido River before Cassino] were too far apart to permit any mutual support."

R. Ernest Dupuy, World War II: A Compact History (New York: Hawthorne Books, 1969), p. 168.

THE PLAN — POOR — strategically unsound (2)   23

"When possibilities of supply and reinforcement, as well as terrain, favor the defense, there exists the chance that in spite of successful landing the battlefield may thus easily become a draining sore in the side of the attackers rather than the opening stages of a destructive campaign against the defender's main forces. This had been the Allied experience at Gallipoli in the first World War, an experience that was partially repeated, for some months in the early part of 1944, in the Anzio operation."

Dwight D. Eisenhower, Crusade in Europe (Garden City, N.Y.: Doubleday & Company, 1948), p. 264.

THE PLAN — POOR — strategically unsound (3)   24

"I agreed to the general desirability of continuing the advance [in Italy] but pointed out [to Churchill, General Sir Henry Maitland Wilson — British CinC Mediterranean — and General Alexander] that the landing of two partially skeletonized divisions at Anzio, a hundred miles beyond the front lines as then situated, would not only be a risky affair but that the attack would not by itself compel the withdrawal of the German front. Military strategy may bear some similarity to the chessboard, but it is dangerous to carry the analogy too far. A threatened king in chess

(over)

reverse of 24

must be protected; in war he may instead choose to
fight! The Nazis had not instantly withdrawn from
Africa or Sicily merely because of threats to their
rear. On the contrary, they had reinforced and fought
the battle out to the end. . . . It was from the
standpoint of costs that I urged careful consideration
of the whole plan. I argued that a force of several
strong divisions would have to be established in Anzio
before significant results could be achieved. I pointed
out also that, because of distance, rapid building up of
the attacking force would be difficult and landing

(more)

---

craft would be needed long after the agreed-upon date
for their release [for the invasion of Normandy].
    "The Prime Minister [Churchill] was nevertheless
determined to carry out the proposed operation
[Anzio]. He and his staff felt certain that the assault
would be a great and prompt success. . . . "

    Dwight D. Eisenhower, Crusade in Europe (Garden
City, N.Y.: Doubleday & Company, 1948), pp.
212-213.

THE PLAN — POOR — strategically unsound (4)   25

    " . . . when he [Lucas] emerged from a planning
session at Alexander's headquarters . . . on January
9, 1944, he wrote in his diary: 'I felt like a lamb being
led to the slaughter. . . . The whole affair has a strong
odor of Gallipoli and apparently the same amateur
[is] still on the coach's bench. . . . ' "

    Anthony Cave Brown, Bodyguard of Lies (New York:
Harper & Row, 1975), p. 418.

Observe that the notes have been gathered into three main groups:

I. Strategic purpose
   A. Allied view
   B. German view
II. The plan
   A. Alexander's conception
   B. Clark's changes in
III. The plan unsound
   A. No provision for exploiting early success
   B. Underestimated Germans
   C. VI Corps too weak
   D. Strategically unsound

That three-part analysis can be adapted to form a strong organizing scheme for the paper by treating "strategic purpose" as the initial category under "the plan":

III. The plan
   A. Strategic purpose
      1. Allied view
      2. German view
   B. Alexander's conception
   C. Clark's changes
IV. The plan unsound
   A., B., C., D., as above

Here is what the actual writing might be like. (Remember that we are supposing two preceding sections: the Introduction and the second section explaining how, when, and why the Anzio operation originated and what part Churchill, Alexander, Eisenhower, and Clark played in the initial conception.)

You might or might not wish to use such section titles in an actual paper.

The opening sentence summarizes the preceding section, tying it and the new one together. Clear transitions are important in a term paper, helping readers to follow the flow of thought.

The second sentence of the paragraph states the immediate topic.

SECTION III: THE PLAN

We have seen how the Anzio operation evolved from the early scheme of a diversionary raid to the final conception of a major assault. The strategic purpose of that assault was simple: to break the deadlock at Cassino. The Allied army had stalled before the Gustav Line, and with it had stalled the strategy of the Italian campaign—to push the Germans out of Italy. If Anzio were completely successful it would mean a militarily cheap capture of Rome and a quick victory in Italy. At the least, or so

Discussion is based on notes 1, 2, and 3. Footnotes are not required as the information is general.

Another example of a careful transition

The quotation requires a footnote. Since it is brief it can be worked into the text. See note 4.

"Then" signals a shift in topic. The rhetorical question identifies the new topic. Rhetorical questions are useful for such a task, but should be employed with restraint. The sentence following the question sets up the two-part structure of the following discussion.

Based on notes 5 and 6. Citation is not necessary because of the general nature of the material.

its planners expected, it would force the German 10th Army to pull back from Cassino by threatening their lines of communication and supply. The Anglo-American forces would resume the march north; the Italian strategy would be back on track.

If Anzio had a clear-cut strategic purpose for the Allies, it also created strategic opportunities for the Germans. Hitler was quick to see the significance for Germany of defeating the VI Corps at Anzio. "If we succeed in dealing with this business down there," he said, "there will be no further landing anywhere."[1] In other words, failure by the Allies at Anzio, Hitler felt, would make them have second thoughts about a cross-Channel invasion of France.

For both sides, then, Anzio came to have great strategic importance. What plan had the Allies developed to achieve their aims? In fact, there were two plans, or rather two versions of the same plan: first, the directive handed down by General Alexander, commanding the Anglo-American 15th Army Group in Italy, to General Clark, commander of the U.S. 5th Army; and second, the modification passed on by Clark to General Lucas, commander of VI Corps, the unit of 5th Army chosen to land at Anzio.

As Alexander and his staff conceived it, the plan was simple: an assault force of two divisions plus several battalions of special troops would land and move in to take the Alban Hills, twenty miles or so inland. At the same time the main force of 5th Army before Cassino would attack across the Rapido and Garigliano rivers and break through into the Liri Valley, by-passing the strong German

The note identifies the primary source. But since the quotation was found in a secondary source that too must be acknowledged.

---

1. Walter Warlimont, <u>Inside Hitler's Headquarters, 1939–1945</u>, trans. R. H. Barry (New York, 1966), p. 410. Quoted in Anthony Cave Brown, <u>Bodyguard of Lies</u> (New York: Harper & Row, 1975), p. 420.

hill-top defense anchored on the abbey of
Monte Cassino.

However, there was a significant change
in that plan as it was passed on by General
Clark to the field commander. Clark's orders
to Lucas read, in part:

Example of an extended quotation
that should be indented. See notes 7
and 8.

> Mission, Fifth Army will launch attack
> in the Anzio area. . . . a) To seize and
> secure a beachhead in the vicinity of
> Anzio. b) Advance on Colli Laziali [the
> Alban Hills].[2]

The brackets indicate that the enclosed
matter was added by the writer of the
paper.

The change in emphasis, hidden in the
innocent preposition "on," seems slight, but
it is significant. As an historian of Anzio
has pointed out:

See note 8.

> What seemed on the surface to be
> perfectly clear—a mission to be executed
> in two logically connected parts—was in
> reality deliberately vague on the second
> portion. The VI Corps was to establish a
> beachhead, but was it then to advance
> toward the Alban Hills or to the Alban
> Hills?[3]

It was not a case of poor writing, of
careless ambiguity. The ambiguity was
intentional; "on" was chosen to give Lucas
tacit authority to use his discretion about
advancing inland. Clark seems to have been
skeptical about the ability of VI Corps to
achieve the objectives Alexander had set for

In other footnotes the publisher is not
named. With government publications,
however, it is useful to know the exact
source.

2. Martin Blumenson, "General Lucas at
Anzio," Command Decisions, ed. Kent Roberts
Greenfield (Washington, D.C.: Office of the
Chief of Military History, Department of the
Army, 1960), p. 330. The primary source is
Fifth Army FO 5, 12 Jan 44. Modern Military
Records Division, The National Archives,
Washington, D.C.

The "Ibids." refer to the title cited in fn.
2. "Ibid." without a page number
signifies "on the same page."

3. Ibid.

The specific nature of the information requires a note, even though there is no quotation. Note 9.

it, and he did not want to force Lucas onto the dilemma of having to disobey orders or else sacrifice his command in a task it was too weak to accomplish. All this was pointed out to Lucas a week before the landing by General Donald W. Brann, the member of Clark's staff responsible for operations.[4]

This is a summarizing paragraph. It emphasizes the essential point of the preceding section and implies a major shift in topic.

About the plan, then, we may conclude that there was a significant shift in emphasis from what Churchill wanted and Alexander directed, on the one hand, to what, on the other, Lucas was ordered to do by his superiors in 5th Army. The shift was from an aggressive action with high expectations to a cautious operation with more limited and, to Clark and his staff at least, more realistic objectives.

SECTION IV: THE PLAN UNSOUND

The opening paragraph links section IV to III, sets up the final sections of the paper, and finally focuses upon the next specific topic.

This alteration in plan obviously bears upon the questions of who was to blame for the failure of the Corps to achieve Churchill's aims and of whether General Lucas was unfairly treated in being relieved a month after the landing. Before we address those questions, however, we must consider a prior one: Was the plan worked out by Churchill and Alexander and modified by Clark a good one?

The short paragraph is justified here. The point itself is of key importance and so is the task of setting up the four-part organization of this section.

The answer is no, and for at least four reasons.

See notes 13 and 14.

For one thing the plan did not provide for exploiting initial success, a flaw several critics have pointed to.[5] Probably no one thought the landing would be so easy, that the Germans would offer no resistance, not only to the actual disembarkation, but during the first forty-eight hours when the troops were establishing the beachhead. The Allies had been pounded badly at Salerno, the

---

4. Ibid., pp. 335–336.

5. See, for example, G. R. D. Fitzpatrick, "Anzio and Its Lessons," Military Review XXXI (April 1951), 100.

main landing in Italy, and perhaps they had
learned the wrong lesson for Anzio.[6]
The failure to provide for early
exploitation involved more than the orders
given Lucas. It involved the assignment of
units to the assault wave. As another writer
has said, the operation erred

See note 12.

> in giving initial priority to a supply
> build-up against counter-attack instead
> of allowing for an early landing of
> elements (e.g., light armoured forces)
> that could have exploited the immediate
> landing more rapidly.[7]

The second flaw in the plan was, in a
curious way, the reverse of the first. If the
planners supposed resistance to the landing
would be stiffer than it was and made no
provision for capitalizing on light
resistance, they seriously underrated the
capacity and the will of the German army to
respond to the threat posed by the beachhead:

See note 16.

> [The Anzio operation] over-estimated the
> consternation that a landing in their
> rear would cause the Germans—overlooking
> the fact that another army of eight
> divisions was uncommitted between Rome
> and the north. Mark Clark, for instance,
> told Truscott that if the force merely
> held a beachhead at Anzio, he believed
> that it would cause the Germans so much
> concern that they would withdraw from
> the southern front.[8]

Clark's superior, General Alexander, was
equally optimistic, believing that a thrust
from Anzio "would so disrupt the German

In fn. 6 a short form is sufficient since
the article has already been cited. The
title is necessary rather than merely the
writer's name because the sources
include two works by Blumenson.

6. Blumenson, "General Lucas at Anzio,"
p. 331.

7. Fred Majdalany, <u>Cassino: Portrait of
a Battle</u> (London: Longmans, 1957), p. 78.

8. Ibid.

See note 11.

defenses that the Fifth Army could move quickly into Rome."[9]

It is not easy to understand why the planners of Anzio supposed that once the VI Corps was on the beach the Germans would fold their tents and disappear. The Allied generals had every reason to respect Kesselring, aptly described as "one of the cleverest defensive commanders in the German Army."[10] He had come very close to defeating the Allies at Salerno and had stopped them cold at the Gustav Line. Neither German commanders nor troops had evinced a tendency to panic when confronted by the unexpected. Threatened from flank or rear, they had fought tenaciously and well.

Perhaps Churchill was the ultimate source of the optimistic belief that the Germans would withdraw from Cassino when threatened by Anzio. Eisenhower, skeptical of the Anzio operation from the beginning, makes a comment about it which hints at an explanation of Churchill's sanguine hopes:

The ellipses mark omission of material from the quotation.

> . . . the landing of two partially skeletized divisions at Anzio, a hundred miles beyond the front lines as then situated, would not only be a risky affair but . . . the attack would not by itself compel the withdrawal of the German front. Military strategy may bear some similarity to the chessboard, but it is dangerous to carry the analogy too far. A threatened king in chess must be protected; in war he may instead choose to fight! The Nazis had not instantly

9. Martin Blumenson, <u>Salerno to Cassino.</u> <u>United States Army in World War II: The</u> <u>Mediterranean Theater</u> (Washington, D.C.: Office of the Chief of Military History, United States Army, 1969), p. 293.

10. R. Ernest Dupuy, <u>World War II: A</u> <u>Compact History</u> (New York: Hawthorne, 1969), pp. 168-169.

See note 24.

withdrawn from Africa or Sicily because of threats to their rear. On the contrary, they had reinforced and fought the battle to the end.[11]

Eisenhower's allusion to chess may hide a criticism of Churchill, who was prone to analogies between chess and war. To the professional military man, such a view of war is theoretical, proper to an armchair, not a battlefield. It assumes that battles are played like games and won or lost by rules. It could lead to a facile belief that once their rear was threatened the Germans would play by the book and withdraw to protect their "king."

See note 10.

At any rate, the optimism was there, "spread throughout the higher levels of the theater command,"[12] and perhaps it contributed to the third flaw in the Anzio plan: the VI Corps simply was not strong enough to achieve its objectives, not certainly if the Germans failed to cooperate by a textbook shortening of their lines.

Eisenhower stressed the weakness of the proposed landing force during the conference at Tunis on Christmas Day 1943 when the Anzio operation was pushed to a decision by Churchill. In the passage just quoted Eisenhower goes on to argue that "a force of several strong divisions would have to be established in Anzio before significant results could be achieved."[13] A more recent military writer has concluded that six divisions would have been needed for a reasonably safe push to take and hold the Alban Hills.[14]

See note 24.

See note 20.

Even though it is a first reference, the writer's name can be omitted from fn. 11 because it is identified in the text.

---

11. Crusade in Europe (Garden City: Doubleday, 1948), pp. 212–213.

12. Blumenson, Salerno to Cassino, p. 322.

13. Crusade in Europe, pp. 212–213.

14. P. Cookley, "Reflections on Anzio," Military Review, XXXIII (October, 1953), 99.

Kesselring, whose professional opinion deserves respect, pointed to the weakness of the VI Corps as the reason for its failure to break out of the beachhead:

> The landing force was initially weak, only a division or so of infantry and without armour. It was a half measure as an offensive that was your basic error.[15]

See note 21.

And a British writer, in an admiring essay on General Alexander, has acknowledged:

> The plan was a good one, but it was conceived on too small a scale. A force of three divisions was not strong enough to open the road to Rome. . . .[16]

See note 18.

It would have been more accurate to say that the "conception" was good (though even that is disputable), not the "plan." A plan "conceived on too small a scale" cannot be construed as "good."

Beyond the failure to gauge accurately the initial German defenses and their subsequent response, beyond the weakness of the VI Corps and the absence of exploitation planning—beyond all this there was perhaps a more fundamental flaw. The Anzio operation was not strategically sound.

An occasional "for one thing . . . , for another" is an easy way of organizing a two-part discussion. But don't overdo it.

For one thing, success depended upon the landing being closely coordinated with a successful crossing of the Rapido River and a breakthrough at Cassino (as it turned out, the attempt to force the river was repulsed with terrible losses). The two operations, however, were sixty miles apart, too far "to permit any mutual support."[17]

See note 22.

A professional historian would track down the source of Kesselring's remark.

15. Ibid., p. 97. No source is given for Kesselring's remark.

16. Nigel Nicolson, "Field-Marshal The Earl Alexander," The War Lords: Military Commanders of the Twentieth Century, ed. Field Marshal Sir Michael Carver (Boston: Little, Brown, 1976), p. 340.

Fn. 17 is a subsequent reference and needs only the writer's name and the page number. No title is required as this is the only work by Dupuy that is cited.

17. Dupuy, p. 168.

For another, the strategy assumed that the enemy would do what you want him to do, always a dangerous assumption, and doubly so when you do not act with sufficient force to compel him to your will. Eisenhower revealed this flaw when he noted the inadequacy of the chessboard analogy. Later in Crusade in Europe, discussing the dangers of the Normandy invasion, he touches again upon Anzio:

> When possibilities of supply and reinforcement, as well as terrain, favor the defense, there exists the chance that in spite of successful landing the battlefield may thus easily become a draining sore in the side of the attackers rather than the opening stages of a destructive campaign against the defender's main forces. This had been the Allied experience at Gallipoli in the first World War, an experience that was partially repeated, for some months in the early part of 1944, in the Anzio operation.[18]

See note 23.

Here, too, there may be a veiled criticism of Churchill's strategic capacities. Gallipoli, a British (largely Australian) landing in Turkey designed to knock the Turks out of the war as Germany's allies, had led to an even bloodier stalemate than Anzio, and ultimately, unlike Anzio, to a British defeat. Gallipoli had been Churchill's idea; the disaster set back his political career for many years.

See note 25.

The parallel to Gallipoli contributed to Lucas' pessimism. His diary for Jan. 9, 1944, two weeks before the landing, contains this:

> The whole affair has a strong odor of Gallipoli and apparently the same amateur is still on the coach's bench.[19]

In fn. 18 neither author nor title is necessary since both are clear from the text.

18. P. 264.

19. Quoted in Brown, Bodyguard of Lies, p. 418. A copy of the Lucas diary is on file at the Office of the Chief of Military History, Department of the Army, Washington, D.C.

This, then, is what a term paper might look like. If you take the time to accumulate notes, if you study them, think about them, and organize them, you will find you have solved the two basic problems of composition: you will have gathered topics of discussion, and you will have organized the paper. The writing itself will be far easier than if you had not done the preliminary work.

As you write, pay attention to transitions between paragraphs and to providing signposts that will point readers in the direction you are going. Your point of view probably will be impersonal. "I," "me," "my" are usually inappropriate in term papers, although certain subjects allow for them. As for tone, it is best to work for objectivity.

THE BIBLIOGRAPHY

The books and articles cited in the sample sections of the paper on Anzio would be listed in a bibliography like this:

Blumenson, Martin. "General Lucas at Anzio." Command
    Decisions. Ed. Kent Roberts Greenfield. Washington,
    D.C.: Office of the Chief of Military History,
    Department of the Army, 1960, 323–348.
    _____. Salerno to Cassino. United States
    Army in World War II: The Mediterranean Theater.
    Washington, D.C.: Office of the Chief of Military
    History, United States Army, 1969.
Brown, Anthony Cave. Bodyguard of Lies. New York:
    Harper & Row, 1975.
Cookley, P. "Reflections on Anzio." Military Review,
    XXXIII (October 1953), 96–103.
Dupuy, R. Ernest. World War II: A Compact History. New
    York: Hawthorne, 1969.
Eisenhower, Dwight D. Crusade in Europe. Garden City:
    Doubleday, N.Y., 1948.
Fitzpatrick, G. R. D. "Anzio and Its Lessons."
    Military Review, XXXI (April 1951), 97–102.
Majdalany, Fred. Cassino: Portrait of a Battle.
    London: Longmans, 1957.

EXERCISE

Selecting one of the topics listed below and using the library resources discussed in this and the preceding chapters, locate twelve to fifteen titles, including both books and articles.

Make two lists of these titles, the first in the proper form for a bibliography, the second in the form for footnotes.

Read at least one of your sources and take half a dozen notes, using 3- by 5-inch cards.

If none of the topics appeals to you, think of one that does. Perhaps, however, you had better clear it with your instructor.

1. The teaching of history in U.S. secondary schools during the past twenty-five years.
2. Major league baseball since 1950.
3. President Franklin D. Roosevelt's attempt to pack the U.S. Supreme Court.
4. The Battle of Iwo Jima.
5. The suffrage movement in the United States.
6. The artist Georgia O'Keeffe.
7. President Harry Truman's dismissal of General of the Army Douglas MacArthur.
8. The killing of Admiral Yamamoto.
9. The sinking of the *Titanic*.
10. The history of jazz or of rock and roll.
11. The life of Edgar Allan Poe.
12. The kidnapping of the son of Charles Lindbergh and the subsequent trial and execution of Bruno Richard Hauptmann.
13. The history of newspaper publishing in the United States.
14. The presidency of Woodrow Wilson.
15. Critical comment on the stories and novels of Flannery O'Connor.
16. The San Francisco earthquake and fire.
17. The career of Charlie Chaplin.
18. The Sacco-Vanzetti case.
19. Greek tragedy.
20. The history of the movies.
21. The Battle of New Orleans.
22. Jean François Champollion.
23. Productions of *Hamlet* in the last fifty years.
24. The history of basketball.
25. The assassination of Georges Jacques Danton.
26. The Teapot Dome scandal.
27. The Eighteenth Amendment.
28. The history of comic strips.
29. The discovery of DNA.
30. Modern theories of the universe.

CHAPTER **52**

Answering
Discussion Questions

Discussion questions require not only that you know the answer but that you organize and express what you know. There is no easy formula for success, but a few general principles will help.

1. READ THE QUESTION CAREFULLY; UNDERSTAND WHAT IT ASKS.

Look for key terms that define and limit the subject and sometimes even indicate how to develop it. Suppose, for example, that in a final exam in American history you are asked: "What were the major military and political consequences of the battle of Gettysburg?" "Consequences of the battle of Gettysburg" establishes the subject; "major," "military," and "political" limit it. "Consequences" reveals as well that the discussion should consist of a series of specifics, and the limiting words suggest how that series might be arranged.

Look below the surface. No question, however detailed, reveals everything the teacher expects; if it did, it would supply the answer. The question on Gettysburg, for instance, does not specify "consequences upon the North" or "upon the South" or "upon North and South"; but you ought to take the absence of such specification as an implicit direction to cover both sides.

Be sure too that you understand what the question does *not* ask. Avoid the haystack answer, heaping together everything remotely connected with the topic and hoping your teacher will be impressed by the size of the pile and will charitably sort it out. Most instructors will be neither impressed nor charitable. They want to see if you can separate wheat from chaff, not to do it for you.

601

2. STUDY THE QUESTION FOR CLUES TO ORGANIZING THE DISCUSSION.

Detailed questions reveal a great deal about how to organize the answer. The one on Gettysburg implies an outline like this:

Major consequences of Gettysburg
    A. Military
        1. . . . .
        2. . . . .
        Etc.
    B. Political
        1. . . . .
        2. . . . .
        Etc.

In developing this outline you still have a lot to decide. Should you begin with the military or with the political effects? How will you arrange the specifics within each section? Will you include consequences upon both the Union and the Confederacy? But while you have much to do, you have been given a clear framework within which to work.

Had the question been phrased less precisely ("Discuss the consequences of Gettysburg," for instance), it would imply a skimpier outline:

Major consequences of Gettysburg
    A. . . . .
        1. . . . .
        2. . . . .
        Etc.
    B. . . . .
        1. . . . .
        2. . . . .
        Etc.

The limiting terms "major," "military," and "political" have gone, and now the writer must fill in "A" and "B" as well as "1" and "2."

The more detailed the question, the easier it is to organize the answer. If teachers sometimes phrase questions for discussion in what seem annoyingly loose terms, they are not lazy. They are challenging you.

3. WORK OUT A STRATEGY.

Once you have read the question carefully and thought about what it implies, plan your attack. In the more detailed of the Gettysburg questions, for example, you would have to consider whether it would be simpler to move from military to political consequences or the other way round. Although you are not bound to follow the order in which the terms appear in the question, that sequence may hint at the teacher's thinking. Next you would have to decide how to arrange the particular consequences within each group: perhaps chronologically, perhaps according to cause and effect, perhaps in an order of relative importance (leading up to the most important is usually best). Often other possibilities of organization are inherent in the subject—for instance, analyzing the effects of Gettysburg in terms of the Union and the Confederacy and developing contrasts where appropriate.

It is not enough, however, to have a plan in your own mind; you need to make the plan apparent to your reader. Phrase the topic sentence concisely and emphatically, ending, if possible, on the key terms that set up your organization. In the case of a detailed question, you will probably find that the topic statement will be little more than a rearrangement of the question:

> There were several major consequences of the battle of Gettysburg, both military and political. [The order of the terms "military" and "political" should anticipate the organization of the dicussion.]

Without being heavy-handed, help readers to follow your strategy. In our example, make sure (1) that they know when you are talking about military results and when you shift to political, and (2) that within each category they can follow you as you move from one result to another:

> There were several major consequences of the battle of Gettysburg, *both military and political. Militarily,* the battle threw back Lee's army and defeated the Confederate strategy of forcing the North to sue for peace by a successful invasion of its territory. *A second military result* was the drain upon southern manpower. While the casualties were about equal, the Union could replace the men it had lost; the Confederacy could not. *Finally,* the victory raised the morale of the Union army, increasing the confidence of the soldiers in themselves and in their officers.
>
> *Politically, too, the battle had great effects. In the North,* Lincoln's position was strengthened. He could take a firmer stand against the doves who wanted peace at any price, even that of allowing the Confederate States to become a new nation. *In the South, on the other hand,* political difficulties increased, especially in relation to foreign powers. Confederate diplomats would now find it impossible to obtain significant financial and political help from England or other European nations.

The italicized words give clear evidence that you have organized your answer, and the evidence is much in your favor, for the most common fault with discussion answers is that they are disorganized. (Of course, you would *not* underline such words in an actual examination answer.)

As the end of the examination period approaches and you are pressed for time, you may need a special strategy. Begin your answer with a brief statement of all your major points and then elaborate these in order as you have time. Then, even if you do not finish, your instructor will see that you understood the question and knew the essentials of the answer. Thus you might begin the Gettysburg discussion like this:

> The battle of Gettysburg had several important consequences. Militarily it threw back Lee's army, defeated the Confederate strategy of forcing peace by invading the North, and raised the morale of the Union forces while lowering that of the South. Politically the battle strengthened Lincoln's position and made it more difficult for the Confederacy to find support abroad.
>
> The defeat of Lee's army was more serious than many people realized. [Develop this point and move on to others as time allows.]

4. PROPORTION DETAIL TO LENGTH.

An essay answer is necessarily limited. An historian could write a book about the consequences of Gettysburg. In fifteen or twenty minutes you can do very little. Make that little cover the ground of the question. The best way to do this is to have your whole answer roughly in mind before you begin to write. Then you will know approximately how much space to allot to each of the consequences you wish to cover, and will not become bogged down in the details of the first so that you never get to the others.

5. BUDGET YOUR TIME.

Discussion questions generally require you to work within a limited period. Teachers may specify a limit ("take twenty minutes"), but even when they do not, they expect you to use common sense. If a two-hour examination consists of four equal questions, allow about thirty minutes for each. Work within the allowance; do not become involved with one topic to the exclusion of others. Most teachers prefer a student to cover all the questions, not to spend the entire period on one or two. It is easier to ramble on for an hour than it is to confine yourself to a twenty- or thirty-minute deadline, selecting what is important and expressing it in a well-organized paragraph or essay.

Use your time allotment for each question wisely. While the bulk should be spent actually writing the answer, take a few minutes at the beginning to study the question and to think out a plan. If your instructor has no objection, work up a quick scratch outline on a blank page of your examination book, marking it "not to be graded." Reserve a few minutes at the end to check over what you have written.

6. PRACTICE BEFORE THE EXAMINATION.

As part of your preparation for a discussion test, make up several likely questions and compose answers, applying the principles we have mentioned and working under conditions similar to those of an exam— a time limit, no notes, and so on. When you have finished, check your answer against your notes and the textbook, incorporating what you missed and correcting what you had wrong. Then rewrite your answer to include these revisions.

In selecting topics for practice, be guided by the instructor's lectures and by your text. Teachers are not trying to strike you out in a test; they will ask about what they stressed in class and the major points of your textbook and other reading. If you are lucky, you may work up a question that appears on the exam. But even if you are not so fortunate, the practice of having written out answers will prove helpful.

Beyond actually composing paragraphs on three or four of the most likely topics, you can also think about how to deal with others, even jotting down a strategy or a scratch outline. This sort of preparatory writing, whether of a complete discussion or of a rough outline, will prove an aid to memory. During the pressure of a test it is usually easier to recall something you wrote down than something you merely read.

EXERCISE

In doing this exercise, assume that you are taking an examination: work within the time limit, use no notes or books (except a dictionary), and consult with no one.

1. On the basis of what you have learned in your English class, discuss the ways of developing an expository paragraph. Take twenty minutes and aim at composing a solid paragraph of about 150 words.

2. Select a topic from another course in which you might have a discussion exam— history, say, or philosophy or sociology. Make up a typical question on that topic and in twenty minutes compose a paragraph answer of about 150 words.

PART NINE

PUNCTUATION

Introduction

THE PURPOSE OF PUNCTUATION

All punctuation exists, basically, to help readers understand what you wish to say. Mostly, marks of punctuation do this by signaling the grammatical or logical structure of a sentence (usually these are the same):

> In the long history of the world men have given many reasons for killing each other in war: envy of another people's good bottom land or of their herds, ambition of chiefs and kings, different religious beliefs, high spirits, revenge.
>
> Ruth Benedict

The colon divides that sentence into its two principal parts: the introductory generalization and the list of specific reasons. The commas within the list mark each single reason. The period closes the total statement.

Less often punctuation marks stress an important word or phrase:

> In 1291, with the capture of the last great stronghold, Acre, the Moslems had regained all their possessions, and the great crusades ended, in failure.
>
> Morris Bishop

Bishop does not need the comma before the closing phrase to clarify the grammar or logic of the sentence. Its purpose is emphatic—to isolate and thus stress the phrase. (The other commas in the sentence, however, function in the more common way, indicating grammatical and logical structure.)

Finally, punctuation may mark rhythm. Listen to this sentence closing an essay on General Robert E. Lee:

> For he gave himself to his army, and to his country, and to his God.
>
> W. K. Fleming

609

The commas separating the coordinated phrases have no grammatical necessity. In such coordinated series, commas are not usually employed with *and.* Here, however, the requirements of a closing sentence—that it be slow and regular in its rhythm—justify the commas.

Of course, these three functions of punctuation often overlap. Sometimes a comma or dash both signals grammatical structure and establishes emphasis. And anytime you put a comma into a sentence to help readers follow its grammar, you automatically affect emphasis and rhythm.

Still, keep in mind that these different reasons for punctuation exist. Asking yourself an unspecified question like "Is a comma needed here?" is not very helpful. Rather you must ask: "Is a comma needed here to clarify the grammar (or to establish a particular rhythm or stress)?" About Bishop's sentence we can answer that the comma before "in failure" is *not* required by grammar but *is* necessary for emphasis.

"RULES" OF PUNCTUATION

It would be nice if punctuation could be reduced to a set of clear, simple directions: always use a comma here, a semicolon there, a dash in such-and-such a place. But it cannot. Much depends, as we have just seen, upon what you want to do. In fact, punctuation is a mixed bag of absolute rules, general conventions, and individual options.

For example, a declarative sentence is closed by a period: that is an inflexible rule. On the other hand, placing a comma between coordinated independent clauses ("The sun had already set, and the air was growing chilly") is a convention and not a rule, and the convention is sometimes ignored, especially if the clauses are short and uncomplicated. And occasionally a comma or other mark is used unconventionally because a writer wants to establish an unusual stress or rhythm (like the commas in the sentences by Bishop and Fleming).

But while punctuation as actually practiced by good writers may seem a melange of rule, convention, and idiosyncrasy, it does not follow that anything goes. To punctuate effectively you must learn when rules are absolute; when conventions allow you options (and, of course, what the options are); and when you may indulge in individuality without misleading the reader. Moreover, you must keep the reader in mind. Younger, less experienced readers, for instance, need more help from punctuation than older, sophisticated ones.

In the discussions of the various punctuation marks that follow, we shall try—as far as it is practical—to distinguish among rules, conventions, and unconventional but possible uses. At times the distinctions may seem a bit confusing. It is no good, however, making up easy rules about how to handle punctuation. Such directions may be clear, but they do not describe what really happens. Instead we must look at what skillful writers actually do. To diminish some of the confusion, just remember that clarity of communication is the one simple "rule" underlying all effective punctuation.

Remember, too, that punctuation is not something you impose upon a sentence after you have written it out. Commas, semicolons, and the other marks are an intimate part of grammar and style. Often mistakes in punctuation do not simply mean that a writer broke an arbitrary rule; rather they signify his or her confusion about how to construct a sentence. To write well, you must punctuate well; but to punctuate well, you must also write well.

THE TWO CATEGORIES OF PUNCTUATION

It is convenient to divide punctuation into two broad categories: the stops and the other marks. *Stops* take their name from the fact that they correspond (though only loosely) to pauses and intonations in speech, vocal signals which help listeners follow what we say. Stops include the period, the question mark, the exclamation point, the colon, the semicolon, the comma, and the dash. We shall study them in the remainder of this chapter.

In the next chapter we shall look at the other marks. These more purely visual signals do not mark pauses (though on occasion some of them signal voice intonations). They include the apostrophe, the quotation mark, the hyphen, the parenthesis and bracket, the ellipsis, and diacritics (marks placed with a letter to indicate a special pronunciation). Along with these marks we shall consider capitalization and underlining (or use of italics), though, in a strict sense, these are not matters of punctuation.

53

Stops

○ **THE PERIOD**

The period is called an "end stop" because it is used at the end of a sentence. More exactly, it closes declarative sentences—those which state a fact, perception, idea, belief, feeling—and it may also close an imperative sentence, or command (though these are often punctuated with an exclamation point).

Ab. **ABBREVIATIONS**

The period is used after many abbreviations: *Mr., Mrs., Ms., Dr.* When such an abbreviation occurs at the end of a sentence, the period does double duty, closing the sentence as well as marking the abbreviation.

Some abbreviations do not take periods: government agencies, for instance, such as the *SEC* (Securities and Exchange Commission) or the *GPO* (Government Printing Office). If you are uncertain about whether a particular abbreviation requires a period, consult a dictionary or an appropriate manual of style.

Not all abbreviations, incidentally, are allowable in composition. Some are perfectly acceptable: *SEC, GPO,* or *Mr., Mrs., Ms., Dr.* (Most professors, though, do not like *Prof.*) Others are not universally accepted. For example, many teachers prefer that instead of *&, i.e., etc.,* and *e.g.,* you write out *and, that is, and so on, for example.* Colloquial, slangy abbreviations are not acceptable at all: *econ* and *polysci* are legitimate enough in conversation, but you should use the full words in composition.

?
.

THE QUESTION MARK

DIRECT QUESTIONS

The question mark (also known as the "query" and the "interrogation point") is used after direct questions. A direct question is always marked by one or some combination of three signals: a rising intonation of the voice, an auxiliary verb inverted to a position before the subject, or an interrogative pronoun or adverb (*who, what, why, when, how,* and so on).

"Yes-no" questions (those answered by "yes," "no," or some variety of "maybe") are always signaled by a rising intonation, which may or may not be accompanied by the inversion of the auxiliary:

You're going downtown?
Are you going downtown?

In speech the intonation alone (without the inverted verb) is often felt to be sufficiently clear. In composition the auxiliary is generally inverted.

Informational questions (those requiring in answer a statement of fact, opinion, belief) do not have the rising intonation and may or may not have the inversion; but they contain, both in speech and in writing, an appropriate interrogative word:

Who is going downtown?
When are you going downtown?

INDIRECT QUESTIONS

Indirect questions do not close with a question mark but with a period. Like direct questions they demand a response, but they are expressed as declarations without the formal characteristics of a question. That is, they have no inversion, no interrogative words, and no special intonation. We can imagine, for example, a situation in which one person asks another, "Are you going downtown?" (a direct question). The person addressed does not hear and a bystander says, "He asked if you were going downtown." That is an indirect question. It requires an answer, but it is expressed as a statement and so is closed by a period, not a query.

RHETORICAL QUESTIONS

A rhetorical question is a variety of direct question and must be closed by a question mark, no matter whether the writer intends to answer it—or to receive an answer—or not. (The notion that a rhetorical question does not require an answer is inaccurate. Rhetorical questions are often asked precisely so that the writer can compose the answer. And even when the writer does not state the answer, he or she expects the reader to supply it.)

QUESTION MARKS WITHIN THE SENTENCE

While question marks are primarily end stops, within the sentence they may signal doubt or uncertainty concerning a particular fact, idea, or feeling:

> It was the Lord who knew of the impossibility every parent in that room faced: how to prepare the child for the day when the child would be despised and how to *create* in the child—by what means?—a stronger antidote to this poison than one had found for oneself.
>
> James Baldwin

Occasionally a question mark in parentheses appears within a sentence after a word to indicate that the writer is uncertain either of the spelling of the word or of the accuracy of the idea. One does this only when no reasonable way exists of checking the spelling or the information. It is abusing a privilege to write: "In 1492 (?) when Columbus discovered America. . . ." Anyone not sure of that date can fairly be expected to look it up.

There is, finally, a problem connected with where a query must be placed relative to a closing quotation mark. But since the problem affects all the other stops as well, we treat it when we discuss the quotation mark in the next chapter.

THE EXCLAMATION POINT

Exclamation points convey emphasis. Most often they close a sentence and signal the importance of the total statement. Used after imperative statements ("Come here!"), they suggest the tone of voice in which such a command would be spoken.

Even more frequently than queries, exclamation points are set within a sentence in order to stress the preceding word or phrase:

> Worse yet, he must accept—how often!—poverty and solitude.
>
> Ralph Waldo Emerson

Interjections are usually followed by exclamation points:

> Bah! you expect me to believe that?

As a device of emphasis the exclamation point is of limited value. Used very occasionally, it can be effective. But like most mechanical means of emphasis, it quickly loses force. It is far better to achieve stress by effective diction and sentence structure.

Period, question mark, and exclamation point

I. Period
Closes all declarative sentence, whether grammatically complete or not

II. Question mark
A. As an end stop
Closes all direct questions, including rhetorical ones
B. Within the sentence
1. May mark a word or construction
2. In parentheses may indicate uncertainty about a matter of fact or belief

III. Exclamation point
A. As an end stop, marks a strong statement
B. Within the sentence, stresses a word or construction

EXERCISE

1. What punctuation do you think closes these sentences? In two of them, internal question marks have been removed. Can you identify where?

A. What difference would it practically make to any one if this notion were true

William James

B. Keep moving   Keep moving   Don't let him set up

A. J. Liebling

C. . . . there was some gift or other—what was it—some essential gift—which the good fairies had withheld

Lytton Strachey

D. The figure and bearing of Clemenceau are universally familiar

John Maynard Keynes

E. All travelers to my city should ride the elevated trains that race along the back ways of Chicago The lives you can look into

Lorraine Hansberry

F. Is the consumer society good for us or hazardous to our health

James Oliver Robertson

2. Compose two sentences closed by a question mark and two closed by an exclamation point. In addition compose one having an internal question mark and one having an internal exclamation point.

:

THE COLON

The colon—along with the semicolon, the comma, and the dash—is an internal stop. That is, it is used only inside a sentence, never at its end.

In modern writing the most common function of the colon is to introduce a specification:

> The first principle from which he [Hitler] started was a value judgment: the masses are utterly contemptible.
>
> Aldous Huxley

> Except for the size of the houses, which varies from tiny to small, the houses look like suburban housing for middle income families in any section of the country: flat, low, lots of wasted space, nothing in the design to please the eye or relieve the monotony.
>
> Sharon R. Curtin

In both those sentences the first portion expresses a general idea (Hitler's "first principle"; "suburban housing for middle income families"). The second portion, introduced by a colon, particularizes the idea.

Sometimes the specification takes the form of a list or series:

> There are three kinds of lies: lies, damned lies, and statistics.
>
> Benjamin Disraeli

Occasionally the usual order is reversed, and the sentence begins with the specific word or phrase, which is followed by a colon:

> Centering: that act which precedes all others on the potter's wheel.
>
> Mary Caroline Richards

But usually the specific follows the general. Such constructions are emphatic. The specification—the key point—is put at the end of the sentence. The effect of the colon, which represents a relatively long pause, is to prepare us for something momentous. The emphasis is seen very clearly in these cases:

> A once-defeated demagogue trying for a comeback, he tried what other demagogues abroad had found a useful instrument: terror.
>
> Wallace Stegner

> Finally, last point about the man: he is in trouble.
>
> Benjamin DeMott

> What distinguishes a black hole from a planet or an ordinary star is that anything falling into it cannot come out of it again. If light cannot escape, nothing else can and it is a perfect trap: a turnstile to oblivion.
>
> Nigel Calder

Notice in all these examples that it is not necessary that the construction following the colon be a complete clause. It can be a phrase or even a single word.

Colons are also used to introduce quotations (really a kind of specification), especially long written ones:[1]

A master expositor, W. K. Clifford, said of an acquaintance: "He is writing a book on metaphysics, and is really cut out for it; the clearness with which he thinks he understands things and his total inability to express what little he knows will make his fortune as a philosopher."

Brand Blanshard

And sometimes rhetorical questions are introduced by a general statement followed by a colon:

The question is: How and to what purpose?

Time magazine

In such a construction it is common practice (but not an absolute rule) to begin the question with a capital letter.

Colon
I. Introduces specifications, often, though not always, in the form of a list or series
II. Introduces quotations, particularly extended written ones
III. Occasionally introduces rhetorical questions

EXERCISE

1. Each of the following sentences originally contained a colon. Where do you think it went? Why do you think so?

   A. The tactics were the tactics of terror strongarm methods borrowed from the Brown Shirts.

   Wallace Stegner

   B. It is some consolation to affirm that one element in conservation is fixed modern English literature.

   Hilaire Belloc

   C. Nor would Calvin do any of the things that normal children did play games, laugh, chat, spend a penny now and then for candy or ice cream.

   Irving Stone

---

1. You can integrate quotations into your text in several ways. These are discussed more fully in the chapter on the research paper (see pages 541–43).

D. The father and son looked a good deal alike lean handsome faces with deep eyes and firm mouths.

John Lardner

E. By the end of 1888 he saw in the gilded labyrinth around him only one figure worth clinging to Mary Vetsera.

Frederic Morton

F. Instantly the book becomes noxious the guide is a tyrant.

Ralph Waldo Emerson

G. Were they right or are we? The answer, I suspect, is Both.

Aldous Huxley

H. Justice Oliver Wendell Holmes once said ''The most stringent protection of free speech would not protect a man in falsely shouting fire in a theater and causing panic.''

Roger Fisher

I. Second marriages may be better or worse than first marriages, but one thing is certain they are quite different.

Leslie Aldridge Westoff

J. Withdrawal, denial, confusion of values these are vague words.

V. S. Naipaul

2. Compose three sentences in which a specification or series of specifications is introduced by a colon.

; THE SEMICOLON

The semicolon has two functions: to separate independent clauses and, under certain conditions, to distinguish the items in a list or series. The first function is by far the more common.

SEMICOLON BETWEEN INDEPENDENT CLAUSES

Independent clauses may be joined either by coordination or by parataxis. In the first case they are linked by a coordinating conjunction (*and, but, for, or, nor, yet, either . . . or, neither . . . nor, both . . . and, not only . . . but*), which is usually preceded by a comma. In the second they are simply run together with no conjunctive word but are separated by a stop, conventionally a semicolon:

Sentimentality and repression have a natural affinity; they're the two sides of one counterfeit coin.

Pauline Kael

Paratactic compound sentences punctuated with semicolons are especially common when the second clause repeats the first:

The New Deal was a new beginning; it was a new era of American government.

Arthur M. Schlesinger, Jr.

Wendell Willkie was publicly and privately the same man; he was himself.

Roscoe Drummond

All of these newcomers—black and white—toiled under some degree of unfreedom; they were bound servants for greater or lesser terms.

Oscar Handlin

Using *and* in such sentences would be subtly misleading, implying a change of thought where none in fact exists.

Parataxis is also effective between clauses expressing a sharp contrast of idea:

Languages are not invented; they grow with our need for expression.

Susanne K. Langer

He [President Calvin Coolidge] knew precisely what the law was; he did not concern himself with what the law ought to be.

Irving Stone

Groups are capable of being as moral and intelligent as the individuals who form them; a crowd is chaotic, has no purpose of its own and is capable of anything except intelligent action and realistic thinking.

Aldous Huxley

Clauses like those could be joined by a comma and *but*. Omitting the conjunction and using a semicolon, however, makes a stronger statement, forcing readers to see the contrast for themselves.

Occasionally even coordinated clauses are separated by a semicolon. This is done at the discretion of the writer and is more common when the clauses are relatively long and complicated, containing commas within themselves. In that case a semicolon more clearly signals the break between them. The following sentence is an example (the Duke of Wellington is commenting with pleasant cynicism upon the capacity of young ladies to endure the absence of lovers gone to war):

They contrive, in some manner, to live, and look tolerably well, notwithstanding their despair and the continued absence of their lover; and some have even been known to recover so far as to be inclined to take another lover, if the absence of the first has lasted too long.

Even when the coordinated clauses are not very long, a semicolon may still replace the more conventional comma if the writer wants a significant pause for emphasis or rhythm:

Children played about her; and she sang as she worked.

Rupert Brooke

So the silence appeared like Death; and now she had death in her heart.

Ford Madox Ford

Run-on   *Run-on sentences*   A run-on sentence occurs when a semicolon has been omitted between uncoordinated independent clauses. Sometimes a comma is used instead (when it is, the error is often called a "comma fault"):

× It was late, we went home.[2]

And sometimes the clauses are simply run together with no stop of any kind:

× It was late we went home.

The most frequent cause of run-on sentences is mistaking the function of conjunctive adverbs—such words as *however, nonetheless, therefore, consequently, even so, on the other hand, for example* (a more complete list is given on pages 673–79). These adverbs do not join clauses grammatically; they only show a relationship between the ideas in the clauses. In this they differ from coordinating conjunctions, which traditionally designate both a grammatical and a logical connection.

The difference may seem arbitrary. The coordinating conjunction *but* and the conjunctive adverb *however,* for instance, can be used almost interchangeably between appropriate clauses. Even so, the first is a conjunction and needs only a comma (or maybe even no stop at all); the second is an adverb and, when it is unaccompanied by a conjunction, requires a semicolon:

It was not late, but we went home.
It was not late; however, we went home.

It would result in a run-on sentence to punctuate it like this:

× It was late, however, we went home.

Run-on sentences may be corrected in several ways, though for any given case one way will probably be best. The simplest solution is to put a semicolon in the proper place. Or the clauses may be joined by an appropriate coordinating conjunction accompanied by a comma (though this stop may be omitted if the clauses are short and simple). Or the two clauses may be recast as two sentences. Finally, the clauses may be kept as parts of the same sentence with one being subordinated to the other, in which case a comma may or may not be needed between them. Thus the run-on sentence × "The search was fruitless, the men were discouraged" can be corrected:

The search was fruitless; the men were discouraged.
The search was fruitless, and the men were discouraged.
The search was fruitless. The men were discouraged.
Because the search was fruitless the men were discouraged.

SEMICOLON IN LISTS AND SERIES

Semicolons are conventionally used to separate all the items in a list or series when any of the items contains a comma. This is done

---

2. Commas are sometimes effective in such cases, the so-called comma link. Comma links are discussed in the next section (see pages 623–24).

because the presence of a comma within one or more items requires a stronger stop to signal the distinction between one unit in the series and another. Look at this sentence about the rise of the Ku Klux Klan in the 1920s:

> There were other factors too: the deadly tedium of small-town life, where any change was a relief; the nature of current Protestant theology, rooted in Fundamentalism and hot with bigotry; and, not least, a native American moralistic blood lust that is half historical determinism, and half Freud.
>
> Robert Coughlan

Even when a comma occurs in only one item, consistency requires that semicolons be used between all the elements of the series:

> He [Huey Long] damned and insulted Bigness in all its Louisiana manifestations: Standard Oil, the state's dominant and frequently domineering industry; the big corporations; the corporation lawyers.
>
> Hodding Carter

SEMICOLON WITH SUBORDINATE CLAUSE

Now and then a semicolon separates a main clause and a subordinate one, a job conventionally assigned to the comma. The stronger semicolon is helpful when the clauses contain internal commas; it more clearly signals the break between the clauses and helps the reader to follow the grammar:

> He [the white policeman] moves through Harlem, therefore, like an occupying soldier in a bitterly hostile country; which is precisely what, and where, he is, and is the reason he walks in twos and threes.
>
> James Baldwin

Semicolon

I. Between independent clauses
    A. Paratactic: semicolon is the conventional stop
    B. Coordinated: comma is conventional
                  semicolon is optional for clarity or emphasis

---

II. In lists and series
    Semicolon between all items when any item contains a comma

---

EXERCISE

A semicolon has been removed from each of the following sentences. Where you think the semicolons should go? Be able to explain your answers.

A. Youth's sense of community is an *ad hoc* thing it is suspicious of institutions and wary of organizations, prizing freedom above system.

Time magazine

B. The American Negro is a unique creation he has no counterpart anywhere, and no predecessors.

James Baldwin

C. The cry was for economic reform the administration settled for a handful of palliatives.

Thurman Arnold

D. We have gone off on a tangent that takes us far away from the mere biological cycle that animal generations accomplish and that is because we can use not only signs but symbols.

Susanne K. Langer

E. By day the hot sun fermented us and we were dizzied by the beating wind.

T. E. Lawrence

THE COMMA

The comma is the most frequent and the most complicated of all marks of punctuation. It is least reducible to rule and most subject to variation, depending upon the need to be clear or emphatic, the preferences of individual writers, and even fashion.

1. COORDINATED INDEPENDENT CLAUSES

Coordinated elements are grammatically identical constructions in the same sentence joined by a coordinating conjunction (*and, but, for, or, nor,* and the correlatives *either . . . or, neither . . . nor, both . . . and, not only . . . but*). Any part of a sentence may be coordinated: two subjects, two verbs, two objects, two adjectivals, two adverbials, two independent clauses.

As a very general rule, two coordinated independent clauses are punctuated with a comma; lesser elements, such as words, phrases, and dependent clauses, are not so punctuated. But exceptions occur each way, depending upon the length and complexity of the constructions. Let's look at several examples.

Two coordinated independent clauses are usually separated by a comma, placed immediately before the conjunction:

It [history] is a story that cannot be told in dry lines, and its meaning cannot be conveyed in a species of geometry.

Herbert Butterfield

When such coordinated clauses are complicated and contain internal commas, the stronger semicolon may be used to separate them, as we saw on page 619. On the other hand when they are short, obviously related and contain no internal commas, the comma between them may be omitted:

> They tried to hold him up against the wall but he sat down in a puddle of water.
>
> Ernest Hemingway

The comma link   A comma link is a comma used between independent clauses that are paratactic—that is, not joined by one of the coordinating conjunctions but simply run together. The semicolon is the conventional mark in such a construction (see pages 618–19), and employing a comma is generally regarded as a fault. Under certain circumstances, however, a comma may be used between paratactic clauses (though it is never obligatory). The clauses must be short and simple and contain no internal stops; the relationship of idea should be immediately clear; and the sentences should move rapidly with only light pauses:

> A memoir is history, it is based on evidence.
>
> E. M. Forster

> The crisis was past, the prospects were favorable.
>
> Samuel Hopkins Adams

When three or more such short, obviously related independent clauses are joined paratactically, comma links are even more frequent:

> Some of the people said that the elephant had gone in one direction, some said that he had gone in another, some professed not even to have heard of any elephant.
>
> George Orwell

> Sheep in the pasture do not seem to fear phantom sheep beyond the fence, mice don't look for mouse goblins in the clock, birds do not worship a divine thunderbird.
>
> Susanne K. Langer

> He becomes more callous, the population becomes more hostile, the situation grows more tense, and the police force is increased.
>
> James Baldwin

The last sentence (about racial tensions in Harlem between white policemen and black residents) illustrates the particular advantage of comma links. By allowing rapid movement from clause to clause, the punctuation reinforces our sense of the inevitability of social cause and effect.

Easy rules about when a comma link is effective and when it is a comma fault do not exist. Certainly long, complicated paratactic independent clauses (especially those containing commas) ought to be punctuated by semicolons, not commas. And even when the clauses are not particularly long and contain no commas within themselves, the relationships among ideas may not be sufficiently close and obvious to allow a comma link. In this sentence, for instance, a semicolon would be clearer:

> ✕ We are overloaded with garbage, in fact we have so much excess garbage that it is being used to make hills to ski on.

For the inexperienced writer the safest course is to use a semicolon between uncoordinated independent clauses unless he or she is very sure that a comma will help the rhythm of the sentence and will not confuse the reader.

As the foregoing discussion suggests, the punctuation of independent clauses is not easily explained in a simple rule. Current practice is summed up in the following table:

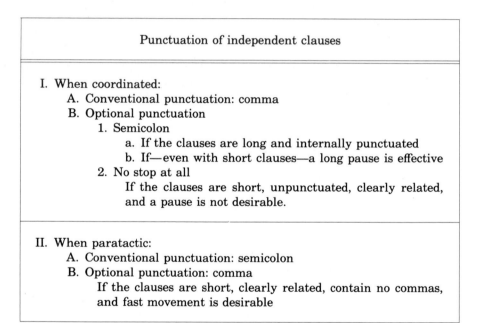

Punctuation of independent clauses

I. When coordinated:
    A. Conventional punctuation: comma
    B. Optional punctuation
        1. Semicolon
            a. If the clauses are long and internally punctuated
            b. If—even with short clauses—a long pause is effective
        2. No stop at all
            If the clauses are short, unpunctuated, clearly related, and a pause is not desirable.

II. When paratactic:
    A. Conventional punctuation: semicolon
    B. Optional punctuation: comma
        If the clauses are short, clearly related, contain no commas, and fast movement is desirable

EXERCISE

1. In the following sentences the stop between the independent clauses has been removed. How would you punctuate them?

A. I stood today watching harvesters at work and a foolish envy took hold upon me.

> George Gissing

B. I could feel the difference between this life and a normal life I could see the difference in my mind's eye but I couldn't satisfactorily express the subtleties in words.

> Richard E. Byrd

C. His prospects had been blighted his family reduced from comparative wealth to abject penury their house and land taken away from them their bank account frozen and devaluated.

Aldous Huxley

D. He could not keep the masses from calling him Lindy but he convinced them that he was not the Lindy type.

John Lardner

E. In a few moments everything grew black and the rain poured down like a cataract.

Francis Parkman

F. The comedy of wit is true to art the comedy of humor is true to life.

Edith Hamilton

G. The railway he had and on the railway everything depended.

Thomas Pakenham

H. Truth is a species of value it is not theoretically useful.

Anthony Quinton

2. Compose four sentences using commas between independent clauses and four using semicolons. Be able to explain why you used the specific stops you did.

,/2

2. THE COMMA WITH COORDINATED ELEMENTS OTHER THAN INDEPENDENT CLAUSES

Two coordinated subjects, verbs, objects, or modifiers are not usually punctuated:

Jack and Jill went up the hill.
NOT Jack, and Jill went up the hill.

We saw them and were surprised.
NOT We saw them, and were surprised.

He picked up his hat and books.
NOT He picked up his hat, and books.

The men were tired and discouraged.
NOT The men were tired, and discouraged.

However, commas may be helpful between the members of such coordinated pairs when the first is long or when the writer wants a pause for emphasis. Thus in the following sentence the comma helps the reader to distinguish the two long predicates that follow the subject ("the twentieth century"):

The twentieth century finds this explanation too vapidly commonplace, and demands something more mystic.

George Bernard Shaw

In the next examples the comma separating two coordinated verbs (while not necessary because of their length) gives the idea more emphasis:

We saw them, and were surprised.

Student

At night we were stained by dew, and shamed into pettiness by the innumerable silences of stars.

T. E. Lawrence

,/3

3. THE COMMA WITH LISTS AND SERIES

A list or series consists of three or more grammatically parallel words or constructions such as three or four subjects of the same verb, say, or three verbs of the same subject, or four or five adjectives modifying the same noun.

The items in a list, or series, may be joined by coordinating conjunctions ("She bought bread and eggs and cheese") or by parataxis ("She bought bread, eggs, cheese").[3] The most common method is to combine parataxis and coordination, linking the last two items with *and, or,* or *but not,* and joining the others paratactically: "She bought bread, eggs, and cheese."

When a list or series is completely paratactic, commas are used between the items:

> Oriental luxury goods, jade, silk, gold, spices, vermilion, jewels, had formerly come overland by way of the Caspian Sea. . . .

Robert Graves

When it is completely coordinated, the commas are usually omitted:

> She was crying now because she remembered that her life had been a long succession of humiliations and mistakes and pains and ridiculous efforts.

Jean Rhys

In the combined method (the most frequent practice), a comma goes between each pair of paratactic elements and is optional between the final coordinated pair, the choice depending upon the preference of the writer or the policy of an editor. The first of these examples uses the comma; the second does not:

> Fifty years ago, when all type was set by hand, the labor of several men was required to print, fold, and arrange in piles the signatures of a book.

Carl Becker

> His plan was to clinch his teeth, shut his eyes, whirl the club round his head and bring it down with sickening violence in the general direction of the sphere.

P. G. Wodehouse

---

3. See page 301 in the chapter on emphasis for a discussion of the different effects of these ways of handling a series.

But whether you choose to place a comma between the final coordinated items or to leave it out, you should follow the same practice consistently in any piece of writing.

Finally about lists and series, remember that, as we saw on page 620, semicolons are conventionally used between all items when any item contains a comma within itself.

Punctuating a series

I. Combined parataxis and coordination: commas and optional comma
   bread, eggs(,) and cheese

II. Completely paratactic: commas
   bread, eggs, cheese

III. Completely coordinated
   A. Conventional punctuation: no stops
      bread and eggs and cheese
   B. Optional punctuation: commas for emphasis or rhythm
      bread, and eggs, and cheese

IV. Series with a comma in one or more items: semicolons
   bread, which she found too moldy; eggs; and cheese

EXERCISE

1. Each of the following sentences contains a list or series. The punctuation has been omitted. Put in appropriate stops where you think they belong:

A. But over against the sorrow the tears the self-generated self-inflicted disasters what superhuman prospects!

> Aldous Huxley

B. Olive oil dried fish chick-peas wine and fruit were in plentiful supply.

> Robert Graves

C. It reverberates with questions anguished meditative alarmed.

> Maynard Mack

D. Joan helped her parents in tillage tended the animals and was skilled with her needle and in other feminine arts.

> George Bernard Shaw

E. Any American college student can run a motor car take a gasoline engine to pieces fix a washer on a kitchen tap mend a broken electric bell and give an expert opinion on what has gone wrong with the furnace.

Stephen Leacock

2. Compose three sentences containing lists and series appropriately punctuated.

,/4

4. THE COMMA WITH ADJECTIVALS

An adjectival is a word, phrase, or clause functioning as an adjective.[4]

Single-word adjectives   Most single-word adjectives are restrictive—that is, essential to the meaning of the nouns they modify.[5] A restrictive adjective is placed after the noun marker, if there is one (*a, an, the, some, this, any,* and so on), and is not punctuated (italics added in the following examples):

The *angry* man sat down abruptly.

However, adjectives are often used in a rather different sense, being either placed before the noun marker (when one is present) and followed by a comma, or after the noun and set off by commas:

Angry, the man sat down.
The man, *angry,* sat down.

They may even be pushed to the end of the clause and preceded by a comma:

The man sat down, *angry.*

In such patterns (especially common with participles acting as adjectives), the word really functions more like an adverb. It tells us something about the action (in this case, how or why the man sat down) rather than about the noun (the man himself). Such "adjectives" are punctuated.

Finally about single-word adjectives: when two or more are used together they are not usually punctuated if they are coordinated. However, should emphasis require it, the second of a pair of coordinated adjectives may be set off by commas:

It [England] always had a peculiar, *and a fond,* relationship with the papacy.

Paul Johnson

When two or more adjectives are run together without conjunctions, they must be punctuated for clarity:

---

4. See pages 738–46 for a discussion of adjectivals.
5. See page 739.

A novel is in its broadest definition *a personal, a direct* impression of life. . . .

Henry James

Participial adjectival phrases Used restrictively, participial phrases follow the noun and are *not* preceded by a comma:

A man *leading a horse* was walking inland from the sea.

W. S. Merwin

Often, however, participial phrases function nonrestrictively. They supply pertinent information about the noun they modify, but not information essential to understanding its meaning in the sentence. Nonrestrictive participals are always punctuated. They may precede their noun; follow it, intruding between it and the verb or remainder of the clause; or be postponed to the end of the clause. In any case they must be followed, set off, or preceded by commas:

Born to lowly circumstances, he came up the easy way.

Samuel Hopkins Adams

Words, *being but symbols by which a man expresses his ideas,* are an accurate measure of the range of his thought at any given time.

Albert C. Baugh

For years he had been blackmailing the rector, *threatening to publish the facts about a certain youthful escapade of his dead wife.*

Robin G. Collingwood

Adjectival clauses Adjectival clauses are less flexible in their positioning than the participial phrase: they must follow their noun. But they too may be either restrictive or nonrestrictive, and they are punctuated accordingly. Restrictive clauses are not punctuated; nonrestrictive ones are set off by commas when they fall inside the main clause, preceded by commas when they fall at the end:[6]

At the apex of the social pyramid, *which was still nominally Republican,* stood the Emperor Augustus.

Robert Graves

All images are symbols, *which make us think about the things they mean.*

Susanne K. Langer

Nonrestrictive clauses are sometimes used in a loose sense, to modify not a single noun but an entire idea. Such clauses are introduced by *which,* placed at the end of the sentence or clause they modify, and always preceded by a comma:

Lenin was cruel, *which Gladstone was not.* . . .

Bertrand Russell

6. This rule reflects current American practice. Sometimes in older usage all adjectival clauses were punctuated without regard to whether they were restrictive or nonrestrictive in meaning.

Comma with adjectivals

I. Single-word adjectives
Restrictive: no comma
The angry man sat down.
Nonrestrictive: comma(s)
Angry, the man sat down.
The man, *angry,* sat down.
The man sat down, *angry.*

II. Participial adjectival phrases
Restrictive: no comma
The man *sitting down* looked angry.
Nonrestrictive: comma(s)
Sitting down, the man looked angry.
The man, *sitting down,* looked angry.

III. Adjectival clauses
Restrictive: no comma
The man *who was sitting down* looked angry.
Nonrestrictive: comma(s)
The man; *who was sitting down,* looked angry.

EXERCISE

1. In the following sentences commas used conventionally with adjectivals have been removed. Replace them where you think they belong. Be able to explain your reasons.

A. The cavalry without which the force of the legion would have remained imperfect was divided into ten troops or squadrons. . . .

Edward Gibbon

B. Looking in the mirror this morning I decided that a man without women around is a man without vanity. . . .

Richard E. Byrd

C. His voice was a high twangy unmusical New England drawl.

Irving Stone

D. Slavery was now regarded by industrialists as a safeguard against the pretensions of the freeborn working classes who could not compete in price against well-organized and well-financed slave labor.

Robert Graves

E. Compared with the forced migrations of our time the Exile was the most trivial affair.

Aldous Huxley

2. Compose two sentences containing nonrestrictive single-word adjectives, two with nonrestrictive participial phrases, and two with nonrestrictive adjectival clauses. Punctuate them according to the conventions discussed in this section.

5. THE COMMA WITH ADVERBIALS

An adverbial is any word or construction used as an adverb.[7] Adverbials are more flexible in their positioning than adjectivals, modify more kinds of words, and convey a wider range of meanings. Consequently their punctuation is especially variable. In the discussion that follows, advice about using commas with adverbials must be understood as loose generalizations, which skillful writers frequently ignore or adapt to their particular need to be emphatic or clear or rhythmic.

Single-word adverbs   When simple adverbs modify verbs, adjectives, or other adverbs, they are not usually punctuated (italics are added in the following examples):

He wept *quietly.*
The people were *extremely* happy.
Everyone was *very deeply* concerned.

Sentence adverbs (those that modify an entire clause rather than any single word) are more frequently punctuated. In composition, sentence adverbs often take the form of connectives, qualifiers, and what may be called "attitudinals" (words like *fortunately* or *unhappily* that express a writer's attitude toward the statement he or she is making). Mostly such words are punctuated, whether in the opening, interrupting, or closing position (italics added):

Further, Hamlet's world is a world of riddles.

Maynard Mack

Unhappily, the gibe has point.

Brand Blanshard

In spite of all these dissimilarities, *however,* the points of resemblance were quite as profound.

Bertrand Russell

But, *luckily,* even at the dreariest moments of our pilgrimage there were compensations.

Aldous Huxley

---

7. See pages 746–54.

I missed that class, *fortunately.*

Student

There is, however, considerable variation in punctuating such sentence adverbs. Some (*however,* for example) are always punctuated. With others (*therefore, luckily, fortunately*) the comma(s)—while probably more often used than not—may be omitted if the writer does not like the pause and feels that clarity does not require it.

When the coordinating conjunctions *and, but, for, or* are used to introduce a sentence, they are *not* punctuated, even though they are serving, for all intents and purposes, just like conjunctive adverbs:

But we stayed.
NOT But, we stayed.

Adverbial prepositional phrases   In the *first position,* when they open a sentence, adverbial prepositional phrases may or may not be punctuated. Much depends upon the conventions regarding specific phrases, upon the writer's own preference, and upon the need for clarity or emphasis. Some idiomatic phrases are usually followed by commas; this is especially the case with those acting as sentence adverbs signaling logical relationship or attitude (*for example, on the other hand, of course*):

For example, in 1913 there was produced in Great Britain seven billion yards of cotton cloth for export alone.

Carl Becker

Less formulaic phrases are often punctuated or not according to the writer's sense of rhythm:

In a crude way, Mickey Spillane is something of an innovator.

Charles J. Rolo

Of Pushkin's shorter stories The Queen of Spades is perhaps the most entertaining.

Rosemary Edmonds

However, if there is any chance that an initial phrase may be misconnected, a comma should always be used. These two sentences, for instance, would be clearer with commas:

In writing these signals must be replaced by punctuation.

In business machines are built to become obsolete within a few years.

In each case the object of the preposition can be misread as grammatically tied to the immediately following word, as if the writers were talking about "writing these signals" and "business machines."

Within a sentence adverbial phrases are punctuated with great variability. What the phrase modifies, where it is placed, what rhythm or emphasis the writer wants are all important. A key consideration is whether or not the phrase is felt as an interrupter—that is, as intruding into the normal grammatical flow of the sentence. If it is, set off the phrase by commas. Interrupting phrases often come between subject and verb:

> Jerusalem, *of course,* contains more than ghosts and architectural monstrosities.
>
> > Aldous Huxley
>
> Barrett Wendell, *in his admirable book on writing,* points out that clearness and vividness often turn on mere specificity.
>
> > Brand Blanshard

But they may come elsewhere:

> And their former masters were, *from the start,* resolved to maintain the old difference.
>
> > Oscar Handlin
>
> Coughlin's activities were clearly, *after Pearl Harbor,* intolerable.
>
> > Wallace Stegner
>
> Newspapermen have always felt superstitious, *among other things,* about Lindbergh.
>
> > John Lardner

In such cases the writer is seeking clarity or emphasis. The option is not so much whether to punctuate the phrase as where to place it. Any of the phrases in the three examples above could be positioned, and more idiomatically, at the end and would then probably not need commas. But placed where they are, they do require punctuation.

At the close of a sentence or clause, adverbial phrases are not generally punctuated:

> The party adjourned *to the kitchen.* . . .
>
> > Herbert Asbury
>
> He was quiet and in-dwelling *from early boyhood on.*
>
> > John Lardner

Final adverbial phrases may be isolated for emphasis, though the technique quickly loses value if overworked:

> They were not men of equal status, *despite the professed democratic procedure.*
>
> > Harry Hansen
>
> And why is this picture an absurdity—as it is, *of course?*
>
> > George Orwell

Adverbial clauses In *initial position,* when they precede the main clause, adverbial clauses are usually punctuated:

If we figure out the answer, we feel devilishly smart; *if we don't,* we enjoy a juicy surprise.

Charles J. Rolo

When the general atmosphere is bad, language must suffer.

George Orwell

A writer has the option of omitting the comma after a short initial adverbial clause if clarity will not suffer. (British writers seem to exercise that choice more often than do Americans):

When he describes the past the historian has to recapture the richness of the moments. . . .

Herbert Butterfield

However, the comma should never be left out if there is any possibility that readers will see an unintended grammatical connection between the last word of the adverbial clause and the first word of the following construction. In the sentence below, for instance, a comma after "sail" would prevent readers from the misstep of thinking the writer is referring to "sail boats":

When you are first learning to sail boats seem to be very cumbersome things.

Adverbial clauses in an *interrupting position* are conventionally punctuated:

The whole thing, *as he himself recognized,* was a clean sporting venture.
P. G. Wodehouse

On occasion, *if no operations were scheduled for the next day,* he would be up early and out on an all-day hunt after getting only one or two hours of sleep.

Ralph K. Andrist

Adverbial clauses in the *closing position* may or may not be punctuated. The primary considerations are clarity and rhythm. A comma generally helps readers follow the grammar, especially before clauses expressing a concession or qualification:

The Supreme Court upheld the conviction, *although the judges could not agree on any one opinion.*

Roger Fisher

Now I seldom cuss, *although at first I was quick to open fire at everything that tried my patience.*

Richard E. Byrd

On the other hand, some writers prefer to omit the comma when the main and the adverbial clauses are both short and unpunctuated within themselves. The comma is often omitted before *because* if the pause might seem overly emphatic:

Comma with adverbials

I. Single-word adverbs
  A. Sentence adverbs: usually punctuated, whether in the initial,
    closing, or interrupting position
      However, the people left.
      The people, *however,* left.
      The people left, *however.*
    But there are exceptions
      Fortunately(,) the people left.
      The people *therefore* left.
  B. Adverbs modifying verbs and other modifiers: not punctuated
    unless they are in an unusual position, when a comma may
    be used for clarity or emphasis
      The people *slowly* left.

      EMPHATIC $\begin{cases} \textit{Slowly,} \text{ the people left.} \\ \text{The people left, } \textit{slowly.} \end{cases}$

II. Adverbial phrase
  A. Initial position: punctuation optional
      On the whole(,) the men were satisfied.
  B. Closing position: not generally punctuated, though comma may
    be used for emphasis
      The men were satisfied *on the whole.*
      EMPHATIC The men were satisfied, *on the whole.*
  C. Interrupting position: punctuation conventionally required
      The men, *on the whole,* were satisfied.
      The men were, *on the whole,* satisfied.

III. Adverbial clause
  A. Initial position: usually punctuated
      When the sun went down, the women left camp.
      OPTION WITH SHORT, CLEARLY RELATED CLAUSES *When the
      sun went down* the women left camp.
  B. Closing position: not usually punctuated, though a comma may
    be used for emphasis or clarity
      The women left camp *when the sun went down.*
      EMPHATIC The women left camp, *when the sun went down.*
  C. Interrupting position: conventionally punctuated
      The women, *when the sun went down,* left camp.

Locke thought traditional theology worthless *because it was not primarily concerned with truth.*

Paul Johnson

On one occasion, however, a following *because*-clause should be preceded by a comma. This is when it comes after a negative statement and is intended as a straightforward explanation of that statement:

They did not elect him, *because they distrusted him.*

Without the comma such a sentence may be read as an ironic assertion that "they did elect him and certainly did not distrust him."

EXERCISE

1. The commas that originally marked the adverbials have been removed from the following sentences. Put them where you think they belong.

A. With the repeal of prohibition the "aspirin age" came to an end.

Carey McWilliams

B. Nevertheless the Supreme Court held that the defendant could *not* be punished for such teaching and urging.

Roger Fisher

C. Similarly the only way to police a ghetto is to be oppressive.

James Baldwin

D. If that is oversimplification it is the kind around which ringing slogans are made.

Time magazine

E. To many his was the only voice that spoke truth.

Wallace Stegner

F. Because the civilization of ancient Rome perished in consequence of the invasion of the Barbarians we are perhaps too apt to think that civilization cannot perish in any other manner.

Alexis de Tocqueville

G. No the ghost is not yet laid although Huey Long is dead these many long years.

Hodding Carter

H. I find too that absence of conversation makes it harder for me to think in words.

Richard E. Byrd

I. In the early days of the century America though it showed evidences of a growing desire for economic isolation still had a frontier.

Thurman Arnold

J. For obvious reasons this like the Nan Britton story was ignored by the newspapers.

Samuel Hopkins Adams

K. Obviously the fact that a neighbor or the community would be disturbed is not always ground in itself to justify prohibiting things from being said.

Roger Fisher

L. At last when the tribute slackened he motioned to his retainers to sweep up the treasure.

Robert Coughlan

M. Today for example we think of the thirteenth century as one of the supremely creative periods of human history.

Aldous Huxley

N. The horses were bred for the most part in Spain or Cappadocia.

Edward Gibbon

O. While he was in the room the atmosphere was electric.

Joy Packer

P. And their former masters were from the start resolved to maintain the old differences.

Oscar Handlin

Q. Beyond the last flutter of actual or possible significance pedantry begins.

Jacques Barzun

R. The hero who marches down these mean streets is at his finest embodiment Raymond Chandler's Philip Marlowe private dick.

Charles J. Rolo

S. On this day she was pretty drunk when she went aboard the ship.

Morris Markey

2. Compose eight or ten sentences containing adverbial phrases or clauses that require punctuation. Vary as much as possible the kinds of phrases and clauses you use and the positions in which you place them.

6. COMMA WITH THE MAIN ELEMENTS OF THE SENTENCE

The main elements of a sentence—the subject, verb, and object—are not separated by commas except under unusual conditions. Very occasionally when the subject is not a single word but a long construction, such as a noun clause, a comma may be put at its end to signal the verb (italics are added in the following examples):

> *What makes the generation of the '60s different,* is that it is largely inner-directed and uncontrolled by adult-doyens.
>
> *Time* magazine

In such a sentence the comma between the subject and the verb may help readers to follow the grammar.

Commas may also be used with the main elements in the case of inversion—that is, when the subject, verb, and object are arranged in something other than their usual order. Sometimes the pattern is object, subject, verb; if the object is a long construction, a comma may be set between it and the subject:

> *What he actually meant by it,* I cannot imagine.
>
> Aldous Huxley

The most frequent kind of inversion in composition occurs with the idiom "I think" ("I suppose," "I imagine," "I hope" are other variations):

The lectures, *I understand,* are given and may even be taken.
Stephen Leacock

Lenin, on the contrary, might, *I think,* have seemed to me at once a narrow-minded fanatic and a cheap cynic.
Bertrand Russell

In this type of sentence the main subject/verb is the "I think," "I understand." The rest (which contains the key idea) is a contact clause acting as the direct object, telling us what is understood or thought. If the sentence were in straightforward order, no comma would be necessary between the main elements:

I understand the lectures are given. . . .
I think Lenin might have seemed. . . .

But when the "I understand" or "I think" is intruded within the noun clause, the subject/verb must be treated as an interrupting construction and set off by commas.

,/7

7. COMMA WITH APPOSITIVES

An appositive is a word or construction which refers to the same thing as another and is (usually) set immediately after it.[8] When appositives are restrictive, they are not punctuated:

The argument *that the corporations create new psychological needs in order to sell their wares* is equally flimsy.
Ellen Willis

In that sentence the clause is in restrictive apposition to the subject "argument"; it specifies "argument," and the noun would be relatively meaningless without it. Notice that the clause is *not* set off by commas. (Sometimes, however, a comma is placed *after* such a clause—though not before—to mark its end and signal a new construction.)

Often appositives are nonrestrictive. In that case they must be punctuated. Usually such appositives follow the noun and should be preceded by a comma (and followed by one if they do not close the sentence):

Poskitt, *the d'Artagnan of the links,* was a man who brought to the tee the tactics which in his youth had won him such fame as a hammer thrower.
P. G. Wodehouse

The newcomers were pagans, *worshippers of Wotan and other Teutonic gods.*
Margaret Schlauch

She was a splendid woman, *this Mme. Guyon.*
W. H. Lewis

8. See Reference Grammar, pages 758–61, for a more complete discussion of appositives.

Appositives occasionally open a clause or sentence, thus preceding the word to which they are in apposition. Then they must be followed by a comma, as in this example where a series of three appositives precedes the subject ("Bishop Andrewes"):

A gifted preacher, a profound scholar, and a great and good man, Bishop Andrewes was one of the lights of the Church of England.

G. P. V. Akrigg

8. COMMA WITH ABSOLUTES

An absolute is a construction that is included within a sentence but is not really a grammatical part of that sentence; it serves as a kind of loose clausal modifier.[9]

Nominative absolutes, the most common kind in composition, may precede, follow, or be intruded into the main clause. In all cases they are punctuated (the absolutes are italicized in the following examples):

The savings of the nation having been absorbed by Wall Street, the people were persuaded to borrow money on their farms, factories, homes, machinery, and every other tangible asset that they might earn high interest rates and take big profits out of the rise in the market.

Irving Stone

The bluffs along the water's edge were streaked with black and red and yellow, *their colors deepened by recent rains.*

John G. Neihardt

The official, *his white shirt clinging with sweat to his ribs,* received me with a politeness clearly on the inner edge of neurosis.

James Cameron

Participial and *infinitive absolutes* are also punctuated:

Allowing for hyperbole and halving the figure, that is still one hell of a pile of pulp.

Pauline Kael

To revert for a moment to the story told in the first person, it is plain that in that case the narrator has no such liberty. . . .

Percy Lubbock

9. COMMA WITH SUSPENDED CONSTRUCTIONS

A *suspended construction* occurs when two or more units are hooked grammatically to the same thing. It is really a form of parallelism, but an unusual or emphatic form, which readers may find difficult. Hence such constructions are often (though not invariably) punctuated:

Many people believed, and still do, that he was taking Nazi money to run his machine.

Wallace Stegner

9. See Reference Grammar, pages 754–56.

Prescott and Parkman were willing, and Motley reluctant, to concede that the sixteenth-century Spaniard's desire to convert American Indians had not been hypocritical.

David Levin

When the idiomatic phrase *more or less* is treated as a suspended construction, it always requires commas to distinguish it from its more common meaning. Usually *more or less* signifies a qualified affirmation, and then is not punctuated:

He was more or less interested. = He was mildly interested.

But when *more or less* is used in a strict disjunctive sense—that is, to mean either more *or* less, but not both—it must be set off by commas:

It is hard to say whether the payment for votes has become more, or less, important.

Ronald P. Dore

10. COMMA WITH DATES AND PLACE NAMES

In American usage, dates are conventionally punctuated like this:

April 14, 1926
April 1926

In European usage the day precedes the month, in which case a comma is unnecessary:

14 April 1926

In those place names that consist of both a local and a larger designation (state, region, province, nation), a comma is placed between the two:

London, Ontario
Kittery Point, Maine

EXERCISE

1. Restore the commas to the following sentences:

A. June saw many prisoners discharged from the sick-ward only a few beds remaining occupied.

Emma Goldman

B. In April 1841 the world's first mass-circulation magazine Grahame's published a story which connoisseurs of the whodunit regard as the first detective story—"The Murders in the Rue Morgue" by Edgar Allan Poe.

Charles J. Rolo

C. Given the way movie companies work it heralds the return of the weepers.

Pauline Kael

D. We the consumers can accept the goods offered to us or we can reject them but we cannot determine their quality or change the system's priorities.

Ellen Willis

E. The present situation is unsatisfactory because the heads of the industry are aggrieved by the "heavyweight critics" whom they accuse of blindness to the virtues and morbid sensitivity to the defects of broadcasting. . . .

Gilbert Seldes

F. A member of the State legislature in 1884 Roosevelt had led in the fight for the passage of four or five bills backed by civic organizations in New York.

Henry F. Pringle

G. Downstairs then they went Joseph very red and blushing.

W. M. Thackeray

H. Her mind being made up the widow began to take such measures as seemed right to her for advancing the end which she proposed.

W. M. Thackeray

2. Write two sentences using appositives which require punctuation, two with nominative absolutes, and two with suspended constructions.

THE DASH

The dash ought not to be confused with the hyphen. It is a longer mark, and on a typewriter is made either by two hyphens (- -) or by a single hyphen with a space on either side (-). The dash, incidentally, is the only stop that may be placed at the beginning of a line.

The dash has no function that is uniquely its own. Instead it acts as a strong comma and as a less formal equivalent to the semicolon, the colon, and the parenthesis. As a substitute for the comma, the dash signals a stronger, more significant pause. *For that reason it should be used sparingly,* reserved for occasions when emphasis is really needed.

1. THE DASH ISOLATING FINAL CONSTRUCTIONS

Dashes force an emphatic pause before the last word or phrase of a sentence:

Our time is one of disillusion in our species and a resulting lack of self-confidence—for good historical reasons.

Barbara Tuchman

So the gift of symbolism, which is the gift of reason, is at the same time the seat of man's peculiar weakness—the danger of lunacy.

Susanne K. Langer

-- /2

2. THE DASH AROUND INTERRUPTING PHRASES AND DEPENDENT CLAUSES

Dashes may set off dependent interrupting constructions such as non-restrictive adjective clauses, adverbial phrases and clauses, appositives, and suspended constructions. In such a use, they create emphasis.

> After graduation from high school—where he [Charles Lindbergh] once wrote an elaborate and not uncomical satire on the finicky methods of his English teacher—he took three semesters in engineering at the University of Wisconsin, where the only thing that seemed to interest him much was shooting (he made the rifle team).
>
> John Lardner

> Occasionally—with a gun in his ribs, another in his back, and a gloating voice saying that in ten seconds he'll be dead—Hammer *does* become a trifle anxious.
>
> Charles J. Rolo

> Rotten logs can also be host to the ghostly glow of slime fungus, a plant that creeps—glowing—over the logs or along the ground.
>
> Ruth Rudner

> Some of those writers who most admired technology—Whitman, Henry Adams, and H. G. Wells, for example—also feared it greatly.
>
> Samuel C. Florman

Notice, in the last example, that dashes are clearer signals of the grammar than commas would be, since the interrupting series contains commas.

-- /3

3. THE DASH WITH COORDINATED ELEMENTS

As we saw with the comma (page 625), coordinated elements are sometimes punctuated for emphasis. Stronger stress can be attained by using dashes:

> We were—and are—in everyday contact with these invisible empires.
>
> Thurman Arnold

> What the youth of America—and their observing elders—saw at Bethel was the potential power of a generation that in countless disturbing ways has rejected the traditional values and goals of the U.S.
>
> *Time* magazine

Coordinated independent clauses are occasionally separated by a dash instead of the usual comma, but it is worth repeating that the dash is not the conventional stop for such a case and should be employed only when emphasis is necessary:

> He was a sad, embittered young man—and well he might be.
>
> Aldous Huxley

Even uncoordinated independent clauses may be punctuated by a dash instead of the conventional semicolon:

Hammer is not just any Superman—he has The Call.

Charles J. Rolo

A town may impose regulation upon the use of trucks which are equipped with loudspeakers—it may, for example, limit the loud playing of music on such trucks.

Roger Fisher

--/4

4. THE DASH INTRODUCING A LIST

The colon conventionally introduces a series of specifics. The dash, however, is employed for the same purpose. The only difference is that the dash is less formal:

In short, says the historian Friedrich Heer, the crusades were promoted with all the devices of the propagandist—atrocity stories, over-simplification, lies, inflammatory speeches.

Morris Bishop

--/5

5. THE DASH AROUND INTRUSIVE SENTENCE ABSOLUTES

An intrusive sentence absolute is a completely independent second sentence which is stuck into the middle of a containing statement without being syntactically tied to it in any way.[10] Such a construction must be clearly marked, but it cannot be set off by commas, semicolons, or colons, since these stops would imply a grammatical connection between it and the containing sentence which does not exist. Parentheses could be used and sometimes are; but they are a little formal for this kind of construction, which is colloquial in tone. Here, then, is the one function which belongs primarily to the dash:

The opening paragraph—it is one of Pushkin's famous openings—plunges the reader into the heart of the matter.

Rosemary Edmonds

He has never, himself, done anything for which to be hated—which of us has?—and yet he is facing, daily and nightly, people who would gladly see him dead, and he knows it.

James Baldwin

He [the psychoanalyst] tells us—and the notion has gained official acceptance to a limited degree—that crime is not so much willful sin as the product of sickness.

Charles J. Rolo

EXERCISE

1. Each of the following sentences originally contained one or two dashes. Restore them to where you think they belong. Explain why you placed them where you did.

10. See Reference Grammar, pages 757–58, for a fuller discussion.

A. It is a serious matter to shoot a working elephant it is comparable to destroying a huge and costly piece of machinery and obviously one ought not to do it if it can possibly be avoided.

George Orwell

B. These moloch gods, these monstrous states, are not natural beings; they are man's own work, products of the power that makes him lord over all other living things his mind.

Susanne K. Langer

C. In one of my books Huckleberry Finn, I think I have used one of Jim's impromptu tales, which he called "The Tragedy of the Burning Shame."

Mark Twain

D. Only a few women over sixty still wore, in 1955, the old Japanese-style clothes blue-dyed cotton short kimono with floppy sleeves that one tied up with a sash for work.

Ronald P. Dore

E. But the Negro-white conflict had and no doubt still has a special intensity and was conducted with a ferocity unmatched by intramural white battling.

Norman Podhoretz

F. He must get them to follow a process of distinguishing, abstracting, and inferring in short, of thinking.

Brand Blanshard

G. This unfortunate accident not a burning love of freedom ended the monarchy.

Robert Graves

H. And all of them investor, farmer, factory worker have their eyes uneasily on the banks.

Wallace Stegner

I. In the Mediterranean the winter is generally short and mild, and the Romans could import unlimited cheap grain from Egypt, Libya, and Tripoli it was not for some centuries that overcultivation made a dust bowl of the whole North African coast.

Robert Graves

J. . . . if assembly lines were to be kept moving, mass purchasing power had to be restored and only employment and wages could do that.

Thurman Arnold

K. But renewal becomes impossible if one supposes things to be constant that are not safety, for example, or money, or power.

James Baldwin

L. Fiction if it at all aspire to be art appeals to temperament.

Joseph Conrad

2. Could any of the dashes you have supplied for the sentences above be replaced by other marks of punctuation?

3. Compose three sentences containing intrusive sentence absolutes and three containing interrupting dependent clauses (adjectival or adverbial), all punctuated by dashes. Compose also three sentences in which a closing specification is set off by a dash.

54

The Other Marks

In addition to the stops, punctuation marks include the apostrophe, the quotation mark, the hyphen, the ellipsis, the parenthesis and bracket, and the diacritics. We shall look at these here, along with the related matters of capitalization and underlining.

THE APOSTROPHE

The apostrophe has three main functions: it marks the possessive form of nouns and some pronouns, the contraction of two words, and the omission of sound within a word. It also appears in the plurals of certain abbreviations.

1. APOSTROPHE TO SHOW POSSESSION[1]

Common nouns   In their singular form common nouns that do not end in *-s* or another sibilant add *-'s* to show possession:

the cat's bowl, the girl's hat, the boy's jacket

Singular nouns with a final sibilant also generally add the *-'s* in modern convention:

the horse's tail, the apprentice's job

However, there is a minor variation of usage in this matter. If such a word has several syllables and the final one is unstressed, some writers and editors prefer to drop the *-s,* using the apostrophe alone to indicate possession:

for appearance's sake OR for appearance' sake

1. See Reference Grammar, page 692, for a discussion of the possessive forms.

645

The issue can often be dodged by using an *of*-phrase:

> for the sake of appearance

> Plural nouns ending in -*s* (the vast majority) add only an apostrophe:

> the girls' books, the mechanics' toolboxes

> Those which do not end in -*s* add -*'s*:

> the men's books, the children's toys

Proper nouns   Proper nouns that do not have a final sibilant follow the same rule as common nouns:

> Sarah's house, Eisenhower's career

With proper nouns ending in sibilants, practice varies. If the noun is monosyllabic, it is conventional to add the full -*'s*:

> Henry James's novels, John Keats's poetry

But opinion differs when proper names have more than a single syllable. Some people prefer -*'s*, some the apostrophe alone:

> Reynolds's paintings OR Reynolds' paintings

However, the -*s* should be omitted from the possessive of names containing several syllables if it would result in an awkward combination of sounds:

> Jesus' ministry NOT ✕ Jesus's ministry
> Xerxes' army NOT ✕ Xerxes's army

When the plural form of a family name is used in the possessive, the apostrophe alone is called for:

> the Browns' house, the Johnsons' boat

Pronouns   Indefinite pronouns form the possessive by adding -*'s*:

> anyone's, anybody's, someone's, everyone's, and so on

The predicative possessive forms of the personal pronouns, however, do *not* use an apostrophe:

> mine, yours, his, hers, its, ours, theirs

Its is especially likely to be misused, probably because of confusion with the contraction *it's* for *it is*. Never use *it's* for the possessive of *it*:

> The cat washed its tail.
> NOT ✕ The cat washed it's tail.

The possessive of *who* is *whose*, not *who's*, which is the contraction of *who is*.

ʾ/2   2. APOSTROPHE TO SHOW CONTRACTION

A *contraction* is the coming together of two or more words with the omission of intervening sounds (in writing, of course, the letters). Contractions are common in speech and are permissible in informal writing, though they should be avoided in a formal style. They are most likely with auxiliary verbs and negative words, and in all cases an apostrophe should be placed in the position of the deleted sound or letter:

He'll go. = He will go.
We would've gone. = We would have gone.
They won't go. = They will not go.

Notice that in the last example several sounds have been dropped, but only one apostrophe is used.

The contracted form of the auxiliary *have*, incidentally, sounds exactly like the unstressed *of*. Because of this confusion such constructions as × *I could of gone* are sometimes seen. That is not in accordance with formal usage and should be avoided. The proper form is: *I could've gone*.

ʾ/3   3. THE APOSTROPHE TO MARK ELISION

Elision is dropping a sound from a word. This often occurs in rapid speech (*goin'* for *going*) and was sometimes done in older poetry (*e'en* for *even*, *ne'er* for *never*), though rarely in modern verse. An apostrophe signals when a sound is elided. Elision is rarely necessary in composition.

ʾ/4   4. THE APOSTROPHE WITH THE PLURAL FORMS OF LETTERS

When letters and numerals are used in the plural, they generally simply add -*s*:

Learn your ABCs.
The 1960s were a period of great change.

There are, however, three exceptions: (1) capital letters in abbreviations with periods, (2) capital letters that might look confusing with a simple -*s* plural, and (3) lowercase letters used as nouns:

The university graduated twenty M.A.'s.
He makes his A's in an unusual way.
Mind your p's and q's.

``

THE QUOTATION MARK

Quotation marks are used with (1) direct quotations, (2) certain titles, and (3) words given a special sense. Quote marks have two forms: double ("...") and single ('...'). Most American writers prefer double quotes, switching to single should they need to mark a quote within a quote. British writers are more likely to begin with single quotes, switching, if necessary, to double. Whether single or double, the quote at the beginning is called an opening quotation mark; the one at the end, a closing.

`"`/1

1. QUOTATION MARKS WITH DIRECT QUOTATIONS

A direct quotation consists of the words actually spoken or written by someone other than the writer. It is distinct from an indirect quotation, which reports the substance of what was said or written but changes the words to fit the context—often altering pronouns and verbs:

DIRECT   She said, "We are not going."
INDIRECT   She said that they were not going.

Direct quotations *must be* signaled by quote marks; indirect quotations *must not be.*

Introducing a quotation   In introducing a quotation the subject and verb of address may precede, follow, or intrude into the quoted matter. The three possibilities are punctuated like this:

She said, "We are not going."
"We are not going," she said.
"We," she said, "are not going."

Notice that the first word of the quotation is capitalized, but that when a quotation is broken—as in the third example—the opening word of the continuation is not capitalized (unless, of course, it happens to be a proper noun or adjective or the beginning word of a new sentence).

Written quotations may be preceded by a comma, or, more formally, by a colon:

Professor Brown writes: "By themselves statistics are rarely enough; they require careful interpretation."

Often written quotations are worked into the text in a smoother manner by an introductory *that*. The *that* requires no stop since it turns the quotation into a noun clause acting as the direct object of the verb; and the first word of the quotation is not capitalized, even if it was so in the source:

Professor Jones writes that "by themselves statistics are rarely enough; they require careful interpretation."

If a quotation is extensive and involves more than one paragraph, it is customary to repeat the opening quote marks at the beginning of each new paragraph. Closing quotes are used only at the end of the final paragraph.

However, extended written quotations are more commonly indented, in which case quote marks are not needed.[2]

Quotation marks in relation to stops   With opening quote marks, a comma, a colon, or any other stop always precedes the quotation mark.

With closing quotes, however, the matter is more complicated. In American usage, commas and periods always come inside a final quote mark; semicolons and colons, outside. This rule applies regardless of whether the stop in question is part of the quotation or not:

She said, "We are not going."
She said, "We are not going," and they didn't.
She said, "We are not going"; they didn't.
She said, "We are not going": why, I wonder?

In the case of question marks and exclamation points, placement depends upon whether the stop applies only to the quotation, only to the sentence containing the quotation, or to both. When the quotation is a question (or exclamation) and the enclosing sentence is a declarative statement, the query (or exclamation point) comes inside the final quote mark:

She asked, "Are we going?"

When the quotation is a statement and the enclosing sentence a question, the query is placed outside:

Did she say, "We are going"?

When, finally, both quotation and sentence are questions, the query is inside the quote mark, where it does double duty:

Did she ask, "Are we going?"

Notice that whether it goes inside or outside the closing quotation, the query (or exclamation point) serves as the end stop; no period is necessary.

2. QUOTATION MARKS WITH TITLES

Some titles of literary works are italicized (in typescript, underlined), others are placed in quote marks. The basic consideration is whether

---

2. The problem of incorporating written quotations into one's text is treated more fully in the chapter on the research paper. (See pages 541–43.)

the work was published or presented separately, or rather as part of something larger (for example, a magazine or collection). In the first case the title is italicized; in the second, set within quotes. In practical terms, this means that the titles of books, plays, and long poems, such as the *Iliad,* are italicized, while the titles of chapters or sections within a book, of short stories, essays, articles in magazines or other periodicals, and short poems are quoted:

> Hemingway's novel *A Farewell to Arms* has been made into a movie.
>
> *A Winter's Tale* is one of Shakespeare's so-called problem comedies.
>
> "A Rose for Emily" by William Faulkner is a shocking short story.
>
> In *Vanity Fair* Thackeray calls one chapter "How to Live on Nothing a Year."
>
> The finest carpe diem poem in English is Andrew Marvell's "To His Coy Mistress."

The titles of movies are italicized, those of television and radio shows are quoted:

> *Robin and Marian* is an unusual and interesting film treatment of the Robin Hood story.
>
> "Truth or Consequences" was popular both on radio and on television.

Notice that the first word of a title is always capitalized. So are the last word and all intervening words except articles (*a, an, the*), short prepositions, and coordinating conjunctions.

" / 3

3. QUOTATION MARKS TO SIGNIFY SPECIAL MEANING

Limited or technical meaning   Sometimes a common word must be used in a special sense that applies only with a limited context. To make the limitation clear, it helps to put the word in quotes:

> Some years later Eton became the first public school—"public" in the sense that students were accepted from everywhere, not merely from the neighborhood.
>
> Morris Bishop

Irony   Irony is using a word in a sense very different from—often opposite to—its conventional meaning. Effective irony depends upon the reader's recognizing the writer's intent. Intention should be clear from the context. Even so, a signal is sometimes advisable. In speech this is given by intonation, as when we speak the word *brave* in a scornful way to mean "cowardly." In writing, the signal may be supplied by quotation marks:

The Indians were therefore pushed back behind ever-retreating frontiers. "Permanent" boundaries were established between the United States and the Indians, tribes were moved out of the United States and established beyond those boundaries. Again and again the boundaries were violated by the whites.

James Oliver Robertson

Citation terms   A citation term is a word used to refer to itself rather than to the object or concept or feeling it conventionally designates. Usually such terms are italicized, but sometimes they are quoted. (They should never be treated both ways.) The following pair of sentences illustrate the difference between the same word used first in its conventional sense and second as a citation term:

A horse grazed in the meadow.
"Horse" is a citation term.

Definitions   When a word is defined, its meaning is sometimes put in quotes, the word itself being italicized:

Other-directed means "accepting and living by the standards of the social group to which one belongs or aspires."

Slang and colloquialisms   It is *not* necessary to place quotation marks around slang or colloquial expressions, apologizing for them, so to speak. If the term says exactly what you want to say, no apology is needed; if it does not, no apology will help.

Hy.

THE HYPHEN

The hyphen has two principal functions. It marks the syllabic division of a word between lines, and it also separates the elements of some compound words.

Hy./1

1. THE HYPHEN TO INDICATE DIVISION OF A WORD

When separating a word between lines, you should always place the hyphen at the end of the upper line, never at the beginning of the new line. The word *supper,* for example, must be divided:

sup- NOT sup
per   -per

Words can be divided only between syllables. Most of us have only a hazy idea of the syllabication of many words, and it is best to consult a dictionary when you must split a word.

Hy./2

2. THE HYPHEN WITH COMPOUNDS

In certain compounds (two or more words treated as one) the hyphen separates the individual words. English does not treat compounds with much consistency. Some are printed as separate words (*contact lens, drawing room, milk shake*); some as single terms (*gunboat, footlight, midships*); and still others are hyphenated (*gun-shy, photo-offset, teen-ager*). Some compounds are treated differently by different writers; *teen-ager,* for example, may appear as *teenager.* You cannot tell how any particular compound is conventionally written without consulting a dictionary or observing how publishers print it.

The examples we just saw are all *conventional compound* words. Another kind exists called the *nonce compound.* This is a construction, usually a modifier, made up for a specific occasion and not existing as a standard idiom. In the following sentence, the first compound is conventional; the other two are nonce expressions:

Old-fashioned, once-in-a-lifetime, till-death-do-us-part marriage. . . .

Leslie Aldridge Westoff

Nonce compounds are always hyphenated.

Hy./3

3. OTHER FUNCTIONS OF THE HYPHEN

Hyphens, finally, have several special applications. When a word is spelled out in composition, the pauses which in speech would separate the letters are signaled by hyphens:

Affect is spelled a-f-f-e-c-t.

If it is necessary to cite inflectional endings or prefixes, they are preceded (or followed) by a hyphen. No space is left between the hyphen and the first or last letter of the cited term:

The regular sign of the plural in English is *-s.*
Anti- and *un-* are common prefixes, while *-ence* is a frequent suffix.

When several different words are understood to be commonly combined with the same final element to form compound words, hyphens are placed after each of the initial elements:

The lemon groves are sunken, down a three- or four-foot retaining wall. . . .

Joan Didion

PARENTHESES

Paren./1

1. TO ENCLOSE PARENTHETICAL MATTER

Parenthetical matter is a word or construction (which may or may not be grammatically related to the rest of the sentence) sufficiently remote in relevance to require a stronger pause than a comma would supply:

> Even for those who can do their work in bed (like journalists), still more for those whose work cannot be done in bed (as, for example, the professional harpooner of whales), it is obvious that the indulgence [of lying in bed] must be very occasional.
>
> G. K. Chesterton

Parenthetical remarks of this sort—which may also be punctuated with dashes—can be a source of interest and variety as well as of necessary information. Moreover, such intrusions loosen the rhythm of a sentence, suggesting more interesting patterns of speech. The effectiveness of parenthetical remarks, however, depends upon their scarcity. Using one in every other sentence costs you whatever advantage the device had, and overused parentheses become an irritating mannerism.

When a parenthetical remark comes inside a sentence, any stop that follows it must be set outside the closing parenthesis:

> In the last act of the play (or so it seems to me, for I know there can be differences on this point), Hamlet accepts his world and we discover a different man.
>
> Maynard Mack

When a parenthetical remark closes a sentence, the period is also placed outside:

> I say only that a considerable number of [TV] set-owners are far from being entirely satisfied (those who are willing to pay and those who hardly ever use their sets).
>
> Gilbert Seldes

However, if an entirely separate sentence is placed within quotes, the period which closes it (or the query or exclamation) must go inside:

> Many of winter's plants are partial to the cleared sides of roads. (The roots—which is what most winter foods are—are not subject to the same kind of road pollution as are the leaves.)
>
> Ruth Rudner

Paren./2

2. PARENTHESES TO ENCLOSE NUMBERS OR LETTERS MARKING A SERIES

When numbers or letters introduce the items in a list they should be put within parentheses to differentiate them from the text:

> We must do three things: (1) study the route thoroughly, (2) purchase supplies and equipment, and (3) hire a reliable guide.

Brack.

BRACKETS

Brackets (which look like this: [ ]) are used in composition to enclose within a quotation any words that are not a part of it. Sometimes a writer needs to explain or comment upon something in the quotation. The sample sentences by G. K. Chesterton and Gilbert Seldes in the section on parentheses contain such editorial additions set within brackets. In the following passage the writer adds a comment within the words spoken by a guide conducting tourists around Jerusalem:

> "This area," he would say as he showed us one of the Victorian monstrosities, "this area [it was one of his favorite words] is very rich in antiquity."
>   Aldous Huxley

Sometimes, too, it is necessary to alter a quotation slightly to fit it into its grammatical context—adding an auxiliary verb or an ending, for instance. Any such addition to the actual quotation must be enclosed in brackets:

> Johnson writes that "monkeys . . . [are] held in great esteem by the tribe."

Finally, brackets are used to enclose parenthetical matter within parentheses (though such a labyrinthine style would be, with rare exceptions [of which this is not one] both unusual and annoying).

THE ELLIPSIS

Ellip.

The ellipsis is a series of three dots, or, under certain conditions, four. It is never five or six or any other number. In composition the principal function of the ellipsis is to mark the omission of material from a quotation.

If the deleted matter occurs within the quoted sentence, three dots are used:

> Dante, someone has remarked, is "the last . . . great Catholic poet."

Notice the spacing: spaces are left between the preceding word and the first dot, between each dot and the next, and between the last dot and the following word.

If the omitted material includes the end of the sentence and/or the beginning of the next one, four dots are used:

> Dante, someone has noted, is "the last great Catholic poet. . . ."

With four dots the spacing is a little different. The first dot, which represents the period of the original statement, is *not* separated from the word it follows, but the spacing between dots remains. Notice too that the ellipsis is placed inside the quote mark.

If the original sentence from which final matter has been dropped was closed by a query or exclamation point, the appropriate stop is placed immediately after the final word and is followed by a standard three-dot ellipsis:

It has been asked, "Was Dante the last great Catholic poet? . . ."

It is considered simple honesty to use an ellipsis to acknowledge that you have omitted something from a passage you are quoting. Of course, the omission must not change the substance of what the other writer said, and if you do alter his or her meaning, the use of an ellipsis will not save you from a charge of dishonesty. The same caution applies to adding explanatory matter within brackets: it must not substantially alter the original meaning.

The ellipsis is also used in dialogue to indicate doubt, indecision, weariness, and so on. In the following sentence, for example, the ellipsis signals not an omission of any words but the trailing off of the voice, suggesting the speaker's uncertainty:

She sighed and answered, "I really don't know. . . ."

Sometimes too a writer will use an ellipsis to imply a conclusion which readers are expected to infer for themselves:

And we certainly know what that remark means. . . .

DIACRITICS

A *diacritic* is a mark placed above, below, or through a letter in order to indicate a special pronunciation. Diacritics are employed because the number of letters in any language is usually fewer than the number of significantly different sounds. Diacritical marks thus supplement the alphabet, enabling a single letter to do the work of two. English, while it certainly has more sounds than letters, has dispensed with diacritical marks except for the diaeresis occasionally seen in words like *naïve* or *coöperate* (meaning that the vowel is to be pronounced as a separate syllable).

But diacritical marks are common in some other languages—the accents grave and acute and the cedilla of French, (` ´ ¸); the umlaut of German (¨); or the tilde of Spanish (˜). When you use a foreign word not yet assimilated into English, reproduce the diacritics that the word has. (If you are typing, it is easiest to put these in afterwards by pencil.)

Under.

UNDERLINING

Underlining is the compositional equivalent to italic type. There are several reasons for underlining a word or phrase.

Under. /1

1. UNDERLINE TITLES

The titles of newspapers, magazines and other periodicals, books, plays, films, and long poems are underlined. Titles of works which were not published separately, but rather as part of something else, are placed in quotation marks (these include magazine and journal articles, short stories, short poems, and also radio and television programs). With newspapers the title is what appears on the masthead: *The New York Times* but *The Times* of London.

Names of ships are also italicized, as are the titles of long musical works (symphonies, tone poems, operas; songs and shorter compositions are referred to in quotes).

Under./2 **2. UNDERLINE FOREIGNISMS**

Any foreign expression that has not been fully assimilated into English should be italicized:

de trop	(French: "unwanted, in the way")
dolce vita	(Italian: "life that is sweet, easy, enjoyable")
Schadenfreude	(German: "malicious joy at the misfortunes of others")

Because English has always been quick to borrow words from other languages and equally quick to anglicize their pronunciation, it is often difficult to tell whether an imported word is still considered foreign and should be underlined. Few of us think of *delicatessen,* say, as a "German" word or *perfume* as a "French" one, and in normal use these would never be italicized. (They are in the preceding sentence because they are citation terms, not because they are foreign.) But in between such fully accepted terms and plainly alien ones like *Schadenfreude,* many words that have become recently popular in English still have a faintly foreign air: *boutique,* for instance, or *détente.* Editors differ on how to treat such terms. If they are not listed in a standard dictionary, it is never wrong to underline them.

Under./3 **3. UNDERLINE CITATION TERMS**

Words used in self-reference are called *citation terms,* and they are usually underlined (less commonly, placed in quotes). In the following pair of sentences, the infinitive *to run* is a citation term in the first, but in the second it has its conventional sense:

To run is an infinitive.
He wanted to run.

Sometimes citation terms are placed in quotes, and then they should not also be underlined. Underlining—that is, italics—however, is the better choice.

Under./4   **4. UNDERLINE FOR EMPHASIS**

Underlining is a legitimate way of achieving emphasis. However, the device must be used carefully and rarely. When emphatic underlining is well done, it has the effect not only of drawing attention to key words, but also of suggesting an actual voice talking to us:

> The cause of pornography is *not* the same as the cause of free speech. There *is* a difference.
>
> Barbara Tuchman

> The church was, in sum, more than the patron of medieval culture; it *was* medieval culture.
>
> Morris Bishop

Cap.   ## CAPITALIZATION

When to use capital letters is a complicated matter; here we shall mention only a few common occasions. You will find more thorough discussions in dictionaries and in style books like *A Manual of Style, 12th Edition,* published by The University of Chicago Press.

Cap./1   **1. CAPITALIZE TITLES**

The first and last words of a literary title should be capitalized, as should all words in-between except articles (*a, an, the*), short prepositions, and coordinating conjunctions:

> *The City of Women*
> *The Call of the Wild*

However, when an article follows a stop in the title (such as a colon or comma), it is usually treated as a second "first" word and capitalized:

> *Charles Dickens, The Last of the Great Men*

Remember that the titles of works published or presented separately (books, magazines, plays, long poems, films) are italicized (underlined), while those published as part of something larger are set in quotes (articles, short stories, most poems, and also television and radio programs).

Cap./2   **2. CAPITALIZE THE FIRST WORD OF A QUOTATION**

The opening word of quoted speech is capitalized, whether it begins a sentence or not. However, when a quotation is broken, the first word of the continuation is not capitalized unless it is a proper noun or adjective or begins a new sentence:

He said, "We liked the movie very much."
"We," he said, "liked the movie very much."

With written quotations capitalization of the first word depends upon whether the quotation is introduced after a stop or is worked into the sentence as a noun clause following *that*. In the former case it begins with a capital; in the latter, it does not, even though it may have done so in the original:

G. K. Chesterton writes: "This is the real vulgar optimism of Dickens. . . ."

G. K. Chesterton writes that "this is the real vulgar optimism of Dickens. . . ."

Cap. / 3

3. CAPITALIZE PROPER NAMES AND ADJECTIVES

A proper name is the designation of a particular person, place, structure, and so on. A proper adjective is a modifier derived from such a name.

Specific people
Harry Jones, Mary Winter, C. S. Lewis

When the name includes a particle, the particle should be spaced and capitalized (or lower cased) according to accepted usage for that name:

Gabriele D'Annunzio
Charles de Gaulle

Nouns, verbs, and modifiers derived from proper names are not capitalized when used in a sense generalized from their origin:

Charles Mackintosh BUT a mackintosh coat
the French language BUT french doors

But if a proper adjective is used in a specialized sense closely related to the name from which it derives, it should be capitalized:

He had a de Gaullean sense of country.

Personal titles
Capitalize these when they are part of a name but not otherwise:

Judge Harry Jones BUT Harry Jones was made a judge.
Professor Mary Winter BUT Mary Winter became a professor.

National and racial groups and their languages

Amerindian	Mexican
Australian	Polish
German	Swahili

Places: continents, islands, countries, regions, and so on

China, Chinese	North America, North American
Europe, European	Manhattan, Manhattanite
the East Coast	42nd Street
New Jersey, New Jerseyan	the North Pole

When a regional name is a common term given specific application (like the Middle West of the United States), an adjective derived from it may or may not be capitalized. Consult a dictionary or style manual for specific cases:

the Far East, Far Eastern history
the Middle West, middle western cities

Structures: names of buildings, bridges, and so on

the Brooklyn Bridge
the Empire State Building

Institutions and businesses

Kearny High School BUT a high school in Kearny
Columbia University BUT a university in the city
the Boston Symphony Orchestra BUT a symphony orchestra
General Motors BUT the motor industry

Governmental agencies and political parties

the U.S. Congress BUT a congressional district
the Supreme Court BUT a municipal court
the Democratic Party BUT democratic countries

School subjects and courses

The subjects you take in college or high school are not capitalized unless they derive from proper nouns (this means language courses only):

anthropology BUT	English
chemistry	French
history	German
philosophy	Latin

Names of particular courses, however, are capitalized since they are, in effect, titles:

biology BUT Biology 201
physics BUT Physics 101

Cap./4

4. PERSONIFICATION

When personified (that is, endowed metaphorically with human qualities) abstractions such as *peace, war, winter* are capitalized. In their
conventional uses they are not:

We had a late spring last year.

Last year Spring arrived reluctantly, hanging her head and dragging her feet.

REFERENCE GRAMMAR

Reference Grammar Contents

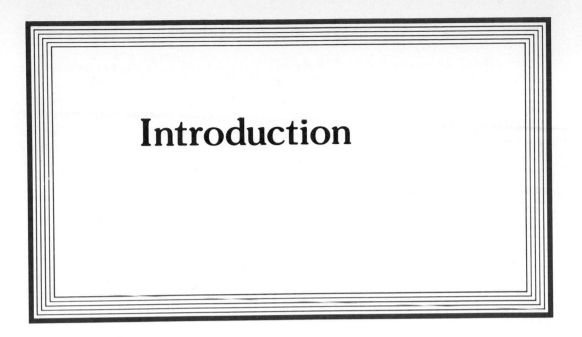

Introduction

The grammar of a language is simply the set of rules by which it works. These are "rules" in a special sense. For one thing, they inhere in the language itself, generalizations of how the language operates, how people actually use it. Grammarians do not dictate the rules; they only describe them, as physicists describe the laws which govern falling bodies.

For another, the rules do not imply approval or disapproval. This follows from our first point: grammarians, committed to observing and systematizing the facts of a language, cannot allow personal preferences to influence what they say any more than physicists can. The rules of grammar, then, do not "rule" that a word or expression is right or wrong; they only say that it exists (or does not exist) and describe its form and functions.

Finally, the rules are not absolute; they change with time. In this they are unlike the laws of physics, which do not alter from one century to another (though our understanding of them may). Rules of grammar grow out of the language. Any language is the creation of many people, and people change, in language habits as in life styles. Grammar necessarily keeps pace. But changes in grammar are slow and may be disregarded in the practical matter of describing a language as it exists at any given moment.

Some people cannot get used to this conception of grammar. They continue to think in terms of "right" and "wrong," in terms of arbitrary laws imposed by authorities. They agree that "I ain't got no money" is "ungrammatical" (even if they say it themselves). As we use the term, however, "ain't got no" is "grammatical" simply because it is a form used by many speakers of English.

Yet the fact that it is grammatical does *not* mean that "ain't got no" is always permissible. In a written composition, the expression violates usage, those narrower rules governing how we should use language in specific social situations. In fact no one who wishes to be well thought of ought to write "ain't got no" in an English theme. The mistake is very real, and calling it an error of usage rather than one of grammar does not lessen its seriousness. What the distinction does do is to draw a useful line between two different sets of rules, and to prevent a short-sighted and misleading view of what grammar is.

The review of grammar that follows has two major sections: surveys first of the parts of speech, and second of the grammar of the sentence—that is, the various constructions which enter into the composition of the sentence.[1]

---

1. In recent years the study of grammar has made much progress and undergone considerable change. Many grammarians are dissatisfied with the traditional, or conventional, grammar of English, which derives from that of Latin. In its place they have proposed more exact analyses of the language, in terms of structural or transformational grammars, for example.

For the moment disagreement exists about how to describe the English language and what terms to use. For example, traditional grammarians distinguish eight parts of speech; others analyze the matter differently and speak of "form classes" and "function words." This book uses the terminology of traditional grammar, not because it is superior, but because it is still the most common currency of grammatical teaching.

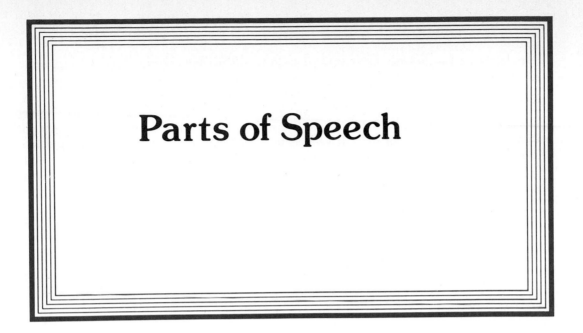

Parts of Speech

Parts of speech are grammatical classes of words determined by how the words function in a sentence (or in smaller groups such as clauses and phrases). Traditional grammarians distinguish eight parts of speech for English:

verbs	adjectives	conjunctions
nouns	adverbs	interjections
pronouns	prepositions	

For convenience we often talk about words as if they were invariably and unalterably nouns and verbs and adjectives and so on. Thus we might say that *eye* is a noun, as in:

His left eye was bruised.

More precisely, we should say that "*eye* usually acts as a noun but may serve as a verb or an adjective, as in these sentences":

They eye strangers suspiciously.

She has eye strain.

Grammarians determine the part of speech assigned to any word by observing the inflections, or endings, that it takes, and the various positions it occupies in phrases, clauses, and sentences. In the first of the three sentences above, the position of "eye" between "his left" and "was" tells us that it is a noun. In the second, the word is marked as a verb by coming after "they" and before "strangers." In the

667

following cases the inflections -*s* and -*d* help to mark the noun and verb functions respectively:

His eyes were bruised.
They eyed strangers suspiciously.

Dictionaries identify the part of speech in which a word commonly acts, and if, like *eye,* it may serve in several capacities, they note that fact as well.

Verbs

INTRODUCTION

The most complicated part of speech is the verb. Traditionally verbs are defined as naming actions or states of being: "People *walk*"; "People *live*." More technically, grammarians describe verbs in terms of the forms they may take and how they fall into patterns with other words. Thus verbs change to show singularity in the third person of the present tense by adding *-s* ("John swim*s*"); they indicate the past tense by adding *-ed* to the infinitive or changing the internal vowel ("People walk*ed*"; "John sw*a*m"); they form a past participle by adding *-ed*, changing the vowel again, or adding *-en* to the past (with or without a second vowel change) ("walk*ed*," "sw*u*m," "br*oken*," "ch*osen*"); and they form a present participle by adding *-ing*. Verbs also characteristically occur before and after certain kinds of constructions.

By changing their forms, verbs signal various states or conditions of reality. These changes occur in six different grammatical categories—person, number, tense, aspect, mood, and voice. In some languages the alterations are made by adding different endings to the verb stem (Latin and Greek); in others the changes are effected by adding other verbs called "auxiliaries" to form a variety of verb phrases.

AUXILIARY VERBS

English uses both methods, though it relies more upon auxiliaries. The auxiliary verbs of English include:

am (is, are, was, were)	do (does)	ought
	have (has, had)	shall
	may	should
can	might	used

could	must	will
dare	need	would

Five of these auxiliaries—all the forms of *be* and *dare, do, have,* and *need*—are full verbs in their own right and may be used alone. The others function only as auxiliaries and are distinguished by the fact that they do not use *-s* in the third person, singular present:

He must go.
He might go.
COMPARE: He has to go AND he does go.

A second mark of the pure auxiliaries is that they do not have conventional past forms, though in a few cases the deficiency has been made good by using some auxiliaries as the past tenses of others:

PRESENT	PAST
can	could
may	might
shall	should
will	would

PRINCIPAL PARTS

To make a verb phrase showing, say, a particular tense, an auxiliary is combined with the appropriate form of the main verb. These forms are called the *principal parts.* Conventional grammar distinguishes four: the *infinitive,* the *past,* the *past participle,* and the *present participle.*

On the basis of how they make their principal parts, verbs are classified as regular, irregular, and defective. In all *regular* verbs the pattern is the same: the infinitive serves as the stem and to it are added *-ed* to form the past, *-ed* to form the past participle, and *-ing* to make the present participle:

INF.[1]	PAST	PAST PART.	PRESENT PART.
(to) walk	walked	walked	walking

Irregular verbs follow different patterns. Most commonly they change their internal vowel, and some add *-en* to form the past participle:

(to) swim	swam	swum	swimming
(to) choose	chose	chosen	choosing

The verb *be* uses for principal parts what were originally different verbs:

---

1. The infinitive is often used with *to,* and this form is called the *periphrastic,* or *marked,* infinitive.

| (to) be | was/were | been | being |

Notice that to make the present participle, both regular and irregular verbs do the same thing: add *-ing* to the infinitive.

Defective verbs are those which do not have all the principal parts. As we have seen, the pure auxiliaries are defective. For example:

| must | _____ | _____ | _____ |
| may | might | _____ | _____ |

If you are unsure of the principal parts of any verb, look in a dictionary. For irregular verbs the principal parts are listed in boldface or in small capitals immediately after the entry (the entry itself is the infinitive):

ring / rang / rung / ringing

If for an irregular verb the past and the past participle have the same form, it is printed only once:

swing / swung / swinging

but it is understood that the one form stands for both parts.

For regular verbs the entry is *not* followed by any principal parts. Their absence is a sign of regularity, and you know that they are formed on the *-ed, -ed, -ing* pattern.

From the principal parts and, usually, an auxiliary are made all the forms of a verb that show differences in person, number, tense, aspect, mood, and voice.

PERSON

Person distinguishes among speaker, spoken-to, and spoken-about. These are called respectively the *first person* ("I"), the *second person* ("you"), and the *third person* ("he, she, it"). English verbs have only one ending for person (*-s* or *-es*) and that only in the third person singular of the present active indicative:

I	stand/go
you	stand/go
he, she, it	stands/goes
we	stand/go
you	stand/go
they	stand/go

An exception is the verb *be*, which has different forms for all three persons in the singular present indicative and for the second person singular in the past:

	PRESENT	PAST
I	am	was
you	are	were
he, she, it	is	was
we	are	were
you	are	were
they	are	were

In addition a distinction is sometimes marked in the future and future perfect tenses by the auxiliaries *shall* and *will*. The rule—of usage, not of grammar—is that to show simple futurity *shall* is used in the first person (both singular and plural) and *will* in the second and third persons. To show determination the pattern is reversed: *will* with the singular and plural forms of the first person and *shall* with both forms of the second and third:

	SIMPLE FUTURE	DETERMINATION
I	shall go	will go
you	will go	shall go
he, she, it	will go	shall go
we	shall go	will go
you	will go	shall go
they	will go	shall go

The rule, however, is often ignored by modern writers. A careful, formal stylist observes it; most others do not, using *will* for all three persons.

NUMBER

Number distinguishes oneness and more-than-oneness. In English only those two distinctions are marked, called the *singular* and the *plural*. (Some languages possess three numbers: singular, one; dual, two; and plural, more than two.)

English verbs do not have an ending for the plural form; the only inflection associated with number is the *-s* or *-es* distinguishing the third person singular.

Again, however, the verb *be* is an exception, both in the present and in the past, as a glance at the forms above shows.

TENSE

Tenses are forms of verbs designating particular intervals of time, and they are the most complicated part of our verb system. Since time may be perceived and analyzed in many different ways, the tense systems of languages often differ. For English, traditional grammar distinguishes six tenses: present, past, future, present perfect, past perfect, and future perfect.

The first three are sometimes called primary tenses. Their point of reference is the moment of speaking or writing. The *present* designates an action or state of being that is going on at the time of speaking:

I *go* to school.

The *past* tense refers to an action or state of being that is now over:

I *went* to school for twenty years.

And the *future* to an action or state of being which is not yet begun:

I *shall go* to college in the fall.

The perfect tenses are more complicated because their reference points in time are not all the same. For the *present perfect* the reference remains the moment of speaking, and the tense designates an interval of time extending from some point in the past to that moment:

I *have gone* to school for three years.

The present perfect implies nothing about the ending of the action or being, which may be over at the time of speaking or which may continue indefinitely into the future.

For the *past perfect* the reference point lies in the past; the tense refers to something that is over in relationship to a particular past time. In other words, the past perfect shows past within the past:

By March of last year he *had earned* five thousand dollars.

For the *future perfect* the reference point shifts to the future, and the tense designates something that will have been completed by a specified point in the future—in other words, past within the future:

By this time tomorrow night I *shall have been* home for two hours.

The present tense is formed from the infinitive, the only inflection being the *-s* or *-es* added to the third person singular indicative. The past tense is formed from the second principal part (called the past, or preterite) with no additions, and the future from the infinitive preceded by *shall* or *will* (these words being selected, in strict usage, according to the convention mentioned on page 672).

All the perfect tenses are made in the same way, the past participle preceded by the appropriate tense of the auxiliary *have*:

PRESENT PERFECT   I have gone.
PAST PERFECT   I had gone.
FUTURE PERFECT   I shall have gone.

Like that of every language, the tense system of English has holes, intervals of time which it does not mark but which occur often enough in thinking and writing to require designation. Idioms develop to fill these needs. The most common is to show future in the past, and the idiom takes the form either of *was/were* + the infinitive of the main verb (with *to*), or of *would* + the infinitive (without *to*):

> In 1540 no one guessed that the seven-year-old Elizabeth *was to become* [or *would become*] one of England's greatest monarchs.

Here the point of reference is in the past (1540) and the action (Elizabeth's monarchy) is in the future with regard to that point.

ASPECT

Aspect concerns the nature of the action named by the verb, specifically whether that action is completed, in process, just beginning, done emphatically, or done repeatedly. Although it is sometimes classed with tense, aspect does not refer primarily to time. Aspect is not as fully developed in English as in some other languages, but it exists and it poses problems.

The most common problem concerns the *progressive* (or durative) *aspect*. Progressive verbs signify an action conceived of as on-going, whether in the past, present, or future. Usually the "on-goingness" is understood with reference to some other event:

> Yesterday, when the phone rang, I *was reading* a book.

Here the progressive pinpoints the action of reading with the event of the phone's ringing. Even when the other event is not actually referred to, it is generally understood in the context. If we hear someone saying to another person, "I was walking the dog," we assume that it is in reference to some specific time or occurrence such as "at two o'clock" or "when you called."

Progressive verbs are formed by taking the present participle (*walking*, for instance) and adding the appropriate tense of *be*:

PRESENT	I am walking
PAST	I was walking
FUTURE	I shall be walking
PRESENT PERFECT	I have been walking
PAST PERFECT	I had been walking
FUTURE PERFECT	I shall have been walking.

There is no commonly accepted name for the aspect of verb forms like those in "I *walk*, I *walked*, I *shall walk*, I *have walked*," and so on. We shall describe such verbs as having *simple aspect*. Exactly how the simple and progressive forms of a verb in the same tense differ from one another is not easy to explain. Generally, verbs of simple aspect refer to a relatively large and indefinite time period and to an action or state of being that is either completed or potential or, if ongoing, is so as a fundamental condition of existence ("I live, I breathe, I go to school"). A professional entertainer, asked what she does for a living, might reply, "I sing," even though she is not at that moment actually singing. As she may have sung professionally for ten years and hope to continue for ten or twenty more, the present time designated by the verb is extensive and inexact. On the other hand, if you are hollering one evening in the bathtub and your mother calls in exasperation, "What are you doing?" you will not answer, "I sing," but rather, "I'm singing," because (even though you have stopped to answer) you are referring to the action going on at the instant of your mother's question.

Foreigners learning English frequently make mistakes with the progressive verbs, applying to them rules of aspect which work in their own language but not in ours.

English verbs also have an *emphatic aspect,* formed for the present and the past (and only those two tenses) by *do/does* or *did* + the infinitive (without *to*):

I *do sing.*
She *does sing.*
We *did sing.*

In speech the emphatic form is often strengthened by placing *so* or *too* between the auxiliary and the main verb:

I *do so sing.*
She *did too sing.*

To show other kinds of aspect, English relies upon idiomatic verb phrases. Thus an action just beginning or about to begin (the *inchoative aspect*) is signaled either by *get/got* + the present or past participle, or by the appropriate form of *be* + *going* + the *to*-infinitive of the main verb:

We'll *get started* on the job tomorrow.
She *is going to start* college in the fall.

Repetitive aspect is shown by a verb phrase using the appropriate tense of *keep* or *keep on* + the present participle of the main verb:

He *keeps going* from job to job.
They *kept on complaining* until something was done.

MOOD

Mood indicates how a speaker thinks of his or her utterance. An utterance may be regarded as a statement (*indicative* mood); a question (*interrogative*); a command (*imperative*); or a possibility, hope, wish, or condition contrary to fact (*subjunctive*).

English verbs do not change very much to show mood. Most sentences are statements, and the verbs, whatever their tense or aspect, are in the *indicative mood*:

> I sing.
> I was singing.
> I did sing.

In the *interrogative mood* the form of the verb itself does not change, but the position of the auxiliary does; it is placed before the subject, while the main verb comes after:

> Is he going?
> Did you go?
> Have the men gone?

Since verbs of simple aspect do not have auxiliaries in the present and past tenses, these forms are not employed for questions in modern English. We never ask:

> × Go you?
> × Went you?[2]

Instead we use a verb phrase with an auxiliary, usually either the progressive or the emphatic form:

> Are you singing?/Do you sing?
> Were you singing?/Did you sing?

These alternatives are not always exact equivalents. For example, while both "Are you singing?" and "Do you sing?" can mean: "Are you planning [*or* scheduled] to sing?" the first can also ask: "Is that noise you are making singing?" and the second: "Are you able to sing?"

In the future and in all the perfect tenses, the simple aspect does have auxiliaries and therefore may serve for questions. The progressive forms also act interrogatively in these tenses (the emphatics exist only for the present and past). Here again the two possibilities create options for subtle distinctions:

> Will you sing?/Will you be singing?
> Have you sung?/Have you been singing?

It is possible to ask questions without relying upon the interrogative mood—that is, without changing the position of the auxiliary verb.

---

2. An exception is the verb *be* in the present and past, where the one and only verb is set before the subject:
   Are you there?
   Were they at home?

Two other signals exist. The first is supplied by the words *who, which, what*:[3]

Who will be there?
What is going to happen?
Which house burned?

The second signal is a particular pattern of intonation, a kind of voice melody created by changes in pitch and slight pauses. Thus in speech we might ask:

Helen is arriving tomorrow?

indicating the question by slightly raising the pitch of the voice and ending the sentence on a rise. In writing, of course, intonation cannot be heard and we have no common symbol for it. A written question in this form must rely exclusively upon the question mark. Such questions, while common in speech, are infrequent in composition, though they occasionally occur in styles that aim at informality. The historian D. W. Brogan, for instance, begins a paragraph about General U. S. Grant with a fragment in this form:

Grant's failures?

The *imperative mood,* used for commands, takes the form of an initial verb without a subject (the sentence may be ended with an exclamation point):

Shut the door.
Come here!

Sometimes the verb is accompanied by a noun or pronoun of direct address, which may either precede the verb or be postponed to the end of the clause:

You come here.   / Come here, you.
Bob, listen to me. / Listen to me, Bob.
Soldier, stand up. / Stand up, soldier.

The imperative mood occurs only rarely in composition—usually as a signpost or transitional formula (and in a less parade-ground tone than "Come here!"):

Let us turn next to the political consequences.
Allow me to explain.

Only traces remain in English of the *subjunctive mood,* though it is more extensive in some other languages. The subjunctive signals

---

3. Relative adverbs like *when, where, why,* however, do require inverted word order: "When are we leaving?"

that the statement is to be understood not as a fact or potential fact, but as a wish, hope, possibility, condition, or even a nonfact. One occasionally sees a subjunctive verb form in sentences like:

If I *were* to go, would you be there?
If he *run* in tomorrow's race, has he any chance?

Notice that how the speaker regards it, *not* whether or not the statement *is* factual, determines the use of the subjunctive. One would use the indicative mood, not the subjunctive, to say "The moon is made of green cheese" because the statement is asserted as a fact, even though it is not.

The only changes in form still used to indicate the subjunctive involve *be* and the third person singular present of other verbs. In the case of *be*, the present forms *am, are,* and *is* may be replaced by *be*, and the past form *was* by *were*:

	INDICATIVE	SUBJUNCTIVE
I	am/was	be/were
you	are	be
he, she, it	is/was	be/were
we, you, they	are	be

In other verbs the *-s* is dropped from the third person singular present:

INDICATIVE	SUBJUNCTIVE
he, she, it *seems*	if he, she, it *seem*

It is also possible to invert the subjunctive *were,* placing it before the subject and dropping *if*:

Were I to go, would you be there?

The same pattern of inversion may also be used with *had* in the past perfect tense:

had I gone to the party = if I had gone to the party

But many of these vestiges of the subjunctive are disappearing. Nowadays one rarely hears or sees *be* instead of *am, are,* or *is.* Careful writers still prefer *were* in place of *was,* though many people are content with "If I was." The *s*-less subjunctive in the third person singular present is still also common, especially in *if*-clauses, but here too writers by no means ignorant of grammar will use the indicative form: "If he runs in the race tomorrow."

There are a few places, however, where the subjunctive form should be retained. One is in traditional idioms, such as:

as it *were*   God *save* America
be that as it may   heaven *help* us
come what may

Another is a *that*-clause of condition or demand:

The university requires that a student *drop* a course within six weeks to avoid failure.

Aside from these two uses, the subjunctive is not a vital issue. Insisting upon *be* is decidedly old-fashioned. Using the other subjunctives is more a matter of tone than of clarity. Now and then a shade of meaning may be helped by a *were* or a subjunctive without *-s*. But usually the conditional or nonfactual nature of the utterance is clearly marked by a word like *if* or by the inversion of *had*. Still, writing "if I were" or "if he run" is a nicety of style, and there is no reason not to observe such subjunctives if you wish. Like chicken soup, they can't hurt.

The possibilities of mood are extended considerably if we include the *modal auxiliaries*. There are ten of these words:

can	ought to
could	shall
may	should
might	will
must	would

The past and present forms of *be* and *have* can also serve as modals.

In a strict sense these auxiliaries are not a matter of mood, but verb phrases employing them have much the same effect. Combined with either the present or perfect infinitive of the main verb (for example, *go/have gone*) the modals express a wide and subtle range of meaning, particularly with regard to future action:

I *can go* tomorrow.	I *was to go* tomorrow.
I *could go* tomorrow.	I *was to have gone* tomorrow.
I *shall go* tomorrow.	I *have to go* tomorrow.
I *should go* tomorrow.	I *had to go* tomorrow.
I *must go* tomorrow.	I *should have gone* tomorrow.
I *will go* tomorrow.	I *would have gone* tomorrow.
I *would go* tomorrow.	I *could have gone* tomorrow.
I *may go* tomorrow.	I *might have gone* tomorrow.
I *might go* tomorrow.	I *ought to have gone* tomorrow.
I *ought to go* tomorrow.	I *go* tomorrow.
I *am to go* tomorrow.	I *went* tomorrow. (= I was to have gone.)

These numerous forms, all "futures" in the sense that they refer to an event yet to occur, convey real differences in meaning. For example, "I am to go" expresses a future action planned or predetermined; "I was to go," a planned future that is now doubtful, even canceled; and "I was to have gone," a planned future definitely canceled. Native speakers use and respond to the subtleties of the modal forms with an intuitive sense of their implications. Foreigners learning English find them very difficult.

VOICE

Voice concerns whether the action named by the verb is determined
by the subject or by something else. When the subject is the determiner,
or actor, the verb is in the *active voice*. When the subject is acted
upon and the action originates elsewhere, the verb is in the *passive
voice*:

ACTIVE   John *threw* the ball.
PASSIVE   The ball *was thrown* by John.

Passives are formed from the past participle by adding as an auxiliary
the appropriate tense of *be*:

PRESENT   is thrown
PAST   was thrown
FUTURE   will (shall) be thrown
PRESENT PERFECT   has been thrown
PAST PERFECT   had been thrown
FUTURE PERFECT   will (shall) have been thrown

In the progressive aspect, passives are formed only for the present
and past tenses. The formula is: the appropriate tense form of *be* +
being + the past participle of the main verb:

PRESENT   is being thrown
PAST   was being thrown

In passive constructions the originator of the action may or may
not be expressed in the sentence. It is in "The ball was thrown by
John," but it is not in "The ball was thrown." In sentences such as
the latter, the agent initiating the action should be clear from the
context.

Passive constructions require more auxiliaries than do active ones.
Moreover they are unemphatic. For those reasons the active voice is
preferable in composition. Avoid such awkward passives as "That course
will be taken by me in the fall." Instead use an active construction:
"I shall take that course in the fall." It is clearer, briefer, and stronger.

Sometimes, however, clarity or emphasis requires the passive. "The
tree was struck by lightning" is better than "Lightning struck the tree"
if the primary subject is the tree.

VERBALS

Most of the verbs we use are finite—that is, limited with reference
to person, number, and tense. We do not say, for instance, × "He *am*
going today," × "We *is* here," or × "I *have gone* tomorrow." Verbs
like *am, is, have gone* cannot be used indiscriminately with any person
or number or tense.

Some verbs can be. These nonfinite forms are called *verbals,* and they include the infinitive, the participle, and the gerund.

The *infinitive* is the stem of the verb; it is the boldface entry in the dictionary: *be, have, take.* It may also occur with *to: to be, to have, to take,* a form called the *periphrastic,* or *marked, infinitive.* With most modal auxiliaries the bare infinitive is used (*I would go, I should go, I must go*); *ought,* however, is an exception (*I ought to go*). Other verbs use the periphrastic infinitive (*I wanted to go, I hoped to go*); a very few may take either (*I helped paint the house* or *I helped to paint the house*).

The infinitive takes different forms to show distinctions in tense, aspect, and voice. These are, however, far less numerous than the forms of the finite verb:

PRESENT SIMPLE ACTIVE INFINITIVE	(to) take
PRESENT SIMPLE PASSIVE INFINITIVE	(to) be taken
PRESENT PROGRESSIVE ACTIVE INFINITIVE	(to) be taking
PRESENT PROGRESSIVE PASSIVE INFINITIVE	(to) be being taken[4]
PERFECT SIMPLE ACTIVE INFINITIVE	(to) have taken
PERFECT SIMPLE PASSIVE INFINITIVE	(to) have been taken
PERFECT PROGRESSIVE ACTIVE INFINITIVE	(to) have been taking
PERFECT PROGRESSIVE PASSIVE INFINITIVE	(to) have been being taken[4]

The *participle* also exists in several forms, but only for the simple aspect. We do not have participles in the progressive.

PRESENT ACTIVE PARTICIPLE	taking
PRESENT PASSIVE PARTICIPLE	being taken
PAST ACTIVE PARTICIPLE	taken
PERFECT ACTIVE PARTICIPLE	having taken
PERFECT PASSIVE PARTICIPLE	having been taken

The *gerund* is the participle used as a noun. Most often it is the present active which is used in this way:

Jogging is good exercise.

Now and then a present passive participle may act as a gerund:

Being congratulated is a new experience for me.

What has become the modern gerund was originally an abstract noun made by adding an ending to the verb, much as we still use a suffix like *-ment* to turn a verb into a noun (*encourage/encouragement*). By chance that ending changed in sound and spelling until it came to be identical with the ending of the present active participle, *-ing.* Thus the two forms fell together, and this has led to the habit of treating

4. These progressive passive forms rarely occur.

the gerund as a kind of semiverb, speaking of its "subject" and describing it, not really accurately, as "the participle acting as a noun." The origin of the gerund explains why its subject is in the possessive case. Just as we say "his encouragement" and not "he encouragement," so we say "His winning was a surprise" and not "He winning was a surprise." Some grammarians prefer to avoid the confusion created by the falling together of the participle and the gerund by simply referring to both as the "-*ing* form."

Verbals are an important resource in shaping sentences and function in several ways:

1. As parts of verb phrases:
   INFINITIVE   He will *go*. / He likes *to go*.
   PARTICIPLE   She is *going*.
2. As nouns:
   INFINITIVE   *To jog* is good exercise.
   GERUND   I like *jogging*.
3. As modifiers:
   PARTICIPLE   The *burning* house could be seen for miles.
   INFINITIVE   They have money *to burn*.

In later sections we shall see in more detail how infinitives, participles, and gerunds function in the sentence.

TRANSITIVE AND INTRANSITIVE VERBS

Aside from being finite and nonfinite, regular and irregular, verbs may be classified in several other ways. An important distinction is between transitive and intransitive verbs, or rather between the transitive and intransitive uses of verbs, since many can act in either way.

A *transitive verb* has an object—that is, a word or construction indicating who or what receives the action named by the verb:

He *loves* candy.   object

They *think* that she will win.   object

An *intransitive verb* does not have an object:

He *walks* to town.
They *sleep* late.

Here "to town" and "late" are adverbial modifiers, not objects naming something being walked or slept.

Many verbs can function in either fashion:

TRANSITIVE   She *drove* the car.
INTRANSITIVE   She *drove* round the corner.

Even verbs commonly used intransitively may be given special transitive meanings:

He *walks* the dog.
They *are sleeping* the long sleep of death.

A special case is provided by three pairs of verbs existing in both a transitive and an intransitive form:

set/sit   lay/lie   raise/rise

The first verb in each pair is transitive; the second, intransitive:

object
She *set* the table.   / They *sit* around the table.

object
They *laid* the blanket on the ground.   / The blanket *lies* on the ground.

object
We *raised* the flag.   / The sun *will rise* about 5:30.

Formal usage requires that the transitive/intransitive distinction be observed for these six verbs. With *lay/lie* the problem is compounded by the fact that *lay* is the form both of the infinitive of the first and of the past of the second:

lay/laid/laid/laying
lie/lay/lain/lying

The accepted formal usage of *lay* is illustrated by these sentences:

TRANSITIVE, PRESENT   They lay the cloth on the table.
INTRANSITIVE, PAST   She lay quietly in the corner.

LINKING VERBS

Linking verbs (also called *copulas*) are a special type of intransitive verb. They have what looks like an object:

John *is* my brother.
My house *is* white.

But actually the object (or "complement") does not receive the action of the verb. Rather it is either another name for the subject—as "my brother" is another name for "John"—or else an attribute characteristic of the subject—as "white" is an attribute of "my house."

In the first case, when the complement is in effect another name for the subject, the complement is called a *predicate noun*. The linking verb acts like an equal sign:

John = my brother

In the second case, when the complement is an adjective naming a quality belonging to the subject, the complement is called a *predicate adjective*.

If it has a complement, the verb *be* is always a linking verb. Other common copulas (or verbs which are frequently used as copulas) include:

seem   sound
become   feel
remain   turn
look

When the complement of a linking verb is a pronoun, formal usage requires the pronoun to be in the subjective case:

It is *I*.
That was *she*.

Informally, however, we generally use the objective form:

It is *me*.
That was *her*.

COMMON ERRORS WITH VERBS

Vb. form

Mistake in verb form   Be sure that you are using the proper principal part and auxiliary. The past participle of *lend*, for example, is *lent*, not × *lended*:

He lent me five dollars.
NOT × He *lended* me five dollars.

The past of *go* is *went*, not *gone*; but the past participle is *gone*, not *went*:

We went to the party.
NOT × We gone to the party.
We had gone to the party.
NOT × We had went to the party.

Principal parts, remember, are given in the dictionary, along with other grammatical information, such as whether a verb is used transitively, intransitively, or in both ways.

Another common mistake in verb form involves the auxiliary *have*. It should never be written *of*:

× We would *of* gone to the party.

In speech we rarely say the full word *have* but instead reduce it to an unstressed sound which is indistinguishable from the sound of *of*. In writing, however, that reduction is properly represented as a contraction:

We would've gone to the party.

Actually even the contraction should be avoided in formal composition (unless you are representing speech); the full auxiliary should be spelled out:

We would have gone to the party.

Tense base

Mistake in base tense. When you write, you must select one tense as a base—that is, to serve as your primary point of reference. Most often it will be the present or the past, though occasionally it may be the future or the present perfect (the past and future perfects are unlikely to serve as starting tenses). Use the present to refer to an activity or state of being that is current, the past for something that is over, the future for what is yet to be, and the present perfect for something extending from a past time to the present moment:

PRESENT	Colleges *are* an important part of American education.
PAST	A hundred years ago colleges *were* less numerous.
FUTURE	By the year 2000 colleges *will be* very different.
PRES. PERF.	Colleges *have changed* considerably in the last fifty years.

A special case in the selection of a base tense is the *historic,* or *narrative, present,* in which a past event is described in the present tense. The short story writer Damon Runyon was fond of using the narrative present:

It is a matter of maybe four years since I see this Joey Perhaps until I notice him on a train going to Boston, Mass., one Friday afternoon. He is sitting across from me in the dining-car, where I am enjoying a small portion of baked beans and brown bread, and he looks over to me once, but he does not rap to me.

Even in more formal situations the narrative present is occasionally used. The historian Hilaire Belloc writes about the early career of Thomas Cranmer (Archbishop of Canterbury under Henry VIII):

At any rate, he becomes Anne Boleyn's Chaplain and is sent to plead the cause of the divorce before the Holy See. . . .

Despite this example, the narrative present is not often employed in serious historical writing. It is, however, when writing about literature. By convention, characters in a story or play are understood to exist in an eternal present (as in fact they do), and are referred to in the present tense:

Hemingway's characters are often alienated Americans, who find little value in the middle-class world. They have lost faith in the America they see around them, indeed in the whole of Western civilization.

Similarly when you refer to a writer as writer use the present, or the present perfect, even though he or she be dead:

Hemingway is [OR has been] one of the most influential novelists of the twentieth century.

But any reference to a writer as a person is cast in the appropriate tense:

Hemingway died in 1961.
Eudora Welty lives in Mississippi. She wrote [OR has written] *Delta Wedding* and many other fine novels and stories.

Tense shift

Tense shift   Whichever tense you select as basic, you should use it consistently. This means, first, that you must continue to employ that tense as long as you are referring to the same time period. Here, for instance, is an unjustifiable shift:

By the year 2000 colleges will be very different. For one thing they × have [SHOULD BE will have] fewer students.

Second, consistency means that when you do refer to a different interval of time you must shift to the appropriate tense:

present   present
I *like* college. I *find* it neither so impersonal nor so difficult as I

past   present   future
feared. I *expect* I *shall continue* to like it.

If the writer had begun that passage using the past as the basic tense, he or she would have had to shift to the past perfect and a "past future":[5]

past   past
I *liked* college. I *found* it neither so impersonal nor so difficult

past perfect   past   "past future"
as I *had feared.* I *expected* that I *should continue* to like it.

Failing to make the appropriate tense shift is especially frequent with the past perfect—that is, when you must refer to a past within a past. In the following example, the reference to the "Colonies" (the

5. For the "past future" see page 674.

United States before 1783) makes it clear that the writer is alluding in the second sentence to a more remote past than he designates in the first sentence; therefore the past perfect is called for:

In nineteenth-century America, colleges were far less numerous. Colleges × began [SHOULD BE had begun] in the Colonies as religious schools.

Tense seq.

Tense sequence A problem with tenses sometimes occur in a dependent clause governed by a main clause. The problem involves sequence of tenses, and the general rule is that when the tense of the main clause is either present or past, that of the dependent clause follows suit:

present   present
He *wants* to go because he *hopes* to find his friend.

past   past
He *wanted* to go because he *hoped* to find his friend.

However, there are many exceptions to the rule. When the main clause is in the future, repeating the future in the dependent clause is awkward. It is preferable to shift to the present:

future   present
He *will go* because he *hopes* to find his friend.

And so too when the main verb is in either the present perfect or the future perfect tense:

pres. perfect   present
He *has gone* because he *hopes* to find his friend.

future perfect   present
He *will have gone* because he *hopes* to find his friend.

If the main verb is past perfect, the subordinate tense is past:

past perfect   past
He *had gone* because he *hoped* to find his friend.

Another exception to the general rule of tense sequence occurs when the main clause and the dependent clause express the result of a hypothetical or nonfactual condition:

dependent: present   main: future
If you *fool* with me, you *will be* sorry.

dependent: past perfect   main: conditional future perfect
If you *had fooled* with me, you *would have been* sorry.

Actually the sequence of tenses here is logical: the result is expressed in a future form because it will—or would—follow the condition stated in the dependent *if*-clause.

Nouns

Traditionally nouns are defined as the names of persons, places, or things. More precisely they may be defined in terms of their inflections— that is, changes in their endings—and their positional relationships with other words. Nouns are inflected for the plural and the possessive forms.

PLURAL FORMS

The great majority of English nouns form their plurals by adding *-s* or *-es*. A few add *-en* instead or change the internal vowel, and a few use foreign plurals or have no plural form at all.

If you are in doubt about the plural of a noun, consult a dictionary. No plural form will be listed if it is the comman *-s* or *-es*. If the plural is one of the other forms, the dictionary will give it with the identifying tag *"pl."*

1. The -s plural This is the most common plural ending:

girl/girls   book/books
boy/boys   sample/samples

A special problem occurs with the plural forms of names, italicized words, citation terms (words used as nouns),[1] letters, and numbers. Names, italicized words, and citation terms add *-s* (or *-es* under the conditions described in a moment) in the conventional manner:

There are four Sams in this class.
We have seven *Pinocchio*s in stock. (Notice that the *-s* is not italicized.)

---

1. A *citation term* is a word used to signify itself rather than its usual referent. In "A horse grazed in the field," *horse* designates the animal; in "*Horse* is a noun," the word is a citation term and refers to itself. Citation terms are indicated by italic type or by being enclosed in quotation marks.

Don't use too many *if*s and *but*s. (Notice that the citation terms are italicized but again not the *-s*.)

A simple *-s* is also the plural ending of numbers and capital letters used as nouns (if the capital plus *-s* is not confusing):

The early 1930s
You should not clutter up a page with (1)s and (2)s.
The town is full of GIs.
He makes his Ns in an odd way.

However, in a few cases the plural form is *-'s*: lower case letters, abbreviations using periods, and capital letters which would be ambiguous in an *-s* plural without an apostrophe:

Mind your p's and q's.
There were several M.D.'s in the audience.
He makes his I's in an odd way.

2. The -es plural   In the following cases *-es* is added to the singular form. Sometimes the *e* is pronounced; sometimes it is silent:

a. Singular nouns ending in *-y* preceded by a consonant. These change the *y* to *i* and add *-es*:

lady/ladies
sentry/sentries

However, proper names ending in *-y* simply add *-s,* as do common nouns ending in *-y* preceded by a vowel:

Harry/Harrys.   guy/guys
Mary/Marys   toy/toys

b. Singular nouns ending in sibilants (*-s, -ss, -ce, -sh, -ch, -tch, -ge, -dge, -x, -ze*):

gas/gases   church/churches
glass/glasses   witch/witches
price/prices[2]   wage/wages[2]
wish/wishes   box/boxes
                   craze/crazes[2]

c. Many singular nouns ending in *-o* preceded by a consonant:

echo/echoes
mosquito/mosquitoes
tomato/tomatoes

---

2. These nouns appear to add only *-s*. However, saying the singular and plural forms out loud reveals that in fact an *-es* sound has been added since the *e* in the singular is silent.

However, some such nouns may take either *-es* or *-s*:

banjo/banjoes OR banjos
cargo/cargoes OR cargos

And some take only *-s*:

dynamo/dynamos
piano/pianos

Nouns ending in *-o* preceded by a vowel add only *-s*:

cameo/cameos
curio/curios

d. Finally, *-es* is the plural form for some nouns ending in *-f* or *-fe* preceded by *l* or by a long vowel (except *oo*). These plurals change the *f* to *v*:

calf/calves   wife/wives
thief/thieves   wolf/wolves

However, when the *-f* is preceded by *oo,* the plural ending is commonly *-s*:

roof/roofs
goof/goofs
hoof/hoofs (*hooves* is also possible)

And words with a final *-f* that have been borrowed from a foreign language (usually French) also add only *-s*:

chief/chiefs
strife/strifes

3. The -en plural   Once a more common plural ending, *-en* remains only in:

child/children
ox/oxen

and in two archaic words:

brethren (an old plural of *brother*)
kine (an old plural of *cow*)

4. Change in the internal vowel

foot/feet   mouse/mice
goose/geese   tooth/teeth

louse/lice   woman/women[3]
man/men

5. *Foreign plurals*   With words borrowed from Latin or Greek the matter of plural forms is confused.

a. Some retain the original Greek or Latin plural:

agendum/agenda   datum/data
alumna/alumnae   erratum/errata
alumnus/alumni   phenomenon/phenomena
criterion/criteria   radius/radii

b. Others are used only with an anglicized plural:

dogma/dogmas
stadium/stadiums

c. And a few have two plurals—the original and an anglicized form:

fungus/fungi OR funguses
curriculum/curriculi OR curriculums
stratum/strata OR stratums

d. Foreign words ending in *-ex, -ix, -yx, -is, -sis* have a plural in *-es.* Those ending in *-is* change the *i* to *e,* and those ending in *-x* may change the *x* to a sibilant *c*:

analysis/analyses   appendix/appendices OR appendixes
crisis/crises   index/indices OR indexes
parenthesis/parentheses   vertex/vertices OR vertexes
thesis/theses

6. *Unchanged plurals*   A few names of animals have no separate form for the plural:

deer/deer
fish/fish (*Fishes,* used in the Bible, is archaic.)
sheep/sheep

However, most nouns designating animals have regular plurals in *-s* or *-es*:

dog/dogs
cat/cats
horse/horses

---

3. *Woman/women* are frequently misspelled. The change of sound occurs in the first vowel, though this is spelled the same (*o*) in both singular and plural. Contrarily, where there is a difference in spelling (*a/e*) there is no difference in sound. A good example of the anomalies of English spelling.

7. Plurals of compound words

a. Some compounds form the plural by adding *-s* to the final word or by making whatever change that word, if it is a noun, would normally make:

blackbird/blackbirds
teen-ager/teen-agers
merry-go-round/merry-go-rounds
chairman/chairmen

b. Compounds in which the headword is a noun followed by a modifying word or phrase usually add the plural to the headword rather than to the end of the compound:

mother-in-law/mothers-in-law
man-of-war/men-of-war
passer-by/passers-by
court martial/courts martial (however, *court martials* is possible)

c. A few compounds denoting weight or measure are used in a plural sense but a singular form:

an *eight-pound* bluefish   a *seven-year-old* child
a *ten-foot* skiff   a *two-week* vacation.

POSSESSIVE FORMS

The possessive inflection is *-'s*. It applies only in the singular and in the plural of those few nouns that do not end in *-s* or *-es*:

child's   children's
woman's   women's

Plural nouns ending in *-s* or *-es*—the vast majority—are not really inflected to show possession. No actual ending is added, but an apostrophe is simply written after the *-s*:

boys'   girls'
students'   wives'

In the case of proper names ending in *-s* (or another sibilant), the apostrophe *-s* is added to monsyllables, and the apostrophe alone to nouns of two syllables or more:

Keats's poetry   Dickens' novels
Joyce's stories   Samuel Clemens' life

In formal usage the *-'s* or *-s'* possessive is restricted to nouns signifying human beings or animals, especially those associated with people. Other living things, inanimate objects, and abstractions show possession by an *of*-phrase:

> the door of the closet NOT × the closet's door[4]
> the concept of honor NOT × honor's concept

NOUN DETERMINERS AND MODIFIERS

In addition to the plural and possessive endings, nouns are also distinguished by their position before the verb and by commonly following words called determiners (*the, a, an, this, that, any, many,* and so on) or single-word modifiers:

> Det.
>
> *A* book lay on the table.
>
> Det.
>
> *Some* books lay on the table.
>
> Det. modifiers
>
> *An old, tattered* book lay on the table.

The determiners *the, a,* and *an* are called *articles,* the first the *definite article,* and *an* and *a* the *indefinite. A* and *an* are used only with singular nouns (*a* before those beginning with a consonant, *an* before vowels).[5] They designate a single unspecified thing named by the noun.

The may be used with singular or plural nouns. In either case it marks a specific one or group of the things named by the noun. "*A* book" designates a book that has not been and is not being specified in any way; "*the* book" refers to a book that has been identified in some way or is about to be:

> A book lay on the table.
> The book on the table is mine.

It is a mistake to use *the* (or such determiners as *this* or *that*) unless the context makes it clear that you mean a particular one or ones.

When you wish to refer to all members of a class or to an unspecified plural number, use the noun (usually in the plural) without an article:

> Books are important.
> Books lay scattered about the room.

---

4. Usually we say *the closet door* or *the table top,* but in such cases nouns like *closet* or *table* are not possessives but merely acting as adjectives.
5. There is some variation in words beginning with *h-*. Most take *a* (*a horse, a house*); a few use either *a* or *an,* though *an* is more common in British usage (*an historian/a historian, an hotel/a hotel*).

The use or nonuse of the articles may be summarized like this:

	SINGULAR	PLURAL
SPECIFIED	*the* book	*the* books
UNSPECIFIED	*a* book	books
	an airplane	airplanes
ALL MEMBERS	_____	books

PROPER NOUNS AND COMMON NOUNS

Proper nouns name particular persons, places, time periods, works of art or literature, and so on:

Abraham Lincoln	the Parthenon
New York City	the Golden Gate Bridge
Friday	Shakespeare's play *King Lear*

Proper nouns are capitalized, even when they are used in an extended sense:

He was the George Washington of his country.
The Cadillac of beers

Occasionally, however, a proper noun passes so completely into the language that it becomes a common noun and is used without capitalization:

a macadam road
a macintosh apple

Common nouns refer to any member or members of a class:

book	street
politician	student

They are not capitalized unless they serve as part of a title or proper name:

The Student Prince
34th Street

ABSTRACT AND CONCRETE NOUNS

Abstract nouns designate a quality, condition, action, or state of being that cannot be directly perceived—that is, seen, heard, touched:

courage	honor
faith	love

Some abstract nouns, but by no means all, do not have plurals and are not used with determiners in the singular. For example, we do not normally say "courage*s*" or "*the* courage" when referring to courage in general.

Concrete nouns refer to something perceptible, which can be seen, heard, touched, and so on:

cow   soup
grass   teacher

MASS AND COUNT NOUNS

These are types of concrete nouns. *Mass nouns* name a material that does not exist in discrete, countable units (not, at least, to the naked eye):

coal   sugar
hydrogen   water

Count nouns designate things that do exist in separate, countable units:

automobile   house
book   town

In the singular, count nouns have a determiner: "*a* girl is," "*the* girl is," "*this* girl is," "*that* girl is," "*some* girl is," "*any* girl is," and so on; but never ×"girl is." And count nouns take a plural ending.

Mass nouns, on the other hand, do not follow a determiner in the singular and are not used in the plural. We do not say:

× *The* water is made of hydrogen and oxygen.
× Gold*s* are valuable.

However, many mass nouns may be used in a special, limited sense referring to a specific kind or amount of the material they designate, and in that case they may be used with determiners and in the plural:

The water is in the white jug.
All sugars harm the teeth.
Some alcohols are poisonous.

An important distinction between mass and count nouns is that they are used with different sets of words signifying quantity or number:

MASS	COUNT
much sugar	*many* books
little sugar	*few* books
less sugar	*fewer* books
least sugar	*fewest* books
a great *deal of* sugar	a great *many* books
a *quantity* (or *amount*) of sugar	a *number of* books

However, some determiners of quantity or number may be used with either mass or count nouns:

some sugar	*some* books
more sugar	*more* books
most sugar	*most* books
a lot of sugar	*a lot of* books

Pronouns

Conventionally defined, a pronoun is a word that stands for a noun. While many pronouns do substitute for nouns, some are also used in ways that cannot be said to refer to any noun: *it,* for example, in "It is going to rain." Perhaps the simplest way of defining a pronoun is that it is a word belonging to one of the classes that follow.

PERSONAL PRONOUNS

Personal pronouns refer to the speaker (first person), the spoken-to (second person), or the spoken-about (third person). They have different forms for case,[1] for singular and plural (in the first and second persons), and for gender[2] (in the third person singular).

	SINGULAR		
	SUBJECTIVE	POSSESSIVE	OBJECTIVE
first	I	my/mine	me
second	you	your/yours	you
third	he, she, it	his/his	they
		her/hers	
		its/its	

1. Case refers to how a noun or pronoun functions in a sentence—that is, whether it is the subject, a possessive, an object of a verb or preposition. In inflected languages like Latin, endings (inflections) are added to the stem of a noun or pronoun to indicate its function. Such languages distinguish varying numbers of cases; Latin, for example, marks five. In its earliest form, English also inflected nouns for case, but only the *-s* of the possessive singular remains. Personal pronouns, however, still preserve three cases: the subjective, the possessive, and the objective.
2. Gender is a grammatical category corresponding to sex. Some languages use distinctive endings to indicate whether a noun, pronoun, or adjective signifies a masculine, feminine, or neuter entity. Gender has disappeared from English nouns and adjectives. It remains only in the third person singular of the personal pronoun.

PLURAL

first	we	our/ours	us
second	you	your/yours	you
third	they	their/theirs	them

SUBJECTIVE FORMS

The subjective pronouns are used as the subjects of verbs:

She said that *she* would be there.
They wanted us to go with them.

In formal usage they are also used as the complements of linking verbs:

It is *I*.
That could be *she*.

In speech, however, and in writing of less formal style, the objective forms are common after linking verbs:

It is *me*.
That could be *her*.

A question about the use of the subjective form sometimes occurs after *than*. When *than* acts as a conjunction introducing a full clause, there is no problem:

Harry's father is taller than *he* is.

But when the clause is elliptical—when, that is, the verb is omitted—many people treat *than* as a preposition and follow it with the objective pronoun. In formal usage the subjective form is required since the construction is regarded grammatically as still a clause, even though an incomplete one:

Harry's father is taller than *him*.
FORMAL USAGE Harry's father is taller than *he*.

POSSESSIVE FORMS

The possessive pronouns have two forms (these are identical for the third person masculine and neuter singular). The forms are not alternatives; they are used in different ways. The first of each pair functions as an attributive adjective, the second as a predicative adjective. An *attributive* stands in front of its noun or occasionally immediately after it; a *predicative* is joined to the noun it modifies by a linking verb. Thus we use *my/mine* like this:

My book is on the table.
The book on the table is *mine*.

When a predicative possessive pronoun ends in *-s,* as most of them do, it does *not* have an apostrophe. In this it differs from the possessive form of nouns:

The book is *Sarah's.*
The book is *hers.*
NOT ⨉ The book is *her's.*

The misuse of the apostrophe is frequent with the possessive *its,* probably because of confusing this possessive with the common contraction *it's* (for *it is*):

The cat licked *its* whiskers.
NOT ⨉ The cat licked *it's* whiskers.
BUT *It's* going to rain.

OBJECTIVE FORMS

The objective forms are used after verbs (except linking verbs) and prepositions:

The doctor saw *me* this morning.
We visited *them* last week.
The people looked to *her* for their salvation.

Note especially the use of the objective pronoun after a preposition, and avoid the following ungrammatical pattern, especially common with the first person singular after *between*:

⨉ between you and *I*

The error is a kind of false elegance; the conventional form is:

between you and *me*

As the complement to a linking verb or verbal—most often some form of *be*—formal usage requires the subjective pronoun:

It was *I* who called.
That is *she* at the door.
It seems to have been *they* who complained.

Informally, however, most people use the objective form after *be*:

It was *me* who called.
That is *her* at the door.
It seems to have been *them* who complained.

THE INDEFINITE USE OF PERSONAL PRONOUNS

You and *they* are often used indefinitely—that is, to refer to someone (or ones) in general rather than to specific individuals:

When *you* first go to college everything is strange.
They are building a new highway around Boston.

The indefinite *you* and *they* are more common in speech than in formal writing. In composition these indefinite personal pronouns should be used sparingly and only in informal situations. *You* especially can get out of hand, leading to vagueness and repetition.

We may also have an indefinite sense of people in general:

We don't like to be criticized.

The indefinite *we* is less colloquial in tone than *you* or *they,* though it should not be overused. Sometimes *we* also serves as a thin disguise for *I,* as when a newspaper columnist expresses his or her opinion by writing:

We think this bill should be defeated.

In college themes, avoid this *editorial we*: *I* is more honest and less pretentious.

REFLEXIVE AND INTENSIVE PRONOUNS

A special form of the personal pronouns ends in *-self* (plural: *selves*):

myself	ourselves
yourself	yourselves
himself [3]	themselves[3]
herself	
itself	

As *reflexives* these are used in the objective function when the object refers to the same thing as the subject. They directly follow a verb or preposition:

He hurt *himself.*
She gave a present to *herself.*

A reflexive pronoun should *not* be used unless its referent is identical with that of the subject:

He hurt *me.*
NOT ✕ He hurt *myself.*

---

3. Forms using the possessive (*hisself, theirselves*) are substandard and should be avoided.

She gave a present to *him*.
NOT × She gave a present to *himself*.

As *intensives* the pronouns in *-self* repeat and emphasize a noun or pronoun:

The woman *herself* said so.
We showed the evidence to the people *themselves*.
I *myself* saw it.

When they repeat the subject of a sentence or clause the intensives may either immediately follow it or be placed at the end of the clause. The first pattern is formal and literary; the second, informal and colloquial:

I *myself* saw it.
I saw it *myself*.

RELATIVE PRONOUNS

The relative pronouns are *who* (*whose*, *whom*), *that*, and *which*. As relative pronouns these words have three characteristics: (1) they introduce a dependent clause; (2) they have a grammatical function within that clause; and (3) they stand for something expressed in the main clause. For example, in

The idea that you suggested is very good,

that is a relative pronoun. It introduces the subordinate clause *that you suggested* (an adjectival construction modifying *idea*); it acts as a noun within that clause (the direct object of *suggested*); and it stands for *idea*.

WHO, WHOM, WHOSE

That and *which* are invariable forms—that is, they have no inflections and do not change no matter what role they play in the clause. *Who* has different forms for the possessive (*whose*) and for the objective (*whom*), and the appropriate form must be selected according to whether the pronoun is the subject of the clause (*who*), a possessive adjective (*whose*), or an object of a verb or preposition (*whom*):

SUBJECT	The man *who* came to dinner stayed two weeks.
POSSESSIVE	This is the fellow *whose* book you borrowed.
OBJECT OF VERB	The man *whom* you expected is not here.
OBJECT OF PREPOSITION	The fellow from *whom* you borrowed the book wants it back.

Considerable confusion is caused by *who/whom* (*whose* is less of a problem). The rule is: Who *when the relative pronoun is the subject of the clause;* Whom *when it is an object.* As an object *whom* may function with a verb, a verbal (infinitive or participle), or a preposition:

OBJECT OF VERB	People *whom* we like are pleasant to be with.
OBJECT OF VERBAL	The people *whom* you wanted to meet cannot come tonight.
OBJECT OF PREPOSITION	The people about *whom* you asked are here.
	The people *whom* you asked about are here.

Notice that when it is the object of a preposition, *whom* may either follow the preposition or open the clause, with the preposition postponed to the end. The first pattern ("about *whom* you asked") is formal and literary; the second ("*whom* you asked about") is informal and colloquial.

A special case occurs when *who* is the complement of a linking verb. Then formal usage requires that it be kept in the subjective case (see the rule on page 698):

They don't know *who* the man is.
NOT × They don't know *whom* the man is.

Whatever its form, *who* (*whose, whom*) opens the clause it introduces (except when *whom* immediately follows a preposition).

SEMANTIC RULES GOVERNING *who, which, that*

As relative pronouns *who, which,* and *that* are used according to certain semantic rules—that is, considerations of meaning. One consideration is whether or not the pronoun designates an animate or inanimate object, and if animate, whether or not it is human:

1. Inanimate objects and abstractions are referred to by *that* or *which*:

   The chair *that* (*which*) she had always used had been pushed into the corner of the room.
   The idea *that* (*which*) you suggested is very good.

2. Human beings are referred to by *who* (*whose, whom*):

   The couple *whom* you met on your holiday seem very nice.

   However, in restrictive clauses (a distinction to be explained in a moment), *that* is still sometimes used for human beings:

   The boy *who* (*that*) lost his hat is here.

Which is not used for human nouns (except in archaic formulas like "Our father *which* art in heaven").

3. Animals are referred to by *that* or *which,* although those closely associated with human beings are sometimes designated by *who,* especially in nonrestrictive clauses:

> She looked fondly at the cat, *who* had curled into a ball of sleep.

A second consideration, applying only to *that* and *which,* depends upon whether the subordinate clause they introduce is restrictive or nonrestrictive. A *restrictive clause* is one that becomes part of the logical meaning of the noun it modifies. For instance:

> All sophomores *who are taking a foreign language* should see their advisers.

The assertion "should see their advisers" does not apply to "all sophomores," but only to "all sophomores who are taking a foreign language." In other words, the *who*-clause is part of the logical subject of the statement, and if we remove it, the sense of the statement changes.

A *nonrestrictive clause* gives pertinent information about the noun it modifies, but does not become part of its essential significance:

> General Grant, *who was an outstanding military leader,* proved to be an ineffective president.[4]

Here removing the *who*-clause would not alter the meaning of the main clause.

With *who,* the restrictive/nonrestrictive distinction makes no difference: the word may serve in either type of clause. With *that* and *which,* however, the distinction is important. *Which* may be used in either restrictive or nonrestrictive constructions. In formal usage *that* functions only in restrictive ones:

RESTRICTIVE   The table *that* you bought has been delivered.
                   The table *which* you bought has been delivered.
NONRESTRICTIVE   The table in the corner, *which* we bought twenty
                   years ago, is one of our favorite pieces.
NOT   ✕ The table in the corner, *that* we bought twenty
                   years ago, is one of our favorite pieces.

In the first of those examples the clause "that (which) you bought" is restrictive because it identifies which table is meant. In the second example the clause "which we bought twenty years ago" is not restric-

---

4. In modern style, nonrestrictive clauses are set off by commas (occasionally by dashes)— that is, a mark precedes them and a mark follows them. Restrictive clauses are not set off.

tive because the table has already been specified by the phrase "in the corner"; hence the clause must use *which*, not *that*.

INTERROGATIVE PRONOUNS

Interrogative pronouns introduce questions. In addition to their function as relative pronouns, *who* (*whose, whom*) and *which* are used as interrogatives, and so too is *what*.

Questions may also be introduced by certain adverbs such as *why, when, where,* and *how.* As interrogatives these differ from the pronouns simply by functioning as adverbs within the clause they introduce rather than as subjects or objects. Some grammarians classify all these terms—both pronouns and adverbs—together as interrogatives or as *wh*-words.

Whether pronoun or adverb, interrogatives introduce both direct and indirect questions. In writing, a *direct question* is signaled by the inversion of the auxiliary verb before the subject ("*Are* you going home?") or by one of the interrogatives ("*Where* are you?" "*Who* isn't here?"). Direct questions are punctuated with a question mark.

An *indirect question* requires an answer just as much as a direct one. It is expressed, however, as a statement and has none of the signals of a direct question:

He asked which movie you went to.

Indirect questions are *not* closed by question marks but by periods.

DEMONSTRATIVE PRONOUNS

When they are used as subjects or objects, *this* and *that* (along with their plurals *these* and *those*) are called demonstrative pronouns. They point out a specific one or group from among many:

That was not a good idea.
This is where we lived.
Those were bad times.
These decided to stay.

The same four words may also act as demonstrative adjectives, modifying nouns rather than acting as nouns:

That idea was not good.
This house is where we lived.
Those times were very bad.
These people decided to stay.

As the subjects of sentences, the demonstratives run the risk of vagueness. The danger is especially likely with *this*, which occurs as a subject far more frequently than the other demonstratives. Especially when *this* refers to the whole of a preceding idea rather than to one specific word, a reader can easily misconnect it:

> The campaign failed because of poor leadership and resulted in the collapse of the government. *This* proved the final disaster.

"This" probably refers to the "collapse of the government," but it could be read as pointing to the lost campaign or to the poor leadership. You can avoid confusion by using *this* as a demonstrative adjective to modify a word that repeats or sums up the concept referred to:

> *This* collapse proved the final disaster.

When *this* and *that* (or their plurals *these* and *those*) are used together to distinguish the relative positions of two things, *this* signifies the nearer object and *that* that farther:

> *This* book in my hand and *that* one over on the desk belong to me.

The same distinction applies with two events in time:

> *That* party last week and *this* one tonight were both duds.

With reference to abstract ideas, *that* looks backward to something already expressed:

> *That* was a fine thing he did.

This may refer backward or point forward, anticipating a new idea:

> Nobody understood her. *This* fact contributed to her sense of isolation.

> *This* is the point to grasp: she felt isolated because nobody understood her.

INDEFINITE PRONOUNS

Indefinite pronouns refer to any one, any two, any several, or all of a class of persons or objects or ideas, but they do not designate any particular member or members of that group. The indefinite pronouns include:

all	everybody ('s)	no one ('s)
another ('s)	everyone ('s)	nothing
any	everything	one ('s)
anybody ('s)	few	other ('s)
anyone ('s)	many	others (others')
anything	most	several

both	much	some
each	neither ('s)	somebody ('s)
each one ('s)	nobody ('s)	something
either ('s)	none	such

As indicated, some of these pronouns have a possessive form; the others do not. Except for *other,* none has a plural. Many of them are also used as adjectives; for example:

PRONOUN	*Either* will go.
	Some will go.
ADJECTIVE	*Either* man will go.
	Some people will go.

In formal usage all forms ending in *-one* or *-body* are construed as singular. With *everyone* and *everybody* this convention causes confusion because logically these words should designate all members of a group. Nonetheless they require singular verbs and should be referred to by singular pronouns:

Everyone has left his seat and is moving to the front of the room.[5]
NOT × *Everyone* have left their seats and are moving to the front of the room.

Even though "everyone" in such a case might refer to forty people, it is treated as singular.

In less formal situations, however, we often allow common sense to override usage, using plural verbs for *everyone* and *everybody* and referring to them by *they, their, them.*

In addition to these indefinite pronouns, the personal pronouns *we, you,* and *they* can also be used indefinitely (see page 699).

RECIPROCAL PRONOUNS

The reciprocal pronouns designate a mutual relationship between two or more individuals or groups. There are two sets of these pronouns, *each other* and *one another*:

Republicans and Democrats never grow tired of slinging mud at *each other.*

---

5. Pronouns in *-one* and *-body* do not have gender and could refer to either male or female. However, it is awkward to acknowledge this fact by writing, for example, "Everyone took his or her hat." The custom is simply to write "Everyone took his hat," allowing the context to make clear that *his* does not have an exclusively masculine significance. However, if the generic *he, his, him* seems offensively sexist, use both feminine and masculine forms. (Often the issue can be evaded by shifting to a plural noun: "People took their hats.")

Both reciprocal pronouns have possessive forms: *each other's, one another's*. The possessive form is always singular:

Harry and Elizabeth dislike *each other's* life style.
NOT × Harry and Elizabeth dislike each others' life style.

In formal usage *each other* is employed for two referents; *one another* for three or more:

The two men dislike *each other*.
The whole family dislike *one another*.

Practically, however, we use the two pronouns interchangeably, without concern for number.

Adjectives

Together adjectives and adverbs are called modifiers. *Modifiers* grammatically connect to other words, called *headwords,* whose meanings they specify or limit. In the phrase *a small boat, boat* is the headword and *small* a modifier making the meaning of *boat* more exact.

Adjectives modify nouns; adverbs modify verbs, adjectives, other adverbs, and in some cases entire clauses.

As a class, adjectives have no distinctive ending, though some are marked by suffixes used to convert other words (usually nouns) into adjectives; for example:

accident*al*   decept*ive*
adventur*ous*   tire*some*

POSITION OF ADJECTIVES

Adjectives may occur either as *attributives* or as *predicatives*. The first is positioned next to the noun it modifies; the second is joined to its noun by a linking verb:[1]

ATTRIBUTIVE   a *small* boat
PREDICATIVE   The boat is *small*.

1. For linking verbs see page 683.

Attributive adjectives may be either restrictive or nonrestrictive. As *restrictive* modifiers they supply essential information which defines who or what is meant by the noun. As *nonrestrictive* modifiers they give information pertinent to the point but not necessary to identify who or what the noun refers to.

This difference in function is marked by differences in position and punctuation. A restrictive adjective precedes its noun with no punctuation (unless it needs to be separated from another restrictive adjective). If the noun has a determiner (*the, a, an, this, that, any, no, some,* and so on), the adjective is placed between the determiner and the noun:

An *empty* house seems cold and uninviting.

Nonrestrictive adjectives are separated from their nouns by a comma. They may come before the noun (in which case they precede any determiner) or after the noun:[2]

Empty, a house seems cold and uninviting.
A house, *empty,* seems cold and uninviting.

COMPARISON OF ADJECTIVES

Adjectives may be compared—that is, changed to show degrees of quality, quantity, or intensity. Three degrees are distinguished: *positive,* which is the basic meaning; *comparative,* which takes the basic sense one step further; and *superlative,* which carries that sense to its final degree.

The positive form is simply the normal adjective (*beautiful, big, small*). The comparative either adds *-er* to the end of the adjective or precedes it with *more* (*bigger, smaller, more beautiful*); and the superlative adds *-est* or uses *most* (*biggest, smallest, most beautiful*).

The choice between the forms in *-er/-est* and those with *more/most* (or *less/least,* used to show negative comparison) depends upon the length of the adjective. Words of one syllable are generally compared with suffixes:

black   black*er*   black*est*
light   light*er*   light*est*
short   short*er*   short*est*

Words of three or more syllables are compared with *more* and *most*:

courageous   *more* courageous   *most* courageous
detrimental   *more* detrimental   *most* detrimental
incongruous   *more* incongruous   *most* incongruous

2. Sometimes a nonrestrictive adjective may come at the end of the clause:
   The girl stood in the silent room, *alone* and *frightened.*
However, not all nonrestrictive adjectives may be positioned this way. It depends upon idiom.

With most adjectives of two syllables either form may be used:

crazy   craz*ier/more* crazy   craz*iest/most* crazy
likely   likeli*er/more* likely   likeli*est/most* likely

However, in particular cases of two-syllable adjectives, one form or the other may be preferred because of idiom or a difficult pattern of sound. *Crazier,* for example, is more idiomatic than *more crazy,* and forms like × *eagerer* or × *oranger* are avoided because of the difficulties of pronunciation.

Whatever the number of its syllables, an adjective should *never* be compared by both methods simultaneously:

× more smaller   × most smallest

A few adjectives are irregular in their comparatives and superlatives:

POSITIVE	COMPARATIVE	SUPERLATIVE
bad, ill	worse	worst
good, well	better	best
far	farther, further	farthest, furthest
little	littler, less, lesser	littlest, least
much, many, some	more	most
top	———	topmost

In formal usage the comparative form, *not* the superlative, is used when comparing two things:

He is the young*er* of two brothers. NOT the young*est*
BUT He is the young*est* of three brothers.

And in formal usage a few adjectives should not be compared at all because their meanings do not admit of degree; for instance:

absolute   perfect
ideal   unique

Something absolute or perfect cannot be more so because perfection and absoluteness allow no room for improvement or addition. This rule of style, however, is often ignored in informal situations, and people do say, "The most ideal thing" or "A more unique place."

DETERMINERS, PARTICIPLES, INFINITIVES, AND NOUN ADJUNCTS

In addition to adjectives, other words function as noun modifiers. These include determiners, participals, and nouns (which, when acting as adjectives, are called *noun adjuncts*). *Determiners* are pilot fish to nouns:

the articles *the, a, an*; the demonstratives *this, that, these, those*; the possessive pronouns *my, your, his, her, its, our, their*; and quantifying words like *any, many, much, some, few,* and so on.[3]

Participals, both present and past, are frequently used to modify nouns, and may function either restrictively or nonrestrictively:

RESTRICTIVE   The *running* boy sped round the corner.
NONRESTRICTIVE *Running,* the boy sped round the corner.

Infinitives (with *to*) may also function adjectivally. They are restrictive and are placed after the noun:

They never had an opportunity *to learn.*
He wants time *to think.*

Noun adjuncts are simply nouns acting as adjectives:

He bought a *London* suit.
They went into the *boat* yard.

All these words differ from true adjectives in that they cannot be automatically compared. Articles, demonstratives, and possessive pronouns are never compared. Nor are noun adjuncts (× *Londoner* suit, × *boater* yard). Some of the quantifying words, however, have comparative and superlative forms (*many/more/most*). Some participles can be compared, and some cannot; it depends upon idiom:

more frightening, most frightening
BUT NOT × more running, most running

Adverbs

Adverbs modify verbs, adjectives, other adverbs, and entire clauses:

MODIFYING VERB   We *almost* went.

MODIFYING ADJECTIVE A *slightly* dirty sweater

MODIFYING ADVERB   She dances *very* well.

MODIFYING CLAUSE   *Certainly* we shall come.

---

3. Some of these words, of course, may themselves act as nouns:
   Many were there.
   These came for you yesterday.

Adverbs describe a wide range of concepts: time, place, manner, degree, number:

TIME	He arrived *late*.
PLACE	They live *here*.
MANNER	People dress *informally*.
DEGREE	She looks *extremely* well.
NUMBER	We called *twice*.

CONJUNCTIVE ADVERBS

A class of adverbs especially important in writing are conjunctive adverbs. They signal how an approaching idea relates to what has gone before, thus helping readers follow the flow of thought. Among the more common conjunctive adverbs are these words and phrases:

as a result	likewise	similarly
besides	nevertheless	still
consequently	nonetheless	therefore
even so	on the contrary	thus
however	on the other hand	yet

Conjunctive adverbs may introduce an entire sentence or an independent clause within a compound sentence:

The attack failed. *However*, the battalion was able to regroup.

We went to the affair; *however*, we were tired and left early.

When they introduce a clause within a sentence, conjunctive adverbs do not join it grammatically to the preceding construction as do true coordinating conjunctions like *and, but, or, for*. This means that a clause linked by a conjunctive adverb is punctuated with a semicolon, not, like a coordinated construction, with a comma:

Some people enjoy mathematics; *however*, others do not.
Some people enjoy mathematics, *but* others do not.

Most conjunctive adverbs can be placed at the beginning of the sentence or clause they introduce, in an interrupting position within that clause, or at its end:

However, others do not.
Others, *however*, do not.
Others do not, *however*.

Generally they are best at or near the beginning, where they most effectively serve their function of alerting readers to the flow of thought.

POSITION OF ADVERBS

In general, adverbs may be positioned more freely than adjectives:

Apparently they did not like his speech.
They *apparently* did not like his speech.
They did not *apparently* like his speech.
They did not like his speech, *apparently*.

Grammatically these sentences are identical; stylistically, however, they differ in emphasis.

Not all adverbs are equally variable in placement. Much depends upon what the adverb modifies, what it signifies (time, place, manner, and so on), and idiomatic restrictions. The possibilities are quite complex, but, briefly, we may say that sentence adverbs (like *apparently* in the example above) are relatively flexible. Those which modify adjectives or other adverbs are not: they precede their headwords:

He was *extremely* impatient.
She sang *very* well.

The moveability of adverbs modifying verbs varies considerably with individual cases. The best rule for any adverb is to position it where it sounds natural and gives the emphasis you want, and to avoid any position where an adverb sounds unEnglish.

A special case of positioning involves the *split infinitive*: an adverb placed between *to* and the infinitive proper:

to quickly go
to suddenly see

In formal writing split infinitives are best avoided. Less formally they are acceptable, and sometimes may be preferable, so long as they sound idiomatic and involve only a single word. Phrases and clauses used to split an infinitive are invariably awkward:

× to on the other hand see

THE *-ly* ENDING

As a general rule adverbs of two or more syllables (excepting some of the conjunctive adverbs) end in *-ly*. Common monosyllabic adverbs (and a few in two syllables) may or may not:

beautifully	better
inherently	fast
rapidly	slow
suddenly	well

Since many of the monosyllabic words also function as adjectives, it may be difficult at times to tell how they are to be construed:

She looks *well.*
She dances *well.*

In the first sentence *well* is a predicate adjective modifying *she*; in the second it is an adverb modifying *dances.*

A few adverbs can be used with or without *-ly*:

I	II
slow/slowly	hard/hardly
quick/quickly	near/nearly
close/closely	late/lately
loud/loudly	high/highly
tight/tightly	real/really
first/firstly,[4]	light/lightly
second/secondly,	
third/thirdly, *and so on*	

The dual forms in column I have no difference in meaning, though the *-ly* adverb is more formal in tone. Those in column II do have different meanings; for example:

He tried *hard.* (= with great effort)
He *hardly* tried. (= with little effort)

We went *near* San Francisco. (= close to)
We *nearly* went to San Francisco. (= almost)

COMPARISON OF ADVERBS

Like adjectives, most adverbs can be compared in the three degrees of positive, comparative, and superlative. Those ending in *-ly* always use *more* and *most.* A few common adverbs of one syllable add *-er* and *-est,* and a few compare irregularly:

beautifully	more beautifully	most beautifully
suddenly	more suddenly	most suddenly
fast	faster	fastest
near	nearer	nearest
well	better	best

4. When it is used as a sequence word to introduce a sentence or clause, *first* is more idiomatic in American English than *firstly.* With the other sequence words, either form may be used. However, once a form has been chosen it should be maintained consistently: "Second. . . . Third," or "Secondly. . . . Thirdly," but not "Secondly. . . . Third" or "Second. . . . Thirdly."

NOUNS AS ADVERBS

It is common in English to use nouns adverbially.[5]

> The building is a *block* long.
> The snow was *knee* deep.
> We went *home*.
> I saw her the next *day*.

Such nouns do not become true adverbs. They cannot be placed in other positions, nor can they be compared. They are simply nouns acting adverbially.

Prepositions

Prepositions are primarily function words. They have no inflections, and they serve to connect words, phrases, or clauses to other words.

Prepositions link a noun (or any term or construction acting as a noun) to another word for purposes of modification. The preposition and the word or construction it introduces (called its object) make up a *prepositional phrase*:

       prep.   object

The book *on the table* is mine.

          prep.   object

They approved *of her going to camp*.

In the first sentence the phrase functions adjectivally, modifying *book*. In the second it is adverbial, modifying *approved*.

Prepositions are numerous in English, and the list changes as new ones are added and older ones become obsolete. Examples of common prepositions include:

5. Adverbial nouns derive from the earlier stages of English, when nouns were inflected to show different cases and the noun in the accusative could be used as an adverb. Today the accusative ending has disappeared, but because the pattern is so convenient, we still use nouns in this way.

about	between	on
above	beyond	to
after	considering	toward
against	for	under
at	in	up
before	into	upon
behind	of	with
beside	off	without

Prepositions may also consist of several words acting as a unit:

according to	because of	in spite of
along with	by means of	outside of
as for	in addition to	with (*or* in) regard to

Some prepositions can function adverbially, and some can merge with verbs, making compound verbs:

PREPOSITION WITH AN OBJECT	He looked *up the street.*
PREPOSITION ACTING ADVERBIALLY	He looked *up.*
PREPOSITION MERGED WITH VERB	He *looked up* the word.

In the first example *up* is part of an adverbial prepositional phrase modifying *looked* and telling us where "he" directed his eyes. In the second, *up* is itself an adverb, also explaining where or how "he" looked. In the third, *up* combines with *look* to make a compound verb meaning to seek information by consulting a reference work such as a dictionary. The difference between the first and third uses become clear when you think about whether you can look up a word in the same way that you can look up a street.

Conjunctions

Conjunctions link words, phrases, or clauses. They come in several varieties: coordinating, subordinating, and correlative.

COORDINATING CONJUNCTIONS

The common coordinating conjunctions (including the correlatives) are:

and	either . . . or
but	neither . . . nor
for	not only . . . but (also)
or	both . . . and
nor	
yet	

These conjunctions join words, phrases, or clauses that are grammatically equal in form and function, and only those that are so equal. Two nouns acting as common subjects of the same verb may be coordinated, or two verbs, two prepositional phrases, adverbial clauses, or independent clauses—two of anything as long as they are in the same form and doing the same job:

TWO COORDINATED SUBJECTS	*John and Mary* went to Boston.
TWO VERBS	John *went* to Boston *but did* not *visit* his aunt.
TWO ADJECTIVES	The book was *torn and tattered.*
TWO PHRASES	The choir sang *at the church and at the reception.*
TWO ADVERBIAL CLAUSES	*Because the plan was well conceived and because the leaders were intelligent,* the community was saved.
TWO INDEPENDENT CLAUSES	*We must succeed, or we shall surely suffer.*

Correlative conjunctions also join the same sorts of things. However, there is a rule of style concerning their use: the same kind of grammatical construction should follow each term of the pair:

Either she will agree or she will not.
She will either agree or will not.

In both those sentences the rule is observed: in the first, full clauses follow each word, and in the second, verb phrases. But the example below breaks the rule because "either" introduces a complete clause while "or" introduces only a verb phrase:

 × Either she will agree or will not.

The rule, remember, is one of style, not of usage or grammar.

If you use a coordinating conjunction to join unequal words, phrases, or clauses, it is a mistake called a *shifted construction.* For example:

 × He likes sailing and to swim.
 × The men said that they would go and when it was time.
 × We went because we were sent for and not out of loyalty.

In the first sentence a gerund (*sailing*) and an infinitive (*to swim*) are falsely coordinated; in the second, an adverbial *when*-clause and a noun

that-clause; and in the third, an adverbial *because*-clause and a prepositional phrase (*out of loyalty*).

In conversation such shifts are commonplace and no great matter.
In more formal situations, however, they should be avoided. The mistake
can be corrected in either of two ways, though in particular cases only
one might supply the appropriate meaning. First, the coordinating conjunction may be retained and one of the constructions changed to make
it equal to the other:

He likes *sailing and swimming* (OR *to sail and to swim*).
The men said *that they would go and that it was time.* (This revision
alters the sense and therefore might not be possible.)
We went *because we were sent for and not because we were loyal.*

Second, it may be possible to drop the coordinating conjunction and
allow the unequal constructions to stand. This solution would not work
with the first example, but it would with the other two:

The men said that they would go when it was time. (Probably the
intended meaning.)
We went because we were sent for, not out of loyalty.

SUBORDINATING CONJUNCTIONS

Subordinating conjunctions are used primarily to join dependent clauses
to main clauses. The common ones are:

after	if	until
although	in order that	what
as	provided that	whatever
as if	since	when
as long as	so (that)	whenever
as though	that	where
because	though	wherever
before	till	while
how	unless	

A subordinate (or dependent) clause is one that acts as a subject,
object, adjective, or adverb, whether in relationship to or within an
independent clause (referred to as the main clause):

We left *before you arrived.*
She said *that she would see us tomorrow.*

In the first example *before you arrived* is an adverbial subordinate
(or dependent) clause modifying the main verb, *left*. In the second,
that she would see us tomorrow is a noun clause acting as the object
of *said* and thus completes the main clause.

Since a clause introduced by a subordinating conjunction is used as a noun or modifier, it is conventionally regarded as unable to stand by itself as a separate sentence. If you punctuate such a construction as a sentence, it is a fragment:

We sat down. × Because we were tired of standing.

Fragments have their value in composition (see pages 107, 283, 314). That value, however, depends upon their being restricted to special situations. Generally a clause introduced by *because* or any other subordinating conjunction should be treated as part of a larger sentence:

We sat down because we were tired of standing.

Occasionally, subordinating conjunctions are used to join grammatically equal words and phrases (though not clauses), in which case they have the idiomatic force of coordinating conjunctions:

Mary, *though* not Lucy, went to college.
He played energetically *though* badly.
I shall leave Wednesday *if* not tomorrow.

In traditional grammar such constructions are regarded as reductions of full subordinate clauses:

Mary went to college, though Lucy [did] not.
He played energetically though [he played] badly.
I shall leave Wednesday if [I do] not [leave] tomorrow.

Interjections

Interjections are words or expressions like *oh, damn, hey, you know* which are inserted into sentences to express emotion or gain attention. They are usually punctuated by a comma or dash and sometimes followed by an exclamation point. Often, though not invariably, interjections come at the beginning of the sentence:

Oh, there you are.
Damn! we missed the bus.
Hey—watch it!
So I said to her—*you know*—that I couldn't get away.

Interjections have no grammatical connection to the sentence in which they occur, and they may stand themselves as separate sentences, though in that case they are not technically interjections. These expressions are common in speech, but they are generally out of place in formal composition.

The Grammar
of the Sentence

The essence of a sentence is a subject and verb, and a minimal sentence needs nothing more:

People were sleeping.

Commonly, of course, sentences do contain more—complements and modifiers and, less often, appositives and absolutes.

In this section we shall review those various elements of the sentence, asking, "What kinds of constructions—that is, words, phrases, and clauses—can serve as subjects, complements, and so on?" We shall not include verbs, however, since they have been sufficiently discussed in the section on parts of speech.

Before we begin the review, we need briefly to define key terms: first, the elements of the sentence, then *phrase* and *clause*.

Definitions

SUBJECTS

The subject names that about which the verb makes an assertion. A subject usually precedes its verb, though for emphasis or for asking questions it is often postponed to a later position.

COMPLEMENTS

The complement is a word or construction that "completes" the action or state of being expressed by the verb. Complements occur only after transitive and linking verbs. After the former a complement may be (1) a direct object by itself, (2) an indirect object plus a direct object, or (3) a direct object plus an object complement; after linking verbs it may be either (4) a predicate noun, or (5) a predicate adjective.

```
                        direct object
1. They gave him ten dollars.

            indirect object   direct object
2. They gave the committee ten dollars.

          direct object  object complement
3. They elected Mary chairperson.

            predicate noun
4. That book is a novel.

          predicate adjective
5. That book is mine.
```

DIRECT OBJECT

A direct object receives the action of a transitive verb. In example 1 it is that which was given.

INDIRECT OBJECT

An indirect object, common after verbs of offering, giving, or telling, indicates who or what received the offering or the message. It precedes the direct object, and in example 2 it names those to whom the money was given.

In traditional grammar the term "indirect object" also applies to a phrasal construction, introduced by the preposition *to* and placed after the direct object:

They gave ten dollars *to the committee*.

More exactly this is called the *periphrastic* (= longer) indirect object.

OBJECT COMPLEMENT

An object complement follows a direct object and tells how the direct object was changed as a result of the action named by the verb. In example 3 "Mary" became "chairperson" by virtue of being "elected." In addition to nouns, adjectives can also function as object complements:

He stained the table *green*.

Verbs which can take object complements form a special class of transitives and are called *factitive* verbs; other examples are *call, consider, make, name, paint, think*.

PREDICATE NOUN

A predicate noun is another name applicable to the subject of a linking verb. In example 4 "novel" is an alternate term that could be used to name or describe "that book."

PREDICATE ADJECTIVE

A predicate adjective designates a quality of the subject, as "mine" is an attribute of "that book" in example 5.

MODIFIERS

Modifiers are words or constructions that limit, enumerate, or somehow describe and clarify the meaning of other terms, called headwords. Modifiers are either adjectival or adverbial in function. The first modify nouns or words acting as nouns; the second modify verbs, adjectives, other adverbs, and entire clauses or sentences.

adverb
The sky *suddenly* darkened and we ran toward the

adjectives
dark, forbidding house.

APPOSITIVES

An appositive is a word that repeats a preceding term or construction. Usually an appositive immediately follows the expression it repeats, as in the following example "John" comes after "brother," the term to which it is in apposition:

My brother *John* has come home.

Occasionally for emphasis an appositive may be postponed to the end of its clause:

My brother has come home, John.

ABSOLUTES

Absolutes are words or constructions within a sentence which do not modify or connect with any single term but rather modify the entire idea expressed in the main clause:

The game over, the fans filed from the stadium.
She flew down the stairs, *her hair streaming behind her.*

Despite the fact that they express complete ideas and are only loosely tied to the main clause of the sentence in which they occur, absolutes are not generally written as separate statements.

PHRASES

Phrases, together with dependent clauses, are functional word groups. Such groups act grammatically just as single words do, serving as nouns and modifiers. They are of the utmost importance in composition; no one can write well who cannot handle the various types of phrases and clauses.

A phrase is a word group that does not contain a finite verb—that is, one limited as to number, person, or tense (see page 680)—though it may have a nonfinite one which, in turn, may even have a subject and complement. There are four kinds of phrases: prepositional, participial, infinitive, and gerundive.[1]

PREPOSITIONAL PHRASE

A prepositional phrase consists of a preposition (*of, to, by, in,* and so on) plus an object. The object may be a noun (with or without modifiers) or one of the other functional word groups acting as a noun:

1. There is also a verb phrase: a main verb plus one or more auxiliaries, like *have been talking* in "They have been talking to the lawyers." However, verb phrases function only as finite verb forms and need not be included in this discussion.

prep.   noun

We are *in the house.*

       prep.   infinitive phrase

He asked *about how to get downtown.*

       prep.   gerundive phrase

They approved *of his going to college.*

       prep.   dependent clause

We were curious *about what was said.*

Prepositional phrases are commonly modifiers and can serve either as adjectivals or as adverbials:

ADJECTIVAL   The book *on the table* is hers.
ADVERBIAL   Her book is *on the table.*

Notice that nothing in the form of a prepositional phrase marks it as adjectival or adverbial. We infer its function from its position.

Certain prepositional phrases have the special functions of showing possession and indicating an indirect object. The first (which uses the preposition *of*) is called the periphrastic (or longer) genitive:

That is the property *of my sister-in-law.*

The second (using *to*) is known as the periphrastic indirect object:

She gave the papers *to me.*

PARTICIPIAL PHRASE

A participial phrase consists of a participal, usually the present or the past, plus one or more words associated with it as modifiers or complement:

part.   modifier

Tiptoeing past the sentry, they entered the castle.

part. complement

Taken prisoner, he was thrown into the dungeon.

Participial phrases function as modifiers, both adjectival and adverbial:

ADJECTIVAL   *Running over the rough stones,* the man fell and twisted his ankle.
ADVERBIAL   The hunters returned to camp, *arriving at sundown.*

INFINITIVE PHRASE

An infinitive phrase is built around an infinitive, usually the present active form, plus one or more words, which may function as a subject of the infinitive, as a complement, or as a modifier:

        subj.   infinitive

They asked *me to go.*

        infinitive complement

She hopes *to study law.*

        infinitive   modifier

You will have *to move quickly.*

Infinitive phrases function within sentences in a variety of ways: as subjects, complements, objects of prepositions, and modifiers:

AS SUBJECT	*To tell her* will not be easy.
	For me to tell her will not be easy.
AS COMPLEMENT	He wanted *to tell her.*
	He wanted *me to tell her.*
AS OBJECT OF PREPOSITION	You are about *to make a serious mistake.*
AS ADJECTIVAL	They have enough money *to buy anything they want.*
AS ADVERBIAL	*To tell the truth,* I don't care.

Infinitive phrases are subject to two special grammatical rules. First, if an infinitive has a subject of its own, that subject must be in the objective case. This rule has significance only for personal pronouns and for *who,* and requires the forms *me, him, her, us, them,* and *whom* to be used as subjects of infinitives:

They wanted *us to tell her.*
NOT ✕ They wanted *we to tell her.*

Second, when an infinitive with a subject of its own acts in its entirety as the subject of a sentence, the phrase is introduced by *for*:

For me to tell her would be difficult.

That second rule applies *only* when the infinitive-phrase subject contains a subject of its own.

When an infinitive acts as the complement to another infinitive, formal usage requires *to,* though in speech *and* is often used:

FORMAL   You must try *to work harder.*
COLLOQUIAL You must try *and work harder.*

Notice finally about infinitive phrases that it is usually the present active infinitive that is used. However, other infinitive forms can also be the center of a phrase, as, for example, the perfect passive infinitive:

To have been snubbed by the family was bad enough, but *to have been completely ignored* was worse.

After most verbs the *to*-form of the infinitive is required. Modal auxiliaries, however, take the bare infinitive:

You must *try a new tack.*

After the verb *help* either form is permissible:

He helped *to paint the house.*
He helped *paint the house.*

GERUNDIVE PHRASE

When a gerund (the *-ing* verbal acting as a noun) is accompanied by other words, the construction is called a gerundive phrase. The other words may relate to the gerund as subject, complement, or modifier:

subject   modifier   gerund   modifier

The king's abrupt dropping of Raleigh was no surprise.

subj.   gerund   complement

His studying German proved valuable in later life.

Notice that the subject of a gerund is in the possessive case: "the king's," "his."

Like gerunds, gerundive phrases act only as substantives—that is, as subjects, complements, or objects of prepositions:

AS SUBJECT	*Our going to Boston* is no concern of yours.
AS COMPLEMENT	Who likes *taking exams*?
AS OBJECT OF PREPOSITION	She worries about *seeing her folks*.

CLAUSES

There are two types of clauses: independent and dependent. *Independent clauses* can stand by themselves as sentences; in fact, a simple sentence is a kind of independent clause. In practice, however, the term is reserved for clauses contained within a larger statement, even though they are not grammatically subordinated to anything in that statement. For example, the following sentence contains two independent clauses connected by the coordinating conjunction *and*:

It was late and *we decided to leave.*

A *dependent clause,* on the other hand, is a *grammatical part* of the sentence in which it occurs and cannot stand by itself as a true sentence. In what follows we shall concern ourselves only with dependent clauses.

A dependent clause consists of a subject and a finite verb; it is commonly introduced by a subordinator, a word marking the construction as dependent and indicating how it relates grammatically to the main clause. In addition, it may also contain a complement and modifiers associated with the subject, verb, or complement. For example, in the sentence "Since very few people were able to see the film, it will be shown again next week," the dependent clause is structured like this:

subordinator	modifier	subject	verb	complement	modifier
Since	very few people	were	able	to see the film	

The subordinator may be a subordinating conjunction (*as, because, since, though,* and so on); a relative pronoun (*that, which, who*); or a relative adverb or adjective (*how, when, where,* and so on):

clause introduced by subordinating conjunction

I'll explain *if you wish.*

clause introduced by relative pronoun

He knows *whom he must see.*

clause introduced by relative adverb

The people are uncertain about *where they should go.*

After verbs that signify saying and knowing, a dependent clause may act as direct object without being introduced by a subordinator. The construction is sometimes called a *contact clause* and is common in speech. In formal writing, however, the clause is generally clearer if it is marked by *that:*

FORMAL   He said *that he would go.*
COLLOQUIAL   He said *he would go.*

According to their grammatical function within the sentence, dependent clauses are characterized as noun, adjectival, or adverbial.

NOUN CLAUSES

Some noun clauses are introduced by *that,* as in "He said that he would go." In such a construction *that* has no grammatical role within the clause but merely marks its noun function. In informal usage, as we have seen, the *that* may be omitted from clauses serving as direct objects of certain verbs.

Other noun clauses are introduced by words that do have a grammatical function within the clause itself—as subjects, objects, or modifiers. Among such subordinating words are relative pronouns, relative adjectives, and relative adverbs:

RELATIVE PRONOUN	We know *who did it.*
RELATIVE ADJECTIVE	He asked *which road he should take.*
RELATIVE ADVERB	*How they escaped* is a mystery.

Noun clauses may serve as subjects, complements, objects of prepositions, or appositives:

AS SUBJECT	*That the mayor had good intentions* is true enough.
AS COMPLEMENT	She knows *whose opinion you prefer.*
AS OBJECT OF PREPOSITION	We shall listen to *whoever will help us.*
AS APPOSITIVE	In the eighteenth century the idea *that all people are created equal* sounded ridiculous to many conservatives.

ADJECTIVAL CLAUSES

Adjectival clauses modify nouns, applying to them complicated ideas that cannot be expressed in a single word. Generally adjectival clauses are introduced by a relative word (they are often called relative clauses). This term may be a pronoun (*that, which, who, whoever*) or a relative adjective or adverb (*which, when, where,* and so on). Whatever its form, the relative word connects the clause to the noun it modifies. At the same time the relative terms acts grammatically within the clause— as subject, complement, object of a preposition, or modifier:

AS SUBJECT	The house *that stands on the corner* has been sold.
AS COMPLEMENT	The man *whom they chose* has done a good job.
AS OBJECT OF PREPOSITION	The book *to which she referred* me has been very useful.
AS ADJECTIVE	The man *whose car you hit* is going to sue.
AS ADVERB	The place *where we live* is very cold.

ADVERBIAL CLAUSES

Adverbial clauses are introduced by one of the subordinating conjunctions, such as *because, since, than, though, when* (see page 716 for a complete list). Adverbial clauses may modify verbs, adjectives, other adverbs, and other clauses:

MODIFYING VERB	He will leave *when he is ready.*
MODIFYING ADJECTIVE	She is smarter *than I am.*
MODIFYING ADVERB	Few people speak more bluntly *than he does.*
MODIFYING ENTIRE CLAUSE	*Because the king was ill,* the ambassadors failed to see him.

Subjects

WORDS AS SUBJECTS

Nouns and pronouns, of course, are the most common kind of single words functioning as subjects. In addition, however, adjectives and verbals (participles, gerunds, and infinitives) may also serve:

ADJECTIVE	*The poor* will suffer most; *the wealthy* always survive.
PARTICIPLE	*The wounded* were left to the mercies of the enemy.[1]
GERUND	*Jogging* is good exercise.
INFINITIVE	*To survive* required all of one's wits.

PHRASES AS SUBJECTS

A phrase, remember, is a functional word-group that does not contain a finite verb (see page 722). Of the four kinds of phrases—prepositional, participial, gerundive, and infinitive—only gerundive and infinitive constructions are commonly used as subjects. Like all phrases, they express as one grammatical unit ideas too complicated to be said in single words.

GERUNDIVE PHRASE AS SUBJECT

Sharing a bed is one of the hardest tests of friendship or even of tolerance.

Reginald Reynolds

Occasionally a gerundive phrase in the subject position is balanced at the other end of the clause or sentence by a gerundive phrase or phrases acting as complement:

---

1. The participle acting as subject is essentially the same as an adjective used this way.

Looking and listening to something with such a group, imaginary or real, means *checking out responses, pointing to particular features, asking detailed questions, sharing momentary excitements.*

Richard Poirier

Gerundive phrases acting as subjects are often placed in anticipatory constructions: the phrase is postponed to the end of the sentence or clause, while the subject position is filled by *it* or, less often, *this, that,* or *there,* words which anticipate the notional subject. (When they function in this way these words are called "expletives.")

It was foolish *taking him to court.*

If the expletive is *this* or *that* the phrase is usually preceded by a comma, colon, or dash:

That was what upset his parents—*his joining the Navy.*

Infinitive phrases are less common as subjects than gerundive ones, at least in the opening position, but they do occur:

To peer into the looking glass is indeed a poetic and fascinating thing. . . .

G. K. Chesterton

To smash something is the ghetto's chronic need.

James Baldwin

More often the infinitive-phrase subject is placed in an anticipatory construction, where it sounds more colloquial and less literary:

It is equally excellent and inconclusive *to say that one must write from experience.* . . .

Henry James

When the expletive is *that, this,* or *there,* the postponed infinitive phrase is generally set off by a colon, comma, or dash:

There is one thing you must remember: *to do nothing without consulting us.*

The infinitive-phrase subjects we have seen thus far have not contained subjects of their own. When the infinitive does have a subject, the entire phrase is preceded by the form word *for*:[2]

For an uninitiated citizen to try to fathom Hollywood argot or the language of the film trade paper Variety is to come unexpectedly upon the tower of Babel.

William Fadiman

---

2. If the subject is a personal pronoun, it must be in the objective form: "They wanted *me to go*"; not × "They wanted *I to go.*"

As a sentence opener, this *for*-pattern is likely to sound old-fashioned. It is usually put into an anticipatory construction:

It has been common all my life *for smart people to perceive in me an easy prey for selfish designs.*

Mark Twain

Like the gerundive construction, an infinitive-phrase subject is sometimes balanced by an infinitive-phrase complement, as in the example above by William Fadiman or in this one:

To regard the dictionary as an "authority," therefore, is *to credit the dictionary writer with gifts of prophecy which neither he nor anyone else possesses.*

S. I. Hayakawa

Using an infinitive phrase as subject is emphatic. Opening directly with the construction is unusual and eye-catching. And even when the sentence begins more conventionally with *it, this, that,* or *there,* postponing the infinitive to the end has the effect of a strong announcement.

CLAUSES AS SUBJECTS

A dependent clause is a functional word group containing a finite verb and acting as a noun or modifier. The noun clause as subject may take two forms. Sometimes it is introduced by the form word *that,* a form grammatically empty in the sense that it performs no function within the clause. Or it may be introduced by a relative word that acts within the clause as subject, complement, prepositional object, or modifier.

That-CLAUSE

That it's rough out there and chancy is no surprise.

Annie Dillard

That John Chaucer was only an assistant seems certain.

John Gardner

Initial *that*-clauses are concise and emphatic, but because they are rather formal, they are often postponed after expletives (*it, there*) in order to create a more colloquial tone:

In our time it is broadly true *that political writing is bad writing.*

George Orwell

There is plenty of evidence *that "democracy" is becoming the same kind of term.*

Richard M. Weaver

CLAUSES WITH FUNCTIONAL RELATIVES

A number of words may serve to introduce a noun clause and at the same time perform a grammatical role within the clause:

who	whosoever	when	whenever
whom	what	where	wherever
whose	whatever	how	however
whoever	which	why	if
whomever	whichever	whether	

In each of the following sentences a noun clause acts as subject, but in each case the relative term serves a different function inside the clause:

Who controls the past controls the future; *who controls the present* controls the past. ("Who" is the subject of the clause.)

C. Vann Woodward

And *what the story does* is to narrate the life in time. ("What" is the direct object of "does.")

E. M. Forster

What Professor Abbot was actually looking for was material for a new life of James Beattie. . . . ("What" is the object of the preposition "for.")

Christopher Morley

Whether Mrs. Harding also knew about her at this time is doubtful. ("Whether" is an adverb.)

Samuel Hopkins Adams

This type of subject-clause is also often postponed after the expletives *it* or *there*:

It may be doubted *whether Harding wanted to live.*

Samuel Hopkins Adams

Complements

The complement is what follows the verb, "completing" the predication the verb begins. A complement may take several forms: a direct object, an indirect object together with a direct object, a predicate adjective, and a predicate noun (see pages 720–21). Because it would be too complicated to illustrate each of these kinds of complements, most of the following examples are simply words or constructions acting as direct objects.

WORDS AS COMPLEMENTS

In addition to nouns and pronouns, adjectives and verbals may serve as complements:

ADJECTIVE   Helen is *shy*. (Here "shy" is a predicate adjective.)
              He helped the *sick*. ("Sick" is an adjective acting as a direct object.)
PARTICIPLE   I prefer *skiing*.
INFINITIVE   I prefer *to ski*.
GERUND   Everyone enjoyed *the skiing*.

PHRASES AS COMPLEMENTS

GERUNDIVE PHRASE AS COMPLEMENT

Gerundive phrases frequently appear after verbs. They are more likely to have objects and modifiers of their own than subjects:

Essentially history for the scissors-and-paste historian means *repeating statements that other people have made before him*.

R. G. Collingwood

Skiing is *sliding down a snowy hill on a pair of expensive boards into a tree*.

Student

Occasionally gerundive-phrase complements do contain subjects. In formal usage that subject is in the possessive form:

It was *all Miss Crawford's doing*.

Jane Austen

However, it is not uncommon to find writers who use the subjective form of nouns (and even of pronouns):

The most extraordinary thing was *the soldiers lining the route*.

Joy Packer

While Packer's sentence is clear, it sometimes happens that failing to use the possessive form for the subject of a gerund can result in ambiguity. This is because English has another construction closely resembling a gerundive phrase but grammatically different. For example, in the sentence "He liked the family living next door," the complement consists of a noun ("the family") modified by a participial phrase ("living next door"). "The family" is the direct object; it is what "he liked"; the fact that they were "living next door" serves to identify them. On the other hand, in the sentence "He liked Susan living next door" we probably cannot construe the complement as a noun modified by a participial phrase, for that would imply that "he" knew several Susans (the phrase merely specifying the one who lived next door). But that is not the writer's meaning; rather the sense is that "he liked" the fact that Susan lived next door. Thus the phrase is a gerundive construction, and a careful writer would have made the grammar—and the sense—clear by writing: "He liked Susan's living next door."

It is good policy, then, to use the possessive form for the subject of a gerund, and particularly to do so when ambiguity may result if you do not.

INFINITIVE PHRASE AS COMPLEMENT

As complements, infinitive phrases are common:

> Mr. Everyman does not wish *to learn the whole truth* or *to arrive at ultimate causes.*
>
> Carl Becker

> In other words, the linguist tries *to find out what kind of social status certain pronunciations, grammatical forms, and expressions have among different groups.*
>
> Neil Postman and Charles Weingartner

Remember that the subject of an infinitive is in the objective form, a fact which requires correct choice when the word involved is a personal pronoun or *who*:

> The plot-maker expects *us to remember,* we expect *him to leave no loose ends.*
>
> E. M. Forster

> You asked *whom to go?*

When an infinitive phrase follows certain verbs, it is used without *to*. Only a few verbs require this idiom: some referring to sense perception (*watch, see, hear, listen to, feel*); others designating compulsion or permission (*make, let, have*):

> We'll have *him paint the ceiling.*
> I watched *him empty the garbage* and heard *him curse.*
>
> Student

> About five years ago I saw *a mockingbird make a straight vertical descent from the roof gutter of a four-story building.*
>
> Annie Dillard

In an informal style the *to* is sometimes dropped in phrases that follow *is* or some other form of the verb *to be*:

The least we can do is *try to be there*.

Annie Dillard

It is not, incidentally, regarded as correct in formal writing to use *and* in place of *to* in such constructions:

× The least we can do is *try and be there*.

One or two verbs may take either the *to*-infinitive or the bare form. These verbs are *help* and, mostly in British usage, *dare*:

I helped *him to fix his car*.
I helped *him fix his car*.

As we saw earlier, infinitive complements are sometimes used to balance infinitive subjects:

But *to define* is also *to differentiate an object, to set forth the features that distinguish it from other objects with which it might be confused.*

Joshua Whatmough

CLAUSES AS COMPLEMENTS

Three types of clauses may serve as complements: those introduced by the form-word *that*; those using a functional relative term (both of which we have seen in the role of subjects); and a contact clause, which does not have a relative word at all.

That-CLAUSES

The activities of Athens east and west suggested *that there was no limit to her ambitions.*

C. A. Robinson, Jr.

People used to think, and some still do, *that Latin should be a universal language.* . . .

Robert A. Hall, Jr.

Occasionally a *that*-clause complement opens a sentence; grammatically, however, such a construction is still a complement of the verb:

That there was danger in such an approach she understood very clearly.

That-clauses may function as object complements, a second object following the direct object after certain verbs (called factitive verbs). An object complement explains how specifically the direct object was changed under the conditions named by the verb. For example, in "They elected Sam president," "president" is an object complement and tells what "Sam" was "elected" to be. Here is an example of a *that*-clause as object complement:

> My three weeks in the Tombs had given me ample proof *that the revolutionary contention that crime is the result of poverty is based on fact.*
>
> Emma Goldman

CLAUSES WITH FUNCTIONAL RELATIVES

In these constructions the relative term does more than introduce the clause: it serves some grammatical function within it:

> We need not ask *what happened next,* but *to whom did it happen.* ("What" is the subject of "happened"; "whom" the object of the preposition "to.")
>
> E. M. Forster

> Now I'd better stop here and explain *how I'm using the word art.* ("How" is an adverb modifying "I'm using.")
>
> Flannery O'Connor

CONTACT CLAUSES

After a small group of common verbs, it is possible (though not necessary) to omit *that* from a noun clause acting as complement.[1] These are verbs signifying perceiving, stating, or engaging in various mental processes. The most frequent are:

hear	sense	pretend	fear	think
feel	understand	prove	hope	imagine
gather	declare	say	know	guess
see	indicate	show	suppose	
seem	mean	decide	tell	

The contact clause is colloquial, suited to an informal style:

> I could tell *he was ready for trouble.*
>
> Student

> I think *I've gone blind or died.*
>
> Annie Dillard

> I gather *it was Mr. Bing's idea of giving the first night customers a fine splash.*
>
> Barbara Tuchman

---

1. Historically this type of clause did not evolve by dropping *that*; its genesis was different. However, it is convenient to speak of the construction as if it were an elliptical *that*-clause.

Objects of Prepositions

WORDS AS OBJECTS OF PREPOSITIONS

A list of prepositions is given on page 714. Besides nouns and pronouns, adjectives and verbals function as prepositional objects:

ADJECTIVE Who worries about *the rich*?
PARTICIPLE She took care of *the injured*.
GERUND They thanked me for *waiting*.
INFINITIVE We were just about *to leave*.

PHRASES AS OBJECTS OF PREPOSITIONS

PREPOSITIONAL PHRASE

A prepositional phrase may serve as the object of another preposition. The idiom is common with verbs of motion:

We ran from *behind the house*.
They went round *by the lilac bush*.
The water came to *within six inches of the house*.
At first we went along *with their plan*.

GERUNDIVE PHRASE

Gerundive phrases are useful as prepositional objects and quite common. Virtually any preposition may take a gerund or a gerundive construction:

It is a new way of *separating the sheep from the goats*.

Ruth Benedict

I can't remember any complaints about *its being un-American*.

Margaret Nicholson

The stock market refused to respond to his sermons about *prosperity being just around the corner*.

Thurman Arnold

(Note that the subject of the gerund, "prosperity," is not in the possessive form, as it usually is in formal usage.)

INFINITIVE PHRASE

Infinitive phrases are not as frequent after prepositions as are gerundive constructions, but they are not rare:

Sporting enthusiasts were indignant over *his attempts to control conditions in the sporting ring.* . . .

Henry F. Pringle

Only the classic beauty of the Athenian girls was left for *the two strangers to admire.*

Lawrence Durrell

CLAUSES AS OBJECTS OF PREPOSITIONS

That-CLAUSE

The noun clause introduced by *that* is not frequently used as a prepositional object, though it occurs with *except* and *in*:

I would have gone to her party except *that I was dating another girl at the time.*

Student

Father Coughlin was lucky in *that he didn't have a real and enduring desperation to play with.* . . .

Wallace Stegner

CLAUSES WITH FUNCTIONAL RELATIVES

These are more common as prepositional objects, especially after *of*:

The program gave a good indication of *what the Klan was all about, or thought it was about.*

Robert Coughlan

Most of *what fills the galleries on Madison Ave.* is simply stuff designed to take advantage of current fads and does not come from an artist's vision or honest creative impulse.

Barbara W. Tuchman

Adjectivals

"Adjectival" is a convenient label for all words and constructions modifying nouns. Adjectivals include adjectives but much else besides.

WORDS AS ADJECTIVALS

Adjectives, of course, are the most common modifiers of nouns, but nouns themselves (called noun adjuncts in this function) and verbals are also used adjectivally:

NOUN ADJUNCT   He was a typical *street* kid.
PARTICIPLE   The armies played a *waiting* game.
INFINITIVE   They have a house *to rent.*

Experienced writers often pile up a series of adjectives as a way of packing meaning into sentences efficiently and economically:

Variegated, virulent, turbulent, literary, inventive, personal, conscienceless and often vicious, the daily newspapers of Paris were the liveliest and most important element in public life.

                                                    Barbara W. Tuchman

Sometimes the adjectives are attached to appositives which precede the subject. In the following example the subject is "he"; the appositive, "sport":

A *pulpy, spluttering, timorous, loose-lipped, dressy country* sport, he loved to loll on the corners, greeting his hundreds of acquaintances with a stale joke and the stock query, "Whaddaya know?"

                                                    Samuel Hopkins Adams

And sometimes the adjectives are gathered on either side of the noun:

A *morbid, gloomy* man, *untaught, unled,* left to feed his soul in grossness and crime, and hard, grinding labor.

                                                    Rebecca Harding Davis

Predicate adjectives may also be piled up. In the following sentence the writer makes not one but six primary assertions about a patroness of the poet Chaucer:

She was, by Chaucer's elegaic account, supremely *modest* yet easily *approachable, refined, temperate, light-hearted,* and *pious* without sterness or coldness.

                                                    John Gardner

PHRASES AS ADJECTIVALS

PREPOSITIONAL PHRASE AS ADJECTIVAL

Prepositional phrases are probably the most common kind of adjectival construction:

The house *on the hill* has a spectacular view *of the river.*

Such phrases are often used in series. Being subordinate constructions, they usually contain secondary rather than primary information:

Afterwards, of course, there were endless discussions *about the shooting of the elephant.*

George Orwell

The passage is from an essay describing how, as a young British police officer in Burma, Orwell had to destroy a rampaging elephant. At this point he is more concerned with the "endless discussions" than with the shooting of the beast.

Yet despite the subordinate nature of the prepositional phrase, skillful writers do use them for major ideas:

This is a story *about love and death in the golden land* and begins with the country.

Joan Didion

The phrase "about love and death in the golden land" is the heart of Didion's idea, giving substance to the abstract term "story."

PARTICIPIAL PHRASE AS ADJECTIVAL

Participial phrases are a very useful form of modification, one beginning writers need to use more. Discussing them is complicated because they may be either restrictive or nonrestrictive and because the nonrestrictive variety is more flexible in its positioning than are adjectivals in general.

First, the distinction between restrictive and nonrestrictive phrases. The following sentence illustrates both:

In the midst of the crowd, *crouching over a smouldering fire,* was a group of Indians *belonging to a remote Mexican tribe.*

Francis Parkman

The first phrase ("crouching over . . .") is nonrestrictive. It tells us something relevant about the Indians, but nothing essential; if we remove it, the basic meaning of the sentence does not change. The second participial construction ("belonging to . . .") is restrictive and essential, identifying the particular Indians the writer focuses on. If we remove this phrase, the sentence sounds silly (to understand why, you need to know that in the context "crowd" means "crowd of Indians"):

In the midst of the crowd (of Indians), crouching over a smouldering fire, was a group of Indians.

Nonrestrictive participial phrases   In modern composition nonrestrictive modifiers—which may be single words or dependent clauses as well as phrases—are separated from their nouns by commas or, less often, dashes. Nonrestrictive modifiers may also be moved about. When they modify the subject they frequently open the sentence, being followed by a comma:

> *Left to themselves,* the facts do not speak. . . .
>
> Carl Becker

> *Having married a young lady of outstanding qualities,* Calvin Coolidge proceeded to put her in her place.
>
> Irving Stone

Alternatively, the nonrestrictive participial phrase may be intruded between the subject and verb. Or it may be postponed to the end of the clause, resulting in a looser, more colloquial tone:

> Words, *being but symbols by which a man expresses his ideas,* are an accurate measure of the range of his thought at any given time.
>
> Albert C. Baugh

> The editor of the Associated Press took over, *being used to such things.*
>
> James Thurber

In any of the four preceding examples the nonrestrictive phrase could be moved to other positions:

> *Left to themselves,* the facts do not speak. . . .
> The facts, *left to themselves,* do not speak. . . .
> The facts do not speak, *left to themselves.* . . .

Such shifts do not alter either the grammatical structure or the logic of the sentence. They do affect emphasis, tone, and rhythm. And sometimes clarity: there is a limit to how far you can remove a nonrestrictive modifier from its noun and still expect readers to make the proper connection

Another limitation on the placement of a nonrestrictive participial phrase occurs when the subject it modifies is a personal pronoun. Then the phrase may precede the subject or come at the end of the clause. But usually it sounds awkward in the interrupting position between subject and verb:

> *Trembling with anger,* he was led from the room.
> He was led from the room, *trembling with anger.*
> AWKWARD He, *trembling with anger,* was led from the room.

Restrictive participial phrases   A restrictive phrase is more fixed: it follows its noun and is not punctuated:

A person *facing danger* does strange things.

As adjectival modifiers such phrases cannot be moved to other positions in the sentence. It is true that this example could have the phrase in different places:

Facing danger, a person does strange things.
A person does strange things, *facing danger.*

But while these variations are perfectly idiomatic, they subtly alter the grammar and the sense. In "a person facing danger" the phrase is adjectival and identifies the individual about whom the predication is made: "a person facing danger" is logically equivalent to "a person who faces danger." On the other hand, in both "facing danger, a person does strange things" and "a person does strange things facing danger," the phrase is really adverbial, not answering the question "who?" but answering the question "when?" "under what circumstances?"[1]

A final point to consider about adjectival participles, both restrictive and nonrestrictive, is the kinds of logical relationships they show. Four are common: condition, cause, attendant circumstance, and specification. The sentence by Carl Becker illustrates *conditionality*:

Left to themselves, the facts do not speak. (*If* they are left to themselves, the facts do not speak.)

Causality is seen in the sentence by Thurber:

The editor of the Associated Press took over, *being used to such things.* (The editor of the Associated Press took over *because* he was used to such things.)

Attendant circumstance is illustrated by the following passage from the English naturalist Gilbert White:

. . . larks rise and fall in large curves, *singing in their descent.*

Their singing is not a cause of the birds' flying in "large curves" and not an effect, but a circumstance that attends upon it.

Finally, participial phrases are used to specify. *Specification* is the primary purpose of the restrictive phrase, as in Parkman's sentence the construction "belonging to a remote Mexican tribe" identifies a particular group of Indians. In the case of nonrestrictive phrases the specification is a bit different. It may be explanatory, as in the following sentence the participles tell us what is meant by "handles well":

1. Not all grammarians would agree that the phrase becomes adverbial, but even if it is construed as adjectival, it is now a nonrestrictive construction and its implications have slightly changed.

The sedan handles well, *steering easily* and *rounding corners and curves with little sway.*

Student

Or the specification may be a way of expanding the topic into a series of accumulating details. In fact, skillful writers sometimes employ the main elements of a sentence to establish the framework of a general idea or situation, which they then elaborate by participial or other modifying constructions:

Joanna, *pursued by the three monks,* ran about the room, *leaping over tables and chairs, sometimes throwing a book or a scriptural text at her pursuers.*

Lawrence Durrell

The main elements and the prepositional phrase "Joanna . . . ran about the room" establish the basic situation, but nothing more. The participles are what render the scene, endowing it with specificity and life. The passage illustrates one of the chief virtues of participles: that they derive from verbs and retain, even in their more abstract uses, something of the force and vitality of verbs.

And participles are economical. They do not require a relative pronoun and an auxiliary as a full clause would:

Joanna, *who was* pursued by the three monks. . . .

And they are able to imply relationships without having to spell them out. The condition, the cause, the fact of attendant circumstance or of specification is implicit in the participle's relation to the word it modifies and does not have to be expressed.

INFINITIVE PHRASE AS ADJECTIVAL

Like participles, infinitives are economical adjectival constructions, allowing a writer to assert more about the subject or object of a sentence without having to repeat the noun:

But we have much *to learn from it.*

E. M. Forster

We seem to be afflicted by a widespread and eroding reluctance *to take any stand on any values, moral, behavioral, or aesthetic.*

Barbara W. Tuchman

Infinitives frequently act as adjectivals with verbs that show compulsion, necessity, capability, purpose, intention, movement toward a goal:

The nature of the fight compelled her *to dramatize the trial as a cruxifiction.*

Carey McWilliams

He is not looking in a dictionary for a key *to open a door to understanding*.
Philip B. Gove

In their passive form infinitives treat the noun they modify as an object receiving the action of the verb-like infinitive:

They were the first books of their kind *to be published in this country*.
Albert C. Baugh

CLAUSES AS ADJECTIVALS

There are three varieties of adjectival clause: those using the relative pronouns *that, which,* and *who* (including the forms *whose* and *whom*); those employing other relative words (*where, when, how, why, while,* and so on); and those having no relative term at all. The first two types are often called relative clauses; the third, the contact clause.

ADJECTIVAL CLAUSES WITH RELATIVE PRONOUNS

These relative clauses are introduced by *that, which,* or *who* (*whose, whom*). While the pronouns may function the same way in the same kinds of clauses, they are not completely interchangeable. The conventions governing their use are complicated. Two considerations apply: (1) whether the noun the relative word stands for designates a human being or something else, and (2) whether the clause is restrictive or nonrestrictive.[2]

Broadly, *who* and its oblique forms are confined to nouns signifying humans and animals associated with man. They may be used in either restrictive or nonrestrictive clauses. Which of the three forms of *who* is correct depends upon how the word functions within the clause. If it is the subject, *who* is necessary ("The boy *who* fell"). *Who* is also required on those occasions—rare for this type of clause—when the word is the complement of a linking verb ("The boy *who* Harry used to be still exists deep within him"). If the *who*-pronoun is the direct object of a verb, the object of a preposition, or the subject of an infinitive, then *whom* is required in a formal style ("The man *whom* you saw"; "The girl to *whom* you gave a ride"; "The friends *whom* I asked to go with me"). If this relative pronoun acts as an adjective, *whose* is the proper form ("The woman *whose* house you bought").

An adjectival clause follows the word it modifies. The relative pronoun introduces the clause, and when it acts as the subject or complement of the verb, it invariably comes at the beginning:

---

2. For "restrictive" and "nonrestrictive" see pages 739–41. For the distinctions governing *that, which,* and *who,* see pages 701–3.

Modern English, especially written English, is full of bad habits *which spread by imitation* and *which can be avoided if one is willing to take the necessary trouble.*

George Orwell

Mr. Hoover, *who was then President,* was an engineer with an engineering mind.

Thurman Arnold

The most striking thing about our present-day civilization is probably the part *which science has played in bringing it to pass.*

Albert C. Baugh

When the relative pronoun functions as the object of a preposition within the clause, the construction follows one of two patterns:

1. Relative word + remainder of clause + preposition
   The idea *that you speak of*
   The idea *which you speak of*
   The man *whom you sold your house to*
2. Preposition + relative word + remainder of clause
   The idea *of which you speak*
   The man *to whom you sold your house*
   NOTE *that* is never used in this pattern.

Beginning with the preposition is more formal and literary; placing the preposition at the end is more colloquial:

That is the fundamental aspect *without which it could not exist.*

E. M. Forster

Spelling reform was one of the innumerable things *that Franklin took an interest in.*

Albert C. Baugh

Now and then, when the relative pronoun is the object of a preposition, a third pattern is possible: subject of clause + preposition + *which* + remainder of clause:

He had squandered monies allocated to a national governors' conference in New Orleans on a riotous, bosomy, party, *testimony concerning which left some Governors red-faced.*

Hodding Carter

This is an unusual construction and a formal one. It occurs only in those rare cases where the prepositional phrase containing the relative pronoun modifies the subject of the clause rather than the verb or complement, as it usually does.

It is also possible to use the relative pronoun as an adjective within the clause. The pattern is most common with *whose,* and the term opens the clause and modifies its subject:

> He couldn't be just a good stone cutter, *whose winged angels were better than other men's.*
>
> R. L. Duffus

Because of its convenience, *whose* is frequently used to refer to nouns that would otherwise take *that* or *which*:

> Many of the marchers carried flaming torches, *whose light threw grotesque shadows up and down Main Street.*
>
> Robert Coughlan

The advantage of the *whose*-clause in such cases is that it avoids the cumbersomeness of an *of-which* construction:

> . . . flaming torches, *the light of which threw.* . . .

Occasionally *which* acts as an adjective within the clause, modifying the subject or sometimes a prepositional object, as in this example:

> Izzy died in New York on February 17, 1938, *by which time his four sons had all become successful lawyers.*
>
> Herbert Asbury

ADJECTIVAL CLAUSES WITH RELATIVE ADVERBS

With nouns of place and time to which the relative adverbs *where* and *when* may be applied, it is possible to introduce adjectival clauses with these words. Such constructions are more idiomatic and economical than their equivalents using a preposition + *that* or *which*. We could say "a place *at which*" or "a time *during which*," but "a place *where*" and "a time *when*" are more concise:

> Finally I fired my two remaining shots into the spot *where I thought his heart must be.*
>
> George Orwell

> October is the bad month for the wind, the month *when breathing is difficult and the hills blaze up spontaneously.*
>
> Joan Didion

CONTACT CLAUSE AS ADJECTIVAL

Contact clauses contain no relative word. The clause simply follows the word it modifies, as in "This is the house *I bought,*" where "I bought" is an adjectival clause modifying "house." Contact clauses are not really elliptical relative clauses—that is, constructions from which the speaker or writer has merely dropped a *that, which,* or *who.* Historically they evolved in a different way, and they cannot be used in every situation where a relative clause would be possible. We shall talk about them as if they were elliptical relative clauses, however, because it is convenient. The most frequent kind of adjectival contact clause is that in which the relative pronoun, if it were present, would be a complement within the clause:

Bret Harte was one of the pleasantest men *I have ever met.*

Mark Twain

In other cases the absent relative term would be the object of a preposition or an infinitive:

Time is but the stream *I go a-fishing in.*

Henry David Thoreau

I stake the time *I'm grateful to have,* the energies *I'm glad to direct.*

Annie Dillard

A contact clause is sometimes possible in place of a relative clause using the adverb *when:*

By the time *we were old enough to choose,* we could not help but doubt.

W. S. Merwin

Contact clauses have a distinctly informal, colloquial flavor. This quality does not at all mean that you should avoid them; simply that they are best when you want a relaxed, conversational tone.

Adverbials

Adverbials include all words, phrases, and clauses which function adverbially—that is, which modify verbs, adjectives, other adverbs, or entire clauses. Because of this greater range of modification, adverbials present more variations than do adjectival constructions. They are also more flexible in positioning and indicate many different relationships of time, place, manner, degree, circumstance, logic, and so on. Adverbials, in short, are complicated, and you might wish at this point to review what was said about adverbs on pages 709–13.

WORDS AS ADVERBIALS

Aside from adverbs proper, nouns, infinitives, and participles may act adverbially:

The people went *home.*
They are coming *tomorrow.*
Everybody left *to watch.*
She looked *longingly* at the picture.

When a present participle such as *longing* functions adverbially, it adds the *-ly* ending and becomes for all practical purposes a regular adverb.

PHRASES AS ADVERBIALS

PREPOSITIONAL PHRASE AS ADVERBIAL

All prepositional phrases look alike, whether adjectival or adverbial. The difference lies in their positioning and punctuation (in speech, pauses and changes of intonation), which signal how the reader is to construe them. Consider these sentences:

The house *on the hill* looks beautiful.
The house looks beautiful *on the hill.*

In form the phrases are identical, but the first is adjectival, identifying the house. The second is adverbial, expressing the condition under which the house is beautiful.

Adverbial prepositional phrases can be placed at the end of the clause, at the beginning, between subject and verb, within the verb phrase, and in other slots as well.

Final position   Final position is the most common, the phrase immediately following the verb if there is no complement, after the complement if one is present:

He appeals *through the senses,* and you cannot appeal *to the senses with abstractions.*

Flannery O'Connor

The execution took place *on August 23, 1927.*

Phil Strong

Opening position   The subject-verb-(complement)-adverb pattern is natural and straightforward, reflecting the usual priorities of how we think. However, placing an adverbial phrase in the final position is not emphatic. When a strong statement is called for, it may be better to shift the phrase to the opening of the clause or sentence, even though this will sound less colloquial. Putting the phrase first stresses the idea it carries:

At this point, I closed my eyes and we knelt together.

Carey McWilliams

Often emphasis attaches to both ends of the statement:

Behind his position he put all of his personal prestige.

John Lardner

In 1928 no one dreamed we were on the verge of a catastrophic depression.

Thurman Arnold

Besides being emphatic an opening phrase may serve to connect a sentence to what has gone before. Describing President Harding's tour of Alaska and the West Coast in 1923, Samuel Hopkins Adams begins the following sentence with a prepositional phrase to signal another stop on the journey:

At Tacoma he had an important message to deliver.

Initial prepositional phrases may or may not be punctuated. Placing a comma after them is a question partly of fashion, partly of individual preference, partly of emphasis, and partly of clarity. Older fashion (so-called close punctuation) used a comma; modern fashion ("open" punctuation) omits it. As for emphasis, the rule is simple: a comma after an initial phrase increases the stress upon it.

When a sentence or clause consists simply of a subject (possibly with modifiers) + verb + prepositional phrase, front-shifting the phrase usually requires that the subject and verb be inverted. Such phrases are not conventionally punctuated:

On the stage was an illuminated scoreboard.

Carey McWilliams

At their backs rose a white marble mantel piece, surmounted by a gold clock and a marble statue of Liberty holding aloft a torch.

Harry Hansen

Adverbial phrases positioned between subject and verb   The end of the clause and the beginning are called "outside" positions. Adverbial prepositional phrases can also go inside the clause. A common slot is between the subject and the verb:

And that, *by every known precedent,* should have been that.

John Lardner

For obvious reasons, this, *like the Nan Britton story,* was ignored by the
newspapers. (This sentence also illustrates a phrase in the opening position.)

Samuel Hopkins Adams

Inserting an adverbial phrase between subject and verb usually
stresses the subject. The phrase is felt to be an interrupting construction
and therefore is generally set off by commas—or dashes—and it is
axiomatic that any comma or other stop throws weight upon the word
preceding it, in this case the subject.

Other inside positions Adverbial prepositional phrases may be set in
a variety of other inside positions, where they are usually punctuated
as interrupters:

There was a moment, *after her return,* when the community was willing
to let bygones be bygones. (The phrase separates a noun and its modifier, the restrictive *when*-clause.)

Carey McWilliams

When, *after two years' incumbency,* Fall resigned from the cabinet to take
open employment with Sinclair, his reputation was still unscathed. (The
phrase separates a subordinating conjunction and the clause it introduces.)

Samuel Hopkins Adams

She seemed to have discovered, *with almost childish delight,* some of the
lighter pleasures of life after she was a grandmother. (The phrase comes
between an infinitive and its object.)

Carey McWilliams

Constant readers must, *after a time,* have come to think of them as The
Typical Catholic Family of the Renaissance. (The adverbial appears within the
verb phrase.)

Robert Coughlan

Indeed Webster was not *in the beginning* sympathetic to spelling
reform. (The phrase interrupts verb and complement, a less common but not unidiom-
atic position. Observe that the writer does not punctuate it.)

Albert C. Baugh

Many prepositional adverbial phrases can appear in all or several
of these positions. The differences are not matters of grammar but of
subtle meaning. For example, placing the phrase first in "*Of course*
he has seen it" affirms the fact of his having seen something; the affirmation would be even stronger if the phrase were punctuated with a
comma. On the other hand, the variation "He, *of course,* has seen it"
emphasizes "he," perhaps in distinction to someone else or perhaps
as a way of hinting at his diligence or nosiness. Such shades of meaning

are subtle but important. An effective style is the total of many, many such subtleties under the control of a good writer.

INFINITIVE PHRASE AS ADVERBIAL

Infinitive phrases do not cover as wide a variety of relationships as prepositional phrases. They are limited primarily to showing reason, purpose, intention, and occasionally subsequence (that is, that one event follows another). Like prepositional phrases, however, infinitive phrases can vary in placement. They are most common at the end or the beginning of a clause, less frequent inside. These adverbial infinitive phrases appear at the end of the clause:

He was only talking *to hear himself talk.* . . .

Mark Twain

Like the bear who went over the mountain, I went out *to see what I could see.*

Annie Dillard

Notice that neither of those sentences has a direct object. In such sentences an infinitive phrase must come at the end. To put it anywhere else would sound un-English. When a clause does contain a direct object, an infinitive phrase can go at either the beginning or the end:

To understand what happened we must go back to Boswell's strict Presbyterian family.

Christopher Morley

Morley's sentence would be as idiomatic if the phrase were at the close.

But it would not be quite as effective. For while such phrases may occur at either end of a clause, a writer does not have a completely free option on where to place them. Depending upon questions of emphasis, context, or tone, one or the other position will probably be better. As with adverbial prepositional phrases, infinitive phrases sound more natural at the close, more literary and emphatic at the beginning. Moreover, one must consider which sequence of ideas best serves one's purpose. Morley, for example, goes on to discuss James Boswell's "strict Presbyterian family"; it is advantageous for him to shift the phrase forward so as to end his sentence on the words that lead naturally into his new topic. Sometimes an initial phrase works in the other direction, hooking the new statement to what has just gone before.

Infinitive phrases may also be placed in interrupting positions:

The case smelled so bad, *to use an inelegant figure of speech,* that Thompson was moved to investigate it.

Phil Strong

Such plants *to operate successfully* had to run at capacity.

Thurman Arnold

But when an infinitive phrase modifies an adjective, it should be kept after the adjective, immediately after if possible:

Too acute *to be taken in by the gross illusion of rationality,* too subtle *to imagine that a home-made abstraction could be a reality,* he derided the academic philosophers. (The phrases modify the adjectives "acute" and "subtle.")

Aldous Huxley

PARTICIPIAL PHRASE AS ADVERBIAL

Customarily participial phrases are treated as adjectivals or as absolutes (we look at absolutes in the next section). Occasionally, however, participial phrases modify the predicate of a sentence rather than the subject or an object. We saw earlier (page 741) that moving the phrase in the sentence "A man *facing death* does strange things" shifts its value from adjectival to adverbial. When it opens or closes the statement like this:

Facing death, a man does strange things
A man does strange things *facing death*

the phrase no longer answers the adjectival question "Who does strange things?" but rather the adverbial question "When—or under what conditions—does a man do strange things?"

Here are two other instances of participial phrases which are essentially adverbial, modifying the entire statement, not just the subject:

Farm prices caved in, *leaving farmers to stagger under a savage burden of debt.*

Arthur M. Schlesinger, Jr.

The station owner may wish that the program would vanish, *leaving him free to sell the time to a local merchant at a much higher profit.*

Gilbert Seldes

CLAUSES AS ADVERBIALS

Adverbial clauses are introduced by one of the subordinating conjunctions listed on page 716. As with adverbial phrases their placement is variable though sometimes restricted by convention. When a clause can be put idiomatically in any of several slots in a sentence, the most effective position will depend upon clarity, emphasis, tone, or the progression of idea that best fits the context.

ADVERBIAL CLAUSES AFTER THE MAIN CLAUSE

Adverbial clauses are most natural after the main clause. Sentences in this pattern are described as "loose"; their tone is conversational rather than literary, matter-of-fact rather than emphatic:

It goes without saying that you will not write a good novel *unless you possess a sense of reality*. . . .

Henry James

The future always looks good in the golden land, *because no one remembers the past.*

Joan Didion

Adverbial clauses set after the main construction may or may not be punctuated, depending upon clarity. Generally a comma is necessary only if its omission might confuse the reader.

ADVERBIAL CLAUSES BEFORE THE MAIN CLAUSE

Placed before the main clause, adverbial clauses sacrifice the naturalness inherent in starting off with the main idea, in putting first things first. On the other hand, such an opening makes for a more emphatic statement, allowing the sentence to build toward its climax, a style described as "periodic." In addition, the order of adverbial clause + main clause may reflect more exactly the logic of events or of a particular idea, especially when the subordinate construction shows precedence, cause, purpose, condition, or concession:

When I left the woods I stepped into a yellow light.

Annie Dillard

Because the civilization of ancient Rome perished in consequence of the invasion of the Barbarians, we are perhaps too apt to think that civilization cannot perish in any other manner.

Alexis de Tocqueville

Most writers still place commas after initial adverbial clauses, particularly if they are of any length and complexity. Occasionally a short, uncomplicated clause that does not contain any internal punctuation is allowed to stand without a comma, as in Dillard's sentence above.

ADVERBIAL CLAUSES IN INSIDE POSITIONS

Adverbial clauses are set inside a main clause less commonly than are single-word adverbs or adverbial phrases. Their added length creates a longer interruption and makes them a greater threat to clarity. But if they are less usual than phrases, interrupting clauses are certainly not rare. They afford a way of controlling emphasis and rhythm in sentences and also of introducing pleasing variation. They are always set off by commas or, less often, by dashes or parentheses. Probably the most frequent slot for interrupting adverbial clauses is between subject and verb:

> My father, *although he has never really said so,* finds it difficult to understand me.
>
> Student

When the subject is a personal pronoun, however, it sounds awkward to split it from the verb by an interrupting clause.

Inside adverbial clauses may be put elsewhere than between subject and verb:

> At last, *when the tribute slackened,* he motioned to his retainers to sweep up the treasure.
>
> Robert Coughlan

> I am informed—*though like Herodotus I make the reservation that it hardly seems credible*—that it is still in the Code of Connecticut that when any three citizens complain that a person is a common scold, that party can and must be ducked, per ducking stool.
>
> Phil Strong

ELLIPTICAL ADVERBIAL CLAUSES

A frequent adverbial construction in English follows this pattern:

Although tired, he decided to go on.

Such a construction is traditionally described as an elliptical adverbial clause, the elements needed to make it a complete clause being, as some grammarians say, "understood":

Although [he was] tired, he decided to go on.[1]

The words that follow the subordinating conjunction may be participles (with or without modifiers and complements), adjectives, a prepositional phrase, or even a finite verb phrase. Like other adverbials, elliptical clauses are variable in placement:

> For a moment—*if only for a moment*—we are utterly estranged.
>
> F. L. Lucas

---

1. Other scholars analyze this construction differently; some, for instance, call it a "conjunction-headed phrase." But while the problem is important to grammarians, it is less significant to students of composition. Whatever we call it and however we analyze it, the construction is common and useful.

Its art treasures had been hidden *when possible,* but many had been looted by Goering and carried off to Germany.

Joy Packer

Other ministers, *while less frantic,* were perhaps no less sure that the Klan was doing God's work.

Robert Coughlan

Elliptical clauses are more economical than full constructions. On the other hand, they diminish the significance of an idea by reducing its relative weight in the sentence. Generally, then, they are best reserved for ideas of secondary or even tertiary value.

Absolutes

Absolutes are a small and special class of constructions which occur as part of a sentence, but which are self-contained in idea and not grammatically tied to the sentence. The common varieties are "nominative," "participial," and "infinitive" absolutes, plus one more which we shall call the "intrusive sentence."

NOMINATIVE ABSOLUTE

This construction came into English as a way of translating a Latin clause called the ablative absolute. In translation, however, the original ablative ending was not used, the noun being placed in the subjective, or nominative, case. The construction consists of participle with a subject and often a complement with attendant modifiers. Frequently nominative absolutes state a cause, a condition, or a preceding event, the effect or subsequent event being expressed in the main clause. In such cases the logic of idea or time is best served by placing the absolute first:

The savings of the nation having been absorbed by Wall Street, the people were persuaded to borrow money on their farms, factories, homes, machinery, and every other tangible asset.

Irving Stone

Her milking finished, the lovely Joanna would gather the cherries which had fallen in the orchard. . . .

Lawrence Durrell

However, writers may prefer to place the nominative absolute—even when it shows cause or condition—after the main construction:

A small village, not a thousand feet away, was blazing in spots, *bits of oil having fallen upon the roofs.*

Theodore Dreiser

His performance is not the kind I care much about—*sincere, vulnerable movie heroes generally not being equipped for much besides trite suffering*— but O'Neal knows how to be emotional without being a slob, and in this unskilled, mediocre piece of movie-making (Arthur Hiller directed) his professionalism shines.

Pauline Kael

(There is, incidentally, another absolute in Kael's sentence. Can you find it?)

In addition to cause and precedence, nominative absolutes may express attendant circumstances—something which follows or accompanies what is stated in the main clause. With this meaning the absolute is more common after the main construction:

It rained steadily, *the clouds resting upon the very tree tops.*

Francis Parkman

They moved upward in the white mists before the sun came out, *the dark gorges aflap with shrieking birds.*

W. S. Merwin

When the participial in a nominative absolute is a form of the verb *to be* (*being, having been*), it is sometimes omitted. The result may be called an "elliptical nominative absolute":

He is medium height, well built, *his tails and starched shirt immaculate.*

Joy Packer

Bolenciecwcz was staring at the floor now, trying to think, *his great brow furrowed, his huge hands rubbing together, his face red.*

James Thurber

From time to time one hears warnings against nominative absolutes as being old-fashioned or too literary. But in fact they are neither. As the examples show, contemporary writers use them—and often writers like James Thurber and Pauline Kael, who cannot be accused of old-fashionedness or pedantry. No doubt nominative absolutes can be misused or so overused that they become an awkward mannerism, but so can any type of clause or phrase. The nominative absolute is still a useful construction, at home in a modern style.

PARTICIPIAL ABSOLUTE

This consists of a participle (usually the present active) together with a complement and/or modifiers; the participle does not have a subject, and this distinguishes the construction from the nominative absolute. Participial absolutes often signal a change of subject by summing up a preceding topic and applying it loosely to a new one:

> *Speaking of daughters,* I have seen Miss Dombey.
>
> Charles Dickens

> *Talking of war,* there'll be trouble in the Balkans in the spring.
>
> Rudyard Kipling

Sometimes, too, they are used to predicate a general condition or concession which applies to the idea expressed in the main clause:

> *Culturally speaking,* the importance of the *Sgt. Pepper* album is that it finally put the Beatles, in the summer of 1967, beyond the shabby treatment or defensive patronization of any of these factions.
>
> Richard Poirier

> . . . the new paperback edition [of *Love Story*] in this country is said to have sold over five million copies. *Allowing for hyperbole and halving the figure,* that is still one hell of a pile of pulp.
>
> Pauline Kael

Participial absolutes should be distinguished from dangling participles, as in a sentence like "*Running down the street,* the Church was on fire." (See page 761 for a discussion of danglers.) On the surface the two constructions seem the same. But the absolute participle expresses an encompassing idea loosely connected to the main point in a conventional relationship which the reader can easily understand. It is, moreover, a deliberate device of style used because of its economy. The dangling participle, on the other hand, implies an unconventional and ridiculous relationship of thought (churches, for example, don't run down streets). Hence it is awkward and, usually, the result of a writer's not being aware of what his or her words imply.

INFINITIVE ABSOLUTE

The infinitive absolute has two forms. One is like the nominative absolute: it has a subject in the subjective case followed by the infinitive, which may have a complement and/or adverbial modifiers. The construction is rare and is used to show action subsequent to the event described in the main clause:

> The men returned to camp, *the search to be continued in the morning.*
>
> Student

> Everyone came out of their houses: *the women to go to the cemetery and take food to the graves of the newly dead, old men to idle by the roadside and to spin their tales to strangers.*
>
> Anne Kindersley

The other infinitive absolute is parallel to the participial absolute, consisting simply of an infinitive (without subject) plus complement and/or modifiers:

> *To revert for a moment to the story told in the first person,* it is plain that in that case the narrator has no such liberty. . . .
>
> Percy Lubbock

INTRUSIVE SENTENCES

More often than one might think, experienced writers place a new, grammatically complete statement inside a sentence without connecting it in any way. Such intrusive sentences are absolute in the sense that they are not tied to the statement which contains them, but unlike other absolutes these take conventional sentence form and can stand alone:

> In one of my books—*Huckleberry Finn, I think*—I have used one of Jim's impromptu tales, which he called "The Tragedy of the Burning Shame."
>
> Mark Twain

> . . . he [Charles Lindbergh] took three semesters in engineering at the University of Wisconsin, where the only thing that seemed to interest him very much was shooting (*he made the rifle team*).
>
> John Lardner

Intrusive sentences are added to explain or justify something a writer has just said. Used sparingly, they suggest an informal, colloquial tone, the speaking voice, as it were, of the writer. They reflect a habit of speech: we often start to say something, pause, insert a new statement by way of explanation, pause again, and return finally to complete the original idea. We clearly signal the intrusion by the pauses and by vocal intonation.

In writing, these signals must be replaced by punctuation, and the intrusive sentence is always cut off by either dashes or parentheses. (In this function the dashes are less formal.) Commas cannot be used around intrusive sentences since they imply that the construction is a grammatical part of the containing sentence, which it isn't. Note, however, that while dashes or parentheses *must* be used around intrusive sentences, they also *may* be used in place of commas around interrupting constructions that are not absolute—nonrestrictive adjectival clauses or participial phrases, for instance, or various kinds of adverbial constructions.

OTHER ABSOLUTES

In addition to the four kinds we have seen, writers occasionally use other constructions and words absolutely:

> *No matter,* we persevered and were rewarded.
>
> Mark Twain

> *Well*—that was the end of that!
>
> Student

Appositives

An appositive is a word, phrase, or clause which signifies the same thing as another term or expression in a sentence but which is not syntactically connected with it. In "John, my brother, is here," "brother" is an appositive—in apposition, as grammarians say, to "John." Often, as in this example, an appositive and the word it relates to are side by side (in fact, "appositive" comes from Latin words meaning "to set near to"). However, appositives do not have to be placed next to the other word; thus we might say "John is here, my brother," adding the appositive almost as an afterthought. The grammatical pattern is the same, even though the appositive has been shifted to the end of the sentence.

The skillful use of appositives is one sign of mature, sophisticated writing. They serve three overlapping functions: to explain or define an important idea, to expand it, or to anchor an otherwise loose modifier. Here are a few examples of *explanatory* appositives:

> This last point—*the relation of characters to the other aspects of the novel*—will form the subject of a future inquiry.
>
> E. M. Forster

He declined a first suggestion, *to become street cleaning commissioner,* because he felt no special qualification for the post.

Henry F. Pringle

Bound to the production of staples—*tobacco, cotton, rice, sugar*—the soil suffered from erosion and neglect.

Oscar Handlin

But the headwaiter, *a man of scant perception,* bowed deferentially and sold them a bottle of whiskey.

Herbert Asbury

Appositives are also very useful for *expanding* a key idea:

It is the trail of an intention gone haywire, *the flotsam of the New California.*

Joan Didion

More important, Bethel demonstrated the unique sense of community that seems to exist among the young, *their mystical feeling for themselves as a special group, an "us" in contrast to "them."*

Time magazine

By using "flotsam" in apposition to "trail" Didion expands her point without having to repeat the meaningless "it is," achieving both economy and emphasis. Similarly the writer for *Time* efficiently enlarges "sense of community" by the appositives "mystical feeling" and "an 'us.'"

Writers often carry this expansive technique further, piling up several appositives in a row:

Harcourt was a noble specimen of the French aristocracy in the days of its highest splendour, *a finished gentleman, a brave soldier, and a skilful diplomatist.*

Thomas Babington Macaulay

He was the pioneer par excellence, *the Bayard of the backwoods, the rough diamond, the epitome of the noble, free soul.*

Thomas Pyles

That beloved President, *Warren Gamaliel Harding, the idol of the man on the street, the apotheosis of the Average American, the exemplar of the triumphant commonplace who lay, stricken, in a San Francisco hotel,* was on the mend.

Samuel Hopkins Adams

Notice in these passages how much is given over to the appositives; in the last it is thirty-one words out of thirty-eight.

Appositives may also precede headwords:

A once-defeated demagogue trying for a comeback, he tried what other demagogues abroad had found a useful instrument: terror.

Wallace Stegner

A member of the State legislature in 1884, Roosevelt had led in the fight
for the passage of four or five bills backed by civic organizations in New
York.

Henry F. Pringle

A common kind of appositive for expanding a key idea is the noun
clause:

Midway, even more than the Battle of the Coral Sea, taught the lesson
*that the gun had been superseded by aircraft as the main weapon of naval
warfare.*

Malcolm Kennedy

He brought with him to America the Hegelian concept *that man might
eventually free himself from the irrationalities of history by mastering nature.*

Samuel C. Florman

In such constructions an abstract term ("lesson," "concept") is, in effect,
an empty container into which is poured the idea expressed in the
appositive clause, filling the abstraction with meaning.

Infinitive phrases may also be used in a similar appositive function:

He declined a first suggestion, *to become street cleaning commissioner,* because he felt no special qualification for the post.

Henry F. Pringle

Appositives, finally, are useful to *anchor down* loose adjectival modifiers—either *which*-clauses that modify the whole of a preceding statement rather than any one word, or phrases far removed from their
headwords. In such cases it is always easy for unwary readers to connect
the constructions to the wrong thing, and a careful writer uses appositives as a way of preventing this:

But now disturbing reports that struck nearer home reached his ears, *matters about which informed circles had been gossiping for months.*

Samuel Hopkins Adams

"Matters" is in apposition to "reports"; remove it and the sentence appears
to say that people had been gossiping about President Harding's ears.

Police headquarters was still on Mulberry Street in 1895, *a gloomy building
with subterranean dungeons where rats and vermin assisted the persuasive
effectiveness of the third degree.*

Henry F. Pringle

"A gloomy building" is in apposition to "headquarters"; without it a reader
might suppose the dungeons to be beneath Mulberry Street.

No Hollywood writer would consider his screenplay complete unless he included a dramatic device to lend it novelty, *a device known as a "gimmick."*

William Fadiman

"Device" is in apposition to "device." The repetition prevents a reader from taking "gimmick" as applying to "novelty."

All these examples of appositives are separated from their headword by commas or dashes. This is because they are nonrestrictive. An appositive can also be restrictive, conveying information essential to our understanding of the headword. In that case it is not punctuated:

Dickens' novel *David Copperfield* is his greatest work.

Johnson *the lexicographer* is not quite the same person as Johnson *the essayist* or Johnson *the tavern companion.*

Murky Modifiers

Adjectival and adverbial constructions create problems when a writer fails to make clear exactly what they modify. These problems are of several kinds: the "dangling modifier," the "misrelated modifier," the "squinting adverb," and the "loose *which*-clause."

DANGLING MODIFIERS

Dang.

Danglers are phrases or clauses that do not hook up grammatically with what they are intended to modify. The idea they relate to, if it is expressed in the sentence at all, is not in a form or position to take the modification. Sometimes the idea is not even in the sentence, either occurring elsewhere in the context or simply being left implicit. In any case, the dangler has nothing it can logically modify, and so is left hanging in the air. The error is most common with adjectival participial phrases, especially when they open a sentence:

Running down the street, the church was on fire.

Being a wandering fish peddler, the man's alibi was worthless.

Understanding its difficulties, the problem can be solved only by us.

Such opening participial constructions are taken to modify the following noun (usually the subject of the sentence), but in the case of a dangler, that relationship is nonsensical, even comic: a church cannot run down a street, an alibi cannot be a fish peddler, a problem cannot understand anything.

Dangling participles can be fixed in either of two ways. First, the main clause can be left alone and the participial construction altered, usually to an adverbial clause or phrase:

When the man ran down the street, the church was on fire.

Since he was only a wandering fish peddler, the man's alibi was worthless.

Because of its difficulties, the problem can be solved only by us.

Second, the participial construction can be allowed to stand and the main clause revised so as to put the modified idea in the subject position, where it can take the modification:

Running down the street, the man saw that the church was on fire.

Being a wandering fish peddler, the man was automatically distrusted and his alibi held to be worthless.

Understanding its difficulties, only we can solve the problem.

Adverbial phrases can also act like dangling modifiers:

On his return from the trip, there was another scene with his father.

However, *in understanding the place of the Klan in American life,* Dr. Evans' significance is less than that of "Colonel" Simmons'.

Such phrases are not danglers to the same degree as "Running down the street, the church was on fire" since they can be construed as loose sentence adverbs. Even so, they are awkward, and for the same reason: there is nothing to take the idea expressed by the phrase—no one, in the first sentence, to "return" from a trip; no one, in the second, capable of "understanding the place of the Klan in American life." The remedy is the same as it is for the true dangler: change the phrase or change the main clause:

When he returned from the trip, there was another scene with his father.

On his return from the trip, he had another scene with his father.

Mis. mod. MISRELATED MODIFIERS

Here the problem is not one of grammar but of position. A misrelated phrase or clause modifies a word in the sentence and that word is in the correct form to take the modification. But the phrase is awkwardly placed, so far removed from its headword that the relationship is obscured. It may even be attracted to another, closer, term so that a reader takes it to modify the wrong thing. This can happen with both adjectival and adverbial constructions. Here are two examples of misrelated adjectival phrases, one prepositional, the other infinitive:

We saw an advertisement in a magazine *about a new kind of beer.*

Only the Empire Impala [a freighter] dropped behind to rescue survivors *to be torpedoed and sunk shortly after daybreak.*

What happens in such cases is that the phrase, rather than directly following its headword, comes after another construction that happens to end with a noun capable of attracting the modifier. In the first example the writer intended us to understand that the advertisement was "about a new kind of beer," but one can suppose that the entire magazine was about the beer. In the second example it is the ship that was torpedoed, but it sounds as though it were the survivors.

To correct a misrelated modifier, the sentence must be revised so as to bring the phrase closer to the word it modifies and break up the attraction between the phrase and the wrong term. Sometimes this can be done simply by shifting one of the constructions before the headword, leaving the other in a clear following position:

In a magazine we saw an advertisement about a new kind of beer.

To rescue survivors only one ship dropped behind, the Empire Impala, only to be torpedoed and sunk shortly after daybreak.

It may also be possible to assure clarity by reducing the shifted phrase to a single word:

We saw a magazine advertisement about a new kind of beer.

Occasionally both phrases can be kept after the headword and simply reversed (though this may lead to an equally awkward progression of thought):

We saw an advertisement about a new kind of beer in a magazine.

(Now it almost sounds as though the beer were in the magazine.)

When the lack of clarity cannot be cured either by front-shifting one of the constructions or by reversing their order, try repeating the headword as an appositive before the second phrase:

We saw an advertisement in a magazine, an advertisement about a new kind of beer.

Repetitiousness can be avoided by using a variant form or a synonym and by adding modifiers:

We saw an advertisement in a magazine, a slick, glossy ad about a new kind of beer.

Only the Empire Impala dropped behind to rescue survivors, a luckless vessel which was to be torpedoed and sunk shortly after daybreak.

Occasionally punctuation may solve the problem:

Once before Richard had been a prisoner in the Tower—surrounded by armed peasants and citizenry.

Henry F. Hutchinson

The dash makes clear that we are to connect "surrounded by" to "Richard," not to "Tower."

When nothing else will do, the sentence may have to be split into two clauses:

Only the Empire Impala dropped behind to rescue survivors, and this ship was to be torpedoed and sunk shortly after daybreak.

Adverbial constructions may also fail to connect clearly to what they modify. A problem sometimes arises when an adverbial phrase follows a complement which it could logically modify and to which it is therefore attracted. Compare these two sentences:

John lost his hat at the beach.
John remembered his date at the beach.

The first presents no difficulty; we read the phrase as an adverb modifying "lost" because we know that a "hat" cannot be "at the beach" as a person can. But the second sentence is ambiguous because "at the beach" could be an adjectival phrase modifying "date" (which might mean, of course, either a girl or an appointment), or it could be an adverbial modifying "remembered." Consequently we do not know whether the meaning is one or another of these possibilities:

1. At the beach John remembered a date that he should have kept someplace else.
2. John remembered someplace else a date he should have kept at the beach.
3. John recollected some time later a date he had once had at the beach.

Here is another example:

A warden's business is to ward the people who are put in his charge without any questions.

Does "without any questions" modify "put" or "to ward"? Probably the latter, in which case the phrase needs to be placed before the adjectival clause:

A warden's business is to ward, without any questions, the people who are put in his charge.

Squint.

SQUINTING ADVERB

A squinting adverb is a special kind of misrelated modifier, an adverbial word, phrase, or clause set between two constructions with the potential for modifying either:

The explanation he gave *at first* sounded reasonable.

There are two interpretations:

1. The explanation he first gave sounded reasonable.
2. The explanation he gave sounded reasonable at first.

As you can see there is no one method of correcting a squinting adverb, or any other kind of misrelated modifier. The best way to avoid the problem is to anticipate it by putting yourself in your reader's shoes, trying to spot places where he or she may make a wrong connection.

Loose which

THE LOOSE *Which-CLAUSE*

At times a nonrestrictive adjectival clause, following the main construction, confuses a reader by being too far from its headword:

There were many reasons why I put off college for two years after graduating from high school, *which no longer seem important.*

The *which*-clause modifies "reasons," but it is uncomfortably far from that noun. The remedy may be to shift the clause forward so that it follows the modified term:

There were many reasons, which no longer seem important, why I put off college for two years after graduating from high school.

If such a revision sounds awkward or if you wish to keep the clause in the final position so as to lead into the next sentence, an alternative solution is to repeat the headword (or its synonym) as an appositive, placing it just before the clause:

There were many reasons why I put off college for two years after graduating from high school, reasons which no longer seem important.

In other *which*-clauses of this kind, the problem may be that they do not modify any single word in the preceding statement, but rather the entire idea. This loose employment of *which* is not uncommon:

I felt bored and out of place at summer camp, *which led me to conclude that it was not for me.*

Student

On the other hand there were no rapes [at the Woodstock Festival], no assaults, no robberies and, as far as anyone can recall, not a single fight, *which is more than can be said for most sporting events held in New York City.*

Time magazine

Loose *which*-clauses like these are not really objectionable; they cause no confusion. However, as a preceding idea grows longer and more diffuse, the meaning of *which* becomes fuzzier and fuzzier. In such a case the solution—if you wish to keep the clause—is to use an appositive. Select a noun or gerund that fairly sums up what you want the *which* to refer to and place it before the clause as an appositive anchor:

I felt bored and out of place at summer camp, a feeling which led me to conclude that it was not for me.

Problems in Agreement

Agreement means that the grammatical form of one word—whether singular or plural; masculine or feminine; first, second, or third person— is determined by another word. Errors in agreement occur when this grammatical relationship between two words is violated. In English, agreement pertains in three situations: between subjects and verbs, pronouns and antecedents, and modifiers and headwords. In addition there is also a kind of agreement that is essentially logical.

Agr. s.-v. ## AGREEMENT OF SUBJECT AND VERB

The rule here is simple: a singular subject requires a singular verb; a plural subject, a plural verb. Applying the rule, however, is sometimes complicated. Errors are likely in the following circumstances.

Agr. s.-v. Comp. ### COMPOUND SUBJECTS

A compound subject consists of two or more members coordinated by *and* or *both . . . and.* Compound subjects are plural without regard to the number of their individual members:

The woman and the child *are* waiting outside.
The women and the children *are* waiting outside.

However, not all coordinated subjects are compound. When two or more subjects are joined by the coordinating conjunctions *or, either . . . or,* or *neither . . . nor,* logic requires that the subjects be regarded separately, and the number of the verb will depend upon that of the individual terms. If both subjects are singular, the verb is singular:

Either John or Mary *is* going to the store.
Neither John nor Mary *is* going to the store.

If both members are plural, the verb is plural:

Either the men or the women *are* able to take care of the children.

Neither the men nor the women *are* able to take care of the children.

If one subject is singular and the other plural, the simplest solution is to let the verb agree with the one closest to it:

Neither the leaders nor the opportunity *was* available.
Neither the opportunity nor the leaders *were* available.

Should such a sentence appear awkward, it ought to be revised, and the compound subject made into two clauses:

The opportunity was not available, nor were the leaders.

Trouble may also be caused by a type of construction which looks like a compound subject but which is grammatically different:

Bill, along with Harry, is going to the convention.

In a sentence of that sort only one term ("Bill" in the example) is the subject, and it alone determines the number of the verb. The other term ("Harry") is the object of a preposition (here "along with"). While logically such a construction means "John and Harry," grammatically it is not "John and Harry." The prepositions *together with, in addition to,* and *besides* are also frequently used in this construction.

Agr. s.-v.
attract.
ATTRACTION

Sometimes a subject is followed by a modifying phrase or clause which happens to end on a noun having a number different from that of the subject and which immediately precedes the verb. In such circumstances it easily happens that the verb is "attracted" to the wrong noun with a consequent error in agreement:

The players named by the coach × *is* to make the trip. (Should be *are.*)

The coach selected by the players × *are* to receive the award. (Should be *is*.)

Agr. s.-v.
ind. pron.

INDEFINITE PRONOUNS IN -ONE AND -BODY

All indefinite pronouns ending in *-one* and *-body* are treated as singular in formal usage, even though their implied meanings may involve a number of individuals:

The people were stunned. While *none was* able to explain exactly what had happened, *everyone was* aware of the disaster.

Everybody knows the truth, but *nobody is* willing to admit it.

Agr. s.-v.
Coll.

COLLECTIVE NOUNS

Collective nouns designate a group of objects or people: *team, crowd, remainder,* and so on. In their singular form these nouns usually take a singular verb and are referred to by singular pronouns, the group being considered as a unit. But if the group is regarded as a number of individuals, then the verb (and pronouns) should be in the plural:

The *team is* leaving tomorrow. (The individuals considered as a single unit.)
The *team are* dressing for the game. (The total number of individual players.)

Once you have established a collective noun as singular or plural in a particular context, you should refer to it consistently, avoiding errors like these:

The team *is* leaving tomorrow. × *They* will arrive at 9:00 P.M.

The team *are* dressing for the game. × *It* will be on the field in twenty minutes.

Agr. pron.

AGREEMENT OF PRONOUN AND ANTECEDENT

Pronouns must agree with the word they refer to (called the antecedent) in number, person, and gender. In English, however, the case of the pronoun is determined by its function, not by the case of the word to which it refers.

Agr. pron.
numb.

AGREEMENT IN NUMBER

A singular noun requires a singular pronoun, and a plural noun requires a plural pronoun:

The *book* you want is on the table. Return *it* by Friday.

The *books* you want are on the table. Return *them* by Friday.

Special problems are posed by collective nouns and by the indefinite pronouns in *-one* and *-body*. In their singular form collective nouns may be treated either as singular or as plural, depending upon the sense you give them. Once you have decided their number, however, be consistent; don't treat them both ways at once:

CONSISTENT   The class *are* gathering *their* books together.
              The class *is* gathering *its* books together.
INCONSISTENT The class *are* gathering *its* books together.
              The class *is* gathering *their* books together.

All forms of the indefinite pronouns ending in *-one* and *-body* are construed as singular in formal usage, even though their implied senses may be plural. In colloquial usage, on the other hand, the antecedents *everyone* and *everybody* are often used with plural verbs and personal pronouns:

FORMAL   Everybody is gathering his books.
COLLOQUIAL Everybody are gathering their books.

AGREEMENT IN PERSON

Agr. pron. pers.

Most often personal pronouns refer to a noun that is neither the speaker nor the spoken-to. Hence the third person is called for:

The *man* was standing on the corner. *He* seemed to be waiting for someone.

With direct address, the second person pronoun is required:

What's the matter, Susan? Has the cat got *your* tongue?

The generic or impersonal *you* (that is, *you* in the sense of "anyone") causes errors in composition, since the implicit sense is third person but formal usage requires that the pronoun be regarded as second person. In speech the impersonal *you* is often used where formal usage would demand a third-person pronoun:

FORMAL   An army *recruit* enters a strange new world. At first *he* thinks nothing makes sense.
COLLOQUIAL An army *recruit* enters a strange new world. At first *you* think nothing makes sense.

AGREEMENT IN GENDER

Agr. pron. gend.

When referring to singular nouns signifying human beings but having no sexual distinction (*doctor, teacher, student,* for example) or to impersonal pronouns like *one, none, anybody,* and so on, you must decide whether to use simply *he, his, him,* or to acknowledge both sexes with *he* and *she, his* and *her, him* and *her*:

Each student must meet with *his* adviser.
Each student must meet with *his* or *her* adviser.

Neither form is completely satisfactory. Using the masculine pronoun alone seems sexist; using both genders is cumbersome, especially if the reference must be repeated:

Each student must see *his* or *her* adviser before *he* or *she* registers for the spring term.

In the past such nouns were referred to simply by *he, his, him,* and it was assumed that the context made clear that no exclusively masculine significance was intended. In such constructions *he* ceases to have gender.

However, if the sense demands explicit identification of both sexes, or if you feel strongly about the sexist implications of the masculine pronoun alone, write both *he* and *she*. But avoid awkward repetitions. Sometimes it is better to turn the noun into a plural so you can use the unisex *they, their, them*:

All students must see their advisers.

Pron. ref.

FAULTY REFERENCE OF PRONOUNS

Faulty reference is not strictly speaking a matter of agreement. Rather the error is that the pronoun is not clear, either because it has no antecedent at all or because it could refer to two or more antecedents. An example of the former case is this passage:

Before taking children on field trips, the school must obtain parental approval. Usually *they* agree.

The initial sentence contains no noun to which "they" can refer. The writer means that the parents usually agree, but since the term *parents*—or some equivalent noun—does not appear in the first sentence, "they" has no antecedent. "Parental" appears, of course. But a pronoun should refer to a substantive—that is, to a word or construction acting as a subject or object. It does not commonly refer to adjectives (like "parental") or to adverbs.

This kind of faulty reference can be mended in two ways: (1) by keeping the pronoun and supplying a proper antecedent, or (2) by substituting an appropriate noun for the pronoun:

1. Before taking children on field trips, the school must obtain permission from the parents. Usually they agree.
2. Before taking children on field trips, the school must obtain parental approval. Usually the parents agree.

The second case of faulty reference involves ambiguity: the pronoun may be read as pointing to either of two possible antecedents, as in:

> Husbands often distrust their wives. *They* magnify problems out of all proportion.

Here "they" can refer to "husbands" or to "wives," and since the second sentence can apply to either, the passage is ambiguous. It could mean that husbands magnify problems (presumably as a result of not trusting their wives), or that wives magnify problems (which is why husbands do not trust them).

Such ambiguity can be resolved in several ways: (1) by making one of the potential antecedents singular and the other plural, if the sense allows this, distinguishing the reference by *he* (*she*) and *they*; (2) by making both singular, if there is a difference in gender, and allowing *he* and *she* to mark the distinction; or (3) by replacing the pronoun with an unambiguous noun:

1. Husbands often distrust their wife. They (she). . . .
2. A husband often distrusts his wife. He (she). . . .
3. Husbands often distrust their wives. Men (women). . . .

Vague this

A special problem of reference occurs with the demonstrative pronouns *this* and *that*, along with their plurals *these* and *those*, and with *such*. When one of these words (most commonly *this*) acts as a subject or object, it often refers not to a single antecedent but to the whole of a preceding idea:

> The man looked furtively over one shoulder and hurried around the corner. *This* aroused her suspicions.

Here the antecedent of "this" is not any one word in the initial sentence, but the whole idea expressed by that sentence.

Some people prefer that *this* not be used at all in place of a noun, especially as the subject of a sentence. On the other hand, good writers often do employ the word in this way without confusion: in the above example it is perfectly clear what "this" means. A better rule is to use *this* (and the other pronouns mentioned) as a subject only when its significance is obvious. If any question exists about its clarity, make *this* an adjectival, modifying a noun that fairly sums up the preceding idea:

> This strange behavior aroused her suspicions.

Agr. mod. ## AGREEMENT OF MODIFIER AND HEADWORD

Modifiers that designate singular or plural quantity must be followed by nouns having the appropriate number. *This, that,* and *one,* for example, are singular; *these, those, many, few,* and so on are plural:

this man   *these* men
that child   *those* children
one book   *many* books

A few quantifying adjectives are indefinite as to number and may be used with either singular or plural nouns:

some tourist   *some* tourists
any woman   *any* women

agr. log. **LOGICAL AGREEMENT**

Logical agreement means that the singularity or plurality of one word requires the same number in another, even though the words have only a semantic relationship and not a grammatical one. In the following sentence, for instance, logical agreement is violated:

Many New York advertising *executives* frequently fly to Chicago on a business *trip*.

Logically, executives take trips.

Close attention to logical agreement, however, is more common in formal usage than in colloquial.

Common Conjunctive Adverbs

Conjunctive (or transitional) adverbs show a relationship in idea between two statements without connecting them in any grammatical way. They signal a writer's flow of thought and prepare readers for turns that flow is about to take. *Therefore* at the beginning of a sentence warns, "Here comes a consequence"; *on the other hand* says, "A contrast will follow"; *besides* prepares us for an idea logically parallel to what has just been said, often a second reason for the same result.

Good writers vary considerably in how often they use conjunctive adverbs. One can call upon them too often, laboring obvious relationships which readers can see for themselves. But a more common mistake among the inexperienced is not employing conjunctive adverbs often enough.

Accordingly, a discussion of the more common of these words and phrases may be useful. They are loosely grouped in terms of the relationships they show. A few expressions function in two or more groups, and some more usually act as coordinating conjunctions—*and, but, for, or*—or as subordinating conjunctions—*so, though.* (The subordinators are best reserved for an informal style.)

The groupings are crude, and you would do well to remember that conjunctive adverbs classed together are rarely exact synonyms. They differ in tone and in shade of thought. As a signal of consequence, for example, *so* is much less formal than *therefore. But, yet,* and *still* all show contradiction, but of subtly different kinds.

The lists, in short, should help you to develop a working vocabulary of conjunctive adverbs. But to use them really well, you need to observe the contexts in which good writers place them.

ADDITION

The new idea parallels what has just been said, a second or third reason or result or example:

But lately I have developed also a sense of destination, or destiny. *And* a sense that if I am to be on quest, I must expect to live like a pilgrim. . . .

Mary Caroline Richards

additionally	furthermore
again	in addition
also	more
and	moreover
and besides	not only . . . but (also)
besides	plus
either	too
further	

When the relationship of addition is between negated statements, two conjunctive adverbs are available:

neither
nor

COMPARISON AND SIMILARITY

The new idea is comparable—often similar—to the preceding one:

One of a woman's jobs in this society is to be an attractive sexual object, and clothes and makeup are tools of the trade. *Similarly,* buying food and household furnishings is a domestic task. . . .

Ellen Willis

by comparison	in similar manner (fashion)
by the same token	likewise
in comparison	similarly
in like manner (fashion)	

CONCESSION, QUALIFICATION

The new topic limits the truth or applicability of what has just been said:

There were, *of course,* certain things to deplore about Bethel.

Time magazine

admittedly	obviously
certainly	of course
doubtlessly	to be sure
granted	true
no doubt	

CONCLUSION

These adverbs signal termination, a final inference in the logical sense, or both:

These examples, *then,* illustrate the first category of restrictions affecting freedom of expression—those which limit the telling of falsehood or the telling of half-truths.

Roger Fisher

and so	in the final (last) analysis
at last	lastly
finally	then
in closing	to conclude
in conclusion	

CONSEQUENCE

The new topic follows from the earlier as a result or an effect:

[The New Deal] violated two basic tenets of the economic fundamentalists. . . . *Hence,* except for a brief period during the first days of the NRA, the New Deal, instead of reassuring business, literally scared it to death.

Thurman Arnold

accordingly	hence
and so	in consequence
as a result	so
consequently	therefore
for this reason	thus

Sometimes the new sentence expresses an unfortunate result that will occur if the preceding condition is not fulfilled:

We must mean what we say, from our innermost heart to the outermost galaxy. *Otherwise* we are lost and dizzy in a maze of reflections.

Mary Caroline Richards

if not
or else
otherwise

CONTRADICTION

Contradiction covers several possibilities: the new statement may assert a contrary fact or idea ("The city was cold and impersonal. *But* it was not dull."); or something that seems to violate reason or logic ("It was late. *But* we didn't leave."); or a strengthening of emphasis ("These reasons were important. *But* others mattered more.").

actually	notwithstanding this
and yet	only

at the same time
but
but yet
even so
even then
for all that
instead
however
nevertheless
nonetheless

on the contrary
quite the reverse
rather
really
still
still and all
to the contrary
though
yet

CONTRAST

The new idea is in opposition to the preceding one:

What he demands is life with flavor and meaning. Though he withdraws from the flow, he does not hide from himself. *On the contrary,* he tries to place himself where he does not feel lost, where he counts for something, above all in such circumstances that his life does not pass him by unnoticed.

Walter Teller

as it happened
by comparison
contrarily
conversely
in comparison

in contrast
on the contrary
on the other hand
on the other side

DIGRESSION

The new topic turns aside, momentarily, from the main point:

The new policy worked. It was not, *incidentally,* absolutely new; it had been suggested before. But this was the first time it had really been effectively pursued.

by the by
by the way
incidentally

EXEMPLIFICATION

The new statement offers a particular instance of the preceding idea:

He is not a great comic writer. Not, *say,* another Fielding or Dickens.

as a case in point
by way of illustration (or example)
for example
for instance

say
take
thus

REASON

The new topic explains or justifies the earlier, giving its reason or cause:

It was hard to believe. *For* no one had ever heard of such a thing.

after all
for

REPETITION

The new statement reasserts the previous idea, often more succinctly or more emphatically:

Few of us take the pains to study the origins of our cherished convictions; *indeed,* we have a natural repugnance to so doing.

James Harvey Robinson

in brief
indeed
in fact

in short
in truth
to repeat

RETURN TO THE POINT

The oncoming sentence swings back to the main topic after a digression:

At any rate the point is first of all to find again the mysteries.

Norman O. Brown

anyhow
anyway
at all events
at any rate
be that as it may

in any case
in any event
to continue
to resume

SERIES

The ensuing sentence is a member of a series, whether first, last, or in between:

There were several reasons for the change in attitude. *For one thing,* the generals had grown increasingly suspicious of their neighbor to the south.

finally
first/second/etc.
for one thing/for another
   (thing)
in the first place/in the
   second (place)/etc.

in turn
last(ly)
the first/the second/etc.
the former/the latter
the one/the other

SHIFT OF SUBJECT

These adverbs tell us, in a general, unspecified way, that a new subject will follow:

Now I'd better stop here and explain how I'm using the word *art.*

Flannery O'Connor

by the way	speaking of
now	thus far
now then	turning (next) to
so far	

SPECIFICATION

The new statement lists the particulars implied in the preceding one:

There are two things to remember: *namely,* to speak carefully and to do nothing.

in specific
namely
specifically

SUMMATION

The new sentence sums up the preceding material:

These, *in short,* were the causes of the Great Depression.

all in all	in summary
in brief	then
in fine	to summarize
in short	to sum up
in sum	

TEMPORAL RELATIONSHIPS

Here the connection is in time rather than in thought. The oncoming idea, fact, or event precedes or follows or occurs simultaneously with what went before:

The bombs fell on the central part of the city. *Later,* the fires spread to the outlying districts.

before	later on
by and by	meanwhile
earlier	now
heretofore	subsequently

hitherto	till now
later	up to (till) now

UNCERTAINTY

These adverbs prepare us for a statement of doubtful truth or applicability:

What was the reason? *Perhaps* he himself didn't really know.

maybe	possibly
perhaps	

A Glossary of Usage

The general nature of usage is discussed on pages 16–17, but here are a few of the common problems you are likely to encounter in your writing. The text treats some of these more fully (and also includes matters not covered here). You ought therefore to consult the index if you do not find help in the Glossary. In addition, dictionaries often comment upon usage, and books exist devoted exclusively to the subject. These include:

Theodore M. Bernstein, *The Careful Writer*
Bergen Evans and Cornelia Evans, *A Dictionary of Contemporary American Usage*
Wilson Follett, *Modern American Usage*
H. W. Fowler, *Dictionary of Modern English Usage*
William Morris and Mary Morris, *Dictionary of Contemporary Usage*
Margaret Nicholson, *A Dictionary of American-English Usage*

Not all these books remain in print, but most of them should be available in your school library.

a, an
> Both *a* and *an* are indefinite articles indicating singularity. They have no difference in meaning. *A* is used with nouns or modifiers beginning with a consonant; *an* is used before vowels. With words beginning with *h-*, usage is idiomatic: some take *an—an honorable man*; others take *a—a history book*; and some take either—*an historian, a historian.*

about for almost
> *About* in the meaning of "almost" is colloquial. In composition *I'm almost done* is more conventional than *I'm about done.*

accept, except

Accept means "to receive"—*They accepted the offer*—or "to admit as true or necessary"—*The teacher accepted his excuse. We accepted our fate.*

Except, as a verb, means "to leave out, exclude," in which sense it is a formal word—*The committee excepted us from the rule.* As a preposition *except* means "excluding, deliberately omitting"—*Everyone went except you.* In speech, though not in composition, *except* is often combined with *for* to make a compound preposition—*Everyone went except for you.*

access, excess

Access means a "way of approach or of admission to"—*We had access to their library.*

Excess, which may serve both as a noun and as an adjective, means "an amount beyond a limit"—*The excess grain was stored in the street.*

actually

As an emphatic term—*Actually you have no right to it*—*actually* has been devalued by overuse and is better avoided.

A.D., B.C.

A.D. (anno Domini, "in the year of our Lord") and *B.C.* ("before Christ") are usually printed in small caps and use periods. *A.D.* precedes the year, *B.C.* follows it—*A.D. 1066, 490 B.C.*

adapt, adopt

Adapt means "to make something serve a new purpose," often by altering it in some way—*We adapted the oar as a mast.*

Adopt means "to take as one's own"—*They've adopted their cousin's child.*

advice, advise

Advice is a noun designating "counsel or information given as a means of help"—*If you want my advice, don't do it.*

Advise is the verb signifying "to give counsel or information"—*I advise you not to do it.*

affect, effect

Affect, as a verb, means (1) "to cause a change in, to produce an effect"—*The war affected the birthrate*—and (2) "to display," often with the implication of falseness or pretense—*He affected an air of learning.* As a noun, *affect* is also a term in psychology meaning, roughly, "a feeling, an emotion."

Effect, as a noun, means "a consequence, result, change"—*The war had an effect upon the birthrate.* As a verb, *effect* means "to cause or bring about"—*She effected her escape.* In this sense it is rather a formal word.

aggravate

Most commonly, *aggravate* means "to annoy or irritate." Purists use the word only in its earlier sense of "adding to or intensifying something bad or undesirable"—*The government's policy only aggravated the crime rate.*

agree to, agree with

Agree is used in idiomatic combination with several prepositions: ____ *with* a person, ____ *to* a plan proposed by others, ____ *on* a plan developed with others, ____ *in* principle, theory, practice. (The last idiom often implies doubt that whatever is being agreed upon will be workable beyond the limits of principle, theory, and so on.)

ain't

Ain't has a long history in English. However, it has been too stigmatized to be used in composition, or even in speech, except humorously, by anyone who wants to be thought educated.

all ready, already

All ready is an adjectival phrase—*The men were all ready to go.*

Already is a temporal adverb meaning "prior to a specified or implied time"—*We had already arrived.*

all right, alright

All right and *alright* are alternate spellings and mean the same thing—"satisfactory" (or, adverbially, "satisfactorily, certainly"). *All right* is the preferred spelling in composition—*The idea seemed all right.*

As an attributive adjective expressing approval—*He's an alright guy*—*alright* (or *all right*) is too slangy for composition.

allusion, delusion, illusion

Allusion means "an indirect or implied reference"—*He made an allusion to their past.*

Delusion signifies the mental state of "being fooled or misled"—*He suffers from delusions of persecution.*

Illusion refers to something in the outer world which fools or misleads us, a "deceptive image or action"—*Magicians create illusions.*

all of

All of in such expressions as *He took all of the cake* is a colloquial redundancy. In composition the *of* should be omitted—*He took all the cake.* With personal pronouns, however, idiom requires *all of*—*He took all of it.*

all together, altogether

All together is an adjectival phrase designating a group considered as a unit—*They went all together*—or, more colloquially—*They all went together.*

Altogether is an adverb meaning "entirely, thoroughly, wholly"—*That's a different situation altogether.*

a lot of

In the colloquial sense of "many or much"—*She has a lot of money*—*a lot of* is legitimate in an informal style, but it should be avoided where a formal tone is called for—*She has a great deal of money.*

alright

See **all right**.

also

Also is a weak word for joining a sentence to what has preceded it. It suggests an afterthought, hastily tacked on, as in *The party was not a success because the weather was cold and damp. Also, people seem preoccupied with the bad news from the front.* Better connectives here would be such words as *besides, moreover, furthermore.*

alteration, alternation

Alteration means "changing in a purposeful way, or the result of being so changed"—*The alteration of his jacket improved the fit.*

Alternation means "occurring by turns"—*Poetic rhythm is the alternation of stressed and unstressed syllables.* The same distinction applies to *alter, alternate* and to *alterant, alternant.*

A.M., P.M.

A.M. (*ante meridiem,* "before noon") and *P.M.* (*post meridiem,* "after noon") may be written in small caps or in lower case (a.m., p.m.). In either case periods are required. The abbreviations are generally used only with numbers, which they follow—*10:15 A.M, 9:30 P.M.*

When clock times are expressed in words, the conventional form is *o'clock*—*ten o'clock in the morning, nine o'clock at night. A.M.* and *P.M.* should not be used with *o'clock.*

The abbreviations for noon is *M* or *M.* (*meridian*), though *twelve o'clock noon* is more common than *12:00 M.* For the other twelve o'clock write *midnight, twelve at night,* or *12:00 P.M.*

among, between

Among implies more than two—*The four thieves divided the loot among them.*

Purists use *between* only with reference to two items—*The two men divided the search between them.* In general usage, however, *between* is often used with more than two items.

When a choice or relation involving a number of items is understood as existing between them as specific individuals, *between* should be used—*The treaty defined the trade relationships between the five countries.*

ampersand

The ampersand (& or &) ought not to be used in composition in place of *and.* However, if it occurs as part of a name, title, or quotation, it should not be changed.

amount, number

Amount is used with mass nouns, that is, nouns designating entities which cannot be separated into numberable units—*a large amount of water, a significant amount of gas.*

Number is used with count nouns, which can be enumerated—*a number of chairs, a large number of people.*

and/or

This expression, loved by lawyers, is rarely necessary in composition. It is preferable to write *A or B,* or *both.* If you feel you do need *and/or,* write it thus, without spaces on either side of the slash.

ante-, anti-

While these prefixes sound the same, they differ in spelling and sense. *Ante-* means "coming before," in either time or space—*antebellum,* "before the war"; *anteroom.*

    Anti- means "opposite or opposed to"—*antitrust laws, antiwar.*

anybody, anyone

These are written as one word and construed as singular—*Has anyone left her* (or *his,* or *his or her*) *jacket in the cloakroom?*

anyway, anyways

Anyway is the preferred form of this adverb in composition. *Anyways* is colloquial.

    Anyway is an efficient connective in an informal style and means "in any case"—*Perhaps they had another appointment. Anyway, they never showed up.* For a formal tone, however, *anyway* may sound a bit too colloquial.

anywheres

This colloquial form of *anywhere* should be avoided in composition.

apiece, a piece

Apiece means "individually, each"—*They took one helping apiece.*

    A piece means "one portion"—*They each took a piece of cake.*

apprehend, comprehend

The pertinent sense of *apprehend* here is either (1) simply "to perceive, that is, to see or hear," in which sense it is a very formal word; or (2) "to grasp mentally but without full understanding"—*The people apprehended the danger, but only dimly.*

    Comprehend means "to understand fully"—*The people comprehended the danger.*

around, round

As prepositions and adverbs these words have the same meaning, though *around* is the common form in American English—*He circled around the house. She turned around.*

    However, as a verb, adjective, and noun, *round* has many meanings of its own—*She rounded the corner. It was an unusual, round object. They played a round of golf.*

    Round is not a clipped form of *around* and should not be written *'round.*

as, like

In formal usage *as* introduces a clause of comparison—*He lives alone, as his sister does. Like* is a preposition and introduces a phrase—

He looks like his sister. With friends like him who needs enemies? With such phrasal constructions, *like* is a perfectly correct word.

Informally *like* is often used, not as a preposition, but as a conjunction—*He lives alone like his sister does.* But in composition *as* is the better choice in such constructions.

as in place of that

As a substitute for *that* in introducing a noun clause, *as* (and *as how*) is a colloquialism. The formal idiom is *I don't know that I care,* not *I don't know as I care.*

assume, presume

Both these words means "to suppose or take for granted," but *presume* implies a greater certainty and may also suggest an unwarranted supposition—*You presume upon our friendship.*

assure, ensure, insure

Assure means "to assert the truth or existence or certainty" of something—*I assure you that we shall be there.*

Ensure means "to make certain"—*Your presence will ensure that we shall be there.*

Insure, originally a variant of *ensure,* now usually means "to promise indemnity for loss in return for payment of a premium."

awful, awfully

As intensives—*It was an awful movie, but we had an awfully good time*—these words are too colloquial for general use in composition.

awhile, a while

Awhile is an adverb meaning "for a short time"—*They stayed awhile and then went home.*

A while is a noun and usually functions as the object of a preposition—*They stayed for a while.*

bad, badly

Bad is an adjective—*a bad actor. Badly* is an adverb—*He acted badly.*

After linking verbs such as *look, seem, feel, bad* is the appropriate word—*She feels bad.*

beside, besides

Beside, a preposition, means "by the side of"—*The house stood beside a small stream.*

Besides has three uses. As a preposition it means "in addition to"—*Besides the men, the women also attended.* As an adverb it means "in addition or also"—*They discussed other solutions besides.* As a conjunctive adverb it joins a sentence to what precedes it and has the meaning "furthermore, moreover"—*The business failed because it was poorly located. Besides, times were hard and people had no money.*

between

See **among**.

between you and me

Between you and me is the grammatical form. *Between you and I* is a false elegance.

bi-

Bi- is a confusing prefix because it is used to mean both "twice a" and "once every two." Thus *biweekly meetings* can signify two per week or two per month. Because of its ambiguity *bi-* is best avoided.

born, borne

Both words are past participles of *bear*. *Born* is generally confined to the meaning "brought into being"—*The idea was born of necessity. He was born in New Hampshire.*

Borne is used in the other senses of *bear,* such as "carry" or "endure"—*The nation had borne its defeat with courage.*

bourgeois, bourgeoisie

Bourgeois is used most frequently as an adjective and means either "characteristic in general of the middle class" or, as a leftist sneer word, "characterized by narrow, middle-class interests and values"—*He has a bourgeois attitude toward individual freedom.* As a noun *bourgeois* designates a member of the middle class.

Bourgeoisie is a noun and describes the urbanized middle class or the social order established by them—*The bourgeoisie dominated Viennese politics.*

bring, take

Bring means "to carry or convey toward the speaker"—*Bring the book when you come.*

Take means "to carry or convey away from the speaker"—*Take the book when you go.*

calculate, reckon

In the senses of "think" or "suppose," *calculate* and *reckon* are too colloquial for composition.

can, may

Can means "able to, capable of"—*I can go tomorrow.*

May means (1) "have permission to" or "seek permission for"—*They said I may go. May I?*—and (2) "have a possible intention of"—*I may go tomorrow.*

In such questions as *May I have more toast? may* is regarded as more polite and precise than *can,* though in practice many people ignore the distinction and say *can.*

cannot, can not

These forms have no difference in meaning. Today the single word is more common.

censor, censure

To *censor* is "to delete words or pictures for moral or political reasons"—*The soldiers' mail was censored.* As a noun *censor* designates anyone who makes such deletions.

To *censure* is "to condemn or disapprove"—*They censured his actions. Censure* as a noun means the act of "blaming or finding fault."

cite, sight, site

Cite is a verb meaning "to quote and refer to." In scholarly work it means not only to quote but to provide the source of the quotation—*She cited your article in her term paper.* The noun is *citation.*

Sight as a verb means (1) "to see," especially something for which one has been looking—*They sighted the ship*—or (2) "to aim"—*They are sighting their guns.* As a noun *sight* means "an extraordinary visual perception"—*The tall ships were a wonderful sight. He was a sight.*

Site means "a place," especially one occupied (or to be occupied) by a building—*The family chose a beautiful site for their new home.* As a verb it means to place or locate a "building, monument," and so on.

climactic, climatic

Climactic means "relating to a climax," that is, a high point, or point of resolution—*The climactic scene occurs in the next-to-last chapter.*

Climatic means "relating to climate"—*There are climatic advantages to living on an island.*

compare, contrast

Compare always involves similarities, whether these are discussed alone or discussed together with differences—*She compared Boston and San Francisco as similar cities.*

Contrast is confined to showing differences—*She contrasted French schooling with our own.*

complement, compliment

Complement, as a verb, means "to complete or fill out"—*The auxiliaries complement the regular police.* As a noun it means (1) "that which completes" and (2) "the full crew of a ship."

Compliment, as a verb, means "to express appreciation or respect or admiration"—*They complimented her on her improvement.* As a noun *compliment* designates "an expression of esteem"—*They paid her a pleasant compliment.*

comprehend

See **apprehend**.

comprise

In a narrow sense *comprise* means "to include, contain"—*The compound comprised seven buildings.* Often, however, its meaning is stretched to "make up, constitute"—*Seven buildings comprised the compound.*

confidant, confident

Confidant is a noun, designating "someone to whom a secret is entrusted"—*I was her only confidant in the matter.*

> *Confident,* an adjective, means "sureness of one's ability or knowl-
> edge"—*I was confident that I could help.*

consensus

> *Consensus* means "general agreement of opinion or belief." In the
> expressions *general consensus* and *consensus of opinion,* the words
> *general* and *of opinion* are redundant.

continual(ly), continuous(ly)

> In formal usage *continual(ly)* means "recurring over and over"—
> *He was continually late for class.*
>
> *Continuous(ly)* means "extending or occurring uninterruptedly"
> in either space or time—*The siren shrilled continuously for an hour.*

could, can

> *Could* suggests a lower degree of possibility than *can. I can go* means
> "I have it in my power to go"; *I could go* means "I have the power
> to go but possibly (or even probably) I won't."
>
> *Could* also functions as the past of *can—Last month I could see
> he was ill.*

could of

> Incorrect for *could've* (*could have*).

consul, council, counsel

> A *consul* is an agent of a foreign government, often concerned with
> matters of business and commerce—*He is the Spanish consul.*
>
> A *council* is a group of people, often elected or appointed, who
> advise, consult, and sometimes make administrative decisions—*The
> town council met last night.*
>
> *Counsel* is "advice"—*Your counsel was very helpful.* A *counsel* is
> also a lawyer engaged in a court case—*The client's counsel addressed
> the judge.*

criteria, criterion

> *Criteria* is the plural, *criterion* the singular. The Greek plural is
> more common in modern English than the anglicized *criterions.*

data, datum

> The Latin plural *data* is used in English both in a plural and in a
> singular (usually collective) sense—*The data are available. This data
> is very useful.* The Latin singular *datum* is rarely heard.

definitely

> *Definitely* in the sense of "certainly" or "clearly"—*She definitely
> agreed*—has been devalued by overuse.

delusion

> See **allusion**.

different from, different than

> In formal usage *different from* is preferred to introduce a phrase—
> *He is different from me.* However, many writers use *different than.*

When the following construction is an adverbial clause, *different than* is the proper choice—*She looked different than I had expected.*

contrast

See **compare**.

disinterested, uninterested

In formal usage *disinterested* means "objective, unbiased"—*A judge should be disinterested.* In that sense the word does not exclude interest in the sense of "concern with" or "curiosity about."

Uninterested means "unconcerned, having no intellectual or emotional curiosity"—*People were uninterested in the issue.*

ditto marks

Ditto marks should not be used in composition or in footnotes or bibliographies. Nor should the word *ditto.*

double negative

Using two negatives in the same construction is substandard—*He didn't want no advice.* Standard English is *He didn't want any advice,* or *He didn't want advice,* or *He wanted no advice.*

However, negating a negative term in order to convey a qualifiedly positive statement is legitimate—*It was not an unfriendly act.* This rhetorical device, litotes, is effective on occasion, but it ought not to be overworked.

due to

Conservative writers use *due to* only to introduce an adjectival phrase—*His illness was due to overwork.* Less formally, *due to* is often used adverbially in the sense of "because of"—*Due to the rain, we didn't go.*

e.g.

Generally this abbreviation should be avoided in composition, the full phrase (*for example*) being written out.

effect

See **affect**.

etc.

In composition *and so on* is usually preferred to the abbreviation *etc.*

Neither the phrase nor the abbreviation should be used except in cases where readers can easily complete the thought for themselves—*He always has an excuse for being late: his clock stopped, his car wouldn't start, he got a phone call, and so on.*

everybody, everyone

In formal usage these impersonal pronouns are construed as singular and take singular verbs and pronouns—*Everybody took his* (or *her,* or *his or her*) *place.* (With regard to *his* or *his or her,* see *he and she.*)

Informally, *everybody* and *everyone* are often treated as plurals, especially in pronominal reference—*Everybody took their places.*

except

See **accept**.

excess

See **access**.

farther, further

Precise writers use *farther* for spatial distance—*The farther shore of the river*—and *further* for more abstract meanings—*It was further alleged that he had fallen asleep on duty.*

fewer, less

In formal usage *fewer* is used with count nouns (things like *car, person, table,* existing in countable units)—*There were fewer people than we had hoped.*

Less is used with mass nouns (*water, air, iron,* which exist in quantities rather than discrete units)—*Less water had leaked in than we feared.*

Informally, *less* is often applied to count nouns.

field

In such expressions as *field of biology, field* is a weak metaphor which says nothing and is better omitted.

finalize

Finalize, used in place of *completed, finished, concluded,* has a jargonistic ring, although the word may be justifiable if you wish to stress an official action—*The general staff finalized plans for the landing.*

first, firstly; second, secondly; and so on

In American usage *first* is preferable to *firstly*. With *second, secondly, third, thirdly,* and so on, the preference, though less pronounced, is still for the form without *-ly.*

flounder, founder

To flounder means "to struggle to find footing, to move clumsily," whether literally or metaphorically—*He floundered through the examination.*

To founder means, in the case of a ship, "to sink," and more generally "to fail, collapse"—*The project foundered because there were too many advisers and not enough helpers.*

for

When *for* is used as a coordinating conjunction, it is usually punctuated with a comma to prevent its being read as a preposition—*The plan was impractical, for the people would not accept any restrictions.*

formally, formerly

Formally means (1) "related to form or structure"—*Formally, the proposal consists of three parts*—or (2) "according to an established pattern or convention"—*They were formally introduced.*

Formerly means "in time past"—*Formerly people were more careful to address one another formally.*

former, latter

Former and *latter* are confined in formal usage to groups of two, though *first* and *second* or *one* and *other* can also be used with twos—*Sarah and Jane left today, the former (the first, the one) for Montreal, the latter (the second, the other) for New Orleans.*

-ful, full

As a suffix *-ful* has one *-l*—*cupful.* The plural is *cupsful.* As a separate word it has two *-lls*—*a cup full*—and the plural is *two cups full.*

general consensus

See consensus.

go for say

As a slang substitute for say—*So I goes, "Why can't you understand?"*—*go* is inappropriate in composition.

good, well

Good is generally an adjective, either attributively—*She is a good golfer*—or predicatively—*She looks good* (that is, "her appearance is attractive").

Well may be used as a predicative adjective in a similar construction—*She looks well* (that is, "she seems in good health"). More often *well* is used adverbially to mean "effectively, carefully, in an accomplished way"—*She looked well but never found the book* (that is, "she searched carefully").

got, gotten

Got and *gotten* are alternate forms of the past participle of *get*—both *I've got very little help* and *I've gotten very little help* are acceptable.

In the simple past *got* is the proper form—*I got very little help.* And when used after *have* in an emphatic sense, only *got* is idiomatic—*I've got to go.*

he and she/his or her

He (*his, him*) has conventionally been used generically to mean either male or female in such constructions as *A student must purchase his books within the next two weeks.* But if precision requires it or if you wish to avoid any taint of sexism, specify both sexes (*his or her books*). Repeated dual references, however, are awkward. Often the problem can be solved by using a plural noun—*Students must purchase their books.*

hisself

A substandard form for *himself.*

hopefully

In the sense of "I hope" or "it is hoped," *hopefully* seems ineradicably established—*Hopefully, the economy will improve in the next quarter.*

Purists object to the usage and prefer to restrict *hopefully* to its narrow adverbial sense of "in a hopeful way"—*The people looked to Lincoln hopefully.*

ideal

In formal usage *ideal* ought not to be compared. Logically, the ideal cannot be more, or less, than what it is.

i.e.

In composition the phrase *that is* is preferable to *i.e.*

if, whether

Whether is more appropriate than *if* in a formal style to introduce interrogative clauses, and clauses conveying uncertainty—*We must ask whether the policy is really to our benefit.*

ill, sick

In American English *ill* and *sick* have much the same meaning, though *ill* is more formal. The distinction that *ill* means "in bad health" and *sick* means "nauseated" is not observed in general usage.

illusion

   See **allusion**.

imply, infer

In formal English *imply* means "to suggest without literally stating." The word is used with reference to a speaker or writer—*The speaker implied that the plan would not work.*

   Infer, which applies to a reader or listener, means "to draw a conclusion"—*The audience inferred that the speaker disapproved of the plan.*

   In less formal contexts *infer* is often used to mean "imply."

incredible, incredulous

Incredible means "improbable, difficult to believe," whether for negative or positive reasons—*The singer gave an incredible performance.* (Note that in such a sentence *incredible* is ambiguous; it could mean "incredibly bad" or "incredibly good.")

   Incredulous applies to the perceiver and means that he or she "does not believe, is not easily convinced"—*They were incredulous when we told them.*

infer

   See **imply**.

inside of

In composition the *of* is not needed—*They were inside the house, not inside of.*

irregardless

The double negation in *irregardless* (*ir-* and *-less*) is redundant. *Regardless* should be used in composition.

its, it's

The possessive of *it* is *its*—*The cat put its paw on the table.*

It's is the contracted form of *it is*—*It may be your cat, but it's my table*.

it is I (he, she, we, they, who)

Most people use the objective pronouns after *it is*—*It is me, it's him*, and so on. The traditional grammatical "rule" that the subjective pronouns should follow *it is*—*It is I, It is she*, and so on—is rarely observed.

However in the question *It's who?* the rule and popular usage coincide.

just

As a vague intensive—*That's just fine*—*just* is a colloquialism.

kind of, sort of

Used as qualifiers, these expressions are colloquial—*She's kind of nice. It's sort of a roadhouse*. In composition *rather* would be more appropriate diction than *kind of* in the first example. In the second *sort of* (or *kind of*) might be used, but the article would come first—*It's a sort of roadhouse*.

lay, lie

Lay is a transitive verb and takes a direct object—*Lay the table for dinner. Lie* is intransitive and does not take an object—*The coat was lying on the floor*.

The problem is complicated by the fact that the past participle of *lie* is *lay*. The principal parts of the two verbs are:

lay	laid	laid	laying
lie	lay	lain	lying

latter

See **former**.

learn, teach

The colloquial use of *learn* in the sense of "teach" is substandard and should be avoided—*She taught me to sail*, not *She learned me to sail*.

lend, loan

In addition to its use as a noun, *loan* is also a verb meaning "to lend." In formal usage *lend* (past and past participle, *lent*) is preferred in the present, though *loaned* is more common in past tenses.

liable

In formal usage *liable* is restricted to situations where the possible consequences are undesirable—*You're liable to get into trouble*, but not *You're liable to have a good time*. With desirable consequences, *likely* is the better word.

lie

See **lay**.

like

See **as**.

loan
  See **lend**.

loose, lose
  These words are often confused because the difference in sound is
  not marked by the spelling. It occurs not in *o, oo,* but in the
  -s's, that in *loose* being silent, that in *lose,* voiced.
     Loose means "free, untrammeled, unconnected," and so on. *Lose*
  means "to misplace" or "to be defeated."

lots of, a lot of
  These are colloquialisms meaning "many, much, a great deal of"—
  Lots of people came—and are better avoided in composition.

majority, plurality
  Majority signifies "more than half." *Plurality* means "the largest
  number, but not a majority," that is, not more than half—*He won
  a plurality of the votes, but not a majority.*

may
  See **can**.

may, might
  Might is used as a past for may—*I may go tomorrow. I might have
  gone yesterday.*
     The two words may also be used with reference to the same future
  action, but they imply different degrees of possibility—*I might go
  tomorrow* suggests greater uncertainty than *I may go tomorrow.*

media
  Media is the common plural of *medium* in the sense of "channel
  of communication." *Media* should not be used to designate a single
  means of communication—*the medium of television,* not *the media
  of television.*
     In the sense of one who communicates with spirits (or claims to)
  the plural is *mediums*—*Three mediums shared the same house.*

might
  See **may**.

need, needs
  Needs is used in the third person singular when something is being
  affirmed—*She needs a new dress.*
     In formal negative statements, *need* is used—*She need not pay
  for it now.* Less formally the idea would be phrased—*She doesn't
  need to pay for it now.*

number
  See **amount**.

numbers
  In composition, numbers are used in dates; addresses; hours (with
  A.M. or P.M.); exact monetary sums (*$7.35*); page references (*p. 37*);

references to acts, scenes, and lines (*IV, ii, 125,* meaning "act IV," "scene ii," "line 125"); measurements, when expressed with conventional abbreviations (*70° F., 6′3″, 25 lbs.*).

In addition, numbers designate sums that would require more than two words—*There were 173 recruits.* Sums requiring two words or one are generally written—*There were five thousand.* However, usage varies in this matter according to editorial policy.

It is awkward to open a sentence with digits—*There were 173 recruits waiting* is preferable to *173 recruits were waiting.*

Written numbers from twenty-one to ninety-nine, and adjectival compounds formed with numbers, should be hyphenated—*He bought a thirty-two-foot ketch,* but *She earned fifty thousand dollars a year.*

of in compound prepositions

Such double prepositions as *inside of, outside of* (though not in the idiomatic sense of "excepting"), and especially *off of,* are redundant in a formal style. The *of* should be dropped.

In some compound prepositions, however, the *of* is necessary—*instead of, in case of, because of, in place of,* and so on.

of for -'ve or have

Of in place of the contraction *-'ve*—*I could of gone*—is substandard.

off of

In composition write *off,* not *off of*—*He fell off the horse,* not *off of the horse.*

O.K., OK, okay

Unless you wish a distinctly informal tone, avoid *O.K.* in composition.

one

One takes a singular verb and is referred to by a singular pronoun.

One strikes many Americans as stilted, especially if it must be repeated—*One cannot always foresee the consequences of one's actions.* Repetition can be avoided by using the third person singular pronoun (either masculine alone or masculine and feminine)—*One cannot always foresee the consequences of his or her actions.*

Using the generic *you* in place of *one* is a possible alternative, but *you* is informal and easily overworked.

only

When *only* modifies a single word rather than an entire clause, it is, in a formal style, placed immediately before that word—*I have only six dollars.*

Informally, *only* often precedes the verb in such constructions—*I only have six dollars.*

Unless clarity suffers, *only* should be positioned where it is most appropriate to tone.

onto, on to

Onto means "to a position on"—*She climbed onto the rock.*

On to means "to continue as far as, to proceed"—*We walked on to the next village. He went on to discuss his trip.*

outside of

In the sense of "physically beyond," *outside of* is colloquial; the *of* should be dropped in composition—*He stood outside the house.*

In the sense of "excepting," the full phrase is necessary—*Outside of Helen, no one guessed the truth.*

perfect

In formal usage *perfect* is not compared since something cannot be more, or less, perfect, most perfect, or least perfect. In speech, however, this logical nicety is often ignored.

phenomena, phenomenon

The Latin plural *phenomena* is used in English.

plenty

The adverbial use of *plenty* in the sense of "extremely, very"—*We were plenty worried*—is colloquial and should be avoided in composition.

plenty of

Plenty of—as in *They had plenty of opportunity*—is at home in an informal style. On more formal occasions such adjectives as *ample* or *considerable* are preferable.

plurality

See **majority**.

P.M.

See **A.M.**

predominant, predominate

Predominant denotes "superior strength or authority"—*Hers was the predominant influence in the family.*

Predominate is a verb with much the same meaning, "to dominate, exercise superior power or authority."

presume

See **assume**.

principal, principle

Principal is primarily an adjective meaning "chief, main"—*The principal point is that the scheme will work.* It has become a noun in the limited sense of "head of a school," a use derived from shortening the phrase *principal teacher.*

Principle is a noun meaning "a comprehensive or fundamental law or conception"—*The principle of self-determination should be upheld by all civilized states.*

prior to

Prior to is an unnecessarily heavy phrase, better replaced by *before.*

proved, proven

Either *proved* or *proven* may serve as the past participle of *prove*— *They have proved* (or *proven*) *capable.* There is no difference in meaning or tone.

provided, providing

Both these forms of *provide* can be used with the meaning "on condition that, with the understanding that"—*I'll go provided* (or *providing*) *they ask me. Provided* is preferable in composition, though in most cases the simpler *if* would be a better choice.

raise, rise

Raise is a transitive verb, taking a direct object—*They raised the flag.*

Rise is intransitive and does not take an object—*We must rise early to catch the tide.*

reason is because

In composition *reason is because* is regarded as a tautology, stating the idea of causation twice. *Reason is that* is preferred—*The reason the city is in difficulty is that it is losing industry.* (The same idea can be stated more succinctly by an adverbial clause—*The city is in difficulty because it is losing industry.*)

reckon

A colloquialism for *think.* See **calculate**.

rise

See **raise**.

round

See **around**.

seem

Unless there is genuine doubt or uncertainty, *seem* is likely to be deadwood. It is meaningful in *They seem to be tired,* but probably dead in *People seem to like television.*

-self pronouns

The reflexives (*myself, yourself, himself, herself, itself, ourselves, yourselves,* and *themselves*) are written as one word. They should be used as objects only when they refer to the same person(s) as the subject— *You will hurt only yourselves.* They should not be used when the object and subject are different—*He gave it to me,* not *He gave it to myself.*

As intensives these pronouns may immediately follow the noun or pronoun they repeat, or they may appear at the end of the clause— *I myself saw it,* or *I saw it myself.* The first pattern is more formal.

shall, will

According to a rule of traditional grammar, *shall* is used in the first person (singular and plural) to show simple future and *will* to indicate determination, force, and so on. In the second and third persons these uses are reversed, *will* showing simple future and *shall,* determination or compulsion.

The rule, however, is generally ignored in contemporary speech and writing, *will* serving for both simple and forced future.

should of

An error for *should've (should have).*

should, would

Should and *would* have many shades of meaning. *Should* often indicates, in all three persons, a future duty or obligation which is uncertain of being fulfilled—*I (you, he, she,* and so on) *should go to the meeting,* meaning "I (you, he, she) have an obligation to go but may not meet that obligation." With reference to the past, *should* similarly designates an unfulfilled obligation—*I (you, he, she) should have gone, and I'm sorry I didn't.*

Would often designates a doubtful, conditional future—*I would go if I thought it would help,* the implication being that "I am doubtful of my presence helping." *Would* may also indicate a potential action in the past which was unrealized because of a doubtful or unforeseen condition—*I would have gone if I had known about the meeting.*

In such first-person uses as *I should* (or *would*) *like to say,* either *should* or *would* may be used with no difference in meaning, though *should* is perhaps a bit more formal in tone.

sick

See **ill.**

sight

See **cite.**

site

See **cite.**

so

As a sentence opener *so* is colloquial. It is better avoided in composition unless an informal tone is important (and even then used sparingly). *Consequently, therefore, as a result* are more appropriate in a formal style. *And so* is less obviously colloquial as an opener than *so,* but it too ought not to be overworked.

So is common and useful within sentences to introduce adverbial clauses of consequence—*It was late so we went home.*

When the adverbial clause conveys purpose rather than strict consequence, it is best to introduce it by *so that*—*We left early so that we would be on time.* Sometimes in such constructions the *so* is used as an intensive within the main clause, the consequence being introduced by *that*—*It was so cold that we left early.*

somebody, someone
> In formal usage these pronouns take singular verbs and are referred to by singular pronouns—*Someone forgot his or her book.*

sort of
> See **kind of**.

split infinitive
> A split infinitive results when an adverbial intrudes between *to* and the verb—*to quickly go.* It is not a heinous crime, and good writers sometimes split infinitives for emphasis or to preserve idiom. As an intruding adverbial becomes longer and more complex, however, the split infinitive becomes increasingly awkward—*To clumsily and lengthily split an infinitive makes for poor writing.*

sure
> Used in the sense of "certainly"—*Are you going? I sure am*—sure is too colloquial for composition.
>
> But in the adjectival sense of "certain," *sure* is perfectly legitimate—*I am sure that I shall go.*

take
> See **bring**.

teach
> See **learn**.

than I, than me
> After *than* formal usage prefers the subjective case of personal pronouns—*She is brighter than I.* But in general usage the objective form is often heard and sometimes seen—*She is brighter than me.*

than, then
> *Than* is a conjunction and introduces a comparison or contrast.
>
> *Then* is an adverb with a range of meanings: "at that time, soon after, following next, in that case, consequently."

that as a loose intensive
> *That* is a common colloquialism in the sense of a loose intensive—*He didn't have that far to go.* The usage should be avoided in composition because it implies an unspecified comparison which might puzzle readers. Such words as *very, so, extremely* are clearer.

that is
> *That is* introduces an explanation or an alternative expression—*The pronouns must be put in the objective case, that is, in the form of me, him, her, us, them.* In such constructions *that is* is usually set off by commas or preceded by a dash and followed by a comma.

their, there, they're
> *Their* is the possessive form of the third person plural pronoun.
>
> *There* is an adverb of place—*The books are there on the table.* It

is often used as an expletive to introduce a kind of anticipatory construction—*There were three books on the table.*

They're is the contracted form of *they are.*

there are, there is

In sentences like *There are three books on the table,* the grammatical subject (*books* in this example), not *there,* determines the number of the verb. Thus the verb is singular in *There is a book on the table.*

this

As the subject of a sentence *this* is useful, neatly linking the statement to what has gone before—*The election was a foregone conclusion. This precluded any sense of drama.*

However, a substantive *this* may seem vague, especially if it refers to the whole of the preceding idea rather than to a single word. In such cases it is better to make *this* an adjective modifying a noun which clearly sums up the preceding point. *This* still functions as a link word but without vagueness—*The election was a foregone conclusion. This inevitability precluded any sense of drama.*

to, too, two

To is a preposition—*The family had gone to town.*

Too is an adverb, meaning (1) "also, besides"—*We took some eggs too*—or (2) "excessively"—*It was too hot.*

Two is an adjective specifying a number—*two children.*

toward, towards

These words are the same in meaning and tone. *Toward* is more common in American usage.

try and, try to

Try to is the appropriate form in composition—*We must try to increase production,* not *try and increase.*

uninterested

See **disinterested**.

unique

In formal usage *unique* is held to be incomparable, that is, something cannot be more, or less, unique; or most, or least, unique.

used to

Used to is always in the form of the past participle, that is, with -*d*—*We used to go,* not *We use to go.*

way

Way in the sense of "much, far"—*He is way ahead*—is a colloquialism to be avoided in composition.

With reference to distance *way* is preferable to *ways*—*We have a long way to go.*

well

See **good**.

where for that

> Constructions like *I read where the housing industry is in trouble* are colloquial. A formal style would use *I read that.* . . .

whether

> See **if**.

who, whom, whose

> *Who* is the subject of a verb—*He is the one who gave the order*—and the complement of a linking verb—*He is who?*
>
> *Whom* is the object of a verb—*She is the one whom you saw*—the object of a preposition—*She is the person to whom you must apply*—and the subject of an infinitive—*Whom to select was a problem.*
>
> *Whose* is the possessive form, used adjectivally—*This is the family whose niece you met in Paris.*

will

> See **shall**.

-wise

> While some established adverbs use *-wise* (*edgewise, lengthwise*), the contemporary tendency to attach *-wise* in wholesale fashion is overdone.

woman, women

> These singular and plural forms are often confused. Where they differ in sound (the *o*s) they do not differ in spelling; contrarily, where they differ in spelling (the final vowel) they do not, in ordinary speech, differ in sound.

would

> See **should**.

would be

> *Would* ought to be used with *be* when the action is imagined—*If I had my druthers I would be slender*—a conditional future—*I would be happy to go*—or uncertain—*It would be possible.*
>
> On the other hand, *would* is awkward and subtly misleading where these conditions do not apply, for example, in a sentence like *That would be John at the door.* If the speaker does not know that it is John, he or she should say *That may be John* or *That could be John.* If the speaker knows that it is John, the appropriate verb is a form of *be*—*That is John at the door.*

would of

> A substandard form of *would've* (*would have*).

Xmas

> Avoid this abbreviation.

you as a generic pronoun

> The generic, or impersonal, *you* means "anyone"—*When you go abroad for the first time you may suffer cultural shock.* It is legitimate

in composition, but it does have a less than formal tone and it must be watched—the generic *you* spreads like dandelions.

you know

The phrase *you know* is used (overused) in conversation to get and hold attention. In that function it has no place in composition.

Subject Index

This index is restricted to particular aspects of composition covered in the text. Words and expressions listed in the Glossary of Usage beginning on page 780 are not generally included. A few exceptions are words listed in boldface for which the discussion in the text may supplement that in the Glossary. The writers whose work supplies examples are listed in a separate index immediately following this one.

Name Index

A listing of the writers whose work is quoted as examples

Reference Grammar Contents